THE
EVOLUTION
OF PRIMATE
BEHAVIOR

The constitution of the ape is hot, and since he is rather similar to man, he always observes him in order to imitate his actions. He also shares the habits of beasts, but both these aspects of his nature are deficient, so that his behavior is neither completely human nor completely animal; he is therefore unstable. Sometimes, when he observes a bird in flight, he raises himself and leaps and tries to fly, but since he cannot accomplish his desire, he immediately becomes enraged. Being similar to man, the (she-) ape also has a menstrual cycle governed by the moon, and since both aspects of its nature are unstable and weak, the creature has no medicinal value.

Saint Hildegard of Bingen   ca. 1150

trans. Jansen, 1952

THE EVOLUTION OF PRIMATE BEHAVIOR

ALISON JOLLY

Rockefeller University

Second Edition

Macmillan Publishing Company
New York

Collier Macmillan Publishers
London

Macmillan Publishing Company
866 Third Avenue, New York, New York 10022

Macmillan Canada, Inc.

Library of Congress Cataloging in Publication Data

Jolly, Alison.
 The evolution of primate behavior.

 Bibliography: p.
 Includes index.
  1. Primates—Behavior.   2. Primates—Evolution.
3. Mammals—Behavior.   4. Mammals—Evolution.
I. Title.
QL737.P9J64   1985   599.8′0451   84-933
ISBN 0-02-361140-5

Printing: 4 5 6 7 8   Year: 7 8 9 0 1 2 3

ISBN 0-02-361140-5

For G. E. Hutchinson

*who once defined ecology as
the study of the universe*

PREFACE

This is a book of philosophy in the sheep's clothing of a freshman text.

I did not start out to write such a book. I merely meant to summarize existing knowledge of primate behavior. That seemed quite enough, given more than ten years' exponential growth of knowledge since the first edition. I had no intention of dealing with the direction of physical causality or the nature of consciousness.

The underlying philosophy is far from original. It is the same reductionist, materialist view that underlies most chemistry or physics texts and that passes for common sense to most scientists. What is, perhaps, original and what forces this view into the open is that my topic is the primates. Primates stand at the hinge of evolution. Primates are to the biologist what viruses are to the biochemist. They are subject to the rules of the simpler discipline; they can be analyzed and partly understood—but viruses are living chemicals, and primates are animals who love and hate and think.

Four developments in the study of primate behavior during the last thirteen years have forced me to be explicit about the assumptions I used to take for granted. First, we have far more data on the ranging patterns, food choice, and population dynamics of wild primates. Ecological principles, as formulated by Evelyn Hutchinson, to whom this book is dedicated, have finally penetrated primate studies. This means we have a far clearer idea than before how environmental pressures shaped primate society and intelligence.

Second, E. O. Wilson in his book "Sociobiology" crystallized the possibility of calculating evolutionary advantages of social behavior. He was certainly not the only one; he built on work by Hamilton, Trivers, and, chiefly, Charles Darwin. Nonetheless his book provoked outcry. This was not really a reaction to his considering the evolutionary advantages of self-interest; we had been through that before with Social Darwinism, although with far less understanding of the evolutionary process. It was outrage at the thought that we might quantify the evolutionary advantages of love. Meanwhile, data on primate kinship and mutual aid began to do just that.

The third influence is exemplified in Donald Griffin's "The Question of Animal Awareness," and "Animal Thinking." Griffin points out that behaviorists cannot have it both ways. They cannot equate the workings of the mind to the workings of the nervous system and then deny conscious mind to creatures with nervous systems much like our own. For him, consciousness is a real property of the complex organization of a brain, just as a virus' capacity to reproduce itself is a property of its complex chemical architecture. Thirteen years' further study of ape language and protocultural transmission leaves purists still arguing over the definition of language, but only by mental gymnastics amounting to sophistry can one doubt ape awareness.

These three currents of thought have clarified some of the links between primate environment, social emotions, and the emergence of mind. The fourth development is the realization that the urgent need in primate studies is conservation. It is sheer self-indulgence to write books to increase understanding if there will soon be nothing left to understand.

This realization has been painful. It began for me in Madagascar, where the tragedy of forest felling, erosion, and desertification is a tragedy without villains. Malagasy peasant farmers are only trying to change wild environments to feed their own families, as mankind has done everywhere since the Neolithic Revolution. The realization grew in Mauritius, where I watched the world's last five echo parakeets land on one tree and knew they will soon be no more. It has come through an equally painful intellectual change. I became a biologist through wonder at the diversity of nature. I became a field biologist because I preferred watching nature go its own way to messing it about with experiments. At last I understood that biology, as the study of nature apart from man, is a historical exercise. From the Neolithic Revolution to its logical sequels of twentieth-century population growth, biochemical engineering of life forms, and nuclear mutual assured destruction, human mind has become the chief factor in biology.

Thus we travel full circle from ecology to ecology. We trace the biological steps from environment to society to the emergence of mind. In the end, one primate's mind determines the environment and, often, the existence of all other species on earth.

AJ

I wish to thank Donald Griffin, Peter Marler, and the Rockefeller University for their hospitality during the preparation of this book. Among those who read and commented on portions of the manuscript are Stuart Altmann, John Fleagle, Robert and Beatrice Gardner, Donald Griffin, Paul Harvey, Sarah Hrdy, Peter Marler, Bob Martin, Emil Menzel, David Premack, Caroline Ristau, Tom Struhsacker, Bob Sussman, and John Terborgh. I am responsible for remaining errors and omissions, but am particularly grateful to have been saved from making even worse ones. Celia Francis and Rosanne Kelly typed the draft and successive strata of corrections. I thank Margaretta, Susan, Morris, and Dickon for putting up with a growing book in the family. Above all I thank my husband, Richard, whose support for the study of primate behavior never falters, although he keeps right on preferring humanity.

CONTENTS

PART ONE

ECOLOGY

CHAPTER 1

WHAT, WHERE, AND HOW MANY?

THE HUMAN RESERVE

Man is a primate. We know this in every detail of our physical form, from our flat fingernails to our blunt big toes, and in the colors we see out of our eyes. Linnaeus (1758) classed man with monkey, lemur, and bat. When T. H. Huxley (1863) drew the anatomy of man and ape, he challenged his hearers to deny their kinship with the chimpanzee.

Does man behave as a primate? No. Chimpanzees, like people, probe twigs into holes; we also analyze moon rock. Chimpanzees, like people, have destroyed their own kind, but one of our species wrote *Medea.* However, the creationist who looks for absolute differences between ourselves and the ape is likely to be disappointed. Our love and hate and fear and curiosity, our birth and death, all our emotions, and even some of our logic stem from the primate past.

On present reckoning we have been placental mammals for more than 100 million years and have been monkeylike, large-brained, possibly group-living creatures for 40 million years. Sometime in the past 5 or 10 million years we became terrestrial hunter-gatherers. Only for the final millennia have some tribes tilled fields, built villages, constructed human life as we now conceive it.

All written history is the chronicle of a *nouveau riche,* abruptly risen to dictatorship of his planet—not what we are, but what we have made ourselves. Our earlier history, the part without writing or artifacts, can only be deduced from what we do and what our primate relatives do—the hieroglyphics of the chimpanzee's whimper and the baby's smile.

The more urgent question in primate evolution is not what it reveals about the past. The real question is what happens in the future. How many kinds of plants and animals will continue to share our planet? As an order of magnitude, perhaps one wild species now becomes extinct every day. By the end of this century, it may be one every hour, and half the world's present wild species will already be dead (Myers, 1979).

How many species will continue in something close to their original habitat? Climax habitats, where nature has approached an equilibrium, may take centuries to reestablish once they have been destroyed. Sometimes they never return to their former state, like cleared areas of Amazon jungle that become hard-baked grassland.

We are approaching a point where the planet will be divided into nature reserves and the human reserve. Only a few wild mammals can live in close proximity to man: rats and mice in houses, bushbabies at the bottom of an African garden, baboons about a garbage dump. Most large mammals cannot. We may be horrified at the slaughter and mutilation of mountain gorillas and quick to condemn the Rwandan poachers who kill them. Few of us, however, want gorillas in our own backyard. We would want them less if rich pasture were forbidden to our own cattle, or if the gorillas demolished our own banana trees, calmly stripping off the leaves and chewing the last shred of pith. Even a tolerant Hindu starts throwing rocks, or seeks out an off-caste monkey trapper, when the sacred langurs invade his field and he faces a choice between the monkeys' survival or his children's.

The most marginal lands and the most crucial watersheds are threatened by the needs of the poor and the greed of the rich. Imagine a forest canopy 80 to 120 feet above ground, with emergent trees towering to 150 feet. The canopy is so dense that little light reaches the forest floor, and there is almost no undergrowth, except where a tree has fallen to make a "clearing"—a solidly braided tangle of vegetation, plants locked together in the light, racing each other upward before the canopy roof shall close again. In the tallest trees, often invisible to a ground-based observer, sit mantled howler monkeys. They sit. And sit. An individual howler rests for 80 per cent of his day. They are black lumps on the branch, without expression, individuals distinguished mainly by the goitrous swellings made by botfly larvae on their throats. They may reward, and threaten, the observer with dramatic howls, or they may simply pick themselves up, drifting from tree to tree by several paths, so it is even difficult to count the animals while the observer stumbles down and up the ravine-slashed forest floor in their wake.

This is not a picture of the selfless heroism of the primatologist. He (or, as likely, she) is usually enjoying himself. It is not even a picture of how he wastes his time, for he ticks "resting" on his checksheet at two-minute intervals, and builds up a quantitative account of the howlers' lack of activity. He will calculate their energy budget, their choosiness over food, and the food available in their environment, so he may contrast results with the many other howler studies from other sites. He even looks forward to a stand-up argument about howlers at the next conference, for such is the peculiar social life of primatologists.

This is, instead, a picture of the primatologist's luck, that he may escape to the wild. He chooses his study site, if he can, where he will not hear the distant grinding of bulldozers or the nearby toc, toc, toc of the peasant's axe. But primates compete for living space and for human food. They carry human diseases. It is clear that in the future most wild primates will live in natural reserves, islanded in the rising sea of the human reserve. It is even clearer that only protected reserves might preserve the richness of natural communities, the remaining fragments of the ecosystems where our own ancestors evolved.

PHYLOGENY

What are primates? We are an order of mammals, difficult to define on any single criterion. Le Gros Clark in 1962 listed, instead, evolutionary trends within the order (see Table 1.1); this grouping seems to be not a thing but a process. Most of the anatomical trends are directly related to behavioral complexity. Four concern the free and precise movement of hands and forelimb, culminating in our own cleverness with tools. Three are the shift from reliance on smell to reliance on vision, and the detailed spatial patterning of a world full of objects. One is cerebral cortex size and complexity, and two, the lengthening of prenatal and postnatal life, which demand prolonged care of our dependent young and allow time for those young to learn the resources of their environment and the manners of their tribe. If there is an essence of being a primate, it is the progressive evolution of intelligence as a way of life.

All these trends long predate the emergence of man and, to a surprising degree, are represented at different levels of development among our modern primate cousins. "Perhaps no order of mammals presents us with so extraordinary a series of gradations as this—leading us insensibly from the crown and summit of the animal creation down to creatures from which there is but a step, as it seems, to the lowest, smallest, and least intelligent of the placental mammals" (Huxley, 1863):

TABLE 1.1. Primate Evolutionary Trends

---

1. Preservation of a generalized structure of the limbs with a primitive pentadactyly and the retention of certain elements of the limb skeleton (such as the clavicle) which tend to be reduced or to disappear in some groups of mammals.
2. An enhancement of the free mobility of the digits especially the thumb and big toe (which are used for grasping purposes).
3. The replacement of sharp, compressed claws by flattened nails associated with the development of highly sensitive tactile pads on the digits.
4. The progressive abbreviation of the snout or muzzle.
5. The elaboration and perfection of visual apparatus with development of varying degrees of binocular vision.
6. Reduction of the apparatus of smell.
7. The loss of certain elements of the primitive mammalian dentition and the preservation of a simple cusp pattern of the molar teeth.
8. Progressive elaboration of the brain affecting predominantly the cerebral cortex and its dependencies.
9. Progressive and increasingly efficient development of those gestational processes concerned with the nourishment of the fetus before birth.
10. Progressive development of truncal uprightness leading to a facultative bipedalism.
11. Prolongation of postnatal life periods.

(Clark, 1960, Napier and Napier, 1967)

Table 1.2 lists 185 primate species. Classification frequently changes at the species level, so the list is far from definitive. Taxonomists even disagree about the major division between lower and higher primates, between the strepsirhines and the rest. There are four great branches of the primate tree: the strepsirhines, the new-world monkeys or ceboids, the old-world monkeys or cercopithecoids, and the apes and men or hominoids.

TABLE 1.2. Taxonomy of the Primate Order

Suborder Superfamily	Strepsirhini Lemuroidea	Malagasy lemurs
Family	*Lemuridae*	
Subfamily	*Cheirogaleinae*	
	Microcebus	mouse lemur
	murinus	western
	rufus	eastern
	coquereli	Coquerel's
	Cheirogaleus	dwarf lemur
	medius	western, fattailed
	major	eastern
	Allocebus	hairy-eared dwarf lemur
	trichotis	
	Phaner	forked lemur
	furcifer	
Subfamily	*Lepilemurinae*	
	Lepilemur	lepilemur
	dorsalis	Nosy Be
	ruficaudatus	redtailed
	edwardsii	Edward's
	leucopus	whitefooted
	mustelinus	mustelinus
	microdon	microdon
	septentrionalis	Montagne d'Ambre
Subfamily	*Lemurinae*	
	Lemur	lemur
	macaco	black
	fulvus	brown
	mongoz	mongoose
	coronatus	crowned
	rubriventer	redbellied
	catta	ringtailed
	Hapalemur	hapalemur
	griseus	gray
	simus	simus
	Varecia	variegated or ruffed
	variegata	lemur
Family	*Indriidae*	
	Avahi	woolly lemur
	laniger	

	Propithecus	sifaka
	verreauxi	Verreaux' or white
	diadema	diademed
	Indri	indri
	indri	

Family	*Daubentoniidae*	
	Daubentonia	aye-aye
	madagascariensis	

Superfamily	**Lorisoidea**	**Lorisoids**

Family	*Lorisidae*	
Subfamily	*Lorisinae*	*lorises*
	Loris	slender loris
	tardigradus	
	Nycticebus	slow loris
	coucang	common
	pygmaeus	pygmy
	Arctocebus	angwantibo
	calabarensis	
	Perodicticus	potto
	potto	
Subfamily	*Galaginae*	*Galagos*
	Galago	galago, bushbaby
	senegalensis	senegal
	crassicaudatus	thicktailed
	alleni	Allen's
	elegantulus	needleclawed
	inustus	inustus
	demidovii	Demidoff's

Suborder	**Haplorhini**	
Superfamily	**Tarsioidea**	

Family	*Tarsiidae*	
	Tarsius	tarsier
	spectrum	spectral
	bancanus	Borneo
	syrichta	Philippine

Superfamily	**Ceboidea**	**New-World Monkeys**

Family	*Callitrichidae*	*tamarins and marmosets*
Subfamily	*Callitrichinae*	
	Callithrix	marmoset
	jacchus	common
	argentata	silvery
	humeralifer	Santarem
	aurita	buffy tuftedeared
	flaviceps	buffyheaded
	geoffroyi	Geoffroy's or white-faced

TABLE 1.2. Taxonomy of the Primate Order *(continued)*

Suberfamily	Ceboidea	New-World Monkeys
	penecillata	black tuftedeared or blackpencilled
	Cebuella	pygmy marmoset
	pygmaea	
	Saguinus	tamarin
	midas	redhanded
	nigricollis	black-and-red
	fuscicollis	saddleback
	mystax	moustached
	labiatus	whitelipped
	imperator	emperor
	bicolor	barefaced
	oedipus	cottontop
	leucopus	whitefooted
	inustus	inustus
	Leontopithecus	lion tamarin
	rosalia	
Subfamily	*Callimiconinae*	
	Callimico	Goeldi's marmoset,
	goeldii	callimico
Family	**Cebidae**	**Cebid monkeys**
Subfamily	*Cebinae*	
	Cebus	cebus, capuchin
	capucinus	whitethroated
	albifrons	whitefronted
	apella	tufted, brown
	nigrivittatus	blackcapped
Subfamily	*Aotinae*	
	Aotus	night monkey, owl monkey
	trivirgatus	
Subfamily	*Callicebinae*	
	Callicebus	titi monkey
	personatus	masked
	moloch	dusky, moloch
	torquatus	widow, yellowbanded
Subfamily	*Saimirinae*	
	Saimiri	squirrel monkey
	sciureus	commmon
	oerstedii	redbacked
Subfamily	*Pithecinae*	
	Pithecia	saki
	pithecia	paleheaded
	monachus	monk
	albicans	white
	hirstuta	hairy
	Cacajao	uakari

	melanocephalus	blackheaded
	calvus	bald
	rubicundus	red
	Chiropotes	bearded saki
	satanus	black
	albinasus	whitenosed

Subfamily	*Alouattinae*	
	Alouatta	howler
	belzebul	black-and-red
	caraya	black
	fusca	brown
	palliata	mantled
	seniculus	red
	villosa	Guatemalan

Subfamily	*Atelinae*	
	Ateles	spider monkey
	paniscus	black
	belzebuth	longhaired
	fusciceps	brownheaded
	geoffroyi	blackhanded
	Brachyteles	woolly spider monkey
	arachnoides	
	Lathothrix	woolly monkey
	lagotricha	Humboldt's, common
	flavicauda	Hendee's, yellowtailed

Superfamily	**Cercopithecoidea**	**Old-World Monkeys**

Family	***Cercopithecidae***	

Subfamily	*Cercopithecinae*	
	Macaca	macaque
	sylvana	Barbary
	nemestrina	pigtailed
	nigra	Celebes black
	togeana	togian
	maurus	moor
	ochreata	ochre
	silenus	liontailed
	fascicularis	crabeating, longtailed
	mulatta	rhesus
	cyclopis	Formosan rock, Taiwan
	fuscata	Japanese
	sinica	toque
	radiata	bonnet
	assamensis	Assamese
	thibetana	Thibetan
	arctoides	stumptailed, bear
	Cerocebus	mangabey
	torquatus	whitecollared, sooty
	galeritus	agile, Tana River
	albigena	graycheeked
	atterimus	black
	Papio	baboon
	papio	Guinea
	anubis	olive

TABLE 1.2. Taxonomy of the Primate Order *(continued)*

Superfamily	Cercopithecoidea	Old-World Monkeys
	cynocephalus	yellow
	hamadryas	hamadryas
	ursinus	chacma
	Mandrillus	mandrill
	sphinx	drill
	leucophaeus	mandrill
	Theropithecus	gelada
	gelada	
	Cercopithecus	guenon
	diana	diana
	dryas	dryas
	salongo	salongo
	neglectus	De Brazza's
	hamlyni	Hamlyn's
	lhoesti	l'Hoest's
	preussi	Preuss'
	mitis	blue, diademed
	nictitans	spotnosed
	petaurista	lesser spotnosed
	erythrogaster	redbellied
	ascanius	redtailed
	cephus	moustached
	erythrotis	redeared
	cambelli	Campbell's
	mona	mona
	pogonias	crowned
	denti	Dent's
	wolfi	Wolf's
	aethiops	vervet
	Miopithecus	
	talapoin	talapoin
	Allenopithecus	Allen's swamp monkey
	nigroviridis	
	Erythrocebus	patas, red hussar
	patas	
Subfamily	*Colobinae*	
	Presbytius	langur, leaf monkey
	aygula	Sunda Island
	melalophos	banded
	frontata	whitefronted
	rubicunda	maroon
	potenziani	Mentawai
	cristata	silvered
	pileata	capped
	geei	golden
	obscura	dusky
	phayrei	Phayre's
	francoisi	Francois'
	senex	purplefaced
	johnii	nilgiri
	entellus	hanuman

Rhinopithecus	snubnosed langur	
roxellanae	golden	
brelichi	brelichi	
avunculus	Tonkin	
Pygathrix	douc langur	
nemaeus		
Nasalis	proboscis monkey	
larvatus		
Simias	simakobu, or	
concolor	Pagai Island langur	
Colobus	colobus	
polykomos	king	
satanas	black	
angolensis	Angola	
guereza	guereza	
badius	red	
Kirkii	Kirk's red	
Procolobus	olive colobus	
verus		

Superfamily	**Hominoidea**	**Apes and Humans**
Family	*Hylobatidae*	
	Hylobates	gibbon
	concolor	black, white cheeked
	hoolock	hoolock
	klossi	Kloss'
	pileatus	pileated
	moloch	silvery, moloch
	muelleri	Mueller's
	agilis	agile
	lar	whitehanded, lar
	Symphalangus	siamang
	syndactylus	
Family	*Pongidae*	
	Pongo	orangutan
	pygmaeus	
	Pan	chimpanzee
	troglodytes	common
	paniscus	pigmy
	Gorilla	gorilla
	gorilla	
Family	*Hominidae*	
	Homo	human
	sapiens	

Napier and Napier, 1967; P. H. Napier, 1976; P. H. Napier and D. Brandon-Jones, personal comments; Petter and Petter-Rousseaux, 1979; and Mittermeier and Coimbra-Filho, 1981.

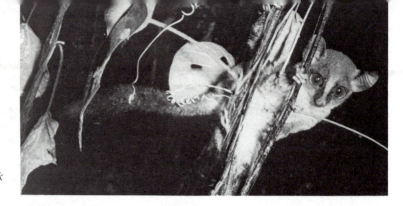

FIGURE 1.1. *Demidoffs bushbaby, one of the smallest primates, among fine lianas in Gabon. (Courtesy C. M. Hladik and P. Charles-Dominique.)*

Strepsirhine comes from the Greek for "turned nose." It includes lorises, bushbabies, and lemurs, whose nostrils are narrow clefts with a damp muzzle like a dog's that connects nose and upper lip. (Figs. 1.1. and 1.2.) Strepsirhines live in Africa, Asia, and Madagascar. They are mainly small-bodied, nocturnal creatures. The potto, the angwantibo, and the slow and slender loris creep stealthily along the nighttime branches, closing a hand on unsuspecting prey. Their cousins, the bushbabies, spring and bounce on elongated hindlegs. Only on Madagascar, which broke free from Africa long before the true monkeys evolved, have strepsirhines evolved into diurnal, troop-living animals. On Madagascar strepsirhines have filled the niches of monkeys and apes.

One genus, the tarsier, is a delight, or a nuisance to taxonomists. (Fig 1.3) In general appearance and habit tarsiers resemble strepsirhines and are often classed with them as Prosimii or premonkeys, as opposed to the Anthropoidea or monkeys and apes. In internal anatomy and reproduction as well as biochemistry and chromosomes, however, tarsiers are higher primates. If tarsiers join the higher primates, the major division in the primate line falls between Strepsirhini and Haplorhini, which means "simple nose." Haplorhine nostrils are surrounded by plain dry skin or fur rather than damp, glandular skin.

The new-world monkeys have been isolated on America almost as long as the lemurs have been on Madagascar. The recency of continental drift and its role in the origin of new-world monkeys is now hotly debated. We are not sure whether new-world monkeys originated from lemurlike ancestors in North America, or monkeylike ancestors from Africa. There is today only one nocturnal form, the owl monkey. Marsupial relatives of the North American opossum fill the niches in South America that the strepsirhines take in Africa and Asia.

FIGURE 1.2. *Variegated lemur, from Madagascar's rain forest, licks its damp muzzle. Lemuriformes and lorisiformes have nose connected to upper lip. (Courtesy P. Coffey and the Jersey Wildlife Preservation Trust.)*

FIGURE 1.3. *A Philippine tarsier clings like a mammalian tree frog. Note the enlarged eye and lengthened heel. (Courtesy P. Wright.)*

One family of new-world monkeys, the tamarins and marmosets, are convergent with birds—brilliant plumaged insect eaters, with trilling voices and a habit of monogamy (Fig. 1.4). The larger new-world monkeys include the manipulative cebus, the resounding howler, and spider monkeys whose diet and social grouping resembles that of chimpanzees. Among all their species, though, none regularly feeds on the ground or has invaded open savannah (Fig. 1.5).

The old-world monkeys fall into two groups: the cercopithecines and the colobines. Colobines are specialized leaf eaters with sacculated stomachs, vats for fermenting cellulose. Cercopithecines are more omnivorous, eating insects and even meat as well as fruit and leaves, and with cheek pouches that hold food to eat later at leisure. Members of five genera of Cercopithecines have colonized the savannah: baboons, macaques, geladas, vervet guenons, and the antelopelike patas monkey (Fig. 1.6). One Colobine has done so as well: the sacred or hanuman langur of India.

FIGURE 1.4. *Emperor tamarin from the upper Amazon with wide set nostrils typical of new world monkeys. (Courtesy P. Coffey and the Jersey Wildlife Preservation Trust.)*

FIGURE 1.5. *An Amazonian red uakari—actually, orange fur and pink bald head. (Courtesy R. Fontaine and Monkey Jungle.)*

FIGURE 1.6. *Patas stands in vigilant posture in the Kenyan savannah. His close set nostrils are typical of old-world monkeys. (Courtesy D. Olson.)*

FIGURE 1.7. *Infant Kloss gibbon on the Mentawai Islands near Sumatra. (Courtesy A. F. Whitten.)*

Apes and men are clearly an offshoot of old-world monkeys. The surviving genera of modern apes are relicts of a wider radiation in the Miocene, when they may have been ecologically distinct from their competitors among the monkeys. Today we have only five ape genera, which are each specialists in their own way. The gibbons and siamang are extreme brachiators that dangle and feed at the ends of swaying branches (Fig. 1.7). They live in monogamous family groups. The great apes are orangutan (Fig. 1.8), gorilla, and common and pygmy chimpanzees. Orangutans are solitary fruit eaters, gorillas are harem-living browsers, and chimpanzees are extrovert parodies of human beings in nearly every capacity—social structure, taste for ripe fruit but not green, mental and emotional development. The little-known pygmy chimpanzee, or bonobo, provides one more comparison, resembling humans in some aspects of their locomotion, communication, and sexual behavior.

FIGURE 1.8. *Adult male Sumatran orangutan. (Courtesy P. Coffey and the Jersey Wildlife Preservation Trust.)*

DISTRIBUTION

Where do primates live? The question, like all others in biology, takes on different meanings at different scales. On the smallest scale, we ask how many species can share the same environment without competing each other out of existence. This will be the subject of the chapters that follow. On a slightly larger scale, we ask how species replace each other in different habitats. On a continental scale, it is a question of the convergence of faunas. Are there regularities between the primate funas of different continents, in spite of their origins from differing stocks? Or are the history and environments of the continents so different that their primates can scarcely be compared (Cody, 1975)?

Primates live in the tropical and subtropical regions of all the southern continents except Australia (see Fig. 1.9.). Only one species lives in more than one continent: the hamadryas baboon, which crosses the Red Sea into southern

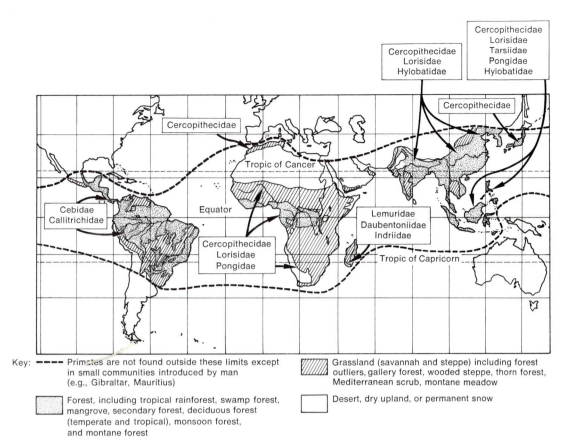

Key: ▬ ▬ ▬ Primates are not found outside these limits except in small communities introduced by man (e.g., Gibraltar, Mauritius)

Grassland (savannah and steppe) including forest outliers, gallery forest, wooded steppe, thorn forest, Mediterranean scrub, montane meadow

Forest, including tropical rainforest, swamp forest, mangrove, secondary forest, deciduous forest (temperate and tropical), monsoon forest, and montane forest

Desert, dry upland, or permanent snow

FIGURE 1.9. *Nonhuman primate distribution. (After Napier and Napier, 1967.)*

Arabia. Two genera cross continents: the African baboons because of the hamadryas, and the Asian macaques, which have a single European and North African species, the Barbary macaque. Grouping Central America with South America, and allowing Madagascar the status of a microcontinent, all other genera are confined to a single continental block.

There are surprisingly similar numbers of species in Africa, Asia, and South America (Table 1.3). Madagascar has fewer, but the number of species is usually closely related to area (MacArthur and Wilson, 1967). By this criterion, even present-day Madagascar has a disproportionately large primate fauna. Six genera and fourteen species of giant lemur died out less than 1,000 years ago, after the arrival of man. If we include these, Madagascar had an enormous number of primates for its size.

In each of the major continents, at least 80 per cent of the primate species live in rain forest. This was probably the original home of the group, and it is not surprising that the most diverse array still live there (Fig. 1.10). More than half the species on each continent can live in drier woodland or woodland savannah (Fig. 1.11). A few are specialized to this habitat, but many are able to live in either moist or dry forest. Few primates, however, range into desertlike environments, and most of those are African, such as the patas monkey and the hamadryas baboon. Some range upward into mountains, and in Asia, where

TABLE 1.3. Primate Faunas by Continent

Madagascar	spp.	S. America	spp.	Africa	spp.	Asia	spp.
Lemuridae		*Callitrichidae*		*Lorisidae*		*Lorisidae*	
mouselemur	3	marmoset	7	angwantibo	1	slender loris	1
dwarf lemur	2	pygmy marmoset	1	potto	1	slow loris	2
hairy-eared		tamarin	10	galago	6		
dwarf lemur	1	lion tamarin	1				
forked lemur	1	callimico	1	*Cercopithecidae*		*Tarsiidae*	
lepilemur	7			macaque	1	tarsier	3
lemur	6			mangabey	4		
hapalemur	2	*Cebidae*		baboon	5		
variegated		cebus	4	mandrill	2	*Cercopithecidae*	
lemur	1	owl monkey	1	gelada	1	macaque	15
		titi	3	guenon	20	baboon	1
		squirrel	2	talapoin	1	langur	14
Indriidae		saki	4	Allen's swamp	1	snubnosed	
woolly	1	uakari	3	patas	1	langurs	3
sifaka	2	bearded saki	2	colobus	6	douc langur	1
indri	1	howler	6	olive colobus	1	proboscis m.	1
		spider	4			Pagai Island	
		woolly spider	1			langur	1
Daubentoniidae		woolly	2	*Pongidae*			
aye-aye	1			chimpanzee	2	*Hylobatidae*	
				gorilla	1	gibbon	8
						siamang	1
						Pongidae	
						orangutan	1
Total Genera-Spp.	12–28		16–52		16–54		13–52

Adapted from Table 1.2.

FIGURE 1.10. *Rain forest primates live in a vertical world of trunk, branch, liana, with separate niches for many different species. De Brazza's guenon, Gabon, is one of the most secretive forest monkeys. (Courtesy A. Gautier-Hion.)*

FIGURE 1.11. *Savannah baboons, Amboseli, Kenya, with impala and zebra. (From S. A. Altmann and J. Altmann, 1970, by permission.)*

there is a broad land connection between south and north, several live in the temperate regions of China and Japan. (Fig. 1.12.)

Madagascar is again peculiar. Only half its species live in the rain forest, the other half in dry forest, and only four species live in both. It seems that Malagasy lemurs have speciated in different environments as an alternative to becoming more adaptable in physiology or behavior within each species.

All three large continents have two main regions of moist forest, with high numbers of endemic species. South America has Amazonia and the south-eastern coastal forests of Brazil. Africa has its massive equatorial forest as well as a western bloc beyond the Dahomey Gap. Southeastern Asia, with the shallow-sea islands of Indonesia, is widely cut off from the other Asian rain forest in south India. On the smaller scale of sympatry within a single forest, rain forests in any continent may have ten or more species living together. Table 1.4 lists some of the best-known primate faunas. Bourliere (in press) points out that primate species' richness is as great as that of the tropical ungulates. The difference is that many ungulates coexist out on the African plains where we can see them, whereas primates mainly flicker in and out of the three-dimensional interlace of rain forest.

During the drier epochs of the pleistocene, the present continuous tracts of rain forest were split into various refuge areas, where the inhabitants di-

FIGURE 1.12. *Tibetan macaque (?) begs near a temple in southern China. (Courtesy D. Hrdy and Anthro-Photo.)*

verged into differing species. When they rejoined, the localized species spread and overlapped. The main moist forest mass of Asia has not, however, been able to rejoin, being cut into peninsulas and islands, which may in part account for both the lower numbers of sympatric species and the possible higher numbers of localized ones.

On the whole, though, the similarities among the three major continents are more striking than the differences. It seems that we can compare the various primate faunas as separate experiments in evolution, which have diversified under many of the same constraints.

WORLD POPULATIONS

There are published estimates for very few world populations of primates. Only 29 species out of the total of 185 are listed in Table 1.5. It is dangerous to publish such figures, because once in print they are quoted forever. If a later worker gives a larger estimate, the government of the countries concerned may quote

TABLE 1.4. Primate Species' Richness

Rain Forest		No. of Species
Africa		
M'Passa, Gabon	Lowland forest	14
Tai, Ivory Coast	Lowland forest	10
Kibale, Uganda	Mountain forest	10
Liboy River, Gabon	Swamp forest	13
Madagascar		
Perinet	Lowland forest	7
Asia		
Kutai, Kalimantan	Lowland forest	8
Kuala Lompat, W. Malaysia	Lowland forest	7
Neotropics		
Cocha Cashu, Peru	Lowland forest	13
Rio Aripuana, Brazil	Lowland forest	11
Raleighvallen-Voltzberg, Suriname	Lowland forest	8
Suframa, Brazil	Lowland forest	7
Barro Colorado, Panama	Lowland forest	5
Woodland		
Africa		
Gombe, Tanzania	Woodland with gallery forest	7
Madagascar		
Berenty	*Didierea* bush and gallery forest	5
Marosalaza	Deciduous woodland	7
Asia		
Polonnaruwa, Sri Lanka	Archeological site	4
Savanna		
Africa		
Bole valley, Ethiopia	Tree savanna	5
Amboseli, Kenya	Tree savanna	3
N'Dioum, Senegal	Gallery forest in Sahelian Zone	2
Fété Olé, Senegal	Sahelian savanna	1
Neotropics		
Masaguaral, Venezuela	Llanos with gallery forest	2

From Bourlière, in press.

TABLE 1.5. Some World Primate Populations

0 to 1,000	1,000 to 10,000	10,000 to 100,000	100,000 to 1M	More than 1M
hairyeared dwarf lemur	Phayre's langur	nilgiri langur	pileated langur	toque macaque
simus hapalemur	moloch gibbon	golden langur	gelada baboon	
aye-aye		Mentawai langur	lar gibbon	
lion tamarin		proboscis monkey	agile gibbon	
bald uakari		Pagai Island langur	black gibbon	
yellowtailed woolly		Mentawai macaque	hoolock gibbon	
woolly spider monkey		Barbary macaque		
liontailed macaque		Japanese macaque		
		Kloss' gibbon		
		pileated gibbon		
		common chimpanzee		
		orangutan		

Petter & Peyrieras, 1970; Coimbra-Filho and Mittermeier, 1977; Mittermeier et al., 1977; Mukerjee in press; Dittus, 1977; Chivers, 1977; Deag, 1977; World Wildlife Fund, 1980; Dunbar, 1977; Oates, 1978; Azuma, personal comment; Rijksen, 1978; Green and Minkowski, 1977; Teleki and Baldwin, 1980.

FIGURE 1.13. *A male golden lion tamarin of eastern Brazil, carrying his infants. World population is 300 to 400. (Courtesy P. Coffey and the Jersey Wildlife Preservation Trust.)*

it as a real population increase and claim there is no need for expensive conservation effort.

The sample is clearly biased toward species with very small populations. They fascinate and dismay zoologists and are easier to survey than those more widely distributed (Figs. 1.13, 1.14, 1.15, and 1.16). A few species with widespread, visible hordes can be put down with confidence as more than a million. The majority of forest-living monkey species are probably far less numerous. World human population just before the invention of agriculture was of the order of 15 million, estimated from humans' large size and sparse distribution and a species range that included all continents (Deavey, 1960; Cohen, 1977).

Suppose, for argument, that the average primate species has a million animals, which is surely an overestimate. That would give a world total of some 180 million wild primates. For comparison, the population of the United States is about 210 million, and India's was 550 million in 1971 and well over 650 million by 1980. World populations of many primate species are no larger than that of a human small town.

FIGURE 1.14. *The yellow-tailed woolly monkey of the Peruvian Andes was not recorded for 50 years until rediscovered in 1974. (Courtesy R. Mittermeier.)*

FIGURE 1.15. *The woolly spider monkey, or muriqi, is the largest South American primate. This is just one protected area with a troop of about 40 animals, in southeastern Brazil. (Courtesy R. Mittemeier.)*

FIGURE 1.16. *The aye-aye of Madagascar probes for insects with skeletal finger. It is killed by villagers as evil luck; there have been three reported sightings in the past 15 years. (Courtesy J. J. Petter.)*

Agencies that must deal with the immediate problems assess the danger to primate populations by more complex criteria than mere numbers. Table 1.6 compares the listings by the Red Data Book of the International Union for the Conservation of Nature, with the Convention on International Trade in Endangered Species (CITES). The official lists overlook some rare species and many distinct subspecies or populations, and perhaps raise undue concern for tarsiers and mouselemurs, but they will be amended as more information becomes available. At present they show some 75 of the 188 primates as currently under threat. As the degradation of both human and animal habitat accelerates, this number can only increase.

TABLE 1.6.   Threatened Primates

	IUCN	CITES
Lemuridae		
lemur, ringtailed	–	I
black	E/–	I
mongoz	V	I
redbellied	–	I
variegated lemur	–	I
lepilemur spp.	E/R/–	I
hapalemur grey	V	I
simus	R	I
hairy-eared dwarf lemur	R	I
dwarf lemur, western	–	I
eastern	V	I
mouselemur, coquerell's	V	I
western	–	I
forked lemur	I	I

Indriidae

indri	E	I
sifaka, verreaux'	E	I
diademed	R/ –	I
woolly lemur	V/ –	I

Daubentoniidae

aye-aye	E	I

Tarsiidae

Tarsiers, Bornean	I/ –	II
spectral	I	II
Philippine	E	II

Callithrichidae

marmoset, silvery	V/ –	II
buffy headed	V	I
buffy tuftedeared	E	I
santarem	I	II
pygmy marmoset	–	II
tamarins, bare-faced	I	I
emperor	I	II
whitefooted	I	I
cottontop	E/ –	I
lion tamarin	E	I
callimico	I	I

Cebidae

squirrel monkey, redbacked	E	I
titi, masked	I	II
bearded		
saki, whitenosed	V	I
uakari, bald	I	I
red	V	I
blackheaded	–	I
howler, brown	I	II
black	I	II
mantled	–	I
spider, blackhanded	–	I/II
woolly, Hendee's	E	II
Humboldt's	V	II
woolly spider	E	I

Cercopithecidae

mangabey, Tana River	E/ –	I/II
mandrill	E	II
macaque, Pagai Island (pigtailed)	I	II
liontailed	E	I
Barbary	V	II
colobus, red	E/R/ –	I/II
black	V	II
olive colobus	R	II
langur, hanuman	–	I
Gee's	R	I

TABLE 1.6.   Threatened Primates (*continued*)

	IUCN	CITES
nilgiri	V	II
pileated	–	I
Mentawai	I	I
proboscis monkey	V	I
Pagai Island langur	E	I
snubnosed langur, golden	R	II
douc langur	E	I
Hylobatidae		
gibbon, agile	–	I
black	I	I
hoolock	–	I
Kloss	V	I
lar	–	I
moloch	E	I
pileated	E	I
siamang	–	I
Pongidae		
orangutan	E	I
chimpanzee, common	V	I
pygmy	V	I
gorilla	V/E	I
All Other Primates	–	II

Total number of species listed as threatened (IUCN) or in imminent danger of extinction (CITES) = 75.

1. IUCN Red Data Book, the official list of threatened species, distinguishes E, endangered; V, vulnerable; R, rare; and I, indeterminate.

   The Convention on International Trade in Endangered Species (CITES) distinguishes Appendix I species in imminent danger of extinction, and Appendix II species liable to extinction if trade is not controlled.

2. Slashes (for example, E/V, I/II) show that different subspecies have been given different status.

3. Nomenclature is given in the lists without authorities. This means that for some species it is not clear which populations the species name is intended to include.

4. Unlisted species include all species in the family Lorisidae, all species in the genera of cebus, owl monkey, sakis, guenon, patas, and gelada, and other species of the listed genera. Nonetheless some of the unlisted species are probably rare, for example, the Tonkin snubnosed langur.

From Wrangham and Dunbar, 1979.

Global populations are only a sum of local populations. It is extremely difficult to extrapolate from a small region to a large one, or even from a richer to a sparser area of the same forest. That is why people are so reluctant to estimate world totals, except when arguing the urgency of conservation measures.

RESERVES AND MINIMUM POPULATIONS

The theory of island biogeography has become one of the most exciting intellectual fields of recent years (MacArthur and Wilson, 1967). This is the theory and study of the relationship between the number of species and the area that supports them when that area is isolated as an island from other habitats where the same species can live. The study of habitat islands shows clearly that the smaller the area, the fewer the species. Some species become too rare to survive (Fig. 1.17.).

FIGURE 1.17. *A Hanuman langur begs from a priest of Shiva in North India. Primates adapt to human habitat wherever they are allowed to. (Courtesy S. Hrdy and Anthro-Photo.)*

In practice, island biogeography turns into the question, how small can a mammalian population be to survive environmental fluctuations, competition from other species, and inbreeding depression? Frankel and Soule (1981) suggest that a conservative estimate of the minimum, bottleneck number to avoid disastrous inbreeding would be an effective breeding population of 50, or a longer-term maintained effective breeding population of 500.

These numbers depend on a series of assumptions. They are offered not as absolute figures but as an indication of when one should become extremely worried. Note that an effective breeding population is not a number of animals. It is a figment of a geneticist's imagination: a group evenly divided into males and females, where each adult contributes an equal number of genes to the next generation. Polygamy or differential success in rearing babies means any real mammalian population must be far larger than its effective breeding analog. What this estimate means is not that you can get away with the minimum, or should give up below the minimum, or even that you can be sure what the minimum is. Frankel and Soule are saying that if you are trying to save an isolated mammalian population in the tens or hundreds, you are fighting from the edge of the cliff. It is dangerous to forget the cliff is there, and worth making even a rough estimate how far it is behind you.

We have seen that several whole species are at this point. Many more, however, live in increasingly isolated pockets of habitat. The best-known wild primates of all are Jane Goodall's chimpanzees of the Gombe Stream in Tanzania. In 1960, when she was beginning her studies, the Gombe Stream was linked to large tracts of woodland eastward and northward perhaps even to Burundi. Today much of this area is cultivated. The chimpanzees of the Gombe Stream will soon be islanded. Their reserve is 32 km^2 (square kilometers)— enough for an estimated 67 to 108 independent adults and adolescents, or 100 to 160 in all. We do not know if this is big enough, when there will be no more migration in or out of the reserve (Goodall, 1983).

The mountain gorilla, *Gorilla gorilla beringei*, is estimated to number 315 to 385 in total. In the three-nation park of the Virunga Volcanos, the population is now about 252 to 285 animals. A rising number of infants and young indicates the population may be recovering from the 1978 series of attacks by poachers, when at least seven gorillas were killed in little more than a year. However, the population before 1978 declined drastically from Schaller's estimate of 400 to 500 individuals in 1960. The park is 120 km^2, much of it volcanic montane heath unsuitable for gorillas, and with the lower slopes under constant pressure from farmers and fuel–wood cutters.

The Virunga Volcano reserve is almost an island. It connects to the forests of Zaire only by a wooded strip less than one kilometer wide, traversed by a road. It is the most important water catchment of Rwanda, and the water table of the area depends on the park's integrity. The density of farming on the land round the park is so great that if the rugged mountains could be cultivated at similar density, they would provide land for 36,000 people, or just three months of Rwanda's annual population growth (Weber and Vedder, in press).

The Virunga park epitomizes the conservationist's dilemma. Clearing it completely would be disastrous for agriculture as well as wildlife. But how do you tell a woman gathering wood that she must not break living trees to heat the meal for her children, or to cook for herself and the baby she expects to bear? How do you convince her that the square kilometer of land just above her homestead, which might indeed support 300 people like her children, is better used supporting two gorillas?

Other primates are threatened by the large, not small, interests. The hardwoods of Southeast Asia and West Africa sell for great profit in Japan and the West. The most primitive of all gibbons is the Kloss gibbon. It lives on the Mentawai Islands off Sumatra, with three other unique primates, including one of the only two monogamous species of old-world monkeys. Siberut, largest in the Mentawai Islands, was totally allocated as logging concessions. This would have destroyed not only the primate population but the way of life of all the 18,000 local people who have lived in harmony with the forest for at least 2,000 years. Fortunately, the government of Indonesia has agreed to make Siberut a biosphere reserve, in view of the island's importance to science, but implementation is slow (Whitten, 1980, 1982).

Meanwhile, as of this writing, the major part of the presumed range of the bonobo or pygmy chimpanzee has been assigned for logging. This includes all areas where the bonobo is still known to exist, including the two sites where bonobo are being studied. The bonobo has a good claim to the status of the animal that most resembles human ancestors in its skeletal structure and contemporary humans in its behavior. At the very least it is the third member of the triumvirate of human, chimp, and pygmy chimp, so close that a biochemist or geneticist can barely distinguish one from another. We do not yet know if the fate of the pygmy chimpanzee (and, for that matter, of the forest dwelling people) will outweigh the price of ebony (Fig. 1.18.), (Susman et al., 1981, Susman, personal communication).

And even presuming that it does, we will not know for 50 or 100 years whether the reserves set aside were large enough for a minimum population to cope in the long term. At least we do know already that it may be crucial whether the long-term stock of the reserve is in thousands or hundreds, which means asking the government of Zaire to save thousands, not hundreds of square kilometers of untouched rain forest.

FIGURE 1.18.   *The pigmy chimpanzee lives among giant ebony trees in Zaire. Is their forest more valuable dead or alive? (Courtesy N. Badrian and Lomako Forest Pygmy Chimpanzee Project.)*

CENSUSES AND BIOMASS

How many primates live in a given area? Counting them is not as easy as it sounds. The only certain method is to spend a year in one spot, learning to recognize individuals and learning how their ranges change and overlap from season to season. This method also provides a valuable check on more extensive census counts.

A larger-scale census is commonly done by the strip method. The observer walks a series of trails chosen to cross a representative sample of habitat and records the animals he sees.

Then he tries to estimate how far he can see, correcting for different visibility through different types of forest. He corrects for different visibility for different age classes or different species; adult males of one sort may come and bellow at him, whereas other animals or species melt into the background. He may even correct for different behaviors; how many animals does he see when they sit still and how many, and at what distance, in panicky flight? If he rarely sees one still, his strip width may be smaller than he thinks, because he misses those that are far enough from him not to panic (Wilson and Wilson, 1974; Green, 1978; Daniels, personal comment). The handbook for actually doing censuses is *Techniques in the Study of Primate Population Ecology*, by the National Research Council, 1981.

Out of all this, he concocts a density figure. Eisenberg (1979) points out that such transect densities are mainly useful as indicators of the *relative* abundance of species within one area. If one knows the average home range size, and it is possible to set transects randomly across the home ranges, one can arrive at absolute estimates (Eisenberg, 1979; Caughley, 1977).

I cannot improve, or even condense Eisenberg's description of the three types of density figures:

1. K density is that value derived from the study of one or more troops for a sufficient time to determine the home range in terms of some area measurement. (A volumetric measure would be better but is as yet impractical). This K density probably is near the carrying capacity for the species in question within the particular environment where the measurements were carried out. Variations in K density within one species from one habitat to the next probably reflect real variations in carrying capacity for the particular habitat.
2. Ecological density results from a study where the number of troops or individuals is expressed in terms of the area which may be termed suitable habitat.
3. Crude density refers to counts expressed in terms of the total base area of the survey whether or not the base area contains suitable habitat for the species in question. Crude density and ecological density become almost synonymous if the habitat is so uniform as to constitute ideal habitat for the species in question. This situation is seldom realized in practice. Thus, crude density estimates tend to be lower than ecological density and the latter lower than K density. All three expressions of density are useful but must be compared separately.

Listing primate densities is a task for a computer file, not a textbook. However, even partial surveying of the literature leads to a few general conclusions. Small-bodied strepsirhines may reach very high concentrations of several hundred animals per square kilometer. The new-world monkeys as a whole have lower population density than the old-world monkeys. A few old-world monkeys reach densities of several hundred, but most populations of both old- and new-world monkeys lie between 10 and 100 per square kilometer. All apes except Muller's gibbon are thinly spread, that is, five or less per square kilometer. This means that a reserve which you hope will hold a viable population of apes should be around a hundred times larger than a reserve to hold the equivalent number of lemurs.

Where productivity of the forest is high, as measured by total rainfall or months of rainrall, or soil type, or vegetation density, there is a higher density of primates. Perhaps not so obvious is that there are more species where there is higher primate density, in the centers of primate distribution. However, at the edges of distribution a few adaptable species may reach high density without sharing the habitat.

To plan reserves for the primates' future, you need data in terms of population. How many animals live in your proposed reserve? Is the population large enough to allow outbreeding and to avoid extinction through random accidents? How much land will a minimum population need and what sort of habitat? You also need to know population density to think about social behavior. How many monkeys does a monkey meet in the course of its lifetime?

If you are interested in ecological relations, *biomass* is a more revealing measure. Biomass is the weight of a given species, or group of species, per unit area. Plant biomass is the amount of standing plants, and primate biomass represents the monkeys and apes that live on them (Eisenberg and Thorington, 1973; Eisenberg, 1980).

Table 1.7 lists some known primate biomasses. In many forests, primates

TABLE 1.7. Biomass Estimates (kg/km²) for 18 Primate Communities

Africa		Asia		Neotropics	
Kibale Forest, Uganda (10 sp)	2217	Polonnaruwa, Sri Lanka (woodland) (4 sp)	2840	Hacienda Barqueta, Panama (3 sp)	4712–5022
Tai Forest, Ivory Coast (10 sp)	1025	Kuala Lompat, West Malaysia (7 sp)	736–1295	Rupununi River, Guyana (4 sp)	1306
Liboy River, Gabon (13 sp)	>595–761	Mean of 5 lowland sites in West Malaysia (6 sp)	750	El Triunfo, Bolivia (4 sp)	704–1129
Mpassa, Gabon (14 sp)	>524–749	Ketambe, Sumatra (6 sp)	727	Cocha Cashu, Peru (11 sp)	367–798
		Kutai, Kalimantan (9 sp)	312	Essequibo River, Guyana (6 sp)	620
				Barro Colorado Island, Panama (5 sp)	418–542
				Ralleighvallen-Voltzberg N.R., Suriname (8 sp)	287
				Samiria, Peru (7 sp)	272
				La Macarena, Colombia (4 sp)	129–233

Rain forest sites, except where otherwise mentioned.
Bourlière, in press.

form a major component of the fauna of which ecologists must take account in these calculations, and plants and predators in their evolution (Bourliere, in press).

The remaining chapters in Part I consider some of the determinants of primate population and biomass and the ways in which different species exist. Chapter 8 returns to the consideration of whole forests and how their species interact as a community.

SUMMARY

Primates are divided into four great groups: the strepsirhines, the new-world monkeys, the old-world monkeys, and apes together with humans. Primates originated in moist tropical forest, where most species still live. Half the species on each continent may range into dry woodland; only Africa and Asia have true savannah-living primates. Twenty-eight of the 185 primate species live on Madagascar; about 50 each in South America, Africa, and Asia.

Most primate species have populations the size of a human small town. At least 8 number less than 1000, and only the most numerous exceed a million animals. Seventy five are listed as threatened by the IUCN or CITES. Inbreeding depression, random fluctuations, and competition for habitat threaten small, isolated populations in the tens or hundreds of animals. Many reserves can support only a few hundred primates, for instance, the Gombe Stream Reserve for chimpanzees in Tanzania, and the Virunga Volcano Park of Rwanda, Uganda, and Zaire. Primates are threatened by habitat loss both to large-scale loggers and to small farmers in need of food and fuel.

Censuses show strepsirhine species may have population densities of several hundred animals per square kilometer of suitable habitat; old- and new-world monkeys tend to have 10–100 animals per square kilometer, and great apes less than five. A reserve to hold similar populations needs to be roughly a hundred times larger for apes than for lemurs.

Rainforest may shelter 5–14 sympatric species of primates, with biomass from 270 to more than 2000 kilograms per square kilometer. Primates thus form an important proportion of the ecosystem, quite aside from our own self-centered interest.

FIGURE 1.19.   *David Greybeard of the Gombe Stream. (Courtesy P. Marler.)*

CHAPTER
2

ECOLOGICAL PRINCIPLES

COMPETITIVE EXCLUSION

What limits primate populations, aside from the influence of man?

Populations of wild animals may be limited by physical factors such as cold, by predation and disease, by the food supply, and by competition among species. Interspecific competition interacts in a complex fashion with the other factors. Even the effects of cold may depend on whether something else is occupying all the warm tree hollows.

A species' way of life may be pictured as a multidimensional space, with each aspect of its requirements measured in a different dimension. This is the *ecological niche* (Hutchinson, 1978).

Ecologists usually divide the limiting factors into those that are density-dependent and those that are not. Cold, for instance, might be considered to cull a certain proportion of the population, or to limit the extension of its range, without reference to the density of the population. Food supply, on the other hand, might have no effect at low population densities but be crucial at high ones. Primate behavior is so variable that many species can find food and refuge in conditions far outside their normal range. Hamadryas baboons can live outdoors in a Russian winter and Japanese macaques range free in Texas. The possible niche of each species is set by absolute physical constraints. However, the observed limits of each species are usually set by interspecific competition for the environment.

One tenet of ecology is the principle of competitive exclusion. This principle, which entered biology in its mathematical form with the work of Gause, can be summed up in Garrett Hardin's words: "Complete competitors cannot coexist" (Hutchinson, 1978). Much of the mathematical treatment revolves around the minimum differences that will allow two species to coexist, to occupy separate niches in the same place. But what are complete competitors? They are species governed by the same limiting factors in a community at equilibrium. Because many species do coexist in a natural community, ecology becomes the study of what limits populations and how populations of different species diversify their requirements.

One caution: in real cases you must specify what you mean by the community at equilibrium and what you mean by primates living in the same place. Spatial separation may occur at very short range. Some primates feed high in the canopy, or in the towering emergent trees; others feed low in the understory. Some stick to the mature trees of primary forest; others run among the tangle of vines and saplings that grow up where a tree fall has let light penetrate to the forest floor. On a larger scale, some prefer dry land forest and others swamp forest, although the two may be so closely intertwined that it still seems on a map that this is living together. Only at the largest scale do you call species *allopatric* rather than *sympatric*. Many closely related species or genera are allopatric; that is, their geographic ranges do not overlap. For instance, tamarins live north of the Amazon; marmosets apparently occupy similar niches south of the Amazon. The minute pygmy marmoset (Fig. 2.1), which clings to the bark of trees, scraping off sap, is different enough to be sympatric, or overlapping in range, with the tamarins (Moynihan, 1976; Eisenberg, 1979).

If there is complete overlap in microspace, the animals differentiate their niches in some other manner, such as specializing in different food (MacArthur and Levins, 1964).

FIGURE 2.1. *Pygmy marmosets' hostile display. (Courtesy M. Moynihan.)*

There is a similar problem with time as with space. What do we mean by a community at equilibrium? If disturbances are so frequent that we accept them as part of the normal scheme of things, we may still call it an equilibrium. Equilibrium communities with frequent large-scale disturbance hold few species—the arctic as opposed to the tropics, or a fast stream with spring floods as against a sluggish pond. Communities with rare disturbance hold most species, the ultimate in species' richness being tropical rain forest. A little disturbance, or localized disturbance, in rain forest, however, actually increases the possibilities of differing environments. A tree fall, a landslide, or a typhoon create a patch of growing trees amongst the mature ones. Such patches of differing habitat multiply the links of food webs; many species depend on forest in different growth stages for their varying needs (Connell, 1978). Thus, equilibrium relates to the generation time of the most important species. In the case of slow-growing rainforest trees, thus means stability over centuries or even millenia.

Some disturbances, whether by local accident or seasonal shift, may reduce populations of many different species to the point where they do not outcompete each other, as in the "paradox of the plankton" (Hutchinson, 1951). There are also "fugitive species," whose habitat is constantly being formed and disappearing. These add to the variegated patterning of an equilibrium community.

With primates, much of the interest lies in guessing how our ancestors evolved from narrow confinement in a particular niche into our present cosmopolitan state.

There is a continuum between the large niche of an adaptable species, able to use a wide range of environments, and the narrow niche of species specialized for a particular part of this range (Gause, 1947; Hutchinson, 1978). We may see this contrast exemplified in the included niche, in which the food, space, and other requirements of one species are included in the niche of another. If the species with the included niche is to survive in an equilibrium situation, it must outcompete the wider-ranging species within its own limited domain. The included species is not only a specialist; it must be a successful specialist. Diamond (1975) analyzed the bird fauna of New Guinea's offshore islands in these terms. He found 13 "supertramp" species that colonized islands recently stripped by volcanic activity and used almost any habitat. In ancient forests, however, the tramps could not compete with the many specialized sedentary birds, "companions in starvation."

Primates as a group may be adaptable, but the fine tuning that lets 14 primates coexist in the rain forest of Gabon is the sign of a community of specialists, whereas tramps like baboons range from the Tibesti Oasis to the Cape of Good Hope, but are outcompeted in true rain forest.

It may be, however, that we emphasize food competition too much, just because food plants hold still to be counted. Often predators are the actual limits to prey, even to the point where two or more prey species that would exclude each other without predation coexist. Darwin (1859) was already aware of this effect. "If turf which has long been mown (and the case would be the same with turf closely browsed by quadrupeds) be let to grow, the more vigorous plants gradually kill the less vigorous, . . . thus out of twenty species growing on a little plot of mown turf (three feet by four) nine species perished, from the other species being allowed to grow up freely. .... The amount of food for each species of course gives the extreme limit to which each can increase; but very frequently it is not the obtaining of food, but the serving as prey to other animals, which determines the average numbers of a species."

The potential ecological niche of a species may be thought of as a multidimensional space whose boundaries are set by climate, spatial and temporal homogeneity, food supply, and predation. The observed niche, in contrast, is a defended space, whose boundaries are intermittently squeezed in by competition from other species.

SIZE

The different meanings of equilibrium when we look at different amounts of area or time gave us one example of the fundamental importance of scale in ecology (Hutchinson, 1971; Hutchinson and MacArthur, 1959). Body size is another case. Size is one of the most pervasive mathematical factors in biology, one of the most intuitively accessible, and one of the most commonly ignored because we take the perspective of our own body size for granted. When Hutchinson asked, "Discuss the influence of man's size on his view of the universe" in a graduate examination, none of the students attempted an answer.

Most biological functions increase as some power of the body size, but this is rarely a 1:1 relation. Basal metabolic rate in mammals, for instance, increases as the three quarter power of weight; an elephant needs relatively less resting energy than a mouse. However, metabolic rate increases rather faster than surface area of same-shape solids, which grows as the two thirds power of the volume.

The general allometric formula is $y = bx^a$, where y is the size of the process under discussion, x is some measure of body size, a is the relation, and b is a constant which gives the starting point of y when $x = 1$. Thus basal metabolism $= bx^{3/4}$ (Kleiber, 1932). This implies that the calories needed for basal metabolism in the food supply of a larger animal may rise only as the three quarter power of weight, with further implications for the size of home range needed to provide those calories.

The next complication is that the allometric proportion is different if you calculate it across a wide phylogenetic range, not within limited phylogenetic groups. The relative brain size jumps from reptile to mammal, and from insectivore to prosimian to monkey, as though the regression lines for species' points were rungs on a sloping ladder. The slope of each rung fits the formula $y = bx^{3/4}$, with the same allometric coefficient as basal metabolism. Climbing to another rung, however, changes the constant b, the level where the line begins. Biologically, each rung represents an *adaptive grade*, a breakthrough in organization, which has meant that each new phylogenetic group has needed a larger brain for its body size. The pattern becomes clear only when plotted against body size within each group. (See Fig. 2.2.)

Hutchinson (1959) showed that for some corixid water bugs and for the beaks of Galapagos finches, a size difference of 1:1.3 seemed enough to allow two closely similar species to coexist in the same place. Ratios of about that magnitude have since been found in other groups, with suggestions that the

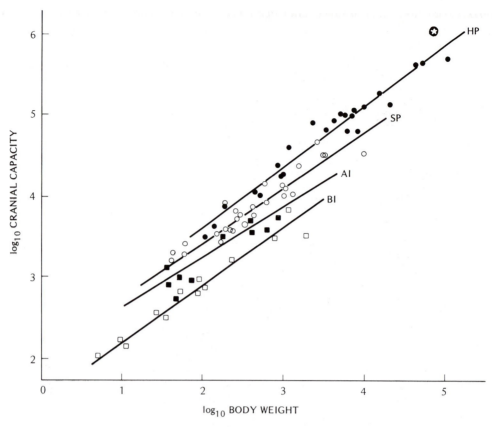

FIGURE 2.2. *Cranial capacity plotted against body weight on logarithmic coordinates for "basal" insectivores (open squares BI); "advanced" insectivores (closed squares, AI); strepsirtrines (open circles, SP) and haplorhines (closed circles, HP). The reduced major axes are shown as lines for each group. The difference between slopes is not significant, but the size of brain for body weight increases between taxonomic groups. Human cranial capacity, shown by the circled star, is not included in the calculation. (Martin, in press.)*

ratios are lower in the tropics than in temperate zones, indicating close species packing. Of course, the biological meaning of such a ratio is a sum of many aspects of the animal. If it is a ratio of beak size, this presumably reflects the size of favored food items. If it is a weight difference, it relates to a complex sum of metabolism, activity, foraging range, and a host of other factors.

Terrestrial mammals have a bimodal size distribution with a clear break between little mammals and large ones. There is no such break among warm-blooded swimming or flying vertebrates. Perhaps the energetic constraints of running and leaping have favored two rather different adaptive complexes (Bourliere, 1975).

Some advantages of small size are concealment from predators, access to small food sources such as the invertebrates and fallen seeds of the lower layers and underground, maximum use of favorable microclimates, and rapid reproduction with large litters, which allows rapid population growth and even genetic adaptation to environmental change (Bourliere, 1975). The chief disad-

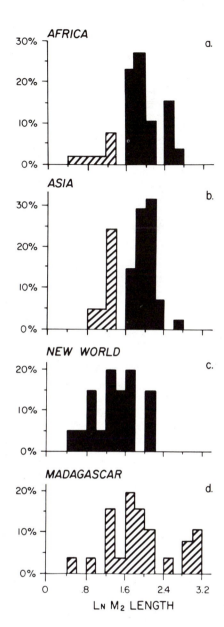

FIGURE 2.3. *Size frequency distributions for primates of each continent. The extinct subfossil giant lemurs are included. The measure is second molar length, which (oddly) is a good indicator of body weight. Strepsirhines shown hatched, haplorhines solid. In Africa and Asia, strepsirhines are the smaller forms; in South America, ceboid monkeys fill nearly the whole spectum; in Madagascar, strepsirhines do (or did). (Fleagle, 1978.)*

vantage of small size is the metabolic cost, and associated high food needs per unit of body weight. "A biomass of 3,000 kilograms of herbivores will consume about 2,000 kilograms of grass per day if made up of voles, and 200 kilograms if composed only of two hippopotamii" (Bourliere, 1979).

If one plots the size distribution of primates on each continent, there is a break between small and large, just as there is among the terrestrial mammals as a whole (Fig. 2.3). The break does not correspond to phylogeny. In Africa and Asia the small forms are the nocturnal prosimians, but in South America the monkeys radiated to cover the whole primate size range, whereas in Madagascar the prosimians did so (Fleagle, 1978).

The real size dichotomy comes, in three of the four continents, between

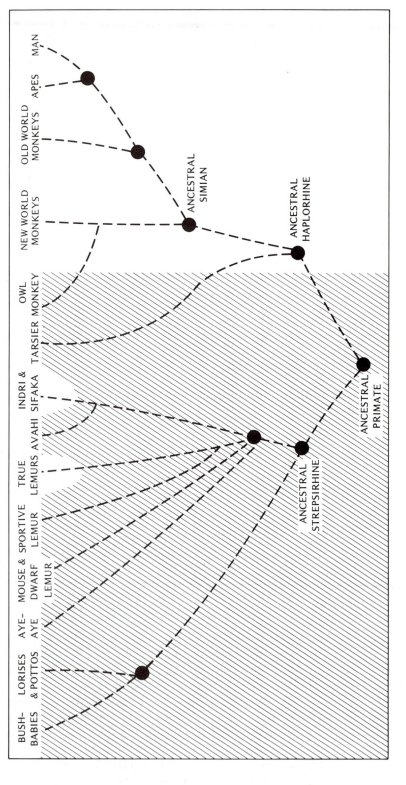

FIGURE 2.4. *Nocturnal and diurnal primates. (Martin, 1979.)*

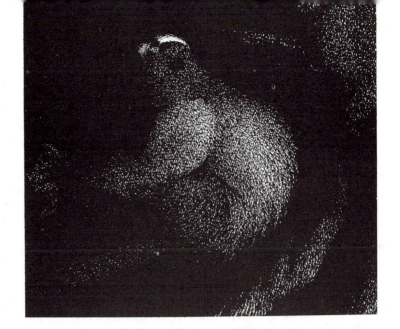

FIGURE 2.5. *The owl monkey, the only new-world nocturnal primate in hostile "arch display." (Courtesy M. Moynihan.)*

nocturnal and diurnal forms (Martin, 1979). (See Fig. 2.4.). The nocturnal primates are, on average, ten times smaller than diurnal primates, averaging 0.5 кg as against 5 кg. Accordingly, they mainly eat high-energy food. The complex of adaptation of small size and nocturnal life also involves feeding on high-energy insects and gums (Fig. 2.5).

Charles Dominique (1978) points out that arboreal frugivores and insectivores of very small body size would be in direct competition with forest birds; "escape" to a diurnal niche may have involved a body size large enough to differentiate bird and monkey.

TIME

Scaling of time, like scaling of size, pervades every aspect of life. Timing, like size and space, can help separate competitors. Even if they compete for the same foods, one can come by day and one by night, or one can visit a fruiting tree early and sporadically, giving way when the stronger species arrives (MacKinnon, 1978; Raemakers, 1978). Even the stages of growth may be separate in time: all lemurs wean their young during the wet season, but the larger ones, with longer periods up to weaning, bear their babies, respectively, at earlier periods in the year. Thus, even the infants of larger species are larger than smaller species infants in any one month (Hutchinson, 1959; Petter-Rousseaux, 1968).

The length of animals' lives increases with body size. There are, again, adaptive grades. A primate has a longer life span for its size than other mammals. As with brain size, we presume these grade differences reflect some change in biotechnology that allows individuals to live longer.

Life history components vary with body size and with phylogenetic grade. Prosimians have a different reproductive strategy from monkeys, with less intimate placentation, shorter gestation periods, and more rapid postnatal growth, even when corrected for body size (Leutenegger, 1976; Rudder, in press).

41   *Ecological Principles*

R AND K SELECTION

Species whose populations grow, then are destroyed, at erratic intervals are selected to maximize their reproductive rate (r) if the intervals are longer than generation length. The evolutionary pressures of r selection are to have short life cycles, many offspring, and small investment in each offspring. In contrast, where the environment is relatively stable and relatively full, so that one's offspring face a predictably competitive environment near its carrying capacity (K), the selective advantage lies with those parents who invest a great deal in each offspring. This is K selection, which selects for few offspring, well equipped for life by long gestation and sheltered, prolonged immaturity (MacArthur and Wilson, 1967; Wilson, 1975). (Fig. 2.6)

This distinction, like most, is a relative one. Primates are K-selected compared to most other animals, but even closely related primates differ between themselves. The savannah-living vervet guenon matures earlier, breeds faster, and dies sooner than its forest congeners, the Sykes and De Brazza guenons (Rowell, 1979). Senegal galagos from South Africa bear twin young twice a year, whereas Allen's galago, of the same body size but living in the stable conditions of the Gabon rain forest, has one young once a year. There is a gradation from the r-selected galago that lives in dry, temperate zone forest to the K-selected races and species in progressively wetter, more tropical, and more stable environments (Martin, 1981).

It is possible to measure the time dimensions of ecological niches, as well as spatial dimensions. Suppose an environmental crisis like a drought occurs on average every five years. A chimpanzee with a forty-year life span sees eight such droughts and a Senegal bushbaby probably only one. The bushbaby's population can suffer a higher percentage loss, with four infants born to each female in the northern races, or 16 in South Africa, during the four years between droughts. Wild chimpanzees' birth interval is about five years, so their females could not afford to loose even one baby each drought. The chimp must have some buffer that allows it to treat the environment as more stable than the bushbaby does.

In spatial terms, the niches of a relative specialist may be included within the niche of a tramp or weed species, if the specialist is better at its specialty. Specialists tend to be more K-selected than weedy, opportunist species of the same body size. Weeds and tramps by definition cope with environments that fluctuate in time and vary in quality. It is an important aspect of the primate

FIGURE 2.6. *In an ideally simple environment, population growth may follow an S curve, determined only by the maximum rate of reproduction (r) and the carrying capacity of the environment (K). Growth of a population of* Paramecium caudatum *fitted to a logistic curve. (Hutchinson 1978, after Gause.)*

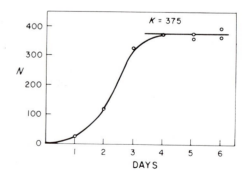

phenomenon that some can override the fluctuations of the environment. Mental agility has in part replaced reproductive agility.

Some of this buffering of the environment depends on individual memory of food source in time of need. Nishida watched savannah-living chimps with a home range of 200 square kilometers, who presumably, like more sedentary forest chimps, knew the botanical resources of their realm. It may, even in primates, depend in part on social memory. The oldest baboons might know which water hole lasts through the kind of drought that comes not in five years but once in a twenty-year generation, and pass this knowledge to their grandchildren (Rowell, 1966).

In conclusion, the mathematical theory of the ecological niche has been far advanced to relate size of food patches to body size, timing of environmental fluctuations to life span and life history, and the number and closeness of species to the rigors of the environment.

Two aspects of ecology are most relevant for students of animal behavior. The first is the contrast between adaptable, opportunist, wide-ranging species, the tramps, and the successful specialists in their narrow niches.

The second is that population ecology has focused on opportunists as r-selected, breeding fast to meet environmental challenge. Evolutionary advances, on the contrary, whether it be air breathing, warm bloodedness, or the human brain, have meant physiological-behavioral opportunism, not simple speeding of reproduction. In fact, they commonly go with slowing of growth and reproduction, because it takes more time to build a more complicated body at a higher adaptive grade. These advances in biotechnology increase the animal's size and life span compared to the patch size of the environment, increasing the range it can travel to find food, increasing the efficiency of finding food, and increasing the means of outliving bad seasons. This is the active expansion of the animal's niche, the conquering of the variations in the environment by means of the *milieu intérieur*, in the phrase of Claude Bernard. The study of the evolution of primate and human behavior is the study of the most recent of these breakthroughs: how our behavioral adaptation has taken over our environment, so far, at least to the advantage of our species' population.

SUMMARY

The competitive exclusion principle states that complete competitors cannot coexist. Animals may differentiate their ecological niches in space on as small a scale as height in the trees, or as large a scale as north or south of the Amazon. If they overlap greatly in space, they may differ in other ways, particularly in choice of diet. The populations may also be limited by culling, which keeps them at such low density that they do not compete—either culling by predators, or impartial disturbances such as landslides and cyclones.

Most aspects of animals' biology are interrelated with their body sizes. Metabolic rate in mammals increases as the three quarter power of body size. Brain size also increases allometrically with body size, but with jumps between different adaptive grades. Life span and the lengths of pregnancy and immaturity also increase with body size, again with jumps between the different primate families.

Ecologists distinguish r-selected and K-selected species. The r-selected species are adaptable, fast-breeding, with multiple offspring, the weeds and tramps that colonize new habitats. The K-selected species live in more stable habitats and breed more slowly, investing more in each offspring so that it may compete in a predictable environment. K-selected species are commonly specialists in narrow, stable niches.

Primate evolution and human evolution is a paradox. We have become larger, with long life and immaturity, and few, much loved offspring, and yet we are more, not less, adaptable. Mental agility buffers environmental change and has replaced reproductive agility.

CHAPTER 3

FOOD AND FEEDING

NUTRITION

Primates, like other animals, need to eat protein for growth and replacement of tissues, carbohydrates and fat for energy, and various trace elements and vitamins for chemicals which they cannot produce themselves.

Proteins come from insect and vertebrate prey and from leaves, particularly young leaves. To some extent, leaves and prey are alternates in a primate's diet. Those animals that can digest leaves have less need of protein from prey, particularly as they also harvest the leaf-fermenting microbes of their guts as an additional source of protein.

Energy is needed for metabolism. Most energy is used in basal metabolism. This means that the body size of the animal is the most important single measure of the amount of energy needed. Secondary factors are the animal's activity and whether it maintains a constant temperature. Many prosimians save energy by letting their temperature drop several degrees at night, and basking in the sun to warm themselves as the day begins. Fruit containing sugar, starch, or oil, is the commonest energy source, but meat and insects are even more concentrated sources.

Vitamins and trace elements mainly come from the normal diet of fruit, leaves, and prey. Vitamin B_{12} is needed by all primates. It is possible that lepilemurs, which seem to eat no animal prey, the usual source of B_{12}, produce it through bacterial action in the hind gut, then eat their own feces. Wild mountain gorilla also consume feces (Charles-Dominique and Hladik, 1971; Harcourt and Stewart, 1978). Various primates have been seen eating earth: gorilla (A. Goodall, 1977), guereza colobus (Oates, 1977), and indri (Pollock, 1977). Perhaps the earth contains minerals needed by these specialized leaf eaters, although analyses have not given clues, or perhaps there is some mechanical aid to digestion.

45

Table 3.1 shows aggregated data for the proportions of leaf, fruit, and prey in primate diets. The data should be treated as indicative only, because the same species may eat widely varying diets in different sites. One comparison can make the point. The black colobus is closely related to the guereza colobus and is probably similar in physiology. It happens that the black colobus has only been studied in a forest where the plants defend themselves by particularly toxic chemicals (McKey, 1978). The black colobus concentrate on seeds (fruit) in this forest, rather than eating their normal diet of less nutritious and equally toxic leaves. In Table 3.1, this difference in the forest environment appears as a difference between species.

A few general indications appear, even with this caveat. Taxonomically, the prosimians include a whole series of extremists: leaf eaters and omnivores who catch no prey; insectivores who eat no leaves. The colobines among the higher primates are also leaf eaters that take no insect food, except as they find it boring inside the leaves or fruit.

There is a wide spectrum of the amount of foliage eaten, but it rarely makes up as much as half the diet. Only six species in Table 3.1 eat as much as 50 per cent leaves, only two eat more than 75 per cent.

Fruit is the staple for most species. Only five species in the Table take less than 25 per cent of their food as fruit, whereas several of the anatomically specialized leaf eaters eat a quarter to half their diet as fruit.

Insects, although important in terms of foraging time and nutrition, make up a minimal percentage of diet. Bushbabies, tamarins, and titis are called insectivores, but only the tiny Demidoff's galago would rate as an insectivore in the wider world of shrews and anteaters.

FRUIT

Fruit eating is almost the hardest category to describe, because it is the norm, at least among diurnal primates. Interest centers on subdivisions of fruit eating. Most primates tolerate a high proportion of green or bitter fruits, which means medium-size food sources at medium distances. Some primates, however, such as the spider monkey and chimpanzee, specialize in ripe fruit. They range widely, because there is usually little ripe fruit on a single tree. Their social groups join and split, often foraging singly or in pairs for these scattered sources.

Others, such as the whitecheeked managabey, specialize on large food sources, especially fig trees (Waser, 1977). Their bands do not split but range erratically through the forest over long distances to search out the superabundance of a fig in fruit. Then they displace other primates that have found the tree first, and settle, sometimes for days, to gorge.

Fig trees may have played a distinctive role in primate evolution (Fig. 3.1) (MacKinnon, 1979). There are some 900 species of *Ficus*, a genus extraordinary in many ways. The fruits are eaten by a large number of vertebrates that are mostly seed dispersers rather than seed predators. There is asynchronous flowering and fruiting among individuals of a population, yet strong synchrony within each tree. Many species of fig coexist in a tropical forest, and yet fig species do not compete for pollinators, for each is fertilized by its own co-

TABLE 3.1. Proportions of Foliage, Fruit, Flowers, and Prey in Primate Diets

	Proportion of Diet			
Species	**Foliage**	**Fruit, Seed**	**Flowers**	**Prey and Gum**
lemur, ringtailed	34	47	7	0
brown	71	25	4	0
indri	57	41	2	0
sifaka, white	41	40	8	0
potto	0	76	0	10
galago, Allen's	0	73	0	25
demidoff's	0	29	0	70
needleclawed	0	80	0	20
senegal				100
tamarin, cottontop	10	60	0	30
owl monkey	30	65	0	5
titi, widow	4	71	0	20
howler, mantled	49	42	10	0
red	15	65	0	20
spider, longhaired	7	83	0	0
blackhanded	20	80	0	0
woolly	0	85	12	2
macaque, longtailed	16	52	5	23
rhesus	19	72	4	2
mangabey, greycheeked	5	61	3	24
Tana River	13	81	2	2
baboon, yellow	8	63	3	10
hamadryas	7	66	22	0
gelada	46	26	0	0
guenon, moustached	8	79	0	10
vervet	12	48	23	17
blue	21	13	44	20
De Brazza	9	74	3	5
crowned	2	84	0	14
redtailed	16	44	15	22
spotnosed	28	61	1	8
talapoin	2	43	2	36
patas	2	52	2	43
langur, banded	37	53	0	0
hanuman	54	37	5	0
nilgiri	71	17	10	0
dusky	48	46	7	0
colobus guereza	82	14	2	0
black	37	56	5	0
red	78	33	8	0
gibbon, lar	34	59	3	10
siamang	45	42	4	8
chimpanzee	28	68	0	4
orangutan	22	62	3	2
gorilla	86	2	2	0

1. Data from Clutton-Brock and Harvey, (1977a and personal comment).
2. This is aggregate data for each species, taken from the mean of all studies. For some site-specific comparisons, see Chapter 8.
3. Only species studied for more than one year are included.

FIGURE 3.1. *Adult male orangutan gorges in a fig tree in Borneo. (Courtesy P. Rodman.)*

evolved fig wasp which acts almost like the active gametes of animals. There is frequently some fig in fruit somewhere, and each fig tree puts colossal energy into the bonanza of fruit that attracts seed dispersers (Janzen, 1979).

The vertebrates that eat figs have at least the precondition for social tolerance: a large clump of food to eat at the same time. If primate ancestors traveled from fig tree to fig tree together, they had the precondition of longer-term social bonding. In such forests as the Krau game reserve in Malaysia, fig trees are still major plants that determine primate movements (MacKinnon, 1980). However, so many vertebrates eat figs that one must also look at smaller, less abundant sources to understand niche differentiation among primate species (Gittins and Raemaekers, 1980).

Eating most fruits presents little challenge to primate ingenuity. An arboreal primate need only reach out a hand, hook in a branch, grab the fruit with the other hand, and munch. Even so, there are differences of manipulative skill. The Madagascar lemurs do not use their hands even to peel a banana, but chew and lick the end (Fig. 3.2.). Monkeys in general peel fruit, and the ground-living macaques and baboons rub grit off their food with a wipe of the fingers.

Blackhanded spider monkeys attack the fruits of *Tocoyena pittieri,* which hang

FIGURE 3.2. *Lemurs do little processing of fruit and do not even peel bananas. This Coquerels' sifaka in northern Madagascar has gouged the ceiba fruit skin with its toothcomb. (Courtesy I. Tattersall.)*

from a stout stem and are protected by a thick green skin. The monkey bites through the skin of a fruit but, finding the pulp green and bitter, leaves it alone. One or two days later, the troop passes by the same way. Opening the fruit has accelerated its ripening. It is full of a sort of sweet jam, which the spider monkey scrapes out to the last drop, spitting the seeds round about. One does not need to invoke foresight in the monkey's performance, but the result favors both monkey and tree over natural, slow ripening with the fruit lost intact on the ground (Hladik and Hladik, 1969).

A few records exist of active sharing of vegetable food between mothers and infants (McGrew, 1975; Silk, 1978; Dare, 1974), among captive juvenile gibbons and douc langurs (Kavanagh, 1972; Berkson and Shusterman, 1964; Schessler and Nash, 1977), and in a captive chimpanzee group between males and young (McGrew, 1975; Silk, 1978, 1979). Among Gombe stream chimpanzees, both mothers and young distinguish easy-to-process and hard-to-process fruit (Fig. 3.3). The mothers gave their offspring easy fruit earlier, but also stopped sharing it earlier than the difficult fruit; they had some notion of which fruits their growing offspring could handle on its own.

One reason that fruit is rarely shared may be that it is not very scarce or

FIGURE 3.3. *Few primates share food. Marmosets and tamarins feed insects to their young; here a Gombe Stream chimpanzee allows her infant to share fruit. (Courtesy C. Tutin.)*

valuable. An adult animal can usually go find another fruit without the stress of begging or social competition with the possessor. The brown lemurs of Mayotte illuminate the link between sharing and snatching. The food they most disputed was the fruit of the houbouhoubou vine *(Saba comorcusis)*, a tough-stemmed object that one animal would scissor off with its teeth and carry away to eat, thus being clearly its possessor. On several occasions, however, the possessor shared the fruit without dispute, usually between two females (Tattersall, 1977).

In short, fruit is easy to eat, at least when ripe, for the trees are selected to offer fruit that tempts seed dispersers. It is not usually worth sharing or begging for individual fruits, but the number of fruit on each tree affects the size of the fruit eater's group. Relatively few animals can feed together in small, dispersed trees, whereas the bonanzas of large fig trees may have favored the original social groupings of primates.

SEEDS

Seed eating is very different from fruit eating. On the whole, fruits are evolved to attract predators that will spread the contained seeds. Plant seeds, in contrast, are selected to repel predators or to go straight through the gut undamaged. They may have hard rinds and nutshells, sticky resins, irritant hairs, or poisonous secondary chemicals. Even if an animal can open such seeds, it may not be worth the time and energy (Raemakers et al., 1980).

An alternative strategy is for the plants to swamp the seed eater by a bonanza of food. Oaks and chestnuts in the temperate zone, durians in the tropics overfeed their predators. There is selection on such plants to concentrate their seed production in a single season, or even from year to year, or in conjunction with other species. Then, although many animals may feed together in such a tree when it does fruit, the lean seasons keep down the total predator population. The plants may also profit from predators' behavior, either complex adaptation, such as squirrels burying nuts, or such simple traits as lion-tailed macaques' habit of eating only half the seeds in the chestnutlike burr of the South Indian durian and dropping the other half (Johnson, 1974).

Still another means of avoiding seed predation is to have seeds that are too small to be worth mammals harvesting, like the winged, windblown seeds of the maple tree or Shorea tree. Grass seeds are similarly small, yet Clifford Jolly (1970) has proposed that feeding on such small objects has played a role in our own evolution. *Ramapithecus*, the earliest fossil hominid, somewhat resembles living gelada baboons. The gelada feeds sitting down, shuffling along on its bottom, picking up grass blades, rhizomes, and seeds with thumb and index finger. Jolly suggested that our short faces, our precision grip, our early terrestriality, and our penchant for sitting upright all stemmed from early small-seed eating. Szalay (1975) took issue. We have long hind legs, geladas long arms. Our incisors and canines probably evolved for biting off meat. Although seed eating may have played a role, it was not overriding for humans as grass-bulb eating was for geladas.

A later fondness for seeds may have contributed to tool use and foresight (Parker and Gibson, 1980). Precisely because plants defend their seeds, it takes strategy to overcome the defenses.

A vervet guenon may pace round the base of a prickly pear cactus until it spots an accessible ripe fruit, and then picks its way upwards with great care, stopping for visible fixes on its route, and making detours to reach its goal. It levers off the fruit gingerly with one finger, then descends with equal caution, sometimes by a different route, to retrieve the fallen prize. The vervet's stalking a prickly pear resembles stalking prey (Rose, 1979).

Blackcapped cebus monkeys go further (Izawa and Mizuno, 1977). One of their staple foods in La Macarena National Park of Colombia is the nut of the black palm or cumare (*Astrocaryum chambira*). The cumare defends its trunk with wicked spines, the petioles of shed leaves, so that the monkeys must leap directly onto the fruit cluster from an adjacent tree. The nuts ripen like small coconuts from a green shell, filled with liquid albumin, to a hard brown husk lined with solid flesh.

At the nut's earliest stage of ripening, the monkey can bite open the base and sip the milk as from a cup (Fig. 3.4). Later in the season, it finds the "co-

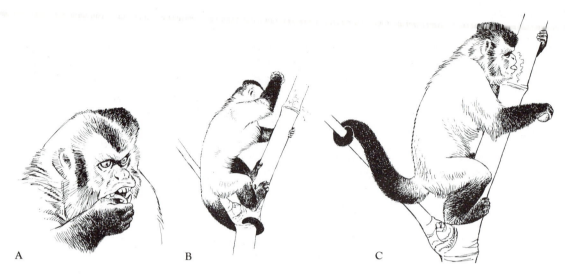

FIGURE 3.4. *Black-capped cebus with unripe palm nuts. A, piercing nut with canine, B and C, tapping and licking the milky contents on bamboo stem. (Courtesy K. Izawa.)*

conut" eye and punctures it with a canine to drink the milk. Then it taps the soft-shelled nut on a bamboo stem internode, licking its yogurtlike contents off the bamboo after every three strikes or so, or from pieces as they split apart. This gentle approach saves most of the soft interior from dropping to the ground.

Still later, the palm nuts are hard-shelled. Then the cebus makes a fulcrum of its hindlegs and tail and bashes the palm nut on a bamboo joint with all its strength (Fig. 3.5). It may take 11 minutes' work and 86 strikes or more; juveniles occasionally abandon the job, but heavier adult males succeed.

This is a local adaptation, absent in other populations of blackcapped cebus. It seems to have developed in La Marcarena because the cumare palm grows

FIGURE 3.5. *A black-capped cebus with a hard, ripe palm nut, bashes it open on bamboo node. (Courtesy K. Izawa.)*

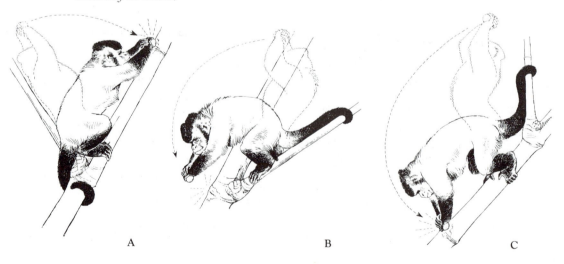

A   B

FIGURE 3.6. *Hammer and anvil stones are used to crack oil palm nuts in Guinea. A, chimpanzee version; B, human version. (Courtesy Y. Sugiyama.)*

there next to many clumps of the hard bamboo, and because the terrain is hilly, which means that there are trees and vines adjacent to the cumare palms at the right height for monkeys to jump to the fruit. At Manu National Park in Peru, the cebus have a different ploy. They investigate the fallen palm nuts, choosing those that have been bored by a beetle, weakening the shell, and only the nuts where the beetle has left some of the meat. These they bash open on branches, or even one against another (Terborgh, in press). Even given these conditions, though, it is only the cebus who have developed nutcracking. Woolly and spider monkeys and bearded sakis in the same forest bite open the green fruit but are baffled thereafter. Cebus of many species in many environments hit food onto branches or cage walls. It seems that the propensity for bashing things pays off (Izawa and Mizuno, 1977).

Chimpanzees in West Africa crack oil palm nuts using two stones, a platform stone with a shallow depression and a pebble tool of half a kilogram or more—a little heavy for a human hand, but probably comfortable for a chimpanzee. (See Figs. 3.6A and 3.6B.) At three sites, there were no other suitable stones within 15 meters, so the hammer and anvil must have been transported under the oil palm. Usually the pebble stone was left properly on the platform stone after use. Sugiyama watched while chimpanzees cracked 18 nuts. It took them 2.8 blows on average, and he found it took him 2.5 blows using the chimp's tools, but he did not crack the meat loose so accurately (Sugiyama and Koman, 1979; Struhsaker and Hunkeler, 1971).

Thus, the extraction of contained seeds may have played a major role in the evolution of manipulation, attention span, and the ability to work toward a goal (Parker and Gibson, 1980).

LEAVES

Leaf eating involves specializations of teeth, gut, and behavior. Most leaf eaters have elongated guts, with fermentation chambers either in the form of a sacculated stomach or else an enlarged caecum, the equivalent of our thumb-sized appendix, with a large colon as well.

Both sorts of enlarged gut allow the leaf eaters to eat large amounts of food. This helps them survive on their relatively un-nutritious diet. Leaf eaters, as a whole, tend to larger body size than fruit eaters. Given that basal metabolism only increases as the three quarter power of body weight, they do not need much more energy to support a large body, and given a large body, they can ferment coarse browse for a longer time in their large gut. Much of basal metabolism supports the resting energy of muscle. Leaf eaters' increased size is a matter of large skeleton and large gut, with not much more muscle, so they probably need even less energy than a fruit eater of the same total weight.

Leaf eaters save energy by sluggish inactivity through most of the day, and gain energy by steady foraging for long periods of time, though with little movement from branch to branch or tree to tree (Clutton-Brock and Harvey, 1977b). Leaf eaters' activity budgets are overwhelmingly devoted to feeding or resting. This indolence is reflected in a smaller brain size than frugivorous primates of the same family and body weight; perhaps leaf eaters are stupid as they seem (Clutton-Brock and Harvey, 1980; Mace et al., 1981).

The two sorts of gut specializations reflect two slightly different approaches to the problem of leaf eating. Hindgut (caecum and colon) fermentors include the Malagasy indriids and, to a small extent, the new-world howlers (Fig. 3.7), as well as such diverse mammals as horse, elephant, and rabbit. Their fermentation is less efficient, and their rate of throughput is faster than for animals with sacculated stomachs.

If a leaf eater needs a relatively large amount of energy, it may do well with hindgut fermentation and a fast throughput, provided it can select the tenderest, most easily fermented leaves and leaf parts. This particularly applies to small animals, which have higher energy requirements per unit weight but less space for fermentation. (The elephant shows that small size does not necessarily go with hindgut fermentation; a large, active animal may also prosper so).

Foregut fermentation, as in colobine monkeys, tends to go with larger body size within a group and with slow throughput that allows time for the bacteria to break down a coarse fibrous diet, such as mature leaves (Parra, 1978). The

FIGURE 3.7. *The mantled howler, a hind-gut fermenter, shown in Barro Colorado, Panama, is the most folivorous of new-world monkeys. (Courtesy C. M. Hladik.)*

record among mammals so far is the sloth, both for conserving its own loco-motor energy and for slothful throughput of a meal from food to feces of 200 to 300 hours, or about ten days (Montgomery and Sunquist, 1978).

Among leaf eaters one finds a gradation from more active, wide-ranging animals that concentrate on new leaves with an admixture of fruit, and those that eat more mature leaves with stress on energy-saving and sedentary habits. The contrast appears over and over in pairs of closely related species—the wide-ranging hanuman langur as against the purple-faced langur in Sri Lanka (Hladik, 1975), the ringtailed against brown lemurs in western Madagascar (Sussman, 1974), and the red colobus against the guereza colobus in East Africa (Clutton-Brock, 1975). It does not appear among primates as a contrast between sympatric fore and hind gut fermentators, because these live in different continents, but may give clues to the habits of the progenitors of each leaf-eating group.

FLOWERS

Many or most primates eat flower parts, much as they eat leaves. A few make flowers a particularly important source of food at seasons when a plant species is flowering in abundance. Mongoose lemurs, for instance, feed on the nectar of silk-cotton trees, lapping the sugary secretion without damaging the plant (Tattersall and Sussman, 1975.) For Coquerel's mouselemurs and forked lemurs, nectar feeding is one of the specializations that lets the animals remain active through the harsh dry winter of western Madagascar. Many plants flower during austral spring, which means that the seed is ready to germinate when the first rains come, providing food for dwarf and lesser mouse lemurs (Hladik et al., 1980; Petter et al., 1975). Flower feeding may also aid animals through a vulnerable stage of life. Soft acacia flowers are ideal weaning foods for young baboons lucky enough to mature in the proper season (Altmann, 1980).

Of course, the effects on the plant of nectar supping and flower chewing are quite different, much like the difference between fruit and seed eating. Sussman and Raven (1978) suggest that there is an ancient coevolution between flowering plants that bear few, strong flowers with copious nectar and their mammalian pollinators. Early primates, marsupials, and bats would have competed for such nectar. Primate pollination is more frequent today in such places as Madagascar, with a relatively impoverished bat fauna. One of the better documented cases of primate pollination is African bushbabies and baobabs. It may be that lemurs as pollinators have played their part in the radiation of Malagasy baobabs and many other endemic Malagasy plants that hold out cups of nectar.

GUM

Eating gum, the sticky sap exuded from the bark of various trees, is a specialty little considered ten years ago. It now appears that many prosimians and a few new-world monkeys are specialized gumnivores (Fig. 3.8).

Senegal bushbabies and forked and dwarf lemurs (Bearder and Martin, 1980;

FIGURE 3.8. *Black-pencilled marmoset on a gum tree, with the slits it has gouged. (Courtesy W. Kinzey.)*

Petter et al., 1975) subsist largely on gum during the dry season when fruit, flowers, and insect food are all scarce. Senegal bushbabies in South Africa eat no fruit or leaves at all, just insects and acacia gum. In the wet forest the needleclawed bushbaby and pygmy marmosets primarily eat gum, and many others make gum a part of their diets (Charles-Dominique, 1977; Moynihan, 1976).

The toothcomb of lemurs and lorises probably evolved as a tool for scraping gum from bark, although it is too fragile to actually bore holes. The prosimians depend on the holes of wood-boring insects for their source of gum (Szalay and Seligson, 1977). The association between small body size, nocturnality, gum eating, and the primitive six-tooth comb suggests that it was a feeding adaptation at the start, later modified in the larger and more social animals as a tool for social grooming. In South America the marmosets also have specialized, procumbent lower incisors and may be able to drill their holes for sap (Moynihan, 1976).

Edible gums are long-chain sugars. Tree species vary in the amount of useful carbohydrates they produce (Hausfater and Bearce, 1976). Gums may also provide calcium, which offsets the low calcium content of insect prey (Bearder and Martin, 1979). Gum-eating animals probably need specialized digestion as well as specialized teeth to cope with this food. Moynihan says, ''Every family . . . of pygmy marmosets seems to own one or more large or thick trees, the trunks and larger branches of which are riddled with small and shallow holes.'' It seems we may have here ancient coevolutionary system, in which particular tree species have shaped the anatomy, physiology, and even social behavior of primate species.

POISON

"The plant world is not colored green; it is colored morphine, caffeine, tannin, phenol, terpene, cavanine, latex, phytohaemagglutin, oxalic acid, saponin" (Janzen, 1978). Once primatologists realized that food is largely poisonous, it transformed our view of habitats; no longer an abundant larder, the jungle was a treacherous place demanding taste discrimination. (Freeland and Janzen, 1974). Plants' chemical defenses cost the plant energy to produce, and the herbivores energy to detoxify, if they can tolerate them at all. There are always trade-offs. At times "a vertebrate may elect to pay a few grams of liver for a large amount of nutrient," particularly in a lean season, when the damage may be repaired later on.

Toxicity varies with the animal. On the whole, more carnivorous and omnivorous mammals do not tolerate secondary compounds so well as leaf eaters. There are also special abilities to cope. Hanuman langurs eat strychnos fruits that would kill macaques, or ourselves. Elephants gobble up to 1,000 Balanites fruits, which have as much diosgenin, a precursor of cortisone and sex hormones, as the yams that we use for birth-control pills. Pigs and all species of monkeys leave the Balanites strictly alone. Vertebrates as a whole may have different tolerances than insects or fungi. It may be that the high tannin content of *Terminalia chebula* and some other ripe fruit (25 per cent to 35 per cent dry weight, compared to 8 per cent for green banana peel) inhibits fungi and thus increases the proportion of fruit eaten by vertebrates (Janzen, 1978).

McKey (1978) has been studying black colobus in a black-water forest in Cameroon, built on ancient beach sand of minimal nutritious value. This forest is generally poisonous; many of its common trees produce drugs, although the chemical analysis has not been completed yet. The trees are mainly evergreen, guarding each leaf as long as possible before replacing it. Here the black colobus eat seeds for 50 per cent of their diet. It seems that they have taken up seed eating because of the high nutritive content of the seeds. Seeds may be no less toxic than leaves of the same tree, but because there is more energy per poison, a seed-eating monkey can survive. The black colobus eat more mature leaves than either the red or the black-and-white colobus of Kibale forest in Uganda. This is probably for want of opportunity to eat new leaves where few of the trees are deciduous. It does, however, increase the black colobus' dependence on seeds for nourishment. They are also highly selective as to the leaves that they eat, avoiding the commoner forest trees.

Glander (1978, 1982) has similarly analyzed the feeding of mantled howlers in Costa Rica. He counted the trees by species in his group's home range and measured the amount of time spent feeding on each. Only four of the fifteen commonest forest species appeared in the list of the fifteen commonest food species. Howlers often eat a small part of the leaf or fruit, the base of the petiole or the pedicel, but not the fruit itself. They drop the rest, which led earlier observers to think that the monkeys are surrounded by an unlimited food supply, so much that they could afford to waste the bulk of each food item. On the contrary, it seems that they must choose carefully the items worth harvesting, or safe to harvest, in a limiting world.

Struhsaker (1978) points out that "when one considers the red colobus diet, nearly half of it comes from so-called toxic families of plants." The combination of inbuilt resistance and great selectivity in the part of leaf or age of leaf eaten allows the red colobus to survive. Again, this has implications for the total available food supply. Struhsaker's Kibale forest in Uganda is one of the richest primate habitats, but even there the monkeys pick and choose the available food.

INSECTS

Insect eating, like gum eating, is probably very ancient in the primate line. Unlike gum eating, it is practiced today not only by small bodied "primitive" primates but by members of every primate line, including chimpanzees and humans.

There are three main techniques for catching insects: the slow stalk, the quick grab, and poking into things.

Stalking a fast insect is difficult. The potto and the slow and slender lorises ooze along a branch, the S-curves of their spine allowing them to move one foot at a time with no shift of balance that would alert their prey (Fig. 3.9.). At last, they clamp a hand on the back of an unsuspecting cockroach, or even a sleeping bird (Walker, 1969). Pottos also specialize in eating repugnant insects that may count on bad taste and smell for protection and be unlikely to fly away (Charles-Dominique, 1977). Cartmill (1974a) points out that forward-facing eyes

FIGURE 3.9. *Slender loris in Sri Lanka is a specialized insect eater. (Courtesy C. M. Hladik.)*

and grasping hands are characteristics of arboreal predators like pottos and chameleons, not arboreal vegetarians like squirrels. Our earliest steps toward primate binocular vision and fine manipulation may have evolved for stalking cockroaches.

Grabbing insects requires speed and precision. Solitary hunters like the bushbabies do this; so do new world squirrel monkeys, African talapoins, and redtailed guenons. These insectivorous monkeys move in troops—sometimes 40 or even 100 squirrel monkeys and talapoins together. They frequently associate with other species, which enlarges their numbers. However, this may be primarily predator defense, rather than an attempt to flush insects out of the leaves. This differs from the mixed flocks of insectivorous tropical birds, each of which may catch a slightly different assortment of the insects they all stir up together (Moynihan, 1960; Struhsaker, 1982).

Finally, in each phylogenetic line, there are pokers and pryers. The most bizarre anatomically is the Malagasy aye-aye, which has bat ears to hear insect crepitations within the trunk, beaver teeth to strip wood and bark, and an elongated, skeletal third finger to extract juicy wood-boring grubs from their holes. The aye-aye is one of a worldwide series of "woodpeckers." True woodpeckers strip wood with their beaks and then harpoon boring insects with a barbed tongue. Where no woodpeckers live their niches are filled by a Galapagos finch that uses a cactus spine for harpoon, by New Zealand huias whose stout-beaked male strips bark so the slender-beaked female can probe, and by a New Guinea marsupial, the long-fingered possum, which is much like an aye-aye (Cartmill, 1974b).

The pokers and pryers among the monkeys are new-world capuchins and old-world blue guenons, mangabeys, baboons, and macaques (Fig. 3.10.). They unroll leaves, search spider webs, chew open dead trunks, prise off bark, and turn over stones. They pick up the insects with clever fingers. South African

FIGURE 3.10. *A Lowe's guenon unrolls a leaf to search for insects. (Courtesy F. Bourliere.)*

baboons and Gabonais chimpanzees slap at stinging forms, such as scorpions, beating them into immobility before eating them. The scorpions that the chimpanzees eat this way are not dangerous; their sting is scarcely worse than a wasp's, but West African baboons eat lethal species by grabbing them in one hand and tweaking off the tail with the other (Thorington, 1967; Moynihan, 1976; Rudran, 1978; Waser, 1977; Marais, 1969; Bolwig, 1959; Hall, 1962; Hladik, 1974).

Chimpanzees spend much of their foraging time in pursuit of insect prey, an indication of the importance of protein in their diet (Hladik, 1973). They pry into leaves and deadwood like the monkeys. They also make use of tools. They strip and trim twigs and grass stems for termite fishing (Fig. 3.11A) They poke their sticks into the exit passages of termite hills; the termites bite and cling to the stick; the chimp extracts them and wipes the stick clean with its rubbery lips. This behavior, learned over years of play and imitation, will be further discussed in the chapters on tool use and on play (Goodall, 1963, 1968, 1971; Jones and Sabater Pi, 1969; Nishida, 1973).

Chimpanzees also "dip" for ants. (Fig. 3.11B) The viciously biting driver ants demand another technique and other tools. The chimp takes a long wand, which it sweeps two-handed through the ant stream, wipes the wand clean with a fast pull-through, and then stuffs ants into the mouth, and chews frantically. If there is no raised log or stone to stand on, the chimp may bend over a sapling and perch on that at a relatively safe distance from the ants. Babies cling tight while their mother is ant dipping, but independent juveniles retreat to a safe distance (McGrew, 1974, 1979).

The sedentary, poke-and-pry approach to small-sized but highly valuable insect prey may thus be linked (like seed eating) throughout the primate line with long attention span, the ability to search for hidden objects, and manipulative skill (Parker and Gibson, 1980).

FIGURE 3.11A. *Flo and Fifi fish for termites from the termite's exit holes in their cement-hard hill, Gombe Stream, Tanzania. (Courtesy H. van Lawick.)*

FIGURE 3.11B. *A chimpanzee in Gabon fishes for tree-ants. (C. M. Hladik.)*

MEAT

Many primates and most people eat and like meat. "The female (greycheeked mangabey) bit the head off a small green snake, which she gripped around the neck, and stuffed the writhing remains into her cheek pouches, grunting contentedly" (Waser, 1977). Similar accounts range from tarsier, consuming a highly neurotoxic Bornean snake, to blackcapped cebus, splintering off fragments of bamboo to reach the small frogs that live in the internodes. Many species eat fledgling birds, and vervets, gibbons, and others pursue and eat adult birds when opportunity arises. Meat eating has been reported in nine of the 11 primate families; but not in indriids or colobinae (Niemitz, 1979; Izawa, 1978; Leck, C. F. 1977; Butynski 1982). Most primates are at least adept enough to bite the head and neck off their prey, although without the innate precision of many carnivores (Steklis and King, 1978).

We scrutinize meat eating with particular interest, because of the widespread view of ourselves as "man the hunter" (Washburn and Lancaster, 1968; King, 1980). Even the attempts to redress the balance by pointing out that most of hunting peoples' food is provided by woman the gatherer does not quite counter the pride of bringing home big game in the anthropological literature and perhaps in fact. Much of the evolution of human behavior has been attributed to hunting—cooperation in the chase, sharing of the spoils, memory of terrain, foresight in planning ambushes, and even language for communicating about game that is out of sight and to be stalked. Do any parallels appear among the primates?

We have already seen that fruits may sometimes be shared. So may meat as highly prized food that is difficult to catch. The sharing reflects the social system of the species. Lion tamarins share insects with their offspring or younger sibs. The tamarin's monogamous family in which father or older young carry and care for the infants is reflected even in the provision of food. Insects are shared more often with juvenile tamarins than fruit, apparently on the same lines that a chimpanzee mother shares difficult-to-obtain fruits. Baboons, on the other hand, with their highly competitive society, share meat chiefly among adult males, and then the troop hierarchy determines who gets a piece. Only rarely does a male move over to let a female feed, or a mother share with her infant. There is only one report of orangutan meat eating. A female completely consumed the carcass of an infant gibbon, taking 137 minutes to do so. (She may have found the gibbon dead, because there was no blood.) An adult male she was consorting with stared at her fixedly for the first hour, but she did not share with him—an index of the antisocial habits of female orangs and either of lethargy or chivalry on the part of the enormous male. Chimpanzee males, in contrast, commonly strip off bits of prey carcass and hand them to begging troop members, including females and young (Brown and Mack, 1978; Strum, 1975; Sugardjito and Nurhuda, 1981; Lawick-Goodall, 1968, 1971).

In old-world monkeys and apes, hunting fairly large prey is primarily a male activity. In zoo lemurs, the dominant females seize and kill such prey as mice and small birds. Other troop members may be allowed to snatch or share portions. This correlates with the usual priority for food in female lemurs over males (Jolly and Oliver, personal observation).

Baboons at Gilgil Ranch in Kenya gradually developed a tradition of hunting (Figs. 3.12A and 3.12B). At first, only the adult males hunted, but then certain females also took to chasing prey, and both females and infants ate prey if they could get a piece, and hunting as a whole increased in frequency. Infants and juveniles at first casually associated with others round a carcass, but once a young baboon had tasted meat, it actively tried to get a share the next time. The size of prey increased over the years of watching, with more hares and large antelope infants such as impala. The males even developed the system of chasing prey in relays like wolves or wild dog (Harding, 1973, 1975; Strum 1975, 1976).

At Gilgil most of the large carnivores have been killed off, so it might seem that there is an unoccupied niche that the primate predators are filling. The growth of the hunting habit at Gilgil may be the process of expanding to fill such a niche. However, that is not the whole story, because at Amboseli, which is uncomfortably full of predators, baboons hunt with about the same frequency as at Gilgil. Another suggestion is that hunting is more likely where there is already a surplus of vegetable food, to give surplus energy for investment in the chase. This is again belied by Amboseli, which has progressively dried up over the last 20 years, and the baboon population, like everything else, has shrunk (Gaulin and Kurland, 1976; Hausfater, 1976).

FIGURE 3.12A. *Infant savannah baboon and vervet at play. (From S. A. Altmann and J. Altmann, 1970, by permission.)*

FIGURE 3.12B. *Adult savannah baboon eating vervet. (From S. A. Altmann and J. Altmann, 1970, by permission.)*

It seems that the baboons actually need meat, either for protein, vitamin B_{12}, or some other factor. Hausfater shows that the seasonal differences in predation rate, with much more hunting during the dry season, follow a lognormal distribution. "A lognormal model is appropriate when the variable of interest (here, rate of predation) is affected by several factors each of which is itself normally distributed and acts multiplicatively, rather than additively, with respect to the other factors." Thus, baboons' need for vitamin B_{12} in the dry season could act multiplicatively with quite different aspects like the shortness of dry season grass, which makes it difficult for prey to hide. Baboons clearly invested effort in stalking unsuspecting vervets, or confronting the charges of a mother gazelle, while other males continued biting at her bleating neonate (Hausfater, 1976).

Chimpanzees, like baboons, prey on both birds and mammals (Nishida et al., 1979; Suzuki, 1975). Some of this is more gathering than hunting, as in McGrew's description of a female Senegalese chimp stripping bark from a tree, finding two bushbabies in a tree hollow, and calmly proceeding to eat them. That population may have a local penchant for bushbabies. (It is a sidelight on primatology that one of the excitements of washing hundreds of chimp feces is the occasional discovery of a whole galago hindleg.) (McGrew et al., 1978, McGrew and Tutin, 1979). Much hunting, however, involves active cornering of colobus and baboons in situations where they cannot escape. Some hunts clearly involve cooperation.

"A group of chimps rested in the shade of a tall tree. In its branches, a juvenile baboon fed alone, separated by some 200 yards from the rest of his troop. Presently Huxley (a mature male) plodded up from the stream toward the peaceful chimpanzee group. About ten feet from the fig tree, he stopped, facing its trunk. To us he seemed unaware even of the existence of the small baboon above. Nonetheless, as though he had in fact given a signal, the other chimps stood up. Two of the males moved to the base of the fig tree; three others stationed themselves under two nearby trees, the branches of which formed an escape route for the baboon. And then, very slowly, with infinite caution, Figan, the youngest of the males present (he was about eight at the time) began to creep toward his quarry . . ." (Goodall, J., 1967).

That hunt ended in screams and frustration. Wrangham describes even more obvious cooperation, this time successful. The chimpanzee Jomeo and adolescent Sherry were both in a tree with a baboon mother Marigold and her disabled infant Mango. Marigold left Mango in a branch fork only two and a half meters from Jomeo, but he did nothing, perhaps intimidated by an adult male baboon who stayed nearby and who had protected Marigold since her infant's injury.

Marigold retrieved Mango and "it appeared that the golden opportunity has been lost. Jomeo moved closer to Marigold, but then young Sherry climbed over and walked along the branch, blocking easy access by the baboon male. Jomeo swung toward the mother baboon; the movements suddenly became faster; she climbed out of the tree, but then both chimpanzees were on her together. They tore the infant off her belly . . . They held the carcass between them, flailing and boxing at the baboons that rapidly gathered to attack them. After four minutes they turned and ran upslope, Jomeo leading with the carcass in his teeth . . . There were no more calls for two hours. It seems that they had rushed away to eat their prey in peace" (Wrangham, personal communication.)

Plooij (1978) watched an attack by all nine males of the Gombe Stream northern community on a family of bushpig. The two adult pigs stood back to back, guarding a juvenile and small piglet. Five chimpanzee females (this was a party of males and estrous females) hung back or climbed nearby trees, while the males edged closer, threatening the pigs with wa-barks and eagle-wra's so loud they even frightened Plooij, and reassuring each other with squeaks and genital touches. Mike, a big male, "then rises, takes a big rock (25x10x10 centimeters) in his right hand, walks bipedally towards the pigs while looking at them, and throws the rock with an underarm throw towards the pigs from a distance of approximately four meters . . . the rock hits the big pig in the front." The pigs did not stampede until male Figan food-barked from behind them, answered by pant-hoots from the others in front. The bushpigs broke back toward Figan to the sound of high piglet screaming, and then Hugo was sitting in a tree with a piglet's head and a foreleg, while Mike and Humphrey begged from him, and Figan loped off with the remains.

The chimpanzee who makes the kill seems to retain some right of ownership, even though he may be subordinate in other circumstances. More dominant males beg a share from him rather than grab (Goodall, 1971; Teleki, 1973). The sharing out follows age and kinship lines, rather than simple dominance (McGrew, 1979).

One macabre aspect of chimp predation is their predilection for brain tissue. In contrast to the chimpanzees' willingness to share most meat, the owner ignores or discourages the advances of others, even though they closely watch him as he consumes the brain—"always a slow, meticulous procedure with a definite undertone of enjoyment" (Teleki, 1973). The process is also fairly complex: peeling back the skin, poking a hole in the cranium, scooping out the contents with a finger, even occasionally wiping the brain case clean with a wad of leaves. Brain tissue is often chewed with a wad of leaves in the mouth, either for flavoring or to "prolong the enjoyment."

Chimpanzees, however, are rare hunters. Teleki (1973) estimates that only about ten successful hunts took place each year among the Gombe chimpanzee community. During the period he watched them, most of their prey were infant baboons, who were also attracted to the banana feeding area (Fig. 3.13). De Pelham and Burton (1976, 1977) suggested that even this rate is higher than normal because artificial feeding increased the amount of hunting to remedy specific nutritional deficiencies in a banana diet. Teleki (1977) strongly refuted this supposition, but it is true that the males often started a hunt when they had just gorged on bananas. The diet, the baboons' proximity, or the social gathering of males triggered several hunts. Knowing that a hunting habit can develop even among baboons, the Gombe chimps might well have been starting an incipient tradition.

Social hunting involves persistence, cooperation, and strategy (Peters and Mech, 1975). Carnivores such as wolves and lions set ambushes, cut corners to relay each other in the pursuit, and track and stalk with great persistence. The

FIGURE 3.13. *A fracas between baboons and chimpanzees. Artificial feeding led to increased interaction, and this perhaps to increased predation. (Courtesy P. Marler.)*

sporadic predation of chimpanzees sometimes has the same qualities. Fruit eating demands none of these except a cognitive map of fruit tree location.

I would add another quality to Peters and Mech's list: speed. The most banal thing about plants is that they hold still. They grow and ripen over days or weeks so that the most critical aspect of timing is to arrive earlier than a more dominant animal or troop who might strip the tree. Hunting does not merely need quick reactions to the prey; cooperative hunting demands simultaneous reaction to prey and fellows and rapid calculation of what each will do next.

Andrew (1962) and others have attributed the long-term increase in mammalian brain size to selection by predators for quicker-thinking prey and by prey for quicker-thinking, more strategically minded predators. The "hunting hypothesis" as one major selective trend in human evolution still stands, even though modified as we have realized the parallel importance of gathering with its tool use and multiple-step attention.

Chimpanzee hunting resembles man's, in its being primarily a male occupation, in the rudiments of cooperation and strategy, and in the sharing of food at the end. All these actions appear sometimes in other contexts, but the complex of interrelated behavior lends strong support to the picture of the evolution of ourselves, or at least one of our own sexes as man the hunter.

SEX DIFFERENCES IN FORAGING

Capturing of large prey seems to be a priority of the dominant sex—male chimps, but probably female lemurs. There are a few other substantial differences between the sexes in foraging techniques. Among blackcapped capuchins in Venezuela, males spent more time foraging and less time resting than females. The males spent more time on the ground, sifting through leaves for invertebrates and fallen palm nuts, or snails. Females tended to break open deadwood or hunt through palm crowns for the same foods and to strip open palm crowns for their pith (Munkenbeck-Fragaszy, in preparation).

Brown capuchins at the study site of Cocha Cashu in Manu National Park, Peru, are similar. Females eat more insects; males rest more and eat more vegetation. There adult males spend more time breaking up wood, and females more hunting through palm crowns (Janson in Terborgh, in preparation).

Female chimpanzees termite-fish and ant-dip more often, in longer bouts, than males. McGrew (1979) points out that this female "gathering" of prey may be a parallel to early human division of labor, for it is compatible with child care. Nursery groups of children play up and down the termite mound, and the females can break off feeding to attend to children's needs and then return to their work.

Finally, the chimpanzees of the Tai forest in the Ivory Coast use hammer stones differently. It is usually the females who open Coula nuts directly in the tree. A male collects 12 or 15 nuts and carries them to a stone anvil, where there is almost always a wooden hammer club at hand. He cracks the nuts and climbs at least 15 meters up the tree for a second round. A female may instead carry a hammer up into the tree, pick, and then store nuts in mouth and hands. She cracks the nuts by a kind of balancing act, hammering on a horizontal branch, holding hammer in one hand, the nut in the other thumb and forefinger, with the rest of the store in mouth and feet, all the while holding onto the tree. To eat, she puts down the hammer on the branch, transfers the uncracked nuts from mouth to hand and the nut meat to her mouth, and so round again. You may conclude, following your own prejudices, either that the female is showing more foresight, more coordination of motor acts, and more economy of effort in that she does not keep climbing up and down trees, or else you may conclude that the male insists on setting up his workshop properly before he starts work. On the whole, it seems likely one should argue for greater female skill or concentration, because either in trees or on the ground the females take fewer blows to open the nuts and obtain more nuts per minute. Besides, only females open another nut, the "Panda" which is so hard that it demands stone hammers, strength, and then precise restraint not to bash it into inedible fragments (Boesch and Boesch, 1981).

WATER

Drinking is a relatively neglected topic in field studies of primates. Many of the forest forms never descend to drink but obtain all they need from dew and the water that is caught in leaf axils or tree hollows. Primates, including ringtailed lemurs, commonly reach a hand into arboreal cups of water and lick the drops from their fingers, but more drastic measures can be adopted if necessary. (Fig. 3.14.). Chimpanzees, as usual, have a tool; they may chew leaves into a sponge, dip the sponge into the water, and then suck it.

The need for water is a question of both physiology and food supply. Some leaf eaters may be able to absorb water released by urea recycling, as in nonprimate ruminants. This could explain why leaf-eating hanuman langurs can range so much farther into the Indian deserts than the omnivorous and otherwise highly adaptable macaques (Bauchop, 1978). Mantled howlers in the dry forest of Costa Rica drink during the wet season but not the dry. This peculiar reversal seems to be caused by the dearth of young leaves during the wet season. Not only do the howlers then eat a diet of drier mature leaves but they may need more water overall to detoxify the poisonous secondary compounds of mature leaves (Glander, 1978).

Arid country primates include groundwater sources in their essential range. A whitefaced capuchin troop came to the ground to drink at the same streambed pool every day for three months, an ideal spot for a predator or a primatologist to wait. Males usually led the troops' progression to the water hole, increasing the chance that a male would meet a predator first (Freese, 1978). Similar male vigilance and leading to water has been seen in toque macaques, hanuman langurs, and baboons (Beck and Tuttle, 1972; Rhine 1975), and a patas male climbed a tree and looked over the area before his females approached (Hall, 1965). In Ethiopia, hamadryas baboons (Figs. 3.15A and 3.15B) dig holes in streambed sand with their hands to reach the water below (Kummer, 1968).

Baboon and patas troops often meet other troops at water holes. This leads to much larger social aggregations than usual, with some skirmishing but also surprising tolerance between otherwise distinct troops (Struhsaker and Gartlan, 1970).

FIGURE 3.14. *A mantled howler reaches a water-filled tree hole. (Courtesy K. Glander.)*

FIGURE 3.15A. *Hamadryas baboons dig for water in a dry streambed. (Courtesy H. Kummer.)*

FIGURE 3.15B. *The result-ing hole. (Courtesy H. Kummer.)*

"Two studies have suggested water as a limiting factor at the edge of primate range, in these cases acting as a density-independent population control. At Gilgil in Kenya, a water supply system put in for cattle allowed baboons to expand into new areas and their population has rapidly increased (Harding, 1976); and at the Waza Game Reserve in Cameroon, water holes dried up in the drought, and baboon and other species died in large numbers (Gartlan, 1975)" (Rowell, 1979).

A third study showed both the importance of water, and the relation between ecology and social factors. At Amboseli one subordinate vervet troop was kept away from the water hole by other vervets. Several of its members died, with little water and a dry season diet high in toxic compounds that demand even more water than usual for digestion (Wrangham, 1981).

Finally, many primates may trade off drinking from standing water sources with obtaining moisture from their food. This does not diminish the importance of water supply. Ringtailed lemurs drink when they can; riverside troops descend once or twice a day in the dry season to drink at the shore. Budnitz (1978) carefully compared the ranging and feeding of a riverside troop and a troop a few hundred meters inland in sparser forest. The inland troop had a home range three times as large as the riverbank troop, yet fine-grained analysis of feeding and searching times, feeding bout lengths, and use of vegetation layers showed very little difference between the troops' food plants. The major difference was that the inland troop traveled regularly to an area where they fed on succulent water-storing vines. Thus, as with the howlers that only drink in the wet season, water needs and supplies interact with all the other aspects of feeding and ranging.

SPECIALISTS AND GENERALISTS

It would be satisfying to conclude this chapter by pointing out differences between primate species that are narrowly specialized to one feeding niche and species that are generally adaptable.

Unfortunately, we have little idea how to go about finding out which is which. Some species are clearly specialized anatomically, such as the colobines with their sacculated stomachs, or the aye-aye with its probing finger. When we look at behavior, though, they nearly all turn out to be more variable than we expect. The aye-aye uses its insect-eating teeth and hand to chisel open coconuts and flick out the pulp. The colobines not only eat fruit as well as leaves but use the same digestive apparatus to ferment mature leaves, or a far-flung diet of new leaves, or even terrestrial herbs.

The gross classifications of Table 3.1 gave a start, with gorilla at one extreme as leaf eater, and demidoff's galago at the other as insectivore. Table 3.2 shows a different approach: the proportion of diet that comes from two, five, or ten plant species.

By this criterion, the most specialized of primates are mature leaf eaters: brown lemur in its tamarind trees, purplefaced langur in Sri Lanka, and gue-

TABLE 3.2. Specialization of Diet by Plant Species

Species	% Diet Top 2 Species	% Diet Top 5 Species	% Diet Top 10 Species
lemur, ringtailed	42	71	93
brown	77	93	100
indri	30	55	73
sifaka, white	28	48	65
howler, mantled	36	51	64
mangabey, greycheeked	34	50	69
Tana River	30	58	79
guenon, vervet	44	61	
blue	16	35	
redtailed	21	37	
langur, banded	9	18	27
hanuman	40	73	96
nilgiri	32	54	67
purplefaced	81	100	100
dusky	14	23	37
colobus, guereza	63	76	88
black	22	44	64
red	45	69	87
gibbon, lar	13	28	44
siamang	18	26	36
orangutan	14	28	42
gorilla	42	53	74

Data from T. H. Clutton-Brock & P. Harvey (1977a and personal comment). These data are species' average, so those animals that have been studied at more than one site have aggregated data. Only species that have been studied for more than one year are included. For site-specific data, see Chapter 8.

reza colobus. The converse, however, does not hold true. Many leaf eaters, including the gorilla, spread their browsing over a fairly wide range of plants, at least in some habitats. The least specialized of all is an anatomical leaf eater, the banded langur.

It is worth looking at the ten species column, to put Table 3.2 into perspective. It seems that the majority of studied primates take half their food from ten species or less of plants, although far more foods are sampled in small quantity.

Finally, we are only beginning to categorize plant parts in detail. Leaves are not just young or old. Frequently, only the petiole or the leaf tip is consumed. The "fruit" may be merely a particular seed coat, meticulously scraped from pulp and seed, while the monkey discards all the rest.

It is possible to point to some species as specialists, typified by the purplefaced langur, which eats no more than five species of plants in the treetops of Polonnarua. We must be wary even there to include the habitat: one troop of purplefaced langurs, whose forest was cut down, foraged on seaside rocks (G. Manley, personal comment). It is much harder to detail the myriad ways of being a generalist.

SUMMARY

Primates eat fruit, leaves, seeds, flowers, gum, insects, and meat in differing quantities. Almost all species eat some fruit, which is a concentrated energy source. Fruiting figs are a common food, which may have favored early primate social groupings in large feeding trees. Plants are adapted to defend their seeds from predation; reaching defended seeds by indirect means, such as opening nutshells, can test a primate's ingenuity.

Digesting mature leaves requires a specialized gut. Leaves provide protein but little energy. Specialized mature leaf eaters are inactive animals with small day and home ranges and relatively small brain size. A few primates lap nectar and may be plant pollinators. Some strepsirhines and callitrichids eat exuded gum. All these plant foods may contain defensive chemicals, which means that herbivores pick out the least toxic plant species or plant parts in the midst of apparent abundance of food. Water is important both for general metabolism and for detoxification. Primates sample a wide range of food, but most take more than half their diet from ten species of plants or even fewer.

Most primates catch and eat insects, either by stalking, by the smash-and-grab approach, or by methodical poking and prying. This last, like seed extracting, goes with long attention span, precise manipulation, and the use of tools. Among Gombe chimpanzees insect gathering is mainly a female occupation. Tai forest female chimpanzees are more efficient than males at opening nuts. Many primates eat meat, and hunting traditions have developed among baboons and chimpanzees. Hunting is mainly a male occupation and involves the rudiments of cooperative strategies and food sharing.

CHAPTER
4

PREDATION, DISEASE, AND DEATH

PREDATION

Primates, like everything else, die from predation, disease, accidents, including falls, and starvation. The final category leads us back to food supply and eventually to a general discussion of demography. Of course, a few wild primates are lucky enough to die of old age.

Having seen what primates eat, we now ask what eats primates. Unfortunately, although we have a wealth of data about food and ranging, we know far less about predation. Usually we argue indirectly, from the alarm calls and other antipredator tactics of the primates themselves.

Primates are prey to carnivorous mammals, (Fig. 4.1) snakes, crocodiles, and eagles. The best statistics come from studies of predators. In records of 46,000 kills by 13 species in the Kruger National Park, 75 were baboons. "Leopards accounted for 58, lions and brown hyenas for seven each, crocodiles for two, and python for one" (Pienaar, quoted by Saayman, 1971). Saayman concludes from this and other studies that leopards are the most significant terrestrial baboon predators, although lions may outdo them in parts of Africa, and at least one adult male rhesus fell to an Indian tigress (Lindburg, 1979). Leopards appear most frequently as killers in the primate literature, both in Africa (Altmann and Altmann, 1970; J. Altmann, 1980; Seyfarth et al., 1980), and in Asia, where they kill langurs and even young orangutans (Eisenberg and Lockhart, 1972; Rijksen, 1978). Jaguars may play the same role in the new world. Smaller carnivorous mammals such as tayras and fossas eat small primates like tamarins and lemurs (Galef et al., 1976; Defler, 1970; Albignac, in press).

Wild bushpigs scavenge the remains of carcasses in the forest. A pig attacked and killed a captive orangutan infant, and wild pigs have attacked human children (Galdikas, 1978; MacKinnon, personal communication).

Snakes, like birds of prey, probably engulf primates on all continents. The

FIGURE 4.1. *A lion eats a subadult or adult male baboon. (From J. Altmann, 1980, by permission.)*

Seyfarths found their vervet troop snake chuttering at an African python with a large lump in its stomach; one of the adult females was never seen again. The activities of Indian pythons are widely known but seem not to have entered the primate literature since Kipling. Fear of snakes may have a partly innate bias in macaques but is much stronger in animals raised in the wild (Huebrer et al., 1979. Pineka et al., 1980).

Eagles and other large raptors carry off primates. The monkey-eating eagle of the Philippines feeds on crab-eating macaques. Martial eagles, black eagles, and crowned hawk eagles catch African vervets (Seyfarth et al., 1980). The Madagascar harrier hawk swoops at lemurs, which scream in chorus as the hawk descends.

The primates' own responses are the clearest sign of the importance of avian predation. Most small monkeys and social prosimians have a frantic air-raid call, which leads the troop to stream lower in the trees. African monkeys are more likely to form groupings of several species together than are primates on other continents. This may reflect a higher density of hawk eagles in Africa (Struhsaker, 1982, and see Chapter 8). The only quantitative data come, again, from studying the predators. Of 141 prey items brought to a single harpy eagle's nest in Guyana, 47 were cebid monkeys—about 30 times what one would expect from relative prey abundance (Izod, 1978, Rettig, 1978, quoted by Struhsaker, 1982).

The difference in group size between nocturnal and diurnal primates is another clue. (Figs. 4.2, 4.3) Nearly all nocturnal mammals, including nocturnal prosimians, forage alone. Nocturnal predators such as owls mainly hunt by sound, by stealth, by ambush. Prey animals do well to be silent. Diurnal predators, particularly hawks, hunt by sight, often with violent pursuit. In these circumstances gregariousness offers some protection (Terborgh, in press).

The most significant predators of all may be other primates, even excluding man. For example, see the section on meat eating, but this time from the point of view of baboon confronting chimp or vervet confronting baboon. The selection pressures by primates may be different from those of other predators. Primates may take account of the positioning or even the personalities of de-

A

B

C

D

FIGURE 4.2. *Tree-hole refuges of lion tamarins, eastern Brazil. (Courtesy A. Coimbra-Filho.) C,D. Tree refuges of lepilemur, southern Madagascar. (Courtesy C. M. Hladik.)*

FIGURE 4.3. *Larger primates sleep on open branches, counting on the troop for protection. Costa Rican howler, with collar for radio-tracking. (Courtesy K. Glander.)*

fenders; the defenders may know potential predators as individuals. One can imagine development of peculiarly primate social strategies of attack and defense (Busse, 1977).

Many primates give alarm calls. Vervet monkeys have several different sorts of call, including "leopard alarm barks," "eagle rraups," and "snake chutters" (Struhsaker, 1967). Vervets respond differentially not only to the calls but to taped playbacks of the calls when other cues are missing (Seyfarth et al., 1980). "Leopard alarms" send them rushing up trees, and "eagle rraups" send them scurrying for cover in or under a bush. "Snake chutters," less menacing, bring the troop out to mob the snake. Several other primates have different calls for aerial and ground predators, for example, red colobus (Struhsaker, 1975), and ringtailed lemurs (Jolly, 1966).

Social group size reflects predator pressure even beyond the nocturnal/diurnal distinction (Bertram, 1978). Clumping may distract the hunter, as schooling fish dazzle their pursuers. A large group of animals has more eyes and ears to spot danger, so they may share vigilance. Green monkeys on the Island of St. Kitts clumped together in the open but spread out when safely under cover (Fairbanks and Bird, 1978). Chimpanzees at Mount Asserik in Senegal travel in larger parties than those seen in East Africa, and nest higher in trees. Parties that do not include an adult male stick to open ground and don't venture into forest, presumably because Mt. Asserik is well populated with lions and leopards (McGrew, 1981).

But how much do primate groups actively defend themselves? And do adult males really play a special role as defenders? These are vexed questions. On the one hand, a whole theory of primate social organization has been based on differences in predator defense; on the other hand, some deny any such thing. Chance and C. Jolly (1970) proposed a fundamental dichotomy between centrifugal monkeys, such as patas, where the females hide at the approach of danger while the male bounces off bushes to lure away the predator, and centripetal monkeys, which draw in toward their "leaders" when in danger. The patas monkeys carry their isolation to an extreme at night. Only one animal, or a mother with dependent young, sleeps in a single thorn tree, and the troop never sleeps two consecutive nights in the same site. They even give birth during the day, unlike most other primates, which minimizes the risk of nocturnal predators sensing the birth (Chism et al., 1983).

It seems, however, that there is great variability in the behavior even of savannah baboons, the typical "centripetal" primate. Several studies have found that adult males position themselves in the front and rear of the troop, in the most protective positions when the troop is alarmed (DeVore and Hall, 1965; Rhine and Westland, 1981; Rhine, 1975), but on some occasions a baboon troop simply gallops in disorder for the trees, with females who carry heavy infants left farthest behind (Rowell, 1966). Marching order seems nearly random (Altmann, 1981), or at best highly variable (Harding, 1977).

Venezuelan wedge-capped cebus have a quite different structure, with the single adult male and dominant female in the best protected positions at the center of the group. Subordinate females foraged at the periphery, and spent most time in vigilance (Robinson, 1981).

Sometimes it is clear that males counterattack. This is not, of course, purely primate; the adult pigs attacked by chimpanzees attempted to protect their piglet (see previous chapter). An adult male baboon attempted to position himself between the crippled baboon infant and the chimps. Baboons have attacked and killed dogs (Stolz and Saayman, 1970) or even leopards (Marais, 1969, Saayman, 1971), and three male patas monkeys ran down a jackal and saved the infant patas in the jackal's jaws (Struhsaker, 1969). Male rhesus both scan the environment from trees and place themselves between the troop and such mild threats as dogs and primatologists (Lindburg, 1977). A 60-strong baboon troop that found a lioness with her cubs on their morning trail mobbed her, screaming and squealing, with volleys of barks from the adult males. After an hour and 20 minutes, the lions had retreated and been harried some 350 meters from the troops' normal route, after which the baboons continued their morning progression (Saayman, 1971).

Various arboreal primates dislodge sticks and dead branches that deter ground predators. Whitefaced cebus persistently work away at deadwood until it falls. Baboons, similarly have dislodged stones from cliff tops and dropped them near observers. These baboons chose relatively large stones—about 600gms—pried them loose and shovelled them underhanded over the cliff side. This defense was primarily a male activity (Hamilton et al., 1975).

The few times that wild chimpanzees have been seen with leopards, they have retreated, ignored the beast, or once thrown sticks from trees. One hand-reared animal charged, but this may have been through surprise and inexperience. (The leopard, perhaps equally surprised, made off.) (Gandini & Baldwin, 1978). However, Kortlandt (1967) placed a stuffed leopard with staring glass eyes, mounted on a moveable trolley, in the path of wild chimpanzees. He filmed the chimpanzees as they advanced, bristling and screaming, and flailing or awkwardly throwing sticks. The stuffed leopard, unlike any real animal in possession of its senses, did not retreat, and held still for heavy clubbing directly on its back.

Kortlandt believes that open-country chimps are more likely to attack than forest chimpanzees. If our own ancestors needed defense in the savannahs they may have used stick clubs for several million years before stone tools appeared in the fossil record, or twirled the powerful deterrent of a broken-off wait-a-bit thorn bush. (Kortlandt, 1978, 1980; Brace, 1979). A great deal depends on local traditions, even in chimpanzees. Jane Goodall (1968) watched the development of branching and rock-throwing in the Gombe stream from miscellaneous flailing of branches in display to accurately hurling stones at visiting dignitaries.

Defense against predators is one factor among many in the evolution of society and culture. The chimpanzees' crude clubs are little different from the sticks used in ant-dipping or in trying to hook down an inaccessible branch (see Chapter 17). Male roles in troop defense may well be less important than inter-male competition in selecting for large body size, large canines, or the hormonal correlates of courage. Predator defense may have acted alongside these other factors in primate evolution or in our own; two or three good reasons for evolving a trait are better than one.

PARASITES AND DISEASE

Although there is a vast medical literature on primates as carriers of human disease, we know relatively little of the effects on wild primates themselves. Viral epidemics have swept through the Amboseli baboons (J. Altmann, 1980). In the Gombe chimpanzees, disease probably accounted for 30 per cent to 40 per cent of the mortality between 1963 and 1973 (see Table 4.1), or more than 50 per cent of the mortality from 1965 to 1980, when three epidemics of respiratory disease and suspected polio struck the population. Some of these epidemics may have started from human contacts and may have been exacerbated by the assembling of chimpanzees for banana feeding. The chimpanzees have no large predators so deaths are either from disease, age, or injury, mainly caused by other chimpanzees (Fig. 4.4) (Tekel, et al., 1976, Goodall, 1983).

The Gombe chimpanzees often catch colds and coughs, which sometimes develop into a serious illness with deep, husky cough, and yellowish mucous. One infant who died was diagnosed as having pneumonia: this suspected pneumonia is one of the commonest fatal diseases. The other major disease started about a month after a human polio outbreak in the area near the park. Ten chimps had obvious symptoms. Four died (two of them shot after they became so paralyzed there was no future but starvation). The remaining six were paralyzed in one or more limbs but have partially recovered. One middle-aged female, Madame Bee, lost so much of the use of one arm that she had no free hand to carry the baby born five years after her illness. It died, but when the next baby was born, the mother began to walk bipedally to support it.

There is a considerable medical literature on primates as vectors of human diseases, particularly malaria, yellow fever, and the dreaded Herpes B menengitis. There are epidemiological studies of the prevalence of antibodies to human disease, but I do not know of studies of their demographic effect on wild, as opposed to human, populations.

Freeland (1976, 1979) suggests that parasite load and disease transmission set an upper limit to the size of social groups. His grey-cheeked mangabey groups acted almost as "islands" for their internal parasites, as found in fecal samples. Each mangabey troop shared many of the same parasite species but differed statistically from adjacent troops. Possibly if the groups were larger, more species of parasites could coexist in the "island" of monkeys, to the monkeys' individual detriment.

TABLE 4.1. Causes of Death in Gombe Stream Chimpanzees

	Baby (0–1)	Infant	Juvenile	Adolescent	Young adult, prime	Middle-aged	Old
Respiratory	gyre[8] FLAME[2] VILLA[2] BEE HINDE[9]	plato[4]		POOCH*[1]	CIRCE[1]	SOPHIE[1] david[1] leakey[12] william	hugo[2] mike[2]
"Polio"	grosvenor[2]		hornby[11] merlin*[12] flint**[9]	mac.d[12] (shot)		j.b.[11]	mcgregor[12] (shot)
Other illness					GILKA[5] NOVA[3] worzle[5]		
Orphaned		SOREMA[2] CINDY[3] ?Jenny's sib[6]	flint** merlin*				
Injury	JANE[2]			POOCH*	sniff[11] godi[1] de[5] charlie[5]	MADAM BEE[9] rix[11]	goliath[2] huxley*[4]
Cannibalized	gandalf[6] OTTA[9] orion[8] GENIE[11] ?melissa's[1] BANDA[3]						
Age							FLO[8] BESSIE[12] MARINA[5]
Vanish-susp. death: cause unknown	wood bee[7] ?Sophie's[9]		BEATTLE[6]		pepe[10] sherry[11]	willy wally hugh? faben[6] OLLY[1]	
Vanish-susp. emigrate, etc.		JAY[3] QUANTRO[10]	LITA[3]	SALLY[2] BUMBLE[6]		JESSICA[3] VODKA[10]	
Lack of care	?patti's[4]						

*, ** Two possible causes of death, or combination of both.

Upper case—♀♀; lower case—♂♂; italics—body seen; ?—sex not known; [1]—Jan, [2]—Feb, etc. Causes of death, known or presumed, are for individuals of Kasakela and Kahama communities, 1963–1980. Month of death or disappearance is indicated.

(*Goodall*, 1982)

FIGURE 4.4. *Wild primates suffer many human pathologies: Ollie, a chimpanzee in the Gombe Stream, has apparent goiter. (Courtesy P. Marler.)*

Short-term behavior may help avoid parasite infection. The ground beneath a baboon sleeping tree is contaminated with fecal parasite eggs which hatch two to eight days after deposition. The baboons at Amboseli, where there are only a few favored sleeping groves, tended to use the same groves only one or two nights in succession, and returned to the same grove after an average of nine days, when the fecal infectivity had returned to baseline (Hausfater and Meade, 1982). Desmond Morris (1967) pointed out that wild primates do not have fleas, whose larvae grow in a den or nest. Purely arboreal primates do not even have to contend with infected ground. Usually they dispose of infectivity by defecating well clear of the branches. Where they do not, as in western dwarf lemurs, it is a specialized, stereotyped scent marking not just sloppiness (Schilling, 1981).

Parasite load varied with social rank in Amboseli baboons (the variation was statistically significant only in males). Oddly, it was the more dominant animals who excreted most parasite ova. Females excreted more parasites in the weeks before and after menstruation than at midcycle, and more when sexually cycling than pregnant or lactating (Hausfater and Watson, 1976).

A quite different parasite also respects motherhood. Botflies cause swellings on the howler monkeys of Barro Colorado and may even lead to death through secondary parasitism by screwworm fly larvae that infect the swellings, then eat through the body wall and organs. Botfly swellings are less common in pregnant and lactating howlers. It is not clear whether females in better health can resist parasites and are therefore more likely to reproduce, whether parasite load can block reproduction, whether pregnancy increases the body's defenses, or all three (Smith, 1977).

Primate reactions to sick troop members seem ambivalent as human reactions. Several baboon mothers at Amboseli increased their care and carrying when their babies were sick (Altmann, 1980). Primate mothers often carry the corpse of a dead infant for days, only abandoning it when it has dried or rotted almost past recognition. On the other hand, the repertoire of response may be limited. Melissa, a young chimpanzee mother, just hugged her baby tighter as it screamed with a broken arm, though her hugs increased its pain (Goodall, 1969).

Jane Goodall tells of the chimpanzees' varying reaction to McGregor's fatal paralysis with polio. Lesser injuries, such as Pepe's and Faben's dangling arms, scared the other chimps at first, but they grew used to them. McGregor died slowly, over ten days. The other males were frightened by him, avoided him, even attacked him, but Humphrey, almost certainly his brother, stayed near and nested by the dying male (Goodall, 1971).

ACCIDENTS

Primates do fall out of trees. (Table 4.2) In a collection of 260 wild gibbon skeletons, a third had healed fractures (Schultz, 1969). This implies that such primates not only have accidents but survive a great many, and we see one-handed baboons and forest chimpanzees coping apparently unhindered (Fig. 4.5). Sometimes, perhaps often in the case of young infants, the accident is fatal. Rix, an adult male chimpanzee of the Gombe southern group, fell and apparently hit his chin on a rock, which snapped his neck upwards and broke it, killing him instantly. Sixteen other chimps stayed with the corpse for four hours. They gave "Wraah" calls, the call of great distress and excitement. The adult males displayed with bristling fur and charging runs, and incidently mated, one after the other, with Melissa, an estrous female accompanying the group. Four younger animals remained more curiously investigating the body. They stared and sniffed, but none dared touch it. Godi, the one young male who, like Rix, moved with the southern Gombe group, was last to lose interest and leave. (Teleki, 1973).

TABLE 4.2. Falls Observed in Gombe Stream Chimpanzees

Context	Mature ♂	Mature ♀	Juvenile ♂	Juvenile ♀	Infant ♂	Infant ♀	Total
Aggressive (fights, chases, etc.)	13	3	4	—	1	—	21
Play	—	—	—	2	10	3	15
Locomotion during feeding, etc.	4	1	3	—	6	1	15
Totals	17	4	7	2	17	4	51
Branch breaks	8	2	2	—	3	1	16
36 falls when distance noted							
Falls over 5 meters	11	1	3	1	7	—	23
Falls over 10 meters	8	1	1	1	2	—	13

(Goodall, 1982)

FIGURE 4.5. *Minor injuries are common in wild primates. One-eyed animals survive without apparent difficulty. Savannah baboon male, Gombe Stream. (Courtesy T. W. Ransom.)*

VIOLENCE

In nearly all long-term primate studies, some animals have been wounded, and probably killed, by conspecifics. Whether this is considered frequent is largely a question of fashion. When the popular primate books stressed violence, primates seemed remarkably peaceful. After primates seemed peaceful for a while, we were shocked to discover them occasional killers.

Males fight, more in some species than others, of course. Ringtailed lemur and white sifaka males fight during their short breeding season, raking downwards with the upper canines to gash flank and thigh. Baboons fight, occasionally to the death. A red howler male wounded a male that tried to move into its troop so badly that it died. Many more species are rarely seen fighting but have a suspiciously high proportion of scar-lipped and ragged-eared males.

Infanticide occurs in many primate species. A broad definition would include killing by members of their own species of immatures at any age from gamete through to the end of parental investment of energy in the child (Hrdy and Hausfater, in press). This includes contraception and abortion, active killing, and neglect leading to death by predation and starvation. Table 4.3 lists explanatory hypotheses for infanticide.

Males of harem-forming species such as Hanuman langurs, redtail and blue guenons, or red howlers sometimes bite and kill infants of a troop they have taken over, the sooner to father their own infants. This has apparently evolved through sexual selection between competing males, and will be discussed in Chapter 12, "Competition." Male chimpanzees kill and eat infants from other communities, with a fascinated attention, grooming and even playing with the

TABLE 4.3. Predictions from Five Explanatory Hypotheses for Infanticide

	Degree of Relationship Between Killer and Victim	Age of Infant	Age and Sex of Killer	Nature of Gains to Killer
1. *Competition for resources*	typically distant	any	adults of both sexes, perhaps especially females	increased availability of resources for killer and killer's kin
2. *Exploitation as resource*	in mammals, usually distant, but cannibalism by kin common among fish and invertebrates	size and vulnerability more important than age	both sexes; any age large enough	typically, a nutritional gain to individual
3. *Sexual selection*	distant	unweaned	adults of sex investing least in offspring—typically males	individual gains increased breeding opportunities
4. *Parental manipulation*	close, at least 0.5	typically unweaned	either sex, but most often the mother	lifetime reproductive success of one (usually the mother) or both parents enhanced
5. *Social pathology*	random	proximity and vulnerability are the only factors	random	none

From Hrdy and Hausfater, in press.

corpse, which contrasts with their straightforward consumption of other prey (Bygott, 1972, Goodall, 1977).

One of the Gombe Stream female chimpanzees killed and ate other females' babies. They may be removing future competitors as well as obtaining meat to eat; that is both resource competition and exploitation of the infants as a resource (Goodall, 1977). There is one case in which fragments of a gorilla infant's bones appeared in the feces of another female of the troop (Fossey, in press). One hopes these females are pathological, but it is possible to argue their case in terms of natural selection. In one sense they are the extreme version of the competition between females, benefiting their own offspring at the expense of others, which goes on in all primate species (again, see Chapter 12).

The elimination of competitors is usually slower and more indirect. Among at least some macaque populations, female infants and juveniles are more likely to die than male infants and juveniles, particularly during periods of food shortage (Dittus, 1975, 1979). Such socially imposed starvation is further discussed in Chapter 11, "Demography," and Chapter 14, "Mothers and Infants."

In wild primates, infanticide by the parents seems extremely rare. The few cases of infant neglect or abuse by orphaned Japanese macaques seem likely to be social pathology rather than parental manipulation to achieve a higher fitness in the long run. In humans, in contrast, infanticide is most commonly either a group phenomenon during warfare, or performed by the parents themselves. Parental infanticide is a widespread and probably ancient practice: in a survey of 99 societies coded in the Human Relations Area Files, infanticide was reported for 84 of them (Whiting et al., 1977; Hausfater and Hrdy, in press). A third of the societies explicitly reported the killing of defective infants, and a third apparently used infanticide to space births. The frequency is highest in hunting, gathering, and fishing societies, where a majority are reported as infanticidal. Infanticide may plausibly have typified hunting and gathering societies throughout the Pleistocene (Birdsell, 1979).

Such parental infanticide is a fairly costly way of diverting parental energy from one child to others that seem more likely to survive, or of saving the mother's own life if she would starve through feeding the child. The nearest parallels are not in primates, but in mice. Pregnant female mice exposed to the odor of a strange male commonly resorb the embryos. This in turn seems to be an evolutionary consequence of male infanticide: the male is likely to kill the litter if he has not been mating with the female. By terminating the pregnancy, she avoids further investment in a litter that is doomed anyhow. The fact that a far higher proportion of malformed human infants than normal ones are spontaneously aborted in early pregnancy suggests that physiological mechanisms reject deformed infants before birth, as parents in many human societies do after birth.

A recently current explanation for infanticide no longer seems valid. We used to attribute it to population regulation for the good of the group, or of the species. Humans often act for the good of their group, and we now need desperately to act for the good of our species. However, there seems no *evolved* mechanism that would let group interest override individual advantage unless the individual were almost indissolubly bound to his or her particular group (see Chapter 9 on levels of organization). Thus, although infanticide, particularly of females, may strongly affect future population size, this does not stand up as an adaptive reason for its evolution.

The conclusions on infanticide, then, are that death seems to be inflicted much more commonly on immature than on adult conspecifics. However, this is not a unitary phenomenon; it may be done by parents or unrelated individuals, for various adaptively advantageous reasons, or through disadvantageous, pathological reasons. In a given case it may be hard to tell whether the infanticide is a benefit or cost to the killer.

All the violence discussed so far concerns individual attacks. The Gombe chimpanzees also give us a parallel to human group conflict. Two communities, the Kasakela group and the Kahama group, ranged together during the 1960s. As the research team slacked off on the banana supply, the two groups gradually separated their ranges and were already distinct by 1972 when Rix of the Kahama group fell from the tree and died (Bygott, 1979).

Between 1973 and 1978, Kasakela males have killed six of seven of the Ka-

TABLE 4.4. Fatal Injuries Inflicted by Gombe Stream Chimpanzees

Year	Name	Age	Nature of Injury	Cause
1965	JANE	3 months	Fractured humerus, bone and tendons exposed: seen dead third day after appearing with injury. Carried by mother.	?
1967	huxley	old	Deep gash from just below eye down to lip. Breaks open repeatedly during 2 month period. Disappears.	Almost certainly during fight with community male.
1967	POOCH	adolescent	Deep wound in groin, does not use leg for walking for 6 months, sexual cycles stop. Disappears during epidemic of pneumonia.	?
1968	rix	middle-aged	Broken neck: dies instantaneously.	Falls from tree during reunion excitement between two parties.
1974	godi	prime	Wounds on face, leg, arm, and ribs. Big gash from lower lip down to chin. Never seen again after attack.	Kahama male attacked by 6 Kasakela males for 10 min.
1974	dé	prime	Deep gash inner left thigh, one toe half off, nails torn from fingers and toes, other wounds. Seen 1 month after attack, incredibly emaciated, unhealed wound in groin. Never seen again.	Kahama male attacked by 3 Kasakela males for 20 min.
1975	goliath	old	Very bad wound lower spine, other wounds on head, and hand. Unable to sit up after attack. Never seen again.	Kahama male attacked by 5 Kasakela males for 18 min.
1975	MADAM BEE	middle-aged	Wounds on head, back, one leg, both hands, and ankle. One toe half off. Dies fifth day after attack.	Kahama female attacked by 4 Kasakela males for 10 min.
1977	charlie	middle-aged	Wounds on head and neck, rump, scrotum, legs, arms, hands, and feet. Found dead in stream third day after attack heard.	Kahama male. Circumstantial evidence of attack by 5 Kasakela males.
1977	sniff	prime	One leg fractured or dislocated (unable to use): bad wounds on face, back, and legs. Seen the following day; subsequently never seen again.	Kahama male attacked by 6 Kasakela males for 19 min.
1976	Melissa's baby	1 week	Neck seems broken, forehead opened and bleeding, some skin torn from upper back. Carried by mother.	Possibly a victim of infant killers Passion and Pom (see text).

Upper case—♀♀; lower case—♂♂.
Injuries which led, or are thought to have led, to death. Kasakela and Kahama individuals, 1965–1980. The three babies killed and eaten by two females of the community (Passion and Pom) are not included.

(Goodall, 1982)

hama males, as well as the matriarch, Madame Bee, in group assaults, and since then have eliminated the entire group. The attackers used hands, feet, and teeth and once even threw a stone at a prostrate victim. They never attacked when two or more Kahama males were traveling together, only when one was alone or with a female. Godi was the first to die—the young male who had lingered by Rix's corpse (Goodall et al., 1979).

The second attack on Dé, observed by E. Mpongo and A. Bandora, took place on February 26, about one and a half months after the Godi attack.

The Kasakela party was small, consisting of Evered, Sherry, Jomeo, and Gigi (in estrus) . . . At 0845 they began to travel purposefully to the south. Once they tensed with hair erect and stared into a tree: two baboons appeared and the chimpanzees gave a few soft grunts and appeared to relax. They then continued their cautious travel, as if they were hunting prey. Suddenly at 0915 the four raced forward, and a moment later there was an uproar of screaming and barking. The observers thought there had been a predation, but the three Kasakela males were found attacking Dé. Calling and displaying nearby were Sniff and Charlie: Little Bee (Kahama female, daughter of Madam Bee) was fully swollen and present with Dé.

After two minutes, Evered and Jomeo left Dé and charged southward, apparently in pursuit of Sniff and Charlie, but the observers stayed to watch the struggle that continued between Sherry and Dé. Sherry stamped on and bit Dé, who was unable to escape as the younger male hit him. . . . he soon stopped struggling and sat hunched over, uttering squeaks, as Sherry continued to assault. Another two minutes passed, and then Dé appeared to bite Sherry, who ran off. Dé managed to climb a tree, but Sherry followed and renewed the attack. Dé began to scream loudly again.

Soon afterward Jomeo and Evered charged back: Jomeo immediately climbed up to join his brother (Sherry) in the attack. Dé managed to jump away to another tree, now very quiet, but the brothers leapt after him and continued hitting and biting their victim. Dé, again screaming, managed to jump away, but the branch he landed on cracked so that he was left hanging close to the ground. Jomeo instantly leapt down, seized Dé's leg and dragged him to the ground. Evered now joined the attack. Dé lay flat on the ground and no longer tried to escape as the three prime males assaulted him, all screaming. Little Bee continued to watch and Gigi began a display, stamping and slapping the ground, and then joining in the attack. The fight became even more vicious, as all four Kasakela chimpanzees hit and stamped again and again on the still prostrate Dé who was uttering a few squeaks. The observers saw the aggressors tear skin from Dé's legs with their teeth, and at some point he was dragged along the ground.

Once again Evered and Jomeo left Dé to charge after the retreating Charlie and Sniff, uttering barks and screams and waving branches. . . . only after the attack had lasted 20 minutes, did Sherry leave his victim to follow the others. Gigi, however, was left near Dé and was not seen again that day.

One and a half hours later, the Kasakela males returned to the site of the attack. As they approached, they called and seemed to listen for a response. They followed the track along which they had dragged their victim and intently smelled the ground and vegetation, particularly where blood was seen. Evered stared into the trees where Dé had been attacked. The party spent about half an hour in that vicinity, but although the observers searched for Dé during that time, they could not find him.

Two months later Dé was found, traveling by himself in Kahama. He was incredibly emaciated with his spine, backbone, pelvic girdle, and anus protruding.

He was lame [because of] a bad unhealed gash on his inner left thigh that prevented normal locomotion: he had great difficulty in climbing and spent much of his time feeding on the ground. Nails were missing from his fingers, a toe was partially bitten off, and part of one ear was torn away. His once large scrotum had shrunk to a fifth of its former size. He was followed for five consecutive days during which he was always alone. He has not been seen since despite intensive searching (Goodall et al., 1978).

SUMMARY

We have little quantitative data on causes of primate death. Leopards and other carnivorous mammals, large snakes, and birds of prey eat primates. Some species have different alarm calls for different classes of predators. Predation by other primates may exert more complex selection pressures. Many primates probably die of disease and parasitism, including human diseases. Diseases accounted for half the known deaths of Gombe chimpanzees. Some primate behavior may help minimize parasite transmission, from defecation patterns to group size. Accidents, mainly falls, are common. Primate males fight, very occasionally fatally, though scarred lips and ears are common. Males of several harem-forming species may kill infants when taking over a new group, and chimpanzee males sometimes eat infants of other groups. The males of one group of chimpanzees have killed the males of a neighboring group one by one. Violence is, however, apparently a rare cause of death compared to disease or even accident. Finally, starvation, which may combine with predation and disease, brings us back to the importance of food supply.

CHAPTER
5

RANGING

DEFINITIONS

A *home range* is the area in which an animal spends its adult life. This excludes long juvenile wanderings, when many mammals emigrate from their birth-place. It also excludes seasonal migrations from summer range to winter range, as well as the occasional foray far outside the usual limits. Leaving aside such exceptional movements, what remains is the home range, where an animal feeds, nests, sleeps (Burt, 1943; Bourlière, 1964). Individual primates sometimes migrate to a new troop, but few primates have regular, seasonal migrations, except people.

A *territory* is a defended portion of the home range (Burt, 1943). Its limits are shown by the owner's aggressive behavior. Pitelka (1949) found that sand-pipers respected each other's areas without overt defense by the owner. He suggested that exclusive use be another criterion of territory.

Core area is a shorthand term for the region in which the animal spends most of its time. When defined quantitatively it may be the number of quadrats (measured map squares) in which the animals were for 50 per cent or 75 per cent of observations. The core area is one of many ways in which we attempt to express differential use of the home range. Home range is never a uniformly crosshatched area on a map. Animals, like humans, have familiar paths from bed to supermarket to local bar, or sleeping tree to feeding tree to water hole.

A few primates, like the gibbon, defend a territory that approximates their whole home range. Many overlap ranges extensively, or even completely, like gorilla groups. The longer you watch, the more overlap you are likely to see, and if you watch in more than one site you may find species like the white sifaka beating the bounds of a defended territory, marked out with smelly scent posts in one forest, yet overlapping without defense of range in another. Furthermore, the more you watch, the more intricate are the ranging patterns that define core area and total home range.

This chapter begins by outlining the major premise of ranging studies: that food is, potentially, a limiting factor, and the cross-species comparisons which support this idea. The great howler debate then shows how hard it is to demonstrate food-limits within one species. Three sections deal with foraging and day range, with the measurement of horizontal range use, and vertical and microhabitat spatial separation. The section on energy budgets treats the attempt to put food intake and energy output into a single balanced equation. The final two sections on knowing one's range and changing range almost revert to anecdote, but these qualities are the bridge between metabolism and intelligence. The question of whether the range is defended as a territory is left until Chapter 7 on relations between social groups, for that becomes the question of whether you fight your neighbors or join them.

IS FOOD LIMITING? CROSS-SPECIES COMPARISONS

The last chapter showed us that we have so little data on predation and disease that we do not know what role they play in limiting most wild populations. On the other hand, we have a mass of feeding and ranging data. Nevertheless, we do not know just how critical it is to feed efficiently, or whether on a given day primates rest to recuperate from the effort of feeding, or from sheer boredom.

We can try to attack the problem of how food supply constrains population size, and individual activity, by analysis of energy budgets within one population, by comparing populations, or by comparing species.

In some ways, it has been most revealing to work at the grossest scale, and to compare ranging behavior between species—day range and home range against the major categories of diet.

A typical primate's typical day begins with waking, stretching, defecating. Little by little it turns to social behavior: grooming and some play among the young. Then off to work: the troop moves in a quick, purposeful fashion to the day's first feeding site and spends the rest of the morning foraging and feeding, perhaps in one place, perhaps in several, or, spread out amoebically, foraging as it moves. At last sated, the primates siesta through the heat of midday. Another bout of feeding in the afternoon precedes the final progression to the sleeping site, more grooming, and sleep.

Of course, any one day may be different. In cold weather the animals eat sporadically all day; in rain they huddle miserably and do nothing; perhaps a fight, a birth, a female in estrus, or an encounter with another troop disrupts the pattern. In study after study, though, the general pattern prevails.

The length of a primate's day range varies from 30 meters for slow loris to 13 kilometers (km) for the hamadryas baboon. As usual, there is a measurement problem. *Path length* is the actual distance an animal moves, counting curlicues; *day range* is some smoothing of path length, probably different for each study. The same animals may travel kilometers on one day and feed in one grove through the next. Even so, regularities emerge. Day range is longer in terrestrial primates than arboreal ones (Fig. 5.1.), and it is shorter in foli-

FIGURE 5.1. *Day range length plotted against average feeding group weight. Figures follow-ing each species name show the proportion of foliage eaten (▲ = nocturnal, arboreal, ■ = diurnal, arboreal, ● = diurnal, terrestrial). (After Clutton-Brock and Harvey, 1977.)*

vores than omnivores (Clutton-Brock and Harvey, 1977b). Surprisingly, it does not change with body weight. The average day range centers on 1.5 km for everything from mouse lemur to orangutan (Martin, 1981).

Such correlations, even with the wide scatter of data, suggest that on the whole primates adjust activity to food supply—leaf versus fruit, tree versus ground. The amount of time spent in each phase varies widely. Cross-species comparison shows that feeding time increases with body weight and decreases with amount of foliage in the diet; amount of time spent moving decreases with amount of foliage in the diet (Fig. 5.2,) (Clutton-Brock and Harvey, 1977b).

This relationship with leaf eating appears even more clearly if we look at pairs of animals in the same forest. Brown lemurs, purplefaced langurs, and mantled howlers are all leaf eaters that take a high proportion of mature leaves in their diet. Their ranges overlap, respectively, with ringtailed lemurs, hanuman langurs, and blackhanded spider monkeys. In each case the mature leaf eater has smaller home range and day range, spends more time resting, and less time either moving or feeding. The contrasting omnivores spend their energy in wide-ranging search for nutritious food (Hladik, 1975; Sussman, 1974).

Home range size is negatively related to the amount of foliage in the diet; the more leaves, the smaller the range. This is intuitively likely: leaves are a more abundant food source than fruit. This negative relationship, however, is weaker than the positive correlation of home range and body size (Clutton-Brock

FIGURE 5.2. *Proportion of day time spent moving against proportion of foliage eaten. (Clutton-Brock and Harvey, 1977.)*

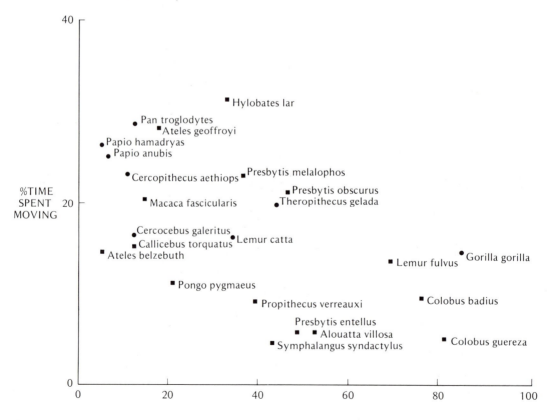

Ecology

and Harvey, 1979). The two relations run counter to each other, for leaf eaters on the whole are larger bodied than frugivores, as discussed in Chapter 2 (Clutton-Brock and Harvey, 1977).

This combination of large body and high population density means that folivores have much higher biomass (total weight of the species per unit area) than do frugivores (Fig. 5.3), and they in turn than insectivores (Hladik and Hladik, 1969; Clutton-Brock and Harvey, 1977a; Eisenberg, 1979). These relationships hold within primate suborders as they do across the order as a whole (Clutton-Brock and Harvey, 1979). New-world monkeys and apes, though, have larger home ranges and much lower biomass than prosimians and old-world monkeys of the same body size.

Foliage eating and body size are merely ways of approaching the central problem: what are the animals' metabolic needs? Home range size should correlate closely with metabolic needs, if there were any way to control for differing environments with their differing supply of food, and if food is indeed limiting in enough cases to shape the species' immediate or evolved behavior.

McNab (1963) established that there is an allometric relationship between

FIGURE 5.3. *Biomass and diet type. The left-hand rectangle in each diagram represents the proportion of leaves in the diet; the central one, fruits and seeds; the right-hand rectangle, insects and other prey. In these forests the species that eat more leaves have higher biomass. (After C. M. Hladik, 1978.)*

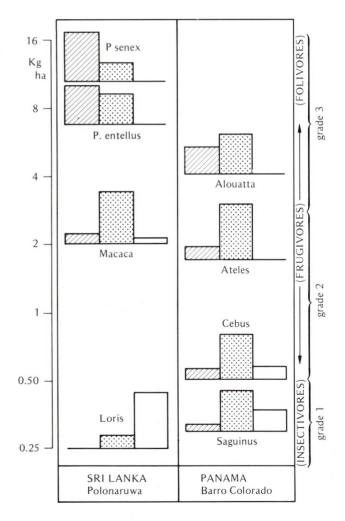

home range and body size in small mammals, Brown (1964) and Schoener (1968) did so for birds, and Turner and colleagues (1968) for terrestrial lizards. Both elevation and slope differ for herbivores and carnivores in each group. Carnivores, being a step up the food chain, need larger home ranges, and large carnivores, whose prey is sparsely distributed, range proportionately more widely.

Mace and colleagues (in press) present a new comparison of galliform and passerine birds, rodents, and squamata (lizards and snakes), avoiding some of the methodological problems of earlier studies, and Harvey and Clutton-Brock do the same for primates (1981). They again find that home range size increases with daily energy expenditure. However, the exponent of the slope is about 2.3 for birds, rodents, and primates, not the predicted 1.0. This means that home range size does not just increase in proportion to metabolic needs, but very much faster. Sensitivity analyses which change the values of various assumptions do not change the result much. Mace and others attribute this to increased patchiness of the environment. As body size increases, relative metabolic needs decrease, but suitable food sources, such as fruit trees worth the detour to reach them, are more widely spaced. The net effect is a relatively much larger range for the big animal—and more to learn about where it is not worth feeding.

Such cross-species graphs are an important step in the right direction, but they need far more data points and a far clearer idea of the range of each species' adaptations for us to understand the components that make up the home range metabolism slope of the whole order.

This in turn leads to a more fundamental challenge. There is real variation in species' behavior in differing environments. Clutton-Brock and Harvey (1981) point out that four of their terrestrial primates, including the ringtailed lemur, have suspiciously small ranges. They rightly attribute this to both primates' and primatologists' affinity for rich habitat by pretty rivers. More data on these species running over waterless crags would see-saw the slope of the graphs. We cannot yet really sort out what are fascinating, precise adaptations to particular environments, what may be historical accident, and what may even be truly random.

IS FOOD LIMITING? THE HOWLER CONTROVERSY

There is an inherent contradiction in our approach. On the one hand we assume that behavior has adapted over evolutionary time to the species' niche. Therefore, we look for success and argue that foraging patterns are intricately attuned to make the most of the environment. However, natural selection is the process of removing failures, or at least forcing them into lower reproductive rates for themselves and their kin. To see evolution in action, we should look for failure—the drought year, the marginal habitat, the relict population whose range is dwindling to zero, and watch to see who dies. Some points on the graphs do not show species' adaptations but ill-adapted populations making the best of where history has stranded them on the margin of their niche.

Even harder to cope with is the possibility that behavior may be neutral. Does fanatic Darwinism blind us to random changes in behavior? Or does concentrating on food lead us to ignore predator-controlled primates banqueting in luxury, who drop half-eaten fruit because they are too full to bother with another mouthful, and who could vary their feeding within wide limits without much affecting selective pressures?

The clearest indication that food is limiting comes from provisioning wild primate troops. In every primate population where people have artificially increased the food supply, animals' weight rises, age at menarche falls, and interbirth interval decreases. Infant mortality may or may not decline; it has in Japanese macques, but because of disease it may have even increased in Gombe chimpanzees (Chapter 11).

Comparison within species should be the most fruitful indication of the importance of food supply. However, Table 5.1. indicates how variable a given index, such as home range size, may be within a species. The size of range per individual clearly decreases from snowy north to clement south of Japan, but there are also a wealth of other factors. It is more useful to concentrate still more, for instance on the well studied populations of howlers.

The most detailed attempt to argue that a primate is not constrained by its

TABLE 5.1. Home Ranges of Japanese Macaques

Area	Troop	Troop Size	Home Range Size (km²)	Size per Individual (km²)	Habitat Conditions	Latitude	References
Shimokita Peninsula							
Northwest	I	31	10.4	0.335	Snowy	41°30'N	Azuma & Ashi-zawa, in prep.
	M	44	18.1	0.411			
	Z	44	26.7	0.607			
Southwest	A	15	2.3	0.153	Snowy	41°10'N	Azuma & Wada, 1970
	O	13	1.75	0.135			
	B	9+	1.3	0.144			Izawa & Nishida, 1963
Tsugaru Peninsula	T	50	10.0	0.200	Snowy	41°10'N	Azuma, 1966, 1967
	Y	20–30	8.5	0.354*			
	M	40–45	11.5	0.272*			
Shiga Heights	C	37	3.66	0.099	Heavy snow	36°40'N	
Hakusan Mountains	T	71	10.25	0.144	Heavy snow	36°10'N	Hayashi, 1970
	Y	23	4.78	0.208			
	K	16	3.84	0.240			
Ryozen	K	44	8.0	0.182	Snow rare	35°20'N	Sugiyama & Ohsawa, 1974
Shodoshima	T	52	4.5	0.087	Snow-free	34°30'N	Yamada, 1966
	K	150–160	7.5	0.049*			
	O	52	8.0	0.154			
	S	130–140	12.0	0.089*			
	I	125–130	15.5	0.122*			
Kawaradake		102	4.2	0.041	Snow-free	33°30'N	Ikeda et al., 1973
Takasakiyama		180	2.71	0.015	Snow-free	33°10'N	Itani & Tokuda, 1954
Yakushima Island	KO	45	0.8	0.018	Snow-free	30°20'N	Maruhashi, 1980

*Mean value calculated from both limits of the troop size range.

(Wada and Ichiki, 1980).

food supply is the work of Coelho and colleagues (1976, 1977a, 1977b, 1979). They estimated 25 Guatemalan howler monkeys and 225 blackhanded spider monkeys living in a five square kilometer study area at the Mayan site of Tikal, Guatemala. Eighty-six per cent of the howlers' diet during July–August 1973 came from the fruit of a single tree species, the ramon (*Brosimum alicastrum*). Ramon were planted by the ancient Maya and are still eaten by Indians today, so many grow among the monuments of Tikal. Coelho calculated that in peak seasons of productivity the ramon alone would support a monkey population 735 times larger than at present, and even in minimal seasons, at least 90 times larger. Besides this, there are at least 100 other tree species, some of which offer primate food.

Coelho concluded that the howlers and spiders at Tikal are not food limited. Their erratic foraging, going right through some ripe ramon trees to feed on others, and their sloppy habit of spitting out seeds, or taking a bite of pulp and dropping the rest, seems genuine waste. If so, it would be silly to look for optimal foraging patterns, or to explain details of troop size, or even the clear antagonism between the two monkey species as competition for food.

Cant (1980) attacked Coelho's calculations. There are fewer ramon trees and the trees are less productive than Coelho assumed. Coelho had used data for particularly rich plots and productivity of mature, not average, trees. Further, the howlers digest only fruit pulp and immature seeds (they spit out mature seeds), but Coelho added in the energy of all ramon seeds. So far, this is an argument over measures of short-term sampling, complicated by the fact that recensuses found even fewer monkeys (Schlichte, 1978; Cant, 1978), which implies even more food available per animal.

The most important of Cant's points is in long-term sampling. Ramon may fruit three times a year in "normal" years, but there is no complete data for a "normal" year. It is known that some fruiting seasons fail. It is probably these periods that set the population limits, for the monkeys may not have time to breed enough to fill their environment before the next period of constraint. A similar mammalian population crash hit Barro Colorado in 1970, when abnormal rain caused the trees to flower early, then set no fruit. Emaciated corpses of howler monkeys, peccaries, porcupines, and sloths lay on the forest floor, outstripping even the vultures' capacity to tidy up, but the spider monkeys survived by raiding the laboratory dining hall (Foster, 1982).

If food is only limiting during intermittent lean seasons, most short-term studies will happen during periods of food abundance and will give clues to the food constraints but not show them in action. Thelma Rowell (1979) generalizes, "the proximal limiting factor of a plant-eating species is likely to be predation in a stable environment, and the majority of primate species are native to tropical forests, the most stable of terrestrial ecotypes. Food supply may (only) become critical when the environment changes, as seems to have been the case at Amboseli for baboon and vervet populations (Struhsaker, 1976)."

However, as we have more and more long term studies, we see more and more instability in the food supply, even within tropical forests. The population of howler monkeys on Barro Colorado seems to have stabilized at around 1200–1350 and between 1974–1979, after the 1970 die-off. Each year there is a season of dietary stress during the latter part of the rainy season, when there is a decrease in both fruit and new leaf production. In several years particular tree species, including the ramon, failed to fruit, throwing the howlers back onto less preferred foods. This is also the period of peak mortality. Of course, the deaths themselves come from parasites and disease, with undernutrition as a contributing factor. This is one of the clearest cases where food is limiting. (Milton, 1982)

A second mantled howler population is probably also food-limited. In the dry forest of Guanacaste, Costa Rica, there is intense competition among female howlers for troop membership, and ultimately for ranges with palatable leaves. (Jones, 1980) Detoxifying mature leaves placed great seasonal stress on the howlers, when they might mainly eat from only one or two tree species for weeks on end. (Glander, 1978a, 1978b, 1980)

Thus, different population studies of the same species may lead one to differing conclusions—particularly if the study is too short to cover a lean year, or if there has been a recent epidemic of disease.

Let us turn, then, to attempts to measure foraging behavior, remembering that for most sites and studies the adaptive explanations are speculations, not cast-iron fact.

DAY RANGE AND SEASONS

How does the day range and time spent moving or feeding relate to availability and choice of food? First, one must distinguish between a *prey* and a *patch*. A *prey* is a food item of which you eat all, or a fixed amount. A mullet for a shark or a leaf petiole for a red colobus might be a prey: one bite is all. A durian fruit for a liontailed macaque or a man for a shark is equivocal: as much like patch as prey, because the monkeys drop bits of durian as sharks drop the odd hand or leg. A treeful of fruit or a shoal of swimmers is clearly a *patch*. At some point it becomes uneconomic to persist in the same place as the ripest fruit is eaten or the swimmers stampede for the beach, so the predator moves on to a new patch.

The theory of optimal foraging concerns the variety of prey that a predator finds worth eating, the trade-off between different kinds and sizes of prey, and the amount of time spent in different patches before they become uneconomic.

In rich habitats, a predator need concentrate on only a few kinds of prey. These are the ones that give greatest return for energy spent—perhaps the size class of insects it catches most readily, or the ripest fruit. In a sparse habitat, or lean times, it should not be so choosy, but accept less preferred food. This reduces the energy cost of travel between widely spaced items of preferred food. This change has been tested experimentally with great tits (a small British bird), offered large and small chunks of mealworm on a conveyor belt. The tits conform fairly well to the quantitative predictions of energy investment and nutritive return, although they keep eating a little too much uneconomic small prey. Perhaps they keep sampling the environment in case of a new bonanza, which would be reasonable behavior in the wild (Krebs, 1978).

Prey diversity, Krebs admits, is a complicating factor. Predators need many different nutrients, but the mathematical models do not handle this yet.

Budnitz' (1978) ringtailed lemurs illustrate the importance of patches of different kinds in a real home range. A riverside troop of 17 animals had a home range of eight hectares, whereas an inland troop of 16 animals ranged over 23 hectares. The riverside group had a prime habitat of continuous, closed canopy tamarind trees, grading to more open canopy away from the riverbank. The inland troop's richest habitat was open canopy forest, which shaded to low, dry bush and scrub. One might predict that the riverside troop would have shorter day range within its small home range, indolently shifting from one fat tamarind tree to the next. On the contrary, the riverside troop moved farther, the animals often retracing their own path as they climbed down the riverbank twice a day. As we mentioned in Chapter 3 on water needs, the inland troop's movements were largely set by the 600-meter hike to its scrub area for succulent water plants—a straight path there and back, with little movement at either end and so little energy to spare that they were thin and scruffy by the time the rains finally came.

Their shorter day range was a reaction to dearth of resources with a minimum length dictated by the distance between patches of food and cover in the woodland canopy and the patches of water source in the dry scrub. The larger home range is no simple function of sparse food supply, but a stepwise function of the need to include patches of different resources.

Modern field studies provide some measure of the preferred foods and some measure of time spent feeding on each food before the troop shifts to another, but the reason for the shift could be prey quantity or quality or patch size. How to sort them out? One solution is to say the field data are not a mere confounding, but a summing of the factors.

Richard (1978) studied white sifaka in Madagascar. She worked with two groups in a relatively rich northern forest and two in the sparse southern spiny desert. In the abundance of the wet season and in the northern forest, the sifaka concentrated on energy-rich prey (fruit and flowers), but shifted more from patch to patch (more tree species, longer day range), and spent more energy doing so (longer day range). That is, in the richer season or forest they put in more investment for higher return. In the harsh conditions of the southern dry season, they cut the investment (day range), shifted to less preferred prey (mature leaves), and even cut the time spent feeding. Because southern sifaka are often visibly in poor condition at the end of the dry season, with sparse fur and wizened muzzles, it may be that they are in negative energy balance, losing weight, but keeping the loss as small as possible, rather than attempt to forage uneconomically for more food.

Homewood's (1978) Tana River mangabeys, however, had an opposite strategy. In the lean season, when not much fruit was available, the mangabeys spent more time foraging, ranged farther, and ate more diverse foods. The result was that they spent about the same amount of time actually eating (not just foraging for food) and ate about the same amount of fruit (their preferred diet) in all seasons.

Homewood's mangabeys, by increasing effort and diversity of diet, ensured adequate return in all seasons. Richard's sifaka, instead, lay low in the harsh season, adjusting not only ranging patterns but reproductive cycles and perhaps even body temperature to make few demands when food was scarce.

Several studies of other primates reveal opposing patterns, but as Clutton-Brock (1977) warns, seasonal changes are a complex subject. Ranging is probably a sum of food availability, composition, and distance, with threshholds where the animals may switch from one approach to another.

One example can make this clear: the versatile red colobus. Red colobus' preferred food is young growth—new leaves and shoots. The Tana River population is a relict, isolated in their stretch of gallery forest since a wetter climatic period when evergreen forest stretched across East Africa. They face a pronounced dry season. In the season of scarcity they eat as high a proportion of new growth as in the rich season, but range much farther to find it, investing more to maintain their income, like the mangabeys with which they share their forest (Marsh, 1980).

Clutton-Brock (1973, 1975) in the Gombe Stream mixed forest found the opposite effect. In poor seasons when there was little new growth, the colobus ranged less widely, and 50 per cent of their diet was mature leaves; when necessary they switched to mature leaves and stayed put. There are many more tree species at the Gombe than at the Tana River. Perhaps at Gombe the colobus could find mature leaves of low toxicity that they could not find at Tana. In other words, the Gombe colobus followed sifaka strategy. In the longest and most detailed red colobus study in the Kibale Forest of Uganda, Struhsaker (1975) found no correlation between day range length and food supply. Thus, the red colobus can apparently play it both ways, or refuse to play at all.

There is no reason to think that ranging strategy has a simple relation to diet. It might well be that in an extremely rich season or area a primate might move little, feed on what comes to hand, and somehow shift more energy into reproduction. In sparser areas it could travel farther to maintain the same food intake (Tana mangabey strategy). In still sparser areas or seasons it might travel less again, conserving energy on the sifaka pattern. One can imagine a final or lemming strategy of desperation—if everything else fails, to travel as far as possible in the hope of food in a new area. The shifts we see are only sections in a curve whose total shape we do not know.

HORIZONTAL HOME RANGE USE

The home range that appears in a published paper is an outline drawn round the day ranges someone observed in the field. As such, the apparent area depends on the length of the study. If you plot the cumulative total of area used during each month's observations, when and if the curve levels off, you can think you have seen most of the range.

Of course, the animals shift over time. Troops grow or dwindle, occasionally merge or fragment. Individuals migrate from troop to troop. Even while the primatologist is summing his observations over a year or two, the primates change the basis of his calculations. Home range and core area are therefore abstractions in the mind of the primatologist, summed up from the hard facts of day range.

Three ways to measure ranging patterns are by drawing linear paths on a map, by checking off troop movements through a grid of arbitrary squares of terrain called quadrats, and by relating troop movements to individual trees or feeding sites. The first, or linear, method can give a quick visual impression of range use: the most densely scribbled areas are most intensively used, and a typical day's movements are clearly seen. However, such spaghetti patterns are not easy to quantify, so people have turned instead to grids of quadrats measured out on the ground and marked by trails or blazed trees.

The first problem here is the size of quadrat to choose: one hectare quadrats would be meaningless for small animals with less than a hectare's range, and the effort of laying out small ones might be wasted for larger animals. If your goal is to compare species, their sizes and ranges almost certainly differ, so the appropriate grid for one species will be too coarse for the other. In short, you need to know your animals and forest *before* choosing your measures.

One of the most detailed attempts to calculate primate feeding in terms of quadrats has been Struhsaker's longterm study of red colobus in the Kibale forest of Uganda. By 1975 he could find no relation of his index of feeding diversity to the dispersion of the food species, or to the percentage of young growth such as new leaves in the monthly diet, or even any relation between food species' diversity and distance traveled. After all that work, the only significant measure was that the colobus' ranging was affected by other troops; they spent most time where they might meet and chase their neighbors. One reason for what Struhsaker, with great restraint, merely called his "counter-intuitive" result was that his plots of food species took no account of phenology, which is the timing of flowering and fruiting. In tropical forests, trees of a single species are often not synchronized with each other and may flower or fruit individually at several-year intervals. This means that a "widespread food species" may in fact have rare individuals providing food for monkeys. Conversely, a rare kind of tree may fruit synchronously, or may have a few adjacent individuals fruiting in a clump and so for the moment is "common." On top of this, individual trees vary enormously in the food they offer. A single fruiting fig may feed all the primates in the area for days, as well as birds, bats, and pigs, whereas smaller trees are stripped in half an hour (Struhsaker, 1975).

Thus, Struhsaker's meticulous use of the quadrat method revealed clearly that the best basis for comparison is the individual food tree. This, in turn, means that primatologists increasingly need botanical backup to estimate the quantity and timing of food offered by each tree.

Waser and Floody (1974), working in the same forest in a later study of greycheeked mangabeys, found that the mangabeys concentrate on big fig trees: superabundant, widely spaced food sources. The trees are far enough apart so that a move to a new source generally means changing quadrat. Here the quadrat system worked, with a stepwise jump when the monkeys traveled.

A simpler means of using quadrats to express ranging patterns is to plot cumulative frequency of use. MacKinnon and MacKinnon (1980) compare data for all five diurnal primates in the Krau Reserve, Malaysia in a six-month study. The core areas, defined as the quadrats where each group spent 75 per cent of its time, are significantly different not only from each other but as proportions of each troop's observed home range.

As shown in Table 5.2, the siamang has a small range, relatively evenly utilized, whereas the lar and the two leaf monkeys each spend three quarters of their time in only a third of their ranges, with forays over a wider area. The macaque is intermediate in coverage but with a total range twice as large as the siamangs.

TABLE 5.2. Home Ranges and Core Areas at Kuala Lompat

	H.R. (ha)	Core Area (ha)	Core Area/HR	Feed/Travel Ratio
Siamang	23.0	16	70%	4.0
Lar gibbon	53.0	19	36%	1.1
Banded langur	28.5	10	33%	1.6
Dusky langur	21.0	7	35%	1.5
Longtailed macaque	46.2	22	48%	2.0

(MacKinnon and MacKinnon, 1980).

The ratio of feeding time to traveling time expresses the same relations in a different way. The heavy-bodied siamang is highly efficient in its use of range: it travels for only a quarter of the time it spends feeding. The light-bodied gibbon darts about (or is displaced by the siamang), passing many trees with no fruit, and travels almost as long as it feeds. Note that "efficiency" in this sense is a judgment. If food limits them both, the more efficient animal feeds longer. If food is not limiting, then even the gibbon may not live merely to eat, and the more efficient animal might be the one that spends more energy gloriously swinging through the trees. There might even be other criteria of feeding efficiency, if a traveling animal was also checking future food sources.

The final type of analysis of range use is being pioneered by people like Egbert Leigh (1978), Annette Hladik (1978, 1980), and Katherine Milton (1978). This involves marking and measuring individual trees and estimating their fruit and leaf production. This is, apparently, the only way we shall in the end arrive at an energy budget.

MICROHABITAT: HEIGHTS, TIMES, AND LOCOMOTION

A horizontal plot of the home range is insufficient to represent the behavior of three-dimensional animals. Each species differs in its use of the layers of forest, and even the most terrestrial primates sleep in the safety of trees or cliffs. The species of a single forest partition their time in the various layers.

Similarly, no plot of spatial use is complete without knowing its time course. The obvious division is between nocturnal and diurnal species. Even those that eat the same food avoid each other's presence and face different predators if one comes by night and the other by day. More subtle differences also exist. Gibbons rise earlier in the morning than sympatric siamang and hurry to the fruit trees to feed. When the siamang arrive, they chase the smaller gibbons, who swing away to another food source. Observing such differences, however, does not prove that they are adaptive in any given situation. The howlers and spider monkeys in the superabundance of Tikal also move and forage at different times of day, without any apparent need (Coelho et al., 1979).

Finally, the habitat even within one tree is not uniform. Light-bodied animals can reach the farthest twigs to eat the terminal buds and new leaves. Lo-

TABLE 5.3. Locomotor Classification

Category/Subtype	Activity	Primate Genera
1. Vertical clinging and leaping	Leaping in trees and hopping on the ground	Avahi, bushbabies, hapalemur, lepilemur, sifaka, indri, tarsier
2. Quadrupedalism		
(i) Slow-climbing type	Cautious climbing—no leaping or running	Golden potto, potto, slow and slender lorises
(ii) Branch-running and walking type	Climbing, springing, branch running	Mouse lemur, dwarf lemur, forked lemur, lemur, all marmosets and tamarins, night monkey, titis, sakis, uakaris, cebus, squirrel, guenons
(iii) Ground-running and walking type	Climbing and ground running	Macaques, baboons, mandrill, gelada, patas
(iv) New-world "semi-brachiation" type	Arm swinging with use of prehensile tail, little leaping	Howler, spider, woolly, woolly spider
(v) Old-world "semi-brachiation" type	Arm swinging and leaping	Colobus, all langurs, proboscis, snubnose
3. Ape locomotion		
(i) True brachiation		Gibbon, siamang
(ii) Modified brachiation	Arm swinging and quad-rumanous climbing	Orangutan
(iii) Knuckle walking	Occasional brachiation, climbing, knuckle walking	Chimpanzee, gorilla
4. Bipedalism	Standing, striding, running	Man

Ape locomotion is defined as what each genus does. It is disputed whether to lump chimpanzee and gorilla with orang. or to separate as terrestrially adapted knuckle walkers.

(Napier and Napier, 1967).

comotion, as well as size, matters here. Table 5.3. shows Napier's classic division of locomotor types.

Locomotion is beloved of anatomists, because limb bones fossilize. It is only recently, though, that anatomy has combined with field study to show locomotion as a means of getting what that particular primate *wants*. Primates use long-distance locomotion in daily travel, fast locomotion in escape, aggression, and play, special feeding locomotion (often dangling from thin terminal twigs by arms and legs) and, for a large part of the day "postural" rather than locomotor activity. That is, they simply stand or sit.

In each of the types, the fastest mode of locomotion is diagnostic. Sifakas spring away from predators, baboons run away on all fours, and gibbons brachiate away. By this criterion, the chimpanzee and gorilla are knuckle walkers (or perhaps knuckle trotters?), as this is their quickest means of progress or flight.

The usual reaction of field observers is surprise that animals can, in fact,

FIGURE 5.4. *The white sifaka is a true "vertical clinger and leaper." The hindlegs propel the sifaka into a straight-line takeoff, then the animal turns in mid-air to land again on its hind feet. (Courtesy D. Attenborough.)*

do *everything*. Sometimes, unsuspected patterns appear. An extreme brachiator, the gibbon, and an extreme vertical clinger and leaper, the sifaka, both feed from the terminal tips of the branches by slinging themselves under the flexible outer twigs (Fig. 5.4.). Their adaptations for fast movement must therefore be seen in the light of their equally specialized feeding behavior (Figs. 5.5A and 5.5B). However, field descriptions do not usually add such a neat new category. Instead, they confuse the picture by pointing out the variation among ages, activities, and even time of day in the way a single species treats its trees

FIGURE 5.5A. *The sifaka dangles to feed. (Courtesy J. Buettner-Janusch.)*

FIGURE 5.5B. *The sifaka walking bipedally. Almost any primate can at times do anything. (Courtesy J. Buettner-Janusch.)*

(Ripley, 1967b). Meanwhile, other species of different locomotor style occupy the same branches.

They may even find apparent lack of adaptation. Struhsaker (1975) says his red colobus is "one of the clumsiest climbers in the forest. Its general mode of locomotion can best be described as suicidal."

Fleagle and Mittermeier (1980) and Mittermeier and van Roosmalen (1981) in their study of eight species of primates in the same Surinam forest conclude that the niche separation of species depends on microhabitat—different heights in the trees, different branch size, and on different choice of foods. Two species that eat the same range of foods must eat in separate places (MacArthur and Levins 1967), but one cannot predict without data which locomotor pattern will get them there. (See Figs. 5.6, 5.7, 5.8A., 5.8B., 5.9., 5.10A., 5.10B., 5.11A., 5.11B., 5.12, 5.13A., 5.13B., and 5.14.)

FIGURE 5.6. *The slow loris is a quadrupedal slow climber. Its hand has become a forceps with reduced index finger. It oozes along the branch and clamps a hand on unsuspecting prey. (Courtesy P. Rodman.)*

FIGURE 5.7. *Quadrupedal branch walkers (here a thick-tailed bushbaby in South Africa) either fling themselves over gaps or gingerly negotiate a passage. (Courtesy S. Bearder.)*

FIGURE 5.8. *A terrestrial quadruped, the vervet, rises to walk bipedally in Kenyan savannah. (Courtesy M. Rose.)*

FIGURE 5.9. *A new-world mantled howler in Costa Rica dangles by its prehensile tail. (Courtesy K. Glander.)*

FIGURE 5.10. *The mantled howler, like other quadrupeds, launches itself over gaps with bent knees and elbows to belly flop on twigs below (compare the leaping sifaka, or the leaping langurs). (Courtesy K. Glander.)*

FIGURE 5.11. *Old-world leaper: the banded langur. (Courtesy J. Fleagle.)*

FIGURE 5.12. *Banded and dusky langurs feed as well as leap with fairly vertical trunk. (Courtesy J. Fleagle.)*

FIGURE 5.13. *The siamang and gibbon are true brachiators. They gain momentum by raising and lowering their center of gravity like a child pumping on a swing. (Courtesy J. Fleagle.)*

FIGURE 5.14. *Orangutans, quadrumanous climbers, can and will do anything. (Courtesy D. Sorby.)*

ENERGY BUDGETS

Attempts to sum all the considerations of this chapter represent *energy budgets* of one's animal. An energy budget includes, on the expenditure side, estimates of the basal metabolism for each age and sex, of the energy spent in each activity—resting, feeding, locomotion, social behavior, and the energy of reproduction. On the intake side, you need estimates of the amount of food consumed and the efficiency of assimilation of the various food types (Bourlière, 1979).

After all this, with all the possible sources of error, you try to relate the animals' energy to the amount of food available in the habitat, which brings in a whole new range of measurement problems.

Time spent in resting, feeding, and moving is recorded in most field studies. It is more difficult to estimate quantity of food eaten although several authors attempt this. Only Nagy and Milton (1979) have measured food assimilation in wild primates.

Nine mantled howler monkeys were captured with tranquilizing darts on Barro Colorado Island, seven of them from one troop. These animals were injected with O_{18} and tritium-labeled water, and a blood sample was taken. They were released to feed normally for a week and then recaptured for long enough to take another blood sample. The amount of isotope lost during the week gives a measure of the amount of CO_2 produced during respiration, and this in turn provides a measure of metabolism and heat production. It was found that for free-living howlers in the dry season field metabolism increased in a one-to-one relation to body weight, not the 0.75 power like basal metabolism.

Nagy and Milton kept three howlers in temporary captivity after the second darting. The monkeys received a measured diet, first of fruit, then of leaves, and the caloric value of their diet and the resulting urine and feces were measured. The captives assimilated 37 per cent and 43 per cent of their food for fruit and leaf, and 34 per cent and 38 per cent of ingested energy was used for metabolism.

Nagy and Milton concluded that a wild howler needs about one kilogram of fresh food per day, or about 15 per cent of its body weight. Table 5.4. shows how this energy is divided.

The most recent census of howlers on Barro Colorado found about 1,340 monkeys, or a biomass of about 4.6 kilograms of howler per hectare, which then would consume about 90 kilograms dry weight per hectare of leaf, fruit, and flowers in a year. This is very close to Leigh's (1975) botanical estimate of the amount consumed, and only one hundredth of the amount produced yearly. Even with the competition of sloths, insects, toucans, and other primates, the howlers are not limited by sheer quantity in most years. They may well be limited by the plants' defenses, particularly of toxic compounds. The problem that is still intractable is measuring what foods are actually available to howlers by anything other than the choosiness of the monkeys.

How closely do the various measures of Nagy and Milton relate to other studies? Coelho's estimates differ (Coelho et al., 1976). The differences of esti-

TABLE 5.4.   Estimated Field Energy Budget for Howler Monkeys on Barro Colorado Island, Panama, During the Dry Season

Avenue		Energy Flux (kJ kg^{-1} day^{-1})	Percent of Input	Percent of Total Metabolism
Input:	food	1,020		
Output:	feces	625	61	
	urine	40	4	
	metabolism	355	35	
	basal	180		51
	maintenance	113		32
	activity	62		17
	(travel)	(8)		(2.3)

(Nagy and Milton, 1979).

mate of energy budget do not, however, alter Coelho's conclusion that the howlers are not food limited at Tikal. Multiplying metabolism by two, or feeding rate by three queries whether Tikal howlers have 1,000 or 10,000 times too much food.

Neither Coelho nor Nagy and Milton consider in detail the energy needed for howler reproduction. The energetic costs of lactation are even greater than those of pregnancy, for one is feeding a larger, more active infant with a less efficient organ. Human mothers do not maintain their own body weight with a caloric intake 23 per cent greater than normal during lactation, but they do maintain weight at 32 per cent greater than normal (Whichelow, 1976 in J. Altmann, 1980). For smaller mammals a lactating female's metabolic rate is 1.5 times "normal," a pregnant female's is 1.25 times, and calculated on her pregnant weight into the bargain (Coelho, 1974).

J. Altmann (1980) uses the human ratios and field data on growth, weights, and normal feeding to calculate the effect of lactation on a mother baboon's feeding time. Altmann has been working with baboons at Amboseli, Kenya, where food supply clearly is limiting, because the water holes and vegetation have been drying up and the wild animal populations have fallen over the past 20 years. If a nonpregnant baboon is just maintaining her body weight by feeding 43 per cent of the time, she cannot provide all an infant's needs past six to eight months of age without cutting into resting time and thus starting on a vicious circle where she spent more energy as well (see Chapter 14).

There is not, so far as I know, any evidence of different energy expenditure in different habitats, although the changes in day range length with season may be indicative. Milton (1978) points out that howlers eat a larger proportion of fruit in some months or some habitats; yet, they remain as inactive as ever. Tikal's howlers rest as much as the Barro Colorado ones, without being energy limited. Perhaps, Milton suggests, the howlers are programmed for sluggishness to cope with periods of low fruit availability.

Among animals, as well as man, there must be some rich and some poor. The very poor do not replace their numbers because of food limits; the rich have a surplus that allows them to reproduce to maximum capacity for the species. Thus, it is only with hindsight of long-term demography that one can really check on the meaning of an energy budget, not simply assume that any one population is "adapted."

KNOWING ONE'S RANGE

One of the chief uses of a wild primate's memory and strategic skill is knowledge of the home range to find the way efficiently from one food source to another. This presents a great challenge to primates whose diet is diverse, which ties in with the larger brain size of omnivores as compared to folivores (Clutton-Brock and Harvey, 1980).

Differences in knowledge of the home range are hard to document. One convincing case is Pollock's (1977) two adjacent groups of indri. A young couple with their first infant, whom foresters said had recently occupied their territory, concentrated their feeding in a few, often visited quadrats of their range. The other group, an older menage with three offspring (indri have a three-year birth interval) stayed to feed even in those quadrats of home range that they visited rarely. Pollock concluded that there was no difference in the ranges as such, but the older couple had a more thorough knowledge of their range.

At the other end of the primate scale, all observers of the great apes conclude they are "excellent botanists . . . (with) detailed spatial memory. Chimpanzees were capable of returning to known food sources precisely and by economical routes, apparently from any direction. (They also) traveled directly to habitat types . . . in which some ripe species were likely to occur" (Wrangham, 1977). Gorillas in the Mount Kahuzi region of Zaire migrate in August and September into bamboo forest, where they dig holes up to 20 centimeters deep for the new bamboo shoots sprouting beneath the ground. They began before any sprouts broke the surface. In August only 20 per cent to 50 per cent of the holes were successful, but by September some 80 per cent were. (No one is sure how they locate the underground shoots; primatologists have a far lower success rate than gorillas.) It seems that this is a migration *to* a food source remembered from the year before, not just *from* a dwindling food supply elsewhere (Casimir, 1975; Goodall, 1979).

MacKinnon (1978) notes "on a single day I followed the subadult male (orangutan) on an impressively economic course that took in six of my (eighteen durian) trees. Even with my map I could not have made a shorter trip to visit so many of the trees! It is quite clear that orangutans do not amble aimlessly around their ranges taking potluck. To plan such routes they must not only have a very good knowledge of what is where but they must also have a good sense of direction, distance, and travel time . . . For such a heavyweight frugivore this is the difference between survival and starvation."

MacKinnon goes on to speculate that apes, and our own ancestors, may have been solitary rather longer than the old-world monkeys and may have been large-bodied frugivores who needed detailed individual memory to forage efficiently (1978). In the early Miocene there were many more ape species than survive today. The monkeys which replaced them could eat a wider variety of foods—their molars evolved as efficient grinders. The apes, meanwhile, may have specialized as more efficient food-finders with larger brains (Kay, 1977).

The largest ranges reported in any primate are about 200 square kilometers for chimpanzees in the savannah of Tanzania's Kasakati Basin and an estimated 500 square kilometers for the driest area of western Tanzania (Kano, in Suzuki, 1979). This is the same order of magnitude as hunter-gathering humans' in dry regions, although the chimp may not know such a large area in the same detail that humans do.

Traditions may contribute to troop survival if stored in individuals' memories. The direction of a day's travel is set in many species without apparent leadership or decision. In gorillas, however, the silverback male decides; other troop members start off, but they look back over one black shoulder. If the silverback does not budge, they subside again. Similarly, in the two–male harems of hamadryas baboons, the older male sets the course. It is the last remaining function of an aging male. Even after the younger partner has taken over both discipline and mating, he defers to the older one on route choice.

A still more striking case was Waser's (1978) aged female mangabey. She was apparently postmenopausal, yet, not only kept up with her troop but even led their movements to unfamiliar swamps in a season of drought. This one animal may bear out Thelma Rowell's (1966) speculation that aged primates are useful to their descendants in just such emergencies, remembering what to do in the sort of drought that comes once in 20 years.

Thus, in both solitary and social primates, individual memories could be selected both to improve day-to-day ranging and to deal with rare crises, before language let group memories pass from generation to generation. Our own next step beyond is told by Birdsell (1979), as an Australian aboriginal hegira. An old man, Paralji, led his tribe on a 600-kilometer trek during the drought of 1943. The first stage, past some 25 water holes, took them to a fallback well that Paralji had visited only once as part of his own initiation rite half a century earlier. When that too failed, they traveled for 350 kilometers on ancient trails between the territories of two tribes, locating water holes by the place names and descriptions Paralji knew from ancient ceremonial song cycles. Man, in his own way, uses his memory of place to survive.

ADAPTING TO NEW RANGE

If there is such a high value in knowing the resources of one's range, do we have examples in the wild of primates spreading to new range and resources? If so, do we have any criteria for judging whether they are successfully enlarging their niche or are merely marginal groups, squeezed out by competition?

We have two sets of data: the changes brought by artificial feeding by primatologists and changes as primates exploit human alteration in the habitat on their own initiative. We do not have records of moves between two "natural" habitats, except what may be inferred about distribution changes in the Pleistocene.

Gorilla, like many of the species that associate with man, frequent secondary forest (Schaller, 1963; A. Goodall, 1979). In Rio Muni they raid forest plantations of sugar cane and manioc and demolish banana and plantain trees to eat the pith, not the fruit (Sabater Pi, 1977). Similarly, the tiny West African talapoins have learned to retrieve manioc that villagers leave soaking in streams, which removes poisonous compounds, but to leave it alone until it is safe (Gautier-Hion, 1970).

More drastic changes come with the artificial transplantation of troops. The Arashiyama West troop of 150 Japanese macaques were moved en masse from Japan to Texas. They promptly began eating the available Texan foods; they efficiently stripped the outer, spiny layer from prickly pear leaves, and leaped from other trees to the flower base of yuccas, avoiding being impaled on the yucca's spiny leaves. After six months they were consuming at least 46 local plant species, about 20 per cent of those available. In Japan, they had been eating 192 species, again 20 per cent of those available. The Texan diet, with extra provisioning, was adequate for weight gain, survival of peripheral males, and a normal birthrate (Clark, T., 1979). A troop of stumptailed macaques released on an island in Mexico adapted with similar rapidity (Estrada and Estrada, 1976).

The success in adapting may, however, depend on intact troop structure, such that related monkeys can observe and copy each other, and try new experiments with some confidence. Kawai (1960) attempted to form a new troop of Japanese macaques by introducing them to each other in a large cage. He released them in a forest where macaques had previously lived and been hunted to death ten years before. The new "troop" began their occupation by a 1.5 kilometer sortie along a ridge top, which let them explore visually. Soon, though, the group splintered into two, and many died of pneumonia and starvation, for they ate only moss and one shrub from their new home even when starving.

Two studies document behavioral changes as wild troops have learned to take advantage of man. Maples and colleagues (1976) say that baboons crop raiding out of the Jadini Forest of Kenya use diversionary tactics. The troops often spread out linearly along the margin between safe but unproductive forest and rich but dangerous cropland. When the conspicuous male sentinels, or an initial raid at one point of the line happened to draw the farmer's attention, other subgroups of baboons took their chance to raid at undefended points. The two-phase alarm bark, "ba-boo," in other baboon troops is given by adult males and provokes caution and retreat. Here, immature males barked when farmers approached, which signaled the farmer's whereabouts to other raiders, who far from retreating, took their chance for a foray.

Groups of vervet monkeys, on the other hand, have learned *not* to vocalize at farmers. Kavanagh (1978, 1980) describes vervets who have invaded secondary forest and clearings in the Cameroon rain forest, far from their normal habitat of forest fringe and tree-savannah. In savannah the monkeys give a loud mobbing call at dogs and other ground predators that cannot reach them up in trees. In the coffee and banana plantations they fall silent and hide when either humans or their dogs approach. These and other changes in male vigilance and general foraging calls show how adapting to new range involves a whole complex of adaptations in troop coordination and predator defense as well as mere food.

Artificial feeding of primate troops is a standard way of taming or habituating them to observations. It was first used with the Takasakiyama troop of Japanese macaques, who are now a great tourist attraction, and has been used with many macaques and chimpanzees. Almost every primatologist's camp has its garbage dump troop, whether they are ringtailed lemurs or Senegal baboons.

In the short term, concentrated sources of food lead to aggression, as every owner of a bird table knows (Balzama et al., 1973; Goodall, 1971). In the long term, providing a surplus of food provokes a population explosion, as happened at Koshima island (see Chapter 11, "Demography"). There may also be changes in social structure. The Takasakiyama troop reached 600 animals before it divided. The largest unprovisioned primate troop was a 200-strong baboon group at Amboseli.

The best documented case is Jane Goodall's chimpanzees (Goodall, 1971; Wrangham, 1974; Teleki et al., 1975). From 1962 onwards Jane provisioned her chimps with bananas, starting with the odd one for a friend, working up through piles of bananas on the ground, then to cement boxes that the chimps soon learned to lever open, and finally a set of sunken boxes filled from underground tunnels and sprung open by switches in the laboratory building. The chimpanzees' aggression to each other and to the researchers over the concentrated banana supply was so great that it swamped the earlier view of their peaceful and casual contacts. Finally after 1968 banana feeding was curtailed, except for an occasional handout to small groups, or for luring individuals up a rope to be weighed.

During the period of provisioning, aggregation size of chimps increased, aggression increased, contact between baboon and chimps increased, and so did preying by chimps on infant baboons. A polio epidemic struck the population, followed by two epidemics of probable pneumonia. The epidemics may have originated from humans; they may have spread through close contact between chimpanzees at the feeding area (Teleki et al., 1975).

Thus, we see in the crop-raiding baboons and vervets how primate behavioral plasticity adapts some troops to life in a world of man. In the Gombe chimps we see how the same sort of change may push a wild population beyond its level of tolerance.

SUMMARY

Comparisons of available data for primates show that day range does not vary with body weight but is shorter in arboreal than terrestial and in folivorous than frugivorous primates. Feeding and moving time is shorter in folivores. Home range is smaller in folivores but larger in heavier primates. Folivores have higher biomass. The same trends appear within families as in the order as a whole, but new-world monkeys and apes, as families, have larger home ranges than old-world monkeys and prosimians. A calculation of average daily metabolic needs of the social group correlates with home range size, with a slope of 2.3 in four different vertebrate orders, which suggests that larger animals range over increasingly patchy environment.

These correlations imply that there is adjustment of ranging behavior to food supply. The data is, however, noisy with a wide scatter of points. Some populations, such as the howlers of Barro Colorado, seem to be food-limited; others, such as the Tikal howlers, seem to have a vast surplus of food. Some primate groups live in range to which they are ill adapted; many may be predator or disease limited.

Foraging strategies may either be to increase investment in lean seasons, visiting more patches for the same amount of prey, or conversely to cut investment, reduce range, and shift to less preferred food, or even just less food. Red colobus can use either strategy in different forests; perhaps many primates may shift strategy.

Horizontal home range use can be measured by maps of day ranges, by percent of time spent in different quadrat squares, and by individual food tree. Quadrat squares have been the most useful quantification up until now, with either a measure of cumulative frequency of use or diversity of use. Because the primates work by tree, not square, the next decade will see more studies tied to estimates of tree leaf and fruit production.

Microhabitat separation can be achieved by height in the tree, or feeding site on branches. Primates of different size can reach different branches, but most anatomical types seem to be able to reach most parts of a tree. The different primate locomotor patterns seem to be adaptations for fast travel or flight, and secondarily feeding adpatations.

Energy budgets provoke arguments. There are now measures of basal metabolism, energy cost of activity, and digestive efficiency for howler monkeys, and a variety of measures and estimates of caloric value and other nutrient availability in common primate foods. Lactation increases a female's needs by one third to one half. We do not have estimates of the amount of food available to a primate, because we do not know their tolerance for secondary compounds, but total forest production is far in excess of arboreal herbivore consumption; most becomes leaf and fruit litter on the ground.

If food is indeed limiting in many environments, there is selective advantage in knowing how to use one's home range efficiently. Such knowledge can be detected in indri and is impressive in the great apes. Primates' adaptability occasionally allows them to exploit new range, particularly in crop raiding, but we do not have examples of range expansion without human interference.

CHAPTER 6

GROUP SIZE AND STRUCTURE

Most primates spend most of their lives in a social group. Their relations to the food and predators of their environment, as well as to each other, are structured by the group.

Group size and structure is to some extent hereditary. Each species seems to gravitate to some norm, whether it is a monogamous pair or a troop of scores of animals. Presumably, social structures have evolved, like anatomical structures, as specific ways of exploiting particular environments. One expects a certain amount of variation in response to the particular habitat of different groups. Part of the advantage of social structure is flexibility to adapt to particular environments. Much of the variation may also be random, not genetic or environmentally adaptive, as accidents of birth, mortality, and local migration change troop composition. We look, then, for general correlations of both evolved and immediate adaptation, while expecting a wider scatter of actual data.

Modern zoological classification of vertebrate societies stems from Eisenberg's (1966) survey of the mammals and Crook's (1965) survey of the birds. They both considered the effects of food supply and predation on grouping and reached many of the same conclusions. Then Crook and Gartlan (1966) attempted to sort primates into different adaptive grades. This was a multicriterion classification: grade 1 included small-bodied, solitary, nocturnal, insectivorous forms, grade 5 was arid-country, harem-groups. Most subsequent treatments have been attempts to dissociate the criteria and to look at one variable at a time (Goss Custard et al., 1972; Milton and May, 1976; Clutton-Brock and Harvey, 1977a).

This chapter will deal with three aspects in turn, after the basic definitions of social group, breeding group, and foraging group. First we measure group *size*. Orangutans and many prosimians forage singly and have small social groups expressed in rare meetings, whereas baboons and macaques may live in troops of a hundred. Group size reflects both predator pressure and food supply. Next we look at social *bonding and migration*. Is a troop a kin group? Which sex is more likely to change troop? Finally, we deal with *breeding structure*. Are there consistent differences between species that form single male or multimale troops or monogamous pairs?

SOCIAL GROUP, BREEDING GROUP, FORAGING GROUP

A *social group* can be defined as one that has a high level of communication between its members, with a steep falling off in the amount of communication between members and nonmembers (Altmann, 1962). This definition may sound unnecessarily abstract. It helps all the same to have a definition that will embrace a community of chimpanzees or a harem of bushbabies, whose members may rarely or never meet in one place at the same time. It also is an objective definition that derives from at least one of our intuitive meanings of society. A troop of zombies that did not communicate with each other would not be a social group.

A *breeding group* is a social group that includes adult males and females. They may not actually be breeding; it may be the wrong time of year, or there may be a famine, or they may not like each other. In primates we use this definition to exclude all-male bands of the same species and solitary individuals. Two female bands have been reported in macaques, and one juvenile gang, but these are not the norm in any primate. Many species, particularly monogamous ones and those that have multimale troops, have no male bands. For them the breeding group and the social group is the same.

A *foraging group* is the number of individuals we see moving together and feeding together, usually the same thing, because so much of a primate's day is spent eating. For many but not all primates, this is the same as the social or breeding group.

A table of primate foraging group sizes is not included because it would be so unwieldy. Its chief morals would be that there is great variability both within and between populations and that many detailed studies focus on very few troops. Instead, I have ruthlessly boiled down the list to the visual format of Table 6.1., which shows the *mean* foraging group sizes reported in various studies made up to 1980. The pattern seems clear and highly consistent within taxa. The majority of primates forage in small groups of less than ten animals. Only in macaques and baboons do many species *average* groups larger than 20. Other large group species are all exceptions in their taxa, with unusual ecological niches.

GROUP SIZE AND PREDATOR PRESSURE

Are there correlates between group size and ecology? We must look separately at predator pressure and at food supply (Crook, 1965; Eisenberg, 1966). Various means of defense against predators are crypsis, flight, attack including mobbing, and distastefulness. More than one defense may be used by the same species. Even a mouse lemur will bite the hand that corners it.

It is generally easier to hide a small group than a large one. Cryptic ani-

TABLE 6.1. Summary of Means of Primate Foraging Group Size

	\multicolumn{8}{c}{Mean Group Sizes}
	1	2–10	11–20	21–30	31–40	41–50	51–100	100
Lemuroidea								
mouse lemur, western	x							
coquerels	x							
dwarf lemur, western	x							
eastern	x							
lepilemur, Nosy Be	x							
whitefooted	x							
lemur, black		x						
brown		x	x					
mongoose		x						
redbellied		x						
ringtailed			x	x				
variegated lemur		x						
hapalemur		x						
avahi		x						
sifaka, diademed		x						
white		x						
Indri		x						
aye-aye	x							
Lorisoidea								
Slender loris	x							
angwantibo	x							
potto	x							
bushbaby, senegal	x							
Allen's	x							
needleclawed	x							
Demidoffs	x							
Tarsioidea								
Tarsier, spectral	x							
Callitrichidae								
marmoset, common		x						
pygmy marmoset		x						
tamarin, redhanded		x						
black and red		x						
saddleback		x		x				
moustached		x						
emperor		x						
cottontop		x						
whitefooted		x						
lion tamarin		x						
Cebidae								
cebus, whitefronted		x	x					
blackcapped		x	x					
olive		x		x				

TABLE 6.1. Summary of Means of Primate Foraging Group Size *(continued)*

	1	2–10	11–20	21–30	31–40	41–50	51–100	100
			Mean Group Sizes					
owl monkey		x						
titi, widow		x						
dusky		x						
squirrel monkey					x	x		
saki, monk		x						
pale-headed		x						
bearded saki, black			x					
howler, black		x						
mantled		x	x					
red		x						
Guatemalan		x	x					
spider monkey, black		x						
black handed		x						
longhaired		x						
wooly monkey, Humbolts		x						
yellow-tailed		x						
Cercopithecinae								
macaque, Barbary			x		x			
pigtailed			x	x	x			
liontailed			x					
crabeating			x	x	x	x		
rhesus			x	x	x	x	x	x
toque			x					
bonnet			x	x				
Assamese			x					
Japanese			x		x	x		x
mangabey, whitecollared		x						
agile			x					
greycheeked			x					
black			x					
baboon, guineae								x
olive			x		x	x	x	
yellow			x		x	x	x	
chacma				x	x	x	x	
hamadryas								
drill				x				
gelada			x					x
guenon, De Brazza		x						
Preuss		x						
blue			x	x				
spotnosed		x	x					
redtailed		x	x	x				
moustached		x						
campbell's			x					
crowned			x					
mona		x						
vervet			x	x				
talapoin							x	
patas			x	x				

Colobinae

	A	B	C	D	E	F
langur, Sunda Is.	x					
banded	x	x				
Mentawai	x					
capped	x					
golden		x				
dusky	x					
purplefaced	x					
nilgiri	x	x				
hanuman		x	x		x	x
proboscis monkey		x				
Pagai Island langur	x					
colobus, black	x	x				
Angola	x					
guereza	x	x				
red		x	x	x	x	
olive		x				

Hominoidea

	Solitary	A	B
gibbon, Kloss'		x	
Muller's		x	
agile		x	
lar		x	
siamang		x	
chimpanzee		x	
pygmy chimp		x	x
gorilla		x	x
orangutan	x		

From Jolly, unpublished survey of primate literature through 1980. Where two different studies, or studies in different seasons or years report different mean group size, x's appear in two or more columns.

mals do well to be solitary or to live in very small groups. The nocturnal prosimians are difficult to spot in the branches for a predator unarmed with headlamps. Among diurnal monkeys De Brazza's guenon is a highly secretive animal. It moves slowly and silently, not leaping about like the other guenon species. It has few vocalizations, but scent-marks branches with musky secretions. It does not drop fruit like a normal guenon. Its group is reduced to a single pair and their young, the only monogamous cercopithecine (Gautier–Hion, 1978).

One revealing comparison is the dusky titi as against the owl monkey at Manu, Peru. Both these small monkeys are about the same size, and both live in monogamous pairs. The titis eat a largely leaf diet and so could perhaps form larger groups than the more insectivorous owl monkey. The titis are diurnal and vulnerable to attack by many birds of prey. They spend much of the day concealed in vine tangles and move stealthily when they move at all. Owl monkeys have to contend only with a small owl and rare carnivores; they bounce noisily through the trees, singing at full moon. It seems likely that the titi monkeys' small groups are a response to predation (Wright, in preparation).

Conversely, an animal that is solitary for other reasons may do well to be cryptic. Orangutans, the only solitary higher primates, are well concealed in rain forest canopy, in spite of their orange fur. You usually find them by listening and looking for the occasional dropped fruit (Gladikas-Brindamour and Brindamour, 1975). The orangutans may be solitary primarily because of their feeding habits, since even one orang may strip a fruit tree. Besides the food constraint, they may have evolved from more social to more solitary life through predator pressure from evolving men. Fossil orang teeth are found in Java, where they may have competed with Javan *Homo erectus*. On Bornea, orangs were the second commonest prey, after wild pig, for the *Homo sapiens* who lived in the Niah caves 35,000 years ago (MacKinnon, 1974).

Larger groups help both in flight and in defense. If there are more eyes, ears, and noses to detect the predator, or more teeth and hands to confront it, a larger group pays off. Terrestrial primates have larger foraging groups than arboreal ones, which may reflect the dangers of life on the ground (Clutton-Brock and Harvey, 1977). (See Fig. 6.1.) Terborgh (in press) concludes that the large groups of little squirrel monkeys and talapoins and the mixed-species feeding groups of many arboreal primates are antipredator devices. Whenever the food supply allows, group size of vulnerable primates will grow. Terborgh is arguing from South American primates. Struhsaker (1982) reaches the same conclusion, on the basis of African observations, so this seems a strong inference.

Distastefulness is rarely considered in primates. Pottos and slow loris, apparently among the most vulnerable, have scrotal glands that smell "like human venereal infection" (Charles-Dominique and Martin, 1972). A house cat killed but did not eat a tarsier, and omnivorous rats actually bump into tarsiers immobile on branches and then move on (MacKinnon and MacKinnon, 1980). Sifakas, so Malagasy villagers have informed me, taste terrible, and lemur slightly better. Mutations that create distaste could spread most rapidly in a population that was divided into kin groups, and mutations that signal distaste by bright and obvious color or form could spread, initially, only in such kin groups (Gittleman and Harvey, 1980). Perhaps we need more research into primate palatability?

GROUP SIZE AND FOOD SUPPLY

The upper limit to the size of a primate foraging group roughly corresponds to the size of a patch of its food (Fig. 6.1). The group could be brought below this upper limit by predators, disease, or other factors, but if it goes for long beyond that limit, some of its members starve.

The lower limit is harder to see. Presumably, each animal is in competition with every other, so why be social at all if that means tolerating close competition? The larger a group, the farther it must range for food, which is an energy cost to each of its members. One answer is defense against predators. Another answer may lie in intergroup competition (Wrangham, 1980). If food comes in discrete, defensible patches, then it is to two or more animals' advantage to band together, if that lets them exclude others. Fruit trees are just such patches. Wrangham proposes that groups defending fruit trees were the origin of primate society.

What aspects of food supply might affect group size? Food is rarely distributed evenly. Prey usually occurs in clumps or patches. We can then mea-

FIGURE 6.1. *The size of the feeding patch determines the size of the animal that can depend on it. A small-bodied, monogamous widow titi in Peru feeding on a vine where one or few berries ripen at a time in each cluster. (Courtesy W. Kinzey.)*

sure the *total* food available, the *size* of an average food patch, average patch *distance* from each other, and patch *predictability* in space and time (that is, the distribution of sizes or distances around the mean). In an ideal mathematical world we would expect each of these aspects to have a different influence on group size. (See Table 6.2.) Of course, we must be prepared to find such large variance of troop size as a result of social, traditional, or microenvironmental influences that the predictions are difficult (or perhaps impossible?) to verify in nature (Rowell, 1979).

Total food available affects the total biomass of primate species. Thus, the centers of tropical rain forest support more primates than the harsher edges of the range. There are always more leaves available than fruit and more fruit than insects, so the biomass of folivores is usually higher than that of frugivores, and higher for frugivores than insectivores. None of this tells us about group size. For that we must turn to food distribution.

When food is distributed in small, separate clumps that require individual skill to find or catch, the animals are likely to be solitary. Insectivorous pri-

TABLE 6.2. Food Distribution and Feeding Group Size

1. Total food availability	low	high
Biomass of primates	low	high
Group size	no consistent effect	
2. Size of food patches	small	large
Biomass of foraging group	small	large
Home range of foraging group	large	small
3. Distance between food patches	near	far
Biomass of foraging group	high	low
Home range of foraging group	small	large
4. Predictability of food patches	stable	uneven, erratic
Foraging group size	uniform	variable
Intergroup behavior	rigid	variable

mates forage alone; there are not enough of most insects to share, even with your best friends. Also, the stalkers of insects, the bushbabies and pottos, would lose by other animals frightening their prey, and insects cannot be herded as wolves herd caribou or, rarely, baboons herd antelope.

Fruit may also grow in small clumps, particularly if the primate likes only ripe fruit, not green. "Small," of course, means small relative to body size. Most fruit trees hold less than a few hours' feed for a 150 kilogram male orangutan, so no wonder orangs are solitary. The small subgroups of chimpanzees and spider monkeys reflect their predilection for ripe fruit. One or a few animals quickly strip a tree (Wrangham, 1980; Klein and Klein, 1975, 1977). A higher proportion of a party of chimpanzees can feed if the party is small than if it is large, and this was particularly true in a harsh dry season when one favored tree species did not fruit (Wrangham, 1977).

If food grows in large clumps, the foraging group may be larger. Several animals can feed in the same tree if they are small-bodied or if they eat both ripe and green fruit. They may then exclude other groups from their tree, or even their fruit grove. If they are close kin, they increase not only their own fitness but their relatives' by such behavior. A plausible origin for the social kin groups that forage together is that the majority of primates are fruit eaters. As Wrangham points out, such patches might be crucial at only one growth season during the year, so long as the sparser, lower nutrient food of the subsistence season occurs in widespread patches, not small clumps that demand individual foraging. Most primates shift from a high fruit diet to relying on young leaves and flowers, or even mature leaves, in the lean season. This is the sort of continuously available resource that allows grouping even if it does not demand it.

Another aspect that matters as well as the size of food clumps is their distance apart in space and time. If they are spaced widely, the troop spends a great deal of energy moving from patch to patch. They can respond by exploiting fewer patches and returning to them more often (which means a smaller troop), or else they can maintain a large troop in a very large range, or some combination of the two. The classic example of decrease in group size is the breakup of Ethiopian hamadryas and anubis baboon troops to feed in small parties on streambed grass or parched acacia bushes. In the hamadryas the small group structure is a tight male-centered harem; in the anubis the groups are diverse assortments of individuals from a larger troop that fissions when times are hard (Kummer, 1968; Aldrich-Blake et al., 1971). The day range of hamadryas is the longest known in primates, so a larger group would perhaps find too little nourishment without walking to exhaustion.

The mangabeys of the Kibale forest offer another example of wide-ranging animals. They specialize in fruit tree bonanzas, that is, very large patches, but travel long distances between them. This is a purer example than the arid country baboons, for it is not confounded with sparseness of total food (Waser, 1977).

Day range length apparently has no relation to individual body size, but it does correlate with the body weight of the foraging group. Folivores have smaller day ranges than frugivores of the same group weight, which may reflect not only total food supply but the distance between patches (Clutton-Brock and Harvey, 1977).

Finally, the "average" food patch may scarcely exist. For some species the sizes or distances of food patches will follow a nice bell curve, centered on the mean, but for others the patches vary erratically from large to small, or even have a two-humped distribution of large and continuous in one season, small and fragmented in another. Such species respond with variable social behavior. Gelada baboons break up into small one-male groups in the dry season, but forage together in hordes of several hundred during the wet. Chimpanzee groups fission and fuse from day to day or hour to hour, hurrying together when someone's food barks announce that 'he is feeding in a favored fruit tree. We may view such variability as intragroup or intergroup behavior, given different biases and circumstances. The ringtailed lemurs who overlapped ranges at one season and spaced out in others never actually associated. Their separate troops just replaced each other in the few tamarind trees in new leaf, like sharing time on a computer (Jolly, 1972; Mertl-Milhollen et al., 1979).

The most relevant parallel for humans is the hamadryas baboons that have a two- or three-tiered social structure. Their harem units link sometimes into bands of perhaps 40 animals when the leaders of many harems confront another band, acting for the moment like the central group of several males in an ordinary savannah baboon troop. Each band takes its own direction as it leaves the hundreds-strong sleeping troop in the morning, and then splits again as each harem unit feeds alone. Only among the hamadryas do we see such hierarchical structuring of social relations.

FEMALE BONDING AND KIN GROUPS

All this talk of group size tells us nothing of group structure. The most common basis of primate society, as for other mammals, is a core of related females. Males born into such female-bonded troops usually migrate to other troops when they reach an age to mate.

Female-bonded troops are not the only form of organization. In several species, females also change troop to live with unrelated females. There are fairly clear ecological differences between the female-bonded and non-female-bonded societies.

Food constraints are relatively more important to females than to males. Females are defined as the sex that puts more investment in each gamete (you need the definition for sea urchins and such where there may be relatively little difference in the size and parental care between eggs and sperm). In mammals the difference in investment is very large, with gestation in uterus and postnatal care of the milky young.

Both males and females need food to grow, but the *limiting resource* for an adult male is access to fertile females. An ounce of energy put into successful courting usually gives him a higher payoff (in terms of more babies) than an ounce of energy put into finding food. A female, on the other hand, has little difficulty in finding a mate to start as many babies as she can bear (in both senses). The critical question for her is her offsprings' survival after all her investment in them. Each extra ounce of food increases her ability to feed her young and to maintain her own good health to raise future young. For adult females, food continues to be the limiting resource (Wrangham, 1980).

When social grouping allows an animal to find more food, the pressure to unite acts more strongly on females than males. Even in the solitary pottos a mother tolerated her daughter sharing the range until the mother moved out (Charles-Dominique, 1977). In the somewhat more gregarious bushbabies, mothers and daughters remain together and groups of females share the same tree nest. Female foraging ranges in most nocturnal prosimians are fairly exclusive, but if you must share a range, better to do so with a daughter who also shares your genes. Then the two of you can keep out more distantly related females (Wrangham, 1980). Of course, these arguments do not apply just to primates. Matriarchy is the common basis of most mammalian societies (Eisenberg, 1966).

When primates moved into the diurnal, frugivorous niche, the pressures for social grouping increased. A fruit tree is a defensible patch with a limited number of feeding sites. Two or three animals could combine to chase away a single one. Four animals could chase out two. With food arranged in tree-sized patches, the pressures are to form a group just larger than your neighbors', up to the maximum size the trees will bear. Because food matters so much to females, it is generally the females who band together. If one must band together, better to do so with close kin, so that the food you share goes to your relatives. In such female bonded species the females not only unite to drive out intruders, but usually associate closely, groom each other, and support each other in intratroop disputes (Wrangham, 1980).

Table 6.3 lists some species where migration between troops has been studied. The distinction between female-bonded and non-female-bonded is (like most distinctions in behavior) relative, not absolute. Occasionally females do migrate even in macaques and savannah baboons, whose normal pattern is a strong central core of related females, with almost all males leaving their natal group. Two species, the bonnet macaques and the hamadryas baboons, indeed have regular female migration. Among non-female-bonded primates males may still change group more often than females. Ideally, we should have long term studies that tell us whether daughters actually remain with their mothers, and the proportion of each sex that migrates. As Table 6.3 shows, we have relatively few such long term studies. Even in the short term one can guess at troop structure by observing day-to-day female interactions. If there is a highly differentiated female dominance hierarchy, and much contact and grooming between females, it is likely the group is female-bonded. In male centered groups like a gorilla harem, females concentrate their attention on the male, and are relatively indifferent to each other (Harcourt, 1979).

Do exceptions illuminate the rule? In which primate species do females move away from their kin? First, the monogamous ones. If a pair sets up together in an exclusive range, both sexes have excluded their parents or have been excluded by them. This is a rare and specialized pattern among mammals but common in birds. Among birds, males commonly stay close to the place where they were born, but females travel farther before or between matings (Greenwood, 1980). Greenwood argues that males in such species defend resources in order to attract mates rather than simply finding and defending mates. This is easier with a food supply that is dispersed fairly evenly throughout one's range, so it is worth defining the whole area rather than food that is clumped in a temporary patch like a fruit tree. Monogamous primates, like monoga-

TABLE 6.3. Female-bonded and Non-Female Bonded Primate Species

FB Species

Species	Females Breed in Natal Group	Males Transfer Between Groups	Differentiated Female Relationships
Ringtailed lemur	?	Budnitz & Dainis, 1975	Jolly, 1966
White sifaka	?	Richard, 1974	Richard, 1974
White-fronted capuchin	?	Oppenheimer, 1968	Oppenheimer, 1968
Squirrel monkey	?	Bailey, pers. comm.	Coe & Rosenblum, 1974
Vervet guenon	Lee, in prep.	Cheney, in press	Struhsaker, 1967
Redtailed guenon	?	Struhsaker, 1977	Struhsaker & Leland, 1979
Blue guenon	?	Rudran, 1978	Rudran, 1978
Gray-cheeked mangabey	?	Struhsaker & Leland, 1979	Chalmers, 1968
Patas	?	Hall, 1965	Hall, 1965
Olive baboon	Moore, 1978	Packer, 1979	Smuts, in prep.
Yellow baboon	?	Hausfater, 1975	Hausfater, 1975
Chacma baboon	?	?	Seyfarth, 1976
Gelada	?	Dunbar & Dunbar, 1975	Dunbar & Dunbar, 1975
Japanese macaque	Itani, 1975	Sugiyama, 1976a	Mori, 1975
Rhesus macaque	Chepko-Sade & Sade, 1979	Sade, 1972; Lindburg, 1969	Sade, 1967
Toque macaque	?	Dittus, 1977	?
Barbary macaque	?	?	Taub, 1978
Crab-eating macaque	?	Angst, 1975	Chance et al., 1977
Hanuman langur	?	Hrdy, 1977a	Hrdy, 1977a
Black-and-white colobus	?	Oates, 1977a	Oates, 1977a

FB species are those in which at least two breeding females travel and forage in the same group, and which meet at least one of the criteria shown. Species for which there are no relevant field data are not included. Females are presumed to breed in their natal group in all FB species, but they have been observed from birth to motherhood only in the species indicated.

Non-FB Species

Species	Males Breed in Natal Groups	Females Transfer Between Groups	Undifferentiated Female Relationships
Mantled howler	Rarely	Jones, 1980	
Red howler	Rarely	Rudran, 1979	
Red colobus	?	Marsh, 1978	Struhsaker & Leland, 1979
Purple-faced langur	Rarely, Manley 1978	Rudran, 1973 Manley, 1978	
Nilgiri langur	Rarely, Oates, 1980	Oates 1980, Moore pers obs.	
Bonnet macaque	Sometimes, Wade 1979, Ali 1981	Ali 1981	
Hamadryas baboon	?	Kummer, 1968	Kummer, 1968
Mountain gorilla	Always	Harcourt, 1978	Harcourt, 1979
Chimpanzee	Pusey, 1979	Nishida, 1979	Goodall, 1968; Nishida, 1979.

Non-FB species are those in which at least two breeding females live in the same group without forming long-term affiliative relationships with particular other females.

(After Wrangham, 1980, Moore, in press).

mous birds, are often territorial; monogamy, territoriality, and female dispersal go together.

Next, there are two genera where females forage alone or in small fluid subgroups: the chimpanzees and various spider monkeys. These species specialize only on ripe fruit (Fig. 6.2). Even in lean seasons they do not shift much to green fruit and leaves, so the gregariousness they might develop in rich fruit trees is broken down when they forage every lady for herself. Among the chimps, the core is the band of related males who violently defend territory while adolescent estrous females change community. Little is known of spider monkey migration, but confrontations between bands of male spiders recall the gang attacks of male chimps, so the systems may have converged to the same patterns. The hamadryas baboons in their arid environment also feed in small shifting patches. Their male-centered harems have the protection of an adult male, but only a few females in direct competition.

Third, there is a collection of leaf-eaters: the two howlers, two langurs, red colobus, and gorilla. Perhaps such large swathes of a leaf-eater's environment are edible that a female is relatively indifferent as to who feeds beside her, kin or nonkin. If we add other species that which fragmentary data suggest a high rate of female migration, including white sifaka and three more langurs, there is quite a strong correlation between leaf diet and female migration.

Another aspect might be that females in these species avoid males who would kill their offspring. Infanticide is fairly widespread among single-male primate groups (see Chapter 12). Gorilla males kill unrelated infants; so do red howlers, many langurs, and red colobus (Fossey, 1979, in press; Marsh, 1979; Struhsaker, personal communication). In one red colobus troop, the females without infants gradually deserted a male who had only five months' tenure. When he was replaced by another male, who lasted at least 21 months, females immigrated into the troop of the secure leader. The process had no relation to good seasons or bad. Even when food was scarce, the troop females did not drive off newcomers. If the newcomers could somehow judge the likely tenure

FIGURE 6.2. *Panamanian red spider monkeys are ripe fruit specialists that forage in small fluid groups, like chimpanzees. (Courtesy R. Mittermeier.)*

of a troop male, social security might outweigh the limitations of the food supply and its distribution meant that established females did not try to keep out the immigrants (Marsh, 1979).

Thus, most of the exceptions agree with Wrangham's rule—all but three. The gelada has orthodox female-bonded groups in an indefensible environment of grassy pastures with no fruit trees; that is, it has a harem with a female core, unlike the hamadryas male-centered harem (Dunbar, 1979; Kawai, 1979; Kawai et al., 1983). The other exception is the bonnet macaque, that seems to have cohorts of related males. It feeds on ripe fruit and leaves, with diet and ranging patterns like all the other, orthodox, female-bonded macaque species (Ali, 1981). The gelada and the bonnets perhaps show that societies may develop their own rationale that defy easy environmental explanation.

The third exception is more revealing, because different troops follow different patterns. Among the red howler monkeys of Hato Masaguaral, in northern Venezuela, most young females, as well as young males, emigrate from their natal group. It seems that they do not leave willingly. Fights and wounding between the adult females coincide with the emigration of daughters. It is likely that females leave because they are driven out. A few daughters are well enough defended by their mothers that they can remain to breed in their mothers' troop; these troops have the female-bonded kin structure. Other emigrants join similar, childless individuals to form new troops. Very few transfer to established breeding troops. The diet and resource base of the newly conglomerated troops is the same as that of neighboring female-bonded troops. This then is an intermediate case where female lineages compete within and between troops in shifting patterns—perhaps not unlike the earliest primate social groups (Crockett, in press).

BREEDING GROUP STRUCTURE

Classification by breeding structure cuts across the earlier considerations of group size and kinbonding. A small foraging unit might breed as a monogamous pair, or be part of a multimale community like the chimps. Slightly larger cohesive troops may have just one breeding pair like the associations of eight or ten tamarins with one pair of breeding adults, or else be a group with one male, three or four breeding females, and three or four young by separate mothers.

Tables 6.4. and 6.5. list and summarize some known breeding structures. Two major points emerge. Some phylogenetic lines are relatively fixed in breeding structure: gibbons are monogamous; only two or three old-world monkeys are so. Marmoset and tamarins are monogamous or polyandrous. Among old-world monkeys the guenons tend to single-male groups; baboons, macaques, and mangabeys have multimale groups. Of 98 species that have even one study of breeding structure, 11 per cent have more than one type of structure. Among genera, 26 per cent of the 38 studied have more than one breeding structure. Breeding structure may have a central tendency in major phylogenetic groups, but it is a fairly labile character on the whole. If we look for its adaptive advantages, we should look on both the large and small scale in phylogeny and in the immediate environment of the study population.

TABLE 6.4. Breeding Group Structure

Lemuroidea

mouse lemur	
western	SM
Coquerel's	P
forked lemur	P
lepilemur	
whitefooted	SM
lemur	
black	MM
brown	MM
mongoose	P SM
ringtailed	MM
woolly lemur	P
sifaka, white	MM
indri	P

Lorisoidea

potto	P
galago	
senegal	SM
thicktailed	SM
Allen's	SM
Demidoff's	SM

Tarsioidea

tarsier	
spectral	P

Callitrichidae

marmoset	
common	P PA?
pygmy marmoset	P
tamarin	
redhanded	P PA?
black-and-red	P PA?
saddleback	PA
moustached	P PA?
emperor	P PA?
cottontop	P PA?
whitefooted	P PA?
lion tamarin	P
Goeldi's marmoset, callimico	P

Cebidae

cebus	
whitethroated	MM
whitefronted	SM MM
blackcapped	MM
night monkey, owl monkey	P
titi monkey	

Cercopithecidae

macaque	
Barbary	MM
pigtailed	MM
liontailed	MM
crab-eating, longtailed	MM
rhesus	MM
Japanese	MM
toque	MM
bonnet	MM
stumptailed, bear	MM
mangabey	
agile, Tana River	MM
graycheeked	MM
black	MM
baboon	
Guinea	MM
olive	MM
yellow	MM
hamadryas	SM
Chacma	MM
olive hamadryas hybrid	SM MM
guenon	
gelada	SM
De Brazza's	P
Preuss'	SM
blue, diademed	SM
spotnosed	SM
redtailed	SM MM
moustached	SM
redeared	SM
Campbell's	SM
mona	SM
crowned	SM
vervet	MM
talapoin	MM
patas, red hussar	SM

Colobinae

langur	
Sunda Island	SM
banded	SM MM
Mentawai	P
silvered	MM
capped	SM
golden	SM
dusky	SM MM
purple-faced	SM
nilgiri	SM MM
hanuman	SM MM
proboscis monkey	MM
simakobu	P SM
colobus	

dusky, moloch	P		Angola	SM
widow	P		guereza	SM MM
squirrel monkey			red	SM MM
common	MM			
redbacked	MM		*Hominoidea*	
saki				
pale-headed	P		gibbon	
monk	P		Kloss'	P
howler			agile	P
black	MM		whitehanded, lar	P
mantled	MM		siamang	P
red	SM MM		orangutan	SM
Guatemalan	MM		chimpanzee	
spider monkey			common	MM
black	MM		pigmy	MM
longhaired	MM		gorilla	SM

P = Pair, SM = Single Male, MM = Multimale, PA = Polyandrous.
Two symbols indicate that different troops or populations have different structures.

TABLE 6.5. Summary of Breeding Structures

	Polyandry Pairs		Single Male		Multi-male	All Three	Unk.	Total
a. Species								
Lemuroidea	4	← 1 →	2		4		17	28
Lorisoidea	1		4				6	11
Tarsioidea	1						2	3
Callitrichidae	11						9	20
Cebidae	5			← 2 →	9		16	32
Cercopithecinae	1		11	← 1 →	18		19	51
Colobinae	1	← 1 →	5	← 6 →	3		12	27
Hominoidea	4		2		2		5	13
Total	28	← 2 →	24	← 9 →	36		87	185
b. Genera								
Lemuroidea	3	← 1 →	1		1	← 1 →	5	12
Lorisoidea	1		1				3	5
Tarsioidea	1							1
Callitrichidae	5							5
Cebidae	3			← 2 →	2		4	11
Cercopithecinae			2	← 2 →	2	← 1 →	2	9
Colobinae		← 1 →		← 1 →	1	← 1 →	3	7
Hominoidea	2		2		1			5
Total	15	← 2 →	6	← 5 →	7	← 3 →	17	55

Numbers in arrows ← x → indicate that different troops or populations of a species or genus have different structures.

Species are listed as unknown if no population's breeding structure has been described; genera if no population of any species has been described. Experience so far suggests that as more populations are studied, more social variability appears. Even to date, 11% of known species and 26% of known genera have variable structure. When the frequency of monogamy versus polyandry is clarified, there will be still more variability.

ONE MALE AND AGE-GRADED MALE GROUPS

Probably the most ancient and still the most widespread breeding group structure is the one or age-graded male group (Eisenberg, 1972). The central male
may tolerate smaller or younger males, often his own offspring (Fig. 6.3.). The
distinction between single and age-graded male groups is not absolute. When
does one consider a son or intruder an adult, particularly if he remains lighter
in body and does not mate, fight, or scent mark like the central male?

Such groups could have two behavioral origins. A male might round up
several unrelated females and keep them away from his rivals. We have seen
that such real harems are rare, so far confirmed in only hamadryas and gorilla.
A much commoner mechanism is for a male to move into a female kin group,
sharing its range and fathering its children.

This social structure was described by Charles-Dominique and Martin (1972)
for Demidoff's galago in the rain forest of Gabon, and for the mouse lemur of
Madagascar. It is also true of Senegal bushbabies in South Africa, whitefooted
lepilemur in Madagascar, and pottos and Allen's bushbabies in Gabon (Bearder
and Martin, 1979; Charles-Dominique, 1977). The prosimians maintain their
contacts by vocal means and by scent marking. The smaller ones in the colder
climates also sleep together; often several females share a nest, and sometimes,
more rarely, a male with females.

FIGURE 6.3. *Patterns of
association of South African
Senegal bushbabies. Large central males each associated with
several females and tolerated
smaller (non-breeding) males.
However, large males were
spaced apart and did not associate with each other. Such a ha-
rem-by-location is the most
likely origin of primate social
structures. (After Bearder and
Martin, 1979.)*

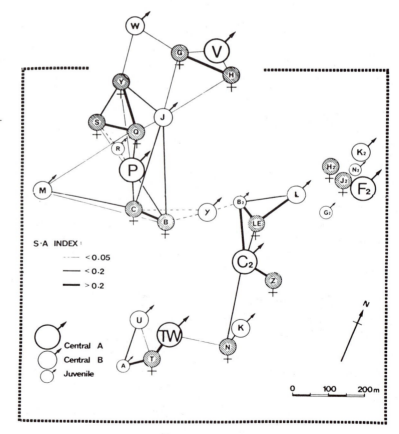

This social structure is a plausible origin for all primate societies (Martin, 1981; Zuckerman, 1932). It could have set the primate pattern of lifelong association between male and female, unlike, for instance, those rodents or mustelids that maintain mutually exclusive ranges and only come together for mating. Also, it is the pattern we see in most nocturnal strepsirhines, which resemble our Eocene ancestors in anatomy and ecological niche. Third, the earliest fossil higher primates were apparently sexually dimorphic.

There is a clear correlation between adult sex ratio and the difference in size between male and female (Clutton-Brock et al., 1977; Leutenegger and Kelly, 1977; Leutenegger, 1978). In monogamous species there is virtually no size difference between the sexes. Among polygynous animals, however, male weight increases in allometric proportion to female weight; that is, in large-bodied species the sexes are proportionately more divergent (Fig. 6.4.). Several species of the oligocene primates of the Fayum were dimorphic, which implies that they, too, had a social system where a male had access to several females (Fleagle et al., 1980, and see chapter 12).

Among the diurnal primates this ancestral structure is very unevenly represented in different taxa, as shown in Table 6.4. So far no diurnal lemurs seem to have one-male groups, except perhaps the mongoose lemur (more on this odd species in the section on monogamous pairs). New-world red howlers, whitefaced cebus, and blackhanded spider monkeys have one-male or age-graded male troops (Eisenberg, 1979), but they are far outnumbered by the monogamous or polyandrous new world species.

FIGURE 6.4. *The single adult male of a Lowe's guenon group in the Ivory Coast, with one of his females and a juvenile. Note the male's much larger size. (Courtesy F. Bouliere.)*

131   *Group Size and Structure*

FIGURE 6.5. *The core of a hanuman langur troop is a related group of females and offspring. There may be one male or several. (Courtesy S. Hrdy and Anthro Photo.)*

In the old world, in contrast, the majority of species share such a structure. It seems to be the norm among the leaf-eating colobines. (Figs. 6.5, 6.6.) Among colobines only the Mentawai Island langur and simakobu have diverged from the ancestral pattern toward localized monogamy; only the hanuman langur and red colobus have true multimale troops in which more than one male may breed. Even they have such forms only in parts of their range, whereas other local populations maintain the ancestral age-graded structure.

The cercopithecines are more prone to multimale troops, but even there the guenon genus overwhelmingly and the baboons in part opt for harem groups. (Fig. 6.7.)

The gorilla among the apes has a classic age-graded structure, with one big breeding silverback (Fig. 6.8.), and the orangutan has a harem by location just like the bushbabies.

Are there particular ecological correlates for this diurnal age-graded male structure? At first sight, no. Such monkeys may be folivorous, frugivorous, or insect eating; their groups may be large or small, and so may their ranges. This is the "normal" category, defined mainly as different from the extremes of multimale or monogamous groups.

FIGURE 6.6. *An all-male band of hanuman langurs. To reproduce, they must usually steal copulations on the sly, or take over a bisexual troop. (Courtesy S. Hrdy and Anthro-Photo.)*

FIGURE 6.7. *Ethiopian Hamadryas baboon male disciplines a female, while others run to their leaders. These are true harems, established by the male. (Courtesy H. Kummer.)*

Group Size and Structure

FIGURE 6.8. *The gorilla group is a harem of unrelated females attracted by the silverback. An elderly silverback, Beethoven, is groomed by Effie, one of his four closely bonded females, as two immatures look on in Virunga Reserve, Rwanda. (Courtesy D. Fossey, and the National Geographic Society.)*

MULTIMALE GROUPS

In a true multimale group (Fig. 6.9.) several of the males reproduce. There are, of course, dubious cases where a young hamadryas steals a copulation behind a bush; yet this leaves the leader still apparently running a harem. The definition is not absolute.

More interesting, many species or even many troops can slide from one male to multimale and back. New-world mantled howlers, old-world hanuman langurs, and the baboons adapt behavior to local ecological conditions and local demography.

Why should males tolerate other males in the group? And why should females tolerate more consumers of the food supply, beyond the one who fathers their offspring? A group of well-armed males may intimidate predators. Terrestrial primates are more likely to have multimale groups, which presumably reflects predator pressure (Clutton-Brock and Harvey, 1977).

FIGURE 6.9. *Part of a ringtail lemur multimale troop on parade. (C. H. F. Rowell.)*

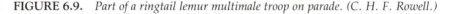

A second reason is intertroop competition. If troop ranges overlap widely, the largest troop is usually dominant and forces the smaller one to retreat when they meet. More males can oust less males. Thus, the optimum troop size is a compromise—small enough to find enough food within a feasible home range, but big enough to be the biggest in the neighborhood. The one pressure is given by the environment, the other only by local demography (Wrangham, 1980).

Why, then, don't all primates develop multimale troops? Perhaps if the food supply is dense and stable enough so that the troop can keep territorial boundaries, they have evolved instead toward less variable troop size and the ability to stand-off neighbors rather than shove them around. It is true that one-male groups are more likely to be territorial than multimale groups, but many multimale groups, from white sifaka to chimpanzees, defend territories, and some single-male groups like patas monkeys do not. The trade-offs here are some of the unsolved questions.

MONOGAMY AND POLYANDRY

Monogamy is odd. Monogamy comprises a definite series of adaptations that channel the possessor toward ever more rigid social structures (Fig. 6.10.). Of all the primate species, only three—the mongoose lemur, the Mentawai langur, and Homo sapiens—have facultative monogamy combined with other possible group structures, and none of these share the anatomical traits of true pair-

FIGURE 6.10. *Indri, largest of the lemurs, live in monogamous pairs in Madagascar's rain forest. (Courtesy J. Pollock.)*

FIGURE 6.11. *Twin mongoose lemurs, Comoro Islands, are one of the three known primates with some monogamous and some polygamous populations. (Courtesy I. Tattersall.)*

bonders. The mongoose lemur (Fig. 6.11.) and Mentawai langur may have temporarily adjusted to life on small rainy islands, which leaves us alone.

Monogamous mammals resemble each other in several ways (Kleimann, 1979). Male and female body size is almost identical. Male and female armory, whether antlers, tusks, or canines, is almost identical in size. Monogamous mammals are frequently territorial, and both sexes take part in banishing the next pair. Female fights female and male fights male, if it comes to fighting. This is rare in the wild, but it is commonly fatal to cage monogamous females together. When Robinson (1979) played titi monkey calls, starting with the female call, in one of his pair's territories, the female titi came down and shook his tape recorder till its transistors rattled and she silenced her "rival." Pairs commonly duet in vocal aggrandizement, the two sexes' calls complementing each other.

Some species in each main primate line are monogamous; in fact, recent work shows a surprising number. Spectral tarsiers are monogamous and territorial, which makes one wonder again about higher primate origins (MacKinnon and Mackinnon, 1980). Avahi and Coquerel's mouse lemur are monogamous among the nocturnal prosimians, indri among the diurnal ones. The owl monkey, titi, and uakari are monogamous in the new world. Perhaps the ancestor of all new-world monkeys was monogamous (Eisenberg, 1979). Only in the old world is the pattern rare, mainly present in gibbons and siamang.

Marmosets and tamarins go further. They have long been thought monogamous, though wild groups seemed surprisingly large and fluid (Dawson, 1978). Now it appears that two or three males may share one breeding female, mating in turn. This is well documented in saddleback tamarins and likely in many others (Terborgh, personal communication, Garber, 1984, Sussman and Kinzey, in press).

Monogamous and polyandrous males contribute to infant care. Siamang males carry their babies and sleep cuddling their juveniles. Marmosets and tamarins carry the twin young until their combined weight equals the adults. Tamarin males catch insects to feed the young. Meanwhile the female is again pregnant as well as lactating. She needs what help she can get, while a group of cooperating males each enhance their reproductive success. Callitrichids may be secondarily dwarfed forms, which depend on gum-eating, and which compete with insect-eating birds, but still retain the large and demanding primate young (Leutenegger, 1980, Ford, 1980, Sussman and Kinzey, in press).

The ecological niche of such species is that of small patches of food, sparsely but evenly distributed and fairly constant through time. There is not much incentive for a female to join other females at these patches; there is not enough to share. On the other hand, the bits of habitat that will support her offer a continuous supply of food that may be worth defending year-round as a piece of geography. For the male, his investment in one female pays off if she needs full-time help to raise his young successfully. He may feed them if he is a bird, carry them if he is a marmoset, or sing about their territory if he is a gibbon. A monogamous male's environment is fairly sparse in food. This may be for reasons that vary from plants' poisonous compounds to interspecific competition. The Mentawai islands are a test case. There even langurs are monogamous, perhaps from the extreme poverty of the Mentawais' leached rain forest soil (Whitten, in press), combined with year-round, if grudging, rain forest productivity. If it were not so sparse, a male might do better to have several wives and let them get on with the child rearing. Monogamy implies that he gains more by investing in one wife who would be vulnerable without him.

PROTOHOMINID GROUPS

What sort of social structure had our own ancestors? We have seen that you cannot extrapolate from one species to the next, or even from one population for the next with much confidence. Start then with the human data. What are we like now?

We are an anatomically polygamous species with a surprising amount of monogamy. Very few groups permit multimale access to the same female; in very few societies do males and females have equal right to adultery. Most societies oscillate between single-male "harems" and monogamy, which puts us in company only with the mongoose lemur and the Mentawai langur.

It appears as though this facultative monogamy narrowed down from earlier, more flexible groupings. Perhaps it came as men and women began to depend on each other for the products of their respective hunting and gathering. Perhaps it came as babies grew more slowly, and women, like nesting birds, needed full-time help from particular males. It may have favored our peculiar prolongation of sexual pleasure to all cycle stages and all seasons, as particular women exerted continuing attraction to particular mates (but we need to know more about pigmy chimpanzees on this score.) If all this happened as we also emerged from the forest, evolving male-female bonds ran counter to the pressure of terrestrial life, which favors multimale bands that can defend their females against predators and against other bands. We simultaneously narrowed down the number of wives a man could keep because he expected to help support them, and embedded such protomarriage in a multimale community (Morris, 1967; Lovejoy, 1981).

This does not tell us whether such a community started from a female-bonded society, like the baboons, or a male-bonded one like the chimpanzees. Either have the behavioral capacity to form consortships, mother-daughter alliances, and male friendships—only the emphasis differs.

What we have done is to develop the capacity to change emphasis. We have opened up our options biologically; then each human society invokes a conscience and moral laws to close off some of the options and enforce particular responsibilities toward each other. We have become surprisingly, but not fully, monogamous of our own free will. We have not wired in such behavior, like the gibbon, but the gibbon lays up no treasures in heaven because it apparently can't help being faithful.

SUMMARY

Foraging groups are those that move and feed together. For most primates, this is the same as the social group of close communication and the male-female breeding group. Some species, however, forage in small parties or alone while having larger social matrices; one, the gelada, has small social groups that forage seasonally in large amorphous herds. Primate foraging groups range from one to several hundred animals.

Animals that hide from predators do well to have small foraging groups. Large groups help in vigilance and active predator defense. Terrestrial primates have larger groups than arboreal ones, perhaps through predator pressure.

The distribution of food affects foraging group size. If total food supply is low, there is low total primate biomass. If the size of food patches is small, the biomass of the foraging group must be small or else its home range must be large. If the food patches are near together, the biomass of the foraging group can be high or its home range can be small. If food patch distribution is stable and even, the foraging groups can be of one size, with rigid intergroup behavior. If food patches are uneven and erratic, troop size varies in space and time, or intertroop behavior varies. Hamadryas have a three-tier social structure in such an erratic environment. There are examples among primates to illustrate each tendency, although one rarely finds a pure comparison where only one aspect of food supply varies at a time.

Social life may have begun when animals, particularly kin groups, united to defend a food patch against nonkin. Food supply is the limiting resource for adult females, and access to mates is the limiting resource for adult males. Groups of related females are the core of most mammalian societies as well as most primate societies. These female-bonded groups share defensible resources such as fruit trees at least at one season, and forage on widespread resources that permit social groupings in other seasons. In several species females are not bonded in kin groups but may migrate to groups of unrelated females. Two of these specialize in very ripe fruit, a food patch so small that females often forage alone. Most others are leaf eaters with widespread food supply. Red howlers are intermediate: daughters of dominant mothers may remain in their natal groups, while other young females emigrate to found new groups with non-kin.

Either males or females or both change troop at adolescence, and sometimes as adults, which largely prevents inbreeding.

The commonest primate mating structure is the single or age-graded male group, in which only one male actively breeds. This is usually formed by a male or males joining a kin group of females. Only gorillas and hamadryas have true harems of unrelated females recruited by the male. A few primates form multimale groups in which several males may breed. This is commonest in terrestrial old-world monkeys, although there are examples in each primate line. Monogamy is a specialised state: monogamous mammals have little size difference between the sexes, are frequently territorial, and share both defense and joint care of the young. Tamarins and marmosets may be polyandrous, with two or three males helping to rear one female's young. Social structure is labile; differing patterns appear within at least 10% of studied species and 25% of studied genera.

We are paradoxical: anatomically polygamous, with a surprising amount of behavioral monogamy, all embedded in multimale communities. The interdependence of particular males and females could have begun with food sharing, shared infant care, and prolonged female receptivity. Terrestrial life in contrast brought pressure for multimale predator defense and interband dominance. We cannot certainly reconstruct whether our own system started from female- or male-bonded groups, but we can see how the two-tiered system of small harems or pairs, within a larger community, could have evolved under the same pressures that produced either pairs and/or multimale groups in other primates. We had the luck to develop both at once, which kept our behavioral options open.

CHAPTER
7

IN-GROUPS AND OUT-GROUPS

In the previous two chapters, we have seen some of the correlates of ranging behavior and home range size and the correlates of group size and structure. One of the chief facts of life for most primate groups is that there are other groups next door, ready to move into their range and explot their food or sleeping sites.

This chapter deals with the relations between groups. In the short term, groups of some species defend clear-cut territories. Other species' groups overlap home ranges, displacing each other wherever they happen to meet. In the short term, then, we can describe a spectrum of more or less territorial behavior.

The depth-of-time perspective that we are now gaining means that this short-term, cross-sectional view is not enough. Animals migrate between social groups. Whole groups fission or sometimes fuse. Next-door groups are not just rivals; they are kin. The second part of this chapter discusses some implications of the kin structure of the neighborhood.

TERRITORY AND DOMINANCE

A *territory* is a piece of geography. The commonest definition is a *defended area* (see Fig. 7.1.). If an animal actively chases its neighbors or struts or sings or scent marks, and if this behavior is victorious in one area of ground, we call the animal territorial (Burt, 1943). Pitelka (1949), however, found that his arctic sandpipers each occupied different parts of the tundra without any obvious defense. He pointed out that the function of territorial behavior is to gain property rights. He therefore proposed another definition: *exclusive use* of an area. One can combine the two definitions to be still more rigorous and demand *successful defense* of an area before calling that area a territory. This book will look

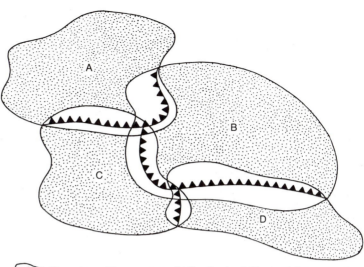

FIGURE 7.1. *An idealized map of home range and territory. (After Bourliere, 1964.)*

A⟩ Boundary of home range, limits of animal A's normal movement

Area of exclusive territory, which conspecifics do not enter

▲▲▲ Defended territorial boundary

at territory from both points of view, because it seems that neither aggression between troops nor long-distance calls and scents necessarily correlate with exclusive ranges.

Davies (1979) takes a broader view. He calls any spacing out of animals territoriality, if it is larger than expected from a random distribution. This would include the troop of crab-eating macaques that melts out of a mango tree as a stronger troop approaches. In this book I call such spacing between groups *dominance* when it has no geographical component.

There is a continuum between dominance and territorial behavior. If troop A wins over troop B wherever they are, that is pure dominance. If troop A excludes troop B from almost all of A's range and wins in clashes one tree inside the A-B boundary, but loses when one tree inside B's range, that is pure territoriality. If A is more likely to win near the center of its range, but less likely to win near the periphery, we are still inclined to call it territoriality, because geographical location influences the outcome. Perhaps, though, such defense does not actually exclude B from the center of A's range but just moves B along and cuts down B's time feeding there. Such transitional behavior allowed the evolution of territoriality itself, and allows species that are territorial in some environments or population densities to slide into dominance relations if circumstances change.

A common measure of exclusive territory is to add up the per cent of home range where other troops are never observed. Thus, Ellefson's gibbons had 90 per cent of the home range to themselves and Shaller's gorillas none. From Chapter 5, on ranging, we can see that this is a highly suspect measure. It depends on the length of study in the first place. Far more important, its significance depends on the amount of time that each troop spends in the overlap zone and the amount of food they find there. More revealing measures would compare the "activity fields" of neighbors; do they concentrate on the boundaries or hardly ever reach them? We can superimpose another map of the "aggression fields." Do encounters and chases take place mainly at boundaries, or randomly through the range? We do not have many primates analyzed in these terms yet, but these measurements will make it easier to distinguish territoriality from dominance (Waser and Wiley, 1979).

Why do animals defend territories? There are *mating territories* like the bower bird's bower; *reproductive territories* like the few inches round a seagull's nest; and *feeding territories* that supply all the animals' needs. A few primates maintain mating territories. Central male Senegal bushbabies allow juvenile and subadult males to feed in their area, but see off adult male challengers. They are presumably defeinding their harem, not their food supply. Many other "solitary" primate males may be doing the same, such as the Allen's bushbabies with a male range three to five times larger than the area that suffices to feed a female Allen's bushbaby.

The commonest primate territories, however, are *permanent feeding territories*. These are most of the group's range for months or years at a time. Let us first consider means of defense—how we know or how the next group knows an area is occupied. Next, we will explore what characteristics of the food supply favor the evolution of territoriality, such that the returns are worth the effort of defense (Burt, 1943).

BATTLES, SONGS, AND STINKS

Primates maintain their spacing in three main fashions. They may physically confront each other, or else they warn each other off with calls or with scent marks.

Physical confrontations are sometimes called battles. Occasionally the combatants do grapple and bite their opponents. The chimpanzees described in Chapter 4 actually killed members of the adjacent group. Ripley (1967), who first documented true primate territoriality, described the fights during encounters between Hanuman langur troops (Fig. 7.2.). However, "battles" are far more often a steotyped display of prowess. Ellefson (1968) watched neighbor lar gibbons swing through the trees, whooping, frequently choosing dead branches for support that cracked and crashed, gibbon and all, into the lower levels of foliage. It was some months before he realized this was not joyous acrobatics but aggression—when one male bit his neighbor in the course of the display. Ellefson reckoned that one gibbon might catch and bite another only 25 times in a lifetime of display every other morning, though male gibbons have many old, healed cuts and scratches.

Sifaka "battles" are a kind of arboreal chess game. The white knights leap to unoccupied branches. Each group faces outwards from their own territorial base. Occasionally defenders of opposite sides arrive back to back on the same branch. The only time I have seen physical contact in such a battle, two sifaka landed on top of each other having aimed for the same empty space. The upper one, finding himself actually touching someone, hesitated, then gave his opponent a quick groom (a couple of short upstrokes with the toothcomb, not the downward canine slash of lemur fighting) before leaping away.

Physical presence then need not be expressed by fighting. It may be ex-

FIGURE 7.2. *Hanuman langur territorial disputes are physical confrontations between the female core of each troop. (Courtesy of S. Hrdy and Anthro-Photo.)*

pressed by any display that the opponent will agree to respect. We will discuss such agreements further in Chapter 10, "Communication."

Effective defense depends on each species' criterion of defense. There is a whole range of such criteria from actual battle, to display battle, to long-distance statements of presence by sound or smell. The long-distance statements need backing by an occasional show of force, but they function for the most part alone.

Indri territoriality is a case in point. In 2,500 hours of watching indri, Pollock (1977) saw only six meetings between troops. These were dramatic enough. The two males leaped forward, springing from trunk to trunk in a vertical clinger and leaper's version of arboreal gymnastics. The females hung back, but wailed their song, much as female gibbons cheer on their males with the "great call." The male indri sang too: a fugue of wails that once lasted nine minutes continuously at pain threshhold for human ears. Then both groups retreated slowly toward the centers of their own territories. For the other 18 months of the study indri groups sang in daily chorus, and indri at least three kilometers from the first callers answered with their own song. It was difficult to *prove* the song territorial. After singing, groups moved off at random, neither forward nor away from the nearest singing neighbors, or as likely did not move at all. The best proof was that indri home ranges only overlapped by 50 meters round the edges of an 18-hectare range. One could argue that the indri song has evolved to an almost ideal territorial defense. The indri have exclusive use of about 90 per cent of their range, at the cost of very little energy. They merely sit and sing, and hardly ever have to do anything else (Pollock, 1975, 1977).

143   *In-Groups and Out-Groups*

Are all such loud calls territorial? That would be too simple. To be sure of the functions of such a call one must try playback experiments. David Attenborough (1961) did so with indri. Frustrated at hearing groups wailing round about him, but invisible, he broadcast their song. A family of indri materialized from the trees, then bellowed at him in high-level alarm.

Approaching a calling tape recorder is the surest sign the call represents territorial defense. Dusky titis in Colombia slept well within their range, tails entwined. On waking, they began to move toward the boundaries, where they called back and forth in bouts with their nearest neighbors. If they met a neighboring group, they might make jumping displays, "long stiff-legged jumps downward into the foliage. The shaking of branches and leaves makes an impressive noise. (But) encounters are usually limited to calling and associated arch postures, piloerection, and tail lashing." In this period, before 8 A.M., the home ranges of two groups overlapped by 17 per cent, but after 8 A.M. they overlapped by only 2 per cent. Five other groups overlapped by only about 2 per cent at any time. These titis had the most exclusive territories measured for any primate (Robinson, 1979).

When Robinson played the titis' "chirrup-pump" call, his groups approached the tape recorder, answering with chirrups, moans, or the loud, complex male-female duet of trills and warbles (Moynihan, 1966). They reacted more when the recorder was near the periphery than at the center of their territory. That is, they are "boundary-oriented" rather than shading out from a home range center in "centripetally focused" fashion (Owen-Smith, 1977). Black-and-white colobus, in contrast, are centripetally territorial. Oates' colobus supplanted other groups significantly more often when they were in the core area of their range than elsewhere and responded more often to playback of roaring in the center than the periphery (Oates, 1977; Waser, 1975, 1977).

Lar and Kloss gibbons, hanuman, and purple-faced and nilgiri langur groups are all reported to approach each other, seeking out their wailing or whooping rivals. Siamang, like dusky titis, approach and shout at a tape recorder, apparently treating a playback of their own voices in the same way as an outside threat (Chivers and MacKinnon, 1977). One cannot extrapolate, however, from one species to the next.

Widow titis, unlike dusky titis, specialize in the forests that grow on almost sterile white sands of the Amazon. One group of widow titis had a home range 40 times as large as that of nearby dusky titis—an area far too large to defend (Fig. 7.3.). The homologous male calls played back to them led to precipitate flight, a duet to answering duets, and then more dignified retreat. Their calls serve for spacing, but not defense (Kinzey and Robinson, 1981, 1983).

One of the best analyzed primates are the greycheeked mangabeys of the Kibale Forest. They "whoop-gobble" over long distances, but they avoid recorded whoop-gobbles instead of attacking (Waser, 1977). This is even true for the Tana River species who live in richer and more evenly dispersed habitat, with home ranges an order of magnitude smaller, where one might expect some form of territoriality. The Tana River mangabeys are fairly indifferent to playbacks. Males whoop-gobbled in answer, but the female cores of the troops did not even shift direction (Waser and Homewood, 1979).

The howler monkeys of Barro Colorado were long thought to be territorial. In Carpenter's (1934) classic account, the first full-scale primate field study, the

FIGURE 7.3. *Dusky titi trills are classic territorial calls, but widow titis, like this one, hear such a call and run (see text). (Courtesy W. Kinzey.)*

howler troops spaced out at low-population density. Each had an area of exclusive range. Their dawn chorus of resounding bellows (Fig. 7.4.) seemed just like the gibbon or indri chorus. Now, at higher population density, howler ranges overlap completely, both in southwestern Panama (Baldwin and Baldin, 1972), and in Barro Colorado itself (Milton, 1980). Their bellowing is largely high-intensity alarm, given to thunder and intrusive primatologists as well as to each other (Carpenter, 1934; Baldwin and Baldwin, 1976). They use it to space out, in that groups are more likely to move away from the nearest calling group in the dawn chorus (Chivers, 1969). However, when they meet, the victor in a confrontation is not the troop nearest a territorial base. There is a dominance hierarchy between troops that depends on the "unity and fighting ability of the coalition of adult males" who roar and even "physically scuffle" until the repulsed troop sedately withdraws (Milton, 1980).

A similar ambiguity arises with scent marking. The scent marks in a sifaka territory in Berenty, a gallery forest in southern Madagascar, form a ring round the boundary. They are true territorial sign posts, like the dusky titi song. Ringtailed lemur in the same forest also scent mark branches, particularly the males who gouge the scent of their forearm glands into branches or saplings, using their horny wrist spurs. The lemur, unlike the sifaka, mark trails throughout their home range. The mark conveys the identity and location of individual males, but it does not unambiguously warn off other troops who merely overpaint the sapling with their own odorous graffiti (Mertl, 1977; Mertl-Millhollen, 1979).

We keep returning to the degree of overlap of range as the final criterion of territoriality. At times this criterion may be too strict when a normally territorial species is squashed together by habitat change, or too loose when a pop-

FIGURE 7.4. *Mantled howlers bellow at dawn, at other groups, at thunder, and at photographers. (Courtesy C. M. Hladik.)*

ulation is expanding into new habitat. Still, this is what we mean by territoriality: to space out conspecifics enough so that some area belongs to each group.

Those that are not territorial may still warn each other away by shouts or stinks. They may still confront each other in physical battles like howlers, and force weak troops to give way like howlers, macaques, and baboons. One troop of rhesus macaques excludes others from a peninsula of Cayo Santiago during the birth season but not the mating season, a case of normal macaque dominance expressing itself as spatial rights (Lauer, 1980). These dominance relations between troops are one end of a spectrum; weaker troops could become too timid to enter the center of a strong troop's range. However, the spectrum may have little in the middle and clearer tendencies at either end. The complex of behavior that is true territoriality may be selected all together; if you are going to defend a property, best to do so wholeheartedly.

146   *Ecology*

FOOD SUPPLY AND DEFENSE

The reasonable assumption is that it is not worth defending a feeding territory that is full of holes. That is, if food supply is very patchy, it would be better to displace other troops from particular patches as they came into flower or fruit, but not defend the unproductive space between patches. If food is fairly evenly dispersed, so that it can be browsed by a "lawn mower" strategy of cropping swathes of whatever grows, then it might be worth defending the whole area (Brown 1964; Brown & Orians, 1970; Milton, 1980).

It is possible to support this hypothesis by taking particular examples. White sifaka (Fig. 7.5A., B.) in a northern forest, with patchy and seasonal food sources, overlapped ranges almost completely. They battled rarely, and at the conclusion of a battle retreated in any direction, without orienting to or from a territorial core. In a southern forest of the "spiny desert," the same species browsed from all the more common trees and found their food sparsely but evenly through their range. There the ranges overlapped little. The sifaka encountered other groups and fought classic battles at the periphery, retreating at the end to their own territories. The territories approximated five hectares, twice as large as the rich gallery forest territories where Mertl mapped sifaka scent marking. It was not the size of area to defend but the food distribution that apparently determined territorial or nonterritorial behavior. Other examples could be taken from the New World. Panamanian cottontop tamarins in a relatively stable, lowland habitat maintained a year-round territory with little overlap. In a more seasonal upland area they ranged on top of each other, like the northern sifaka (Dawson, 1979).

The two species of titi monkeys that responded in opposite manners to the playbacks of their calls also use their home ranges in different fashions. Dusky titis, the territorial form, are normal forest primates who eat about 70 per cent fruit and, otherwise, mainly leaves. The widow titis in their white sand forest also ate 70 per cent fruit, although they ranged over an area 40 times as large to find it. Trees in such nutrient-poor regions defend their leaves by many poisonous secondary compounds. The party of widow titi supplemented the fruit with insects, not leaves. Presumably, it is not efficient for them to guard such a huge space as permanent territory, and, presumably, the patchiness of fruiting trees and insects means most of the area is not worth defending (Kinzey and Gentry, 1979). Wide ranging species of the Old World may grade from one end of the spectrum to the other within a single species, such as the vervets, black-and-white colobus, hanuman langurs (Kavanagh, 1981; Oates, 1977; Dunbar and Dunbar, 1974; Jay, 1965; Ripley, 1967).

Finally we can look at savannah baboons. The baboons as a whole lie at the dominance end of the spectrum. Large troops generally displace small troops, with no ritualized long-distance communication and little reference to geography. Their habitat is extremely patchy, not in food, which is largely grass, but in water holes and sleeping trees or cliffs. Altmann (1974) pointed out that the ranging behavior of such creatures is constrained by the most limiting resources. Amboseli baboons overlap range more than at Nairobi because water holes and sleeping groves are more clumped at Amboseli. Thus, the Nairobi baboons can have exclusive core areas without actually defending them (Devore and Hall, 1965). On the other hand, some chacma baboons actively de-

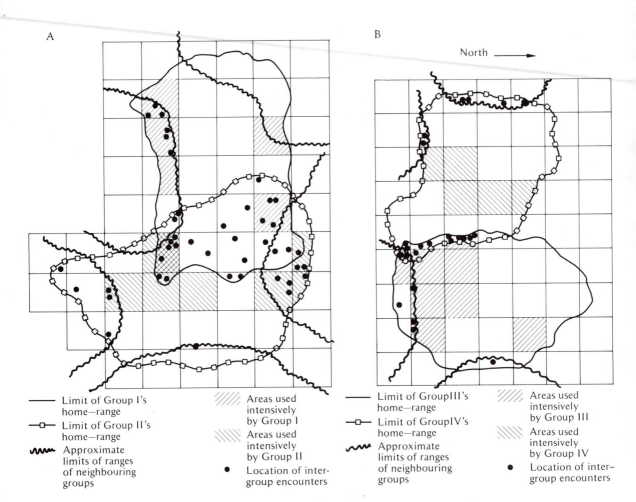

A

B

North ⟶

—— Limit of Group I's
home—range

—□— Limit of Group II's
home—range

〜 Approximate
limits of ranges
of neighbouring
groups

▨ Areas used
intensively
by Group I

▨ Areas used
intensively
by Group II

● Location of inter-
group encounters

—— Limit of GroupIII's
home—range

—□— Limit of GroupIV's
home—range

〜 Approximate
limits of ranges
of neighbouring
groups

▨ Areas used
intensively
by Group III

▨ Areas used
intensively
by Group IV

● Location of inter-
group encounters

FIGURE 7.5. A,B. *Intergroup spacing in white sifaka. The squares are 50-meter by 50-meter quadrats. (A) In a northern forest, two groups had exclusive use of only 46 per cent and 43 per cent of their home range. The core areas, the quadrats where they spent 75 per cent of their time, largely lay in zones of overlap. During intergroup encounters, groups did not necessarily retreat toward the center of the range. (B) In a southern forest, the white sifaka defends classical territories, with exclusive use of 90 per cent of the range, central core areas, and stinkfights at the boundaries, in which each group faces outward from its base like chessmen and retreats afterwards toward the central core. (After Richard, 1978.)*

fend desert water holes and even boundaries of swamp (Hamilton et al., 1976). Just as sifaka, with their evolved, stereotyped territorial marking, still have some nonterritorial populations, so baboons, which seem to have no particular inter-troop code beyond simple bullying, still can sometimes use bullying to defend space.

The sifaka, tamarin, titi, and baboons illustrate two points: food distribution molds territoriality, and territoriality is highly labile behavior. It may differ between closely related species, or even between populations of the same species. One group of primates that seem to have rigidly fixed territorial behavior is the gibbons, but for the others it is dangerous to extrapolate from one population to another.

Three cross-species correlations could test the hypothesis that food distribution and defensibility determine territoriality: the relation between home range size and day range (Brown and Orians, 1970), the relation between territoriality and folivory, and the relation between territoriality and monogamy.

There are two attempts to deal with the relation between day range and home range size: Martin (1981) and Mitani and Rodman (1979). The two make an instructive comparison, because they are mathematically interchangeable and based on much of the same data, but have different pictures of behavior and opposite conclusions.

Mitani and Rodman calculated an index D, which is the ratio of average day range to the diameter of the home range, if the home range is idealized to a perfect circle (Table 7.1.). They pictured their territorial primates camped

TABLE 7.1.   Index of Defendability

$D<1$	$D\geq1$
Territorial Species	
	indri
	dusky titi
	vervet guenon
	hanuman langur
	black-and-white colobus
	siamang
	lar gibbon
	redtail guenon
	ringtail lemur
	white sifaka
	lepilemur
Nonterritorial Species	
gorilla	brown lemur
patas	talapoin
rhesus macaque	chacma baboon
hanuman langur	mongoose lemur
bonnet macaque	squirrel monkey
orangutan	red howler
greycheeked mangabey	
gelada	
mantled howler	
yellow baboon	
black-and-white colobus	
red colobus	
chimpanzee	
olive baboon	

Defendability represents mean daily path length divided by diameter of the home range.

(After Mitani and Rodman, 1979).

somewhere in the center, then sallying out to meet challenges on the borders, so that the diameter or radius of the range gives the relevant measure of defensibility. (We have seen that dusky titi and lar gibbon behave like this.) They found a highly significant effect: all territorial populations had D equal to or greater than 1, so they could travel from center to edge and back in an average day. Nonterritorial populations had D of widely varying quantities. They might move much or little compared to range diameter.

Martin's index, the Range Traversing Index, is the ratio of day range to circumference of home range idealized to a perfect circle, in other words, $RTI = D/\pi$. Martin chose this version because he wanted to know whether territorial primates could patrol their frontiers to monitor intruders. Mertl-Millollen's sifaka spent 60 per cent of their time in the outer 40 per cent of their range (with no apparent food or feeding differences), so at least one troop does just that. However, Martin suspected that such boundary patrols were hardly ever substantiated. He found no correlation between RTI and either territoriality or nonterritoriality. He concluded that it is perfectly possible to defend a territory by long-distance calls and little movement.

Why are the conclusions so different? Two reasons: a discrepancy of data that should be trivial, and a discrepancy of point of view. The data discrepancy is a question of taking Pollock's summary of ranging for Indri or breaking it down by study site; breaking it down gives Mitani and Rodman's version. Data discrepancy should give shudders to anyone dealing with cross-species correlations, because such multispecies comparisons are meant to transcend the noise in the data. A trivial decision about which figures to use for a species that happens to have extreme behavior can affect one's whole view of the conclusion. Without indri, Martin's figure is just like Mitani and Rodman's.

The second conclusion is that biologists should keep referring back to the meaning of their mathematics. Martin is very likely right that most territorial species do not and need not move round the circumference. Mitani and Rodman may also be right that radial movements are frequent and important. What looked then like an argument over mathematically equivalent indexes has turned into a prescription for reporting one's fieldwork differently, to see whether animals actually move in and out or round and round when defending their territories.

The next test might be a relation between territoriality and folivory. One expects a diet of mature leaves to be more evenly spread and thus more worth defending than other diets. However, the per cent of any sort of foliage in the diet has no relation to territoriality. The ground grazers—gorilla, gelada, and hamadryas and savannah baboons—all lack territorial boundaries. The choosers of new leaves—red colobus and mantled howler—live in patchy environments like the species that eat fruit. Furthermore, many territorial species do not specialize in leaves at all. The conclusion (contrary to what I wrote in the first edition of this book) is that leaf eating is not a strong guide to territoriality. Evenness of food supply and *mature* leaf eating may be so, but this must be confirmed habitat by habitat, as for Richard's southern sifaka.

Is there a relation between monogamy and territoriality? Several populations of marmosets and tamarins are territorial. One species of titi is, and another is not. Avahi and indri are, but mongoose lemurs are not. All gibbons so far described are territorial, and so are Mentawai langurs. On balance, monogamy seems to be correlated with territoriality, but loosely so.

On the other hand, multimale troops are rarely territorial. The only one on Mitani and Rodman's list is the ringtailed lemur, which is a dubious case. Ringtails have exclusive core areas and intertroop aggression, but the ranging and range overlap varies from season to season, so they are best considered intermediates like the Nairobi baboons (Mertl-Millhollen, et al., 1979; Jolly, 1966). This fits Wrangham's (1980) argument that multimale troops evolved as means of dominating neighbors and forcing them out of food patches in the short term.

Finally, does territoriality run in phylogenetic lines, even though labile at the population level? Gibbons and siamang are the clearest case. Many langurs also have exclusive territories and whooping or honking territorial calls: purple-faced, nilgiri, dusky, banded, Mentawai, and some but not all hanumans. Some but not all colobus are territorial. There may be such a tendency throughout the colobinae, correlated with their anatomical specializations for mature leaf eating. A similar tendency could surface in the indriids, again correlated with mature leaf eating and, in this case, monogamy. The baboon-macaque tribe are rarely, if ever, strictly territorial.

Thus, the environmental correlations may be there, mixed with phylogenetic bias, but the clearest conclusion is that they must still be checked out case by case.

THE NEIGHBORHOOD

The discussion of territoriality versus dominance concerns only the competitive aspects of group relations. Primates do not merely shove each other out of the food patch (dominance) or out of the area (territory). They also migrate, mix, inherit, and, perhaps, even cooperate.

Two main questions are, how much interchange do we see between troops and how much can we judge by genetic studies they are related. Can we reconcile the two approaches and judge that their affiliative behavior or lack of it somehow relates to genetic distance? This is one of the most fascinating fields of current primatology, but we do not have data yet for unequivocal answers.

All long-term studies report migration between troops, either male, female, or both. In macaques and baboons, essentially all adolescent males shift troops, and adult males may do so too. The rates of exchange are probably similar in other primates.

This means that a young macaque "knows" it is related to its mother and most other females of its troop. However, it may also be related to neighboring juveniles. One of the commonest reports of friendly troop interactions is play between juveniles—sometimes even while their elders engage in territorial chases, like the vervets of Amboseli (Struhsaker, 1967). Adult male vervets behave more amicably toward their natal group or toward a group in which they have previously fathered infants than to more distantly related groups (Cheney and Seyfarth, in press).

Wild Barbary macaques run the gamut of possible interactions from aggressive troop displacements to spending hours or days in joint foraging. Troops sometimes walked through each other with individual aggression but no apparent change in either troop's movements (Deag, 1973). A few other primates feed amicably together, like the geladas. Waser and Homewood (1979) report "a certain insouciance" in the intertroop encounters of Tana mangabeys. They sometimes joined and fed together for hours at a time. Perhaps this reflects the close kinship between groups in isolated patches of gallery forest by the Tana River.

These are exceptions, though. The vast majority of primate groups avoid each other, sometimes with active aggression, and some with titilike standoffs at the boundaries. The balance of interactions are competitive, not cooperative.

SUMMARY

A territory is a geographical area, defended by the owners, or exclusively used by the owners, or both. Intertroop conflicts that do not refer to geographical location are dominance behavior, not territory defense. There is a continuum from pure dominance to pure territoriality. Ideally, we should measure activity fields and aggression fields, not simply overlap of ranges, to understand territoriality. Territories may be defended for mating, for reproduction, or for food supply.

Intertroop conflicts include fighting, ritualized visual battles, vocalizations at short or long distance, and scent marking. Any of these may function as dominance spacing or as territorial defense. Playbacks of loud vocalizations elicit approach from territorial species and usually retreat from nonterritorial ones. Some territorial species like the dusky titi are boundary-oriented in that they respond more strongly to calls played at the edge, not the center of their territory. Others respond more strongly to challenges near the center of their range, which may be an intermediate state between dominance and strict territoriality.

Feeding territories presumably evolve when the food supply is both defensible and worth defending. Several examples support the idea that evenly distributed food supply is more defensible than a patchy food supply. However, it is difficult to support the hypothesis by a general correlation between folivory and territoriality, because ground grazers and eaters of new leaves are generally not territorial. Defensibility is more straightforward. Territorial forms' average day range commonly exceed the diameter of their home range, so they can be frequently present at any boundary, though they may not range in circles round the periphery. Nonterritorial forms may or may not have average day ranges longer than home range diameter.

Territoriality correlates with phylogeny. All gibbons and many colobines and indriids are territorial, with scattered examples from other groups and virtually none from macaques and baboons. However, the general tendencies do not predict the behavior of particular populations: in several species, some populations are territorial and others not so.

Most interactions between primate troops are characterized by competition or avoidance, not cooperation. This is explicable because members of female-bonded troops are more closely related to each other than they are to members of neighboring troops. This is so even if the next troop only recently split from their own, because troops fission along genealogical lines. The exception are monogamous and solitary primates whose offspring become their neighbors. Such parents occasionally cede territory to their offspring. Neighboring vervet troops are less aggressive to troops where they have close kin.

Males who have immigrated into female-bonded troops may take the lead in intertroop defense, thus aiding the group as a whole as well as their own offspring. Intertroop competition may thus be one selective pressure for within-troop cooperation.

CHAPTER 8

INTERSPECIFIC RELATIONS

Much of ecology is the study of interspecific relations. Questions such as how many species live in a given environment, which species, and how they interact are far too broad to treat in one chapter. Neither is it possible to list the permutations of interactions even among primate species. This chapter will necessarily select a few things to say, leaving much more unsaid. It will focus on case studies of five rain forests. This at last fleshes out what you actually see when you go out to watch primates. (It is tempting, but too long, to have a comparative section on dry woodlands or savannahs, or temperate zones; perhaps students would like to do that for themselves.)

FIVE FORESTS: THE TREES

Most species of primates live in tropical rain forest, as we have already said. This was probably the original home of the order. Five of the best studied and longest studied primate faunas live on Barro Colorado Island in Panama; by the oxbow lake of Cocha Cashu in the Manu Park of Peru; by the Makokou Field Station of Gabon; in Kibale Forest, Uganda; and at Kuala Lompat in the Krau Game Reserve of Malaysia. All five have in common, as research, that they are team efforts, involving many scientists for many years. One person is not enough to tackle the job of understanding rain forests or even part of a primate's lifetime therein.

All five areas are classed as rain forest. Rain forest is not a homogeneous habitat; neither does it stay the same forever. The structural changes in rain forest, however, take place in a time scale that stretches far beyond primate or human generations. If a tree falls, or a farmer clears a little patch of field, the forest regenerates in orderly fashion. First, bushes, and then fast-growing trees shoot up. These trees are called weed or scar species, or pioneers—typically

Cecropia in tropical America, *Musanga* in Africa, and *Macaranga* in Asia. These trees have a short life span. They die off in about 20 years. Beneath them a second-stage crop of shade-tolerant species has grown, festooned with the lianas that start from the ground at the same time as their principal host tree. Such trees may not sprout in large clearings where fire and sunlight have killed the soil fungi. Many later-stage seedlings depend on mycorrhiza, the association of root with symbiotic fungus, to absorb minerals from the poor soil. The succession of trees above ground depends on an equally complex succession of fungal symbients (Janos, 1980a,b). The second stage of a tree rises in the typical tall rain forest pattern of growth (Fig. 8.1.), but its species in turn have a limited life span of a mere 100 years. Some of the saplings beneath them are the third-stage giants, which mature over centuries. They may have begun to grow as early as the pioneers, but their seedlings and saplings can wait for decades without growth in the densely shaded understory, until the death of a tree above lets them move toward the canopy (Hallé et al., 1978).

When a tree falls or a farmer comes with his machete to clear a manioc field, the forest responds in various ways. If the clearing is extensive, the whole process goes back to start over from grasses, if at all. If it is only a small gap, the forest regresses to some earlier stage and heals the scar with the relatively rapid growth of light-demanding trees. If, by chance, the tree fall occurs over a sapling of a later stage, it may even advance the succession, like weeding in a garden.

Gaps and tree falls are natural processes. (See Fig. 8.2.) The patches of rapidly growing forest provide special habitats for some primates and a source of fruit and new leaves for many others. The architecture of such a gap begins from a dense tangle of bushes and creepers, with an interlace of twigs near ground level when the gap is newly made. The pioneer species as they grow provide horizontal branches with both food and pathways at fairly low level. When they are eventually replaced by the high canopy trees, continuous foliage is far above ground. Most rain forest mammals, birds, and insects feed in the canopy. Below that lies a region of vertical trunks and obliquely fanned branches with wide spaces between—short of food and difficult to traverse unless you fly or glide.

Forests differ. Some grow on shallow soils where the trees cannot spread to make single-layered canopy, or in swamps with a different set of species. Some grow on hillsides where the treetops overlap like roof tiles. Some are cut by roads or rivers, where the bankside trees bend obliquely toward the light. Some have dry seasons, such that the long-term growth is overlaid by yearly cycles. What are the particular characteristics of our five rain forests?

Barro Colorado Island has more than 2,500 millimeters of rainfall a year (Fig. 8.3.), and Cocha Cashu in one year had 2,080 millimeters. Kuala Lompat in the rain shadow of Malaysia's main range averages 1,980 millimeters, Makokou, 1,700 millimeters in one year, and Fort Portal near Kibale averaged 1,475 millimeters over 52 years. The dewpoint and evaporation matter, as well. Kibale, at medium altitudes, remains humid even with slightly lower rainfall than the others.

Only Kuala Lompat has about the same amount of rain in each month. Barro Colorado and Cocha Cashu have marked wet and dry seasons; in some years it may not rain at all in February, March, or April. The ground bakes and

FIGURE 8.1. *Vertical structure of rain forest at Ipassa, near Makokou, Gabon, along a transect 90 × 5 meters. The horizontal projection is enlarged to 90 × 10 meters to include trunks of trees standing behind the profile. Note the scale, the concentration of horizontal branches high in can-*

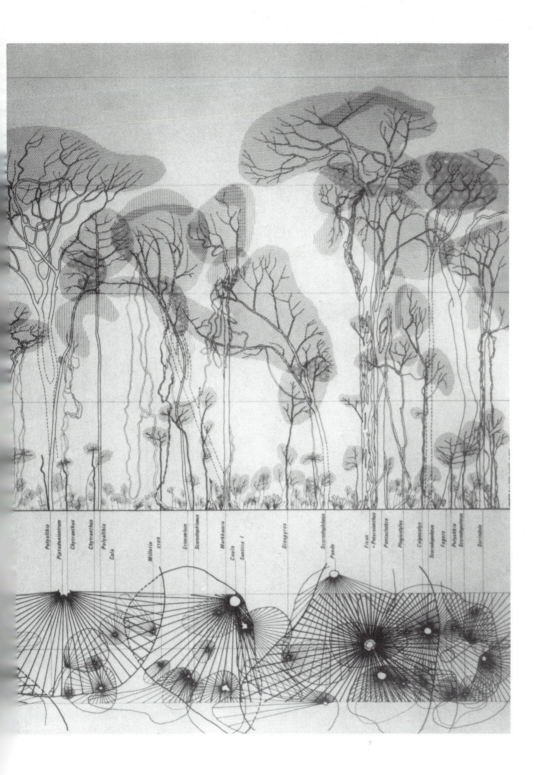

...opy and emergent layers, with vertical trunks below, and the dearth of small branches at mid-level, with liana tangles connecting understory and canopy. (Courtesy A. Hladik.)

157 *Interspecific Relations*

FIGURE 8.2. *A section of old forest revealed by tree fall, Barro Colorado. (By permission of the Smithsonian Institution Press, from* The Ecology of the Tropical Rainforest, *E. G. Leigh, A. S. Rand, and D. M. Windsor, eds., p. 69 (drawing by Daniel Glanz) © Smithsonian Institution, Washington D.C. 1982.)*

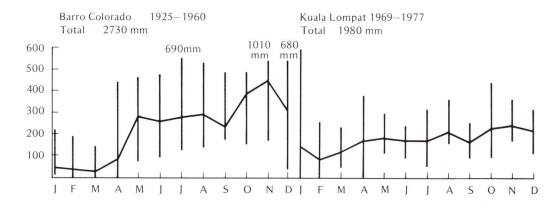

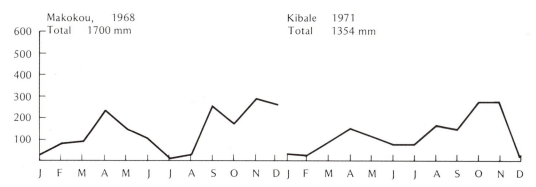

FIGURE 8.3. *Seasonal distribution of rainfall in four of the forests. Although they are all "rain forest," seasonal changes markedly affect primate food supply.*

cracks toward the end of the dry season. The two African forests have a double pattern of rainfall, with two wet and two dry seasons each year. Further, the range of variation is marked from year to year: it would be an experience to do field work in Barro Colorado during the November of the 1,000-millimeter rainfall—more precipitation than an average 12 months' in London.

Table 8.1. gives tree heights, densities, and species diversity. Barro Colorado was islanded by the rising waters of Gatun Lake when the Panama Canal was formed. (You can sit on the steps of the Smithsonian Laboratory watching freighters and cruise ships sail between oceans past the little landing dock.) Its old forest canopy rises to 30 meters, emergents to 40. Just beside the laboratory buildings, however, the forest is regrowing from a regime of slash and burn that ended some 80 years ago. The howlers of Lutz Ravine, wild primates studied since Carpenter's pioneering work of 1932, come there to eat the leaves of colonizing *Cecropia* trees as well as the abundant figs. Lutz Ravine has tree density and diversity as great as the fully mature forest, although the canopy is not quite so high (Milton, 1980). The tall trees have the typical rain forest structure of vertical unbranched trunks leading up to an obliquely vertical fan of branches.

Manu National Park, in contrast, is a huge area of virgin jungle in the headwaters of the Amazon. The study site in the meander plain of the Manu river encompasses high ground forest even taller than that of Barro Colorado,

TABLE 8.1. Forest Structure and Diversity

Forest	Criterion cm. Girth at Breast Height	No Trees/ ha.	No spp./ ha.	% 10 Commonest Spp.	Continuous Canopy Height, m.	Emergents Height, m.	
Barro Colorado							
Old forest	≥60	168–214	52–58	42–45	8–30	40	Milton, 1980
Lutz Ravine	≥60	200–208	58–63	42–55	→25		
Cocha Cashu							
High-ground forest					30–35	50–60	Terborgh, in press
Kuala Lompat	≥50	200–300	107	34	15–40	80	Raemaekers, 1981
	≥60	155–230					
	≥91	83–124					
Makokou	≥31	427			15–30	45	Hladik, 1978
	≥95	119	50				
Kibale			37.5			ridge 30 valley 55	Struhsaker, 1976

and short, thick swamp forest, filled with the prop roots of *Ficus trigona,* "as if the whole swamp were enveloped in the ramifying tentacles of some sinister creation of the vegetable kingdom." There are also a series of early forest stages on land newly made by the river (Terborgh, in press).

Asian forests are the tallest in the world and the most diverse in species. Kuala Lompat is no exception. The general look of the forest resembles Barro Colorado magnified in both scale and complexity; though a portion of the study site has been selectively logged.

The two African forests differ sharply. Kibale is a mosaic of ridge top and valley, where some sections have been selectively logged. The valley forest is not much lower in stature than the mature sections of Barro Colorado, but somewhat less diverse in species composition, which is reasonable at 5,000 foot altitude. Finally, Makokou has the typical rain forest shape and height but with so much denser understory that primate-watching is much harder than on Barro Colorado or in Malaysia. Thus, these five forests are directly comparable in structure, each with a mosaic of succession stages, and with greatest height and tree species diversity at Kuala Lompat.

FIVE FORESTS: THE PRIMATES

Barro Colorado has five species of primates; Cocha Cashu, 11 to 13; Kuala Lompat, seven; Kibale, 11; and Makokou, about 12, differing in various micro-habitats (see Table 8.2). Being drier and higher does not inhibit primate diversity at Kibale.

At Barro Colorado there are mantled howlers, blackhanded spider monkeys, whitethroated capuchins, and the rufus-naped tamarin, a subspecies of the cottontop. The one nocturnal primate of the New World is not a prosimian, but the owl or night monkey. Cocha Cashu has black spider monkeys, red howlers, the whitefronted capuchin that resembles that of Barro Colorado, and also the much stockier brown capuchin. Infrequently seen are monk sakis and woolly monkeys in a forest 30 kilometers downstream. Squirrel monkeys, dusky titis, owl monkeys, emperor and saddlebacked tamarins, rare Goeldis monkeys, and pygmy marmosets complete the list.

At Kuala Lompat live siamang and lar gibbons, dusky and banded langurs, and longtailed macaque, with sporadic visits by pigtailed macaques who mainly live on higher ground beyond the study site. Slow loris occupy one nocturnal primate niche.

The primates of Kibale are chimpanzee, red colobus and black-and-white colobus, the blue, redtail, and l'hoests' guenons, and the greycheeked mangabey. Olive baboons range from the fringe into the forest proper. Three nocturnal prosimians—the potto, Demidoff's galago, and a larger galago—have yet to be studied at Kibale. (See Figs. 8.4A,B, 8.5, 8.6A,B,C,D, and 8.7A,B.)

The primates of the various forests near Makokou include five nocturnal prosimians, the little green talapoin, four guenons, two mangabeys, the black-and-white guereza colobus, drills that are forest-loving baboons, and two now transient or extinct apes, the gorilla and chimpanzee. One hundred and twenty mammalian species live at Makokou, so the habitat must be finely divided.

Some, but not all of these primates are clearly differentiated by weight. In

TABLE 8.2. Social Group Structure, Range, and Territory

	Mantled Howler	Red Spider	Whitefaced Cebus	Cottontop Tamarin	Owl Monkey
Barro Colorado					
Group size	14	fluid	15	6–9	3
Group structure	MM	MM	SM–MM	Pr.	Pr.
Male:Female Ratio	1:2–4		1:1–5		
Core of Troop	Females?	Males?			
Migrants	Both?	Females?			
Pop. Dens./Km²	80	1	16	4–6	
HR Size, ha.	40		90		
HR Overlap	Complete				Small?
Boundary defense	No				
Participants in conflicts	Males	Males?	Males and females		
Nature of encounters	Avoid-aggr.	Aggr.	Aggr.		Aggr.
Adult Male Loud call used in spacing	Yes	No	No		
Countercalling	Yes	No	No		

	Siamang	Lar	Dusky Langur	Banded Langur	Longtailed Macque	Pigtailed Macaque
Kuala Lompat						
Group size	3.0	3.5	10.3	9.03	17	35
Group biomass (Kg)	26	17	51	57	39	190
Groups structure	Pr.	Pr.	MM	MM	MM	MM
Male:Female ratio	1:1	1:1	1:2–4	1:3	1:3	
Core of troop	Both	Both	Females?	Females?	Females?	Females?
Migrants	Both	Both	Males?	Males?	Males?	Males?
Pop. density/Km²	4.5	6	31	74	39	
Hr size, на.	28	53	29	21	46	100 +
HR Overlap	None	Almost none	Slight	Extensive	Moderate	
Boundary defense	Yes	Yes	?	Yes	?	
Areas of Excl. use	Large	Large	Large	Small		
Participants in conflicts	Both	Both	Males	Males	?	
Nature of encounters	Aggr.	Aggr.	Avoid.	Avoid.	Mild	
Adult Male or Female loud call used in spacing	Yes	Yes	?	Yes	No	
Countercalling	Yes	Yes	No	Yes	No	

	Red Colobus	Guereza	Mangabey	Redtail Guenon	Blue Guenon	L'Hoest's Guenon
Kibale—Compartment 30						
Mean Group size	50	9	15	30–35	24	10–15

Group structure	MM	SM	MM	SM	SM	SM
Male:Female ratio	1:2	1:3–5	1:2	1:9	1:10	
Core of group	Males	Females	Females	Females	Females	
Migrants	Females	Males	Males	Males	Males	
Pop. density/Km²	300	10–100	9	140	40–50	
HR size, ha.	35	16	410	24	61	
HR Overlap	Complete	Large	Large	Small	Small	
Boundary defense	No	Yes	No	Yes	Yes	
Areas of excl. use	No	Perhaps	Yes	Yes	Yes	
Participants in conflicts	Males	Males	All	All	All	
Nature of encounters	Variable	Aggr.	Avoidance	Aggr.	Aggr.	
Adult Male loud call used in spacing	?	Yes	Yes	?	Yes?	No
Countercalling	Yes	Yes	Yes	No	No	

	Potto	Angwantibo	Allen's Galago	Needle-clawed Galago	Demidoff's Galago
Makokou, Nocturnal Species					
Group size	1	1	1	1	1
Group structure	Pr. SM		SM		SM
Male: Female ratio	1:2	1:1	1:4	1:0.8	1:1
Core of Group					
Migrants					Males
Pop. density/Km²	8–10	2	15–20	15–20	50–80
HR size, ha. Female	7.5		10		0.8
HR size, ha. Male	9–40		30–50		0.5–2.7
HR Overlap, Female	Small				Small
HR Overlap, Male	Small		Small		Small
Boundary defense					
Participants in conflict					
Nature of encounters	Avoidance				Avoidance
Scent important in spacing	Probably	Probably	Probably	Probably	Probably
Countercalling	No	No	Perhaps	No	No

	Crested Mangabey	De Brazza Guenon	Talapoin
Makokou, Riparian Species			
Group size	10	3–6	70–100
Group structure	SM	Pr. SM	MM
Male:Female ratio	1:1–3	1:1	1:2
Pop. density/Km²	8–13	28–38	40–90
HR. size	200	10	120
HR Overlap	Large	Small	Small
Boundary defense	No		No
Participants in conflicts			
Nature of encounters	Join	Variable	
Adult Male loud call used in spacing	?	Yes ?	No
Countercalling	No	Rare	

TABLE 8.2 Social Group Structure, Range, and Territory (*continued*)

	Guereza Colobus	Mandrill	Spotnosed Guenon	Crowned Guenon	Moustached Guenon
Makokou—Mature Rain forest					
Group size	10	200	20	15	10
Group structure		MM	SM	SM	SM
Male:Female ratio			1:2–4	1:2–4	1:2–4
Pop. density/Km^2			20–40	20–25	20–30
HR size, ha.		500	55–80	55–100	20–50
HR Overlap					Small
Boundary defense					
Participants in conflicts					
Nature of encounters			Avoidance		
Adult male loud calls used in spacing			Yes	Yes	No
Countercalling			Yes		No

	Black Spider	Red Howler	Brown Capuchin	Whitefronted Capuchin	Monk Saki	Squirrel Monkey
Cocha Cashu						
Group size	Variable	6	10	14	5?	30
Group structure	MM	SM	MM	MM	?	MM
Male:Female ratio						
Core of troop						
Migrants						
Biomass Kg/Km^2	175	180	104	84	2	48
Pop. dens/Km^2	25	30	40	35	2?	60
HR overlap			moderate	large		large
Boundary defense						
Nature of encounters			mild	avoid aggr.		mingle
Loud calls, probably spacing			no	yes		no
HR size, ha			70	>150		>250

	Owl Monkey	Dusky Titi	Emperor Tamarin	Saddleback Tamarin	Pygmy Marmoset
Cocha Cashu					
Group Size	4	3	4	5	5
Group structure	Pr.	Pr.	Pr.	Pr.	Pr.
Male:Female ratio					
Core of troop					
Migrants					
Biomass Kg/Km^2	17	17	4	4	1
Pop. dens/Km^2	24	24	12	16	5
HR overlap			none	none	

Boundary defense		
Nature of encounters	aggr.	aggr.
Loud calls, probably spacing		
Hr size, ha	20	20

Barro Colorado data from Oppenheimer, 1968, Hladik and Hladik 1969, Eisenberg 1979, Smith 1977, Carpenter 1935, Kuala Lompat data from Mackinnon and Mackinnon 1980, Gittins and Rae-maekers 1980, Curtin 1980, Aldrich Blake, 1980, Kibale data from Struhsaker and Leland, 1979, Struhsaker pers. comm., Makokou data from Gautier-Hion, 1979, Quris 1975, Gautier 1969, Gautier-Hion and Gautier 1974, Gautier Hion and Gautier 1978, Gautier-Hion 1971, Charles-Dominique 1977, Cocha Cashu data from Terborgh, in press.)

FIGURE 8.4. *Allen's bushbaby (A), and anowantibo (B) are two of the five nocturnal pro-simians at Makorow, Gabon. Both these species prefer small supports, but the bushbaby leaps among vertical saplings and lianas of the high forest, while the angwantibo creeps about the tangled thickets left by tree falls. (Courtesy A. R. Devez and P. Charles-Dominique.)*

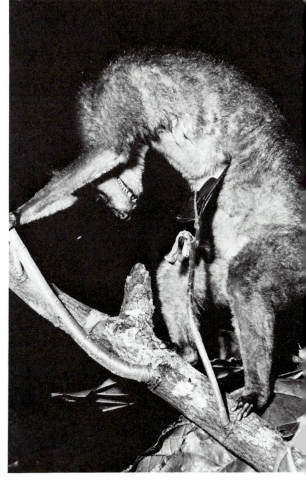

A

B

A

B

C

D

FIGURE 8.5. *Guenons of the Makokou region. All but the de Brazza form mixed species troops. (A) de Brazza's (courtesy of B. L. Deputte), (B) spotnose (courtesy J. P. Gautier), (C) moustached (courtesy J. P. Gautier), (D) crowned (courtesy J. P. Gautier).*

FIGURE 8.6. *Talapoin mother, infant, and juvenile, sleeping in tangled branches by a river at Makokou. When disturbed, they leap in the water and swim away. (Courtesy A. R. Devez.)*

FIGURE 8.7. *A. Siamang dangles to feed at Kuala Lompat. (Courtesy D. J. Chivers.) B. The lighter gibbon dangles in the same posture and eats much of the same food, but travels two to three times as far in a day as the siamang. (Courtesy D. J. Chivers.)*

A   B

many the males differ more from the females than they do from adjacent species. It will be complicated to predict how body weight affects niche difference. The statistical generalization that large primates are more likely to depend on leaves and small ones on insects does not predict the habits of any one primate (see Table 8.3). On the other hand, we would expect primates of similar weight and diet to be in close competition, so we may have to look for other niche differences. One clear difference appears between the forests: the new-world primates are smaller in body size and much smaller in biomass.

TABLE 8.3. Biomass, Folivory, and Insectivory

	Biomass Kg/Km²	Rank	% Leaves	Rank	% Insect	Rank
Barro Colorado						
geoffrey's tamarin	30	1	10	1	30	4
whitethroated capuchin	42	2	15	2	20	3
blackhanded spider	180	3	20	3		2
mantled howler	440	4	40	4	0	1
Cocha Cashu						
Goeldis monkey	0.5	1	small			
pygmy marmoset	1	2	small			
monk saki	2	3				
saddleback tamarin	4	4	0		58	2–3
emperor tamarin	4	5	0		58	2–3
dusky titi	17	6	high			
owl monkey	17	7			low	1
squirrel monkey	48	8	0		82	6
whitefronted capuchin	84	9	0		64	4
brown capuchin	104	10	0		76	7
black spider	175	11				
red howler	180	12				
Kuala Lompat						
pigtailed macaque	?	1	small	1		
lar gibbon	30	2	29	3	0	1
siamang	39	3	36	4	15	5
crabeating macaque	89	4	25	2	3	4
dusky langur	172	5	66	6	0.5	2
banded langur	406	6	46	5	1	3
Kibale						
mangabey	60	1	6	1	19	3
blue guenon	126–139	2	21	3	20	4
redtail guenon	328	3	16	2	22	5
black-and-white colobus	64–570	4	76	5	0	1
red colobus	1,760	5	75	4	3	2
Makokou riverine						
de Brazza	30	1	9	2	5	2
talapoin	60	2	2	1	36	3
mangabey	100	3	14	3	3	1

DIET AND LOCATION OF FOOD

How much do the primates overlap with each other in food species? For plant food species, Struhsaker (1982) gives blue guenon 22 per cent of diet; redtail guenon, 18 per cent; mangabey, 16 per cent; black-and-white colobus, 12 per cent; and red colobus, least at 8 per cent. If anthropod prey were included, the overlap would be even less. He points out, however, that this has little meaning unless one distinguishes which foods are abundant. If there are many trees of a species in flower or fruit within the troop's home range, they can shift to another when competition appears. If it is a rare, but popular food, the two groups may associate to feed simultaneously. (See Figs. 8.8, 8.9, 8.10, 8.11, 8.12, and 8.13.)

Hladik and Hladik (1969) first pointed out a relationship for the primates of Barro Colorado that is roughly true in the other forests. The biomass of folivorous primates is greater than that of omnivores, and that of omnivores, in turn, more than the insectivores. Because the food base of leaves is greater than fruit, and fruit in turn more than insects, this may seem a tautology. If one only examines the five forests, it is not always true. Makokou in particular has many frugiorous species, and a great diversity of fruit (Emmons et, al. 1983). One needs to take other animals into account, such as the colossal biomass of sloths that are vegetating in the canopy of Barro Colorado. One needs to discount particular species like the dusky titi of Cocha Cashu, which is small-bodied and not particularly common, but eats a great many leaves. Hladik's generali-

FIGURE 8.8. *At Cocha Cashu, Manu Park, Peru, a brown cebus and a white-fronted cebus forage in a Scheelia palm, one of their dry season staples. Only the brown has the strength to crack intact palm nuts for itself. (Courtesy C. Janson.)*

FIGURE 8.9. *Adult male brown cebus strips palm pith at Cocha Cashu, Peru, showing massive jaw muscles. (Courtesy C. Janson.)*

FIGURE 8.10. *Female white-fronted cebus at the more delicate task of unrolling palm fronds to look for insects. (Courtesy C. Janson.)*

FIGURE 8.11. *Squirrel monkey troops at Cocha Cashu keep traveling until they find a bonanza of figs. (Courtesy C. Janson.)*

FIGURE 8.12. *Saddlebacked tamarins, meanwhile, patrol tiny, well-monitored home ranges for small trees in fruit. (Courtesy C. Janson.)*

FIGURE 8.13. *Emperor tamarins are more likely to travel on vines and to forage for insects among the leaves. (Courtesy C. Janson.)*

zation holds true in many forests, both dry and wet, but it is a deduction that will not work in reverse to confidently predict what will be found in an unstudied area.

A second generalization is the distinction between species with small, well-known ranges, and those that range widely between erratically rich food sources. One way to measure the difference is to record sizes of feeding tree. The biggest bonanza trees are figs (Terborgh even christens one species a superfig). There is a good correlation between depending on large fig trees and size of range. The Kibale mangabeys or Cocha Cashu squirrel monkeys travel far and wide to rich sources, whereas animals like the tamarins actively avoid such places. Again, though, simple rankings of dependence on figs, or of plant food diversity, do not correlate perfectly with any given measure of ranging for all forests.

Neither does the degree of territorial defense differentiate life styles. The two tamarins of Cocha Cashu are highly territorial and invariably aggressive toward neighbors. They feed by choice in small fruit trees, too small to interest their far-traveling competitors. They also glean insects. It seems a classic pattern of small, dispersed food supply correlating with defended territory. Equivalent fruit and insect gleaning may correlate with territoriality in the blue monkey of Kibale, or the nocturnal Allen's galago of Makokou. However, in Kuala Lompat, it is not the small insectivorous animals who are territorial. It is the gibbon and siamang, respectively specialists in ripe fruit and new leaves. Like the tamarins, gibbons feed on relatively small, scattered food sources; for their body size, they travel a good deal between food sources, and they depend on knowledge of their range.

The height within the trees is relatively one easily measured indication of niche difference. Just as the animals of each forest eat different foods, so they eat them at different heights. This may, again, be means to an end; as Struhsaker (1982) says, "Primates go where the food is." Whether one considers height in trees an indication of active avoidance or just an index of slight differences in diet, it reveals differences in species habit.

Finally, we are only just coming to see the importance of the critical season. We know there have been sporadic die-offs in Barro Colorado during years of adverse rainfall. In Kuala Lompat, there are marked changes from year to year. In Cocha Cashu, five primates eat a fairly similar diet of fruit and insects during the wet season. During the dry season, when fruit is short, they reveal their niche separation. Brown capuchins open palm nuts and extract tree pith, which allows them to remain in a small home range. Whitefronted capuchins range far and wide to ripe fig trees, and squirrel monkeys even more so. Meanwhile, the tamarins continue to pick over their tiny ranges for fruit and insects in small, scattered units.

In sum, within each forest we can see that each monkey has its own way of life. We can see many coherent patterns—a correlation between leaf eating, biomass, and lethargy, or between small-patch feeding and territoriality. However, we are far from predicting just how the niche space will be divided by any given group of primates.

POLYSPECIFIC ASSOCIATIONS AND ANTAGONISMS

A much more revealing index of the species' relations is to look at how the animals treat each other. In many forests, primates supplant each other; they also commonly feed and travel in mixed groups.

At Barro Colorado, spider monkeys harass howlers and have occasionally kidnapped howler babies. This may be species-specific behavior, because the two species conflict at Tikal as well. Spiders also fight cebus at Barro Colorado.

At Kuala Lompat, siamang displace lar gibbons, who for the most part retreat from a fruit tree as siamang enter (Raemaekers, 1979; MacKinnon, 1977). This is not surprising as the siamang are heavier and stronger. It also fits the gibbon strategy of a quick bite and much travel, whereas the siamang strategy of efficient travel between trees in a small range fits with a need to feed in the tree when they reach it, not leave it to upstart lars. Two other species that conflict in both food and space are the crab-eating macaque and the banded langur. The macaques occasionally oust the langurs from feeding and sleeping sites. The macaques are smaller-bodied, but more numerous and more aggressive (MacKinnon and MacKinnon, 1980).

There is little polyspecific association at Kuala Lompat. At Kibale, in contrast, mangabeys displace blue guenons, and red colobus displace chimpanzees, but the associations are far more striking than the antagonisms.

Struhsaker (1982) has summarized the available data on primate association (Table 8.4). He points out that the absolute percentage of association is prob-

TABLE 8.4. Polyspecific Associations

Location and Source	% of all Primate Contacts that were Polyspecific (n = Total no. Contacts)	% of all Social Groups Contacted that were in Polyspecific Associations (n = Total Contacts with Social Groups)	% Social Groups Contacted that were in Polyspecific Association (n = Number Contacts with Social Groups)
Africa			
Kanyawara, Kibale Forest, Uganda (Struhsaker, 1975, 1978a, and unpubl. data)	50.9% (234 in 44 censuses)	73.6% (394)	red colobus, 64.6% (144) black-and-white colobus, 37.1% (35) redtailed guenon, 90.7% (107) blue guenon, 94.7% (75) l'Hoest's guenon, 30.8% (13) greycheeked mangabey, 78.6% (14) chimpanzee, 16.7% (6)
K 13, 12, 17 (selectively felled forest), Kibale Forest, Uganda (Struhsaker, 1975)	25.6% (82 in 11 censuses)	41.9% (105)	red colobus, 36% (25) black-and-white colobus, 29.2% (48) redtailed guenon, 60% (20) blue guenon, 77.8% (9) greycheeked mangabey, 100% (2) savannah baboon, 0 (1)
Ngogo, Kibale Forest, Uganda (Struhsaker, unpubl. data, 1975–1978)	47.1% (87 in 25 censuses)	65.4% (133)	red colobus, 78.6% (42) black-and-white colobus, 33.3% (3) redtailed guenon, 79.1% (43) blue guenon, 70% (10) l'Hoest's guenon, 0 (2) greycheeked mangabey, 85.7% (14) savannah baboon, 0 (6) chimpanzee, 0 (13)
Magombero, Tanzania (Struhsaker and Leland, 1980)	62.5% (16 in 5 days)	78.6% (28)	Gordon's red colobus, 90% (10) Angola black-and-white colobus, 88.9% (9) blue guenon, 55.6% (9)
Idenau, Cameroon (Gartlan and Struhsaker, 1972)	31.6% (95)	57% (151)	spotnosed guenon, 43.3% (4) redeared guenon, 96.3% (27) mona guenon, 75% (4) crowned guenon, 100% (20) l'Hoest's guenon, 11.8% (17) cherry-capped mangabey, 60% (5) drill, 25% (8) chimpanzee, 33.3% (3)
Southern Bakundu, Cameroon (Gartlan and Struhsaker, 1972)	57.7% (324)	79.6% (671)	spotnosed guenon, 76.1% (222) redeared guenon, 78.4% (222) mona guenon, 94.7% (169) crowned guenon, 96.2% (26) drill, 18.8% (32)
Four sites in Gabon (Gautier and Gautier-Hion, 1969)	49.3% (144)	70.9% (251)	spotnosed guenon, 76.7% (60) moustached guenon, 88.4% (59) crowned guenon, 77.6% (49) de Brazza guenon, 0 (7)

			talapoin, 47.4% (57)
			greycheeked mangabey, 85.7% (14)
			agile mangabey, 100% (2)
			black-and-white colobus, 33.3% (3)

S.E. Asia

Krau Reserve, W. Malaysia (Struhsaker, unpubl. census data; group densities from Mackinnon and Mackinnon, 1978)	0 (25 in 7 days)	0 (25)	siamang, 0 (1) lar gibbon, 0 (1) crabeating macaque, 0 (8) dusky langur, 0 (5) banded langur, 0 (9)
Ketambe, Gunung Leuser Reserve, N. Sumata (Rijksen 1977)	19% (1,017)	34% (1,250)	orangutan, 0.3% (4)[1] siamang, 28.6% (7) lar gibbon, 33.3% (6) banded langur, 0 (0) crabeating macaque, 37.5% (8) pigtailed macaque, 0 (1)
Ketambe, Gunung Leuser Reserve, N. Sumatra (Struhsaker, unpubl. data, 1979)	13% (23 in 9 days)	26.9% (26)	orangutan, 0.3% (4)[1] siamang, 28.6% (7) lar gibbon, 33.3% (6) banded langur, 0 (0) crabeating macaque, 37.5% (8) pigtailed macaque, 0 (1)

S. America

Ventura, N. Colombia (Bernstein et al., 1976)	approximately 14.8% (108)	18.8% (173)	longhaired spider monkey, 14.3% (77) woolly monkey, 20% (45) red howler, 28.6% (21) whitefronted cebus, 50% (13) whitefooted tamarin, 0 (17)
Peninsula Vicente La Macarena, Colombia (Klein and Klein, 1973, 1975, 1976)	no data	no data	longhaired spider monkey, ? 30–35% (150) red howler, ? 30–35% (150) whitefronted cebus, 59.5% (200) squirrel monkey, 70.4% (169)
El Duda, La Macarena, Colombia (Struhsaker, unpubl. data, 1974)	18.6% (43)	32.7% (52)	longhaired spider monkey, 27.3% (11) woolly monkey, 0 (2) red howler, 25% (12) whitefronted cebus, 31.8% (22) squirrel monkey, 80% (5)
Surinam (Mittermeier, 1977; Mittermeier and Van Roosmalen, in press)	no data	no data	black spider monkey, 3.7% (136) red howler, 0.96% (106) black-capped cebus, 6.7% (30) white-fronted cebus, 30.2% (139) squirrel monkey, 45.7% (92) black bearded saki, 33.3% (36)–23.9% pale-headed saki, 13% (23) redhanded tamarin, 3.2% (126)

[1] Orangutan were observed for 1,553 minutes. During this period they were with another species (lar gibbon) for less than five minutes.

Data is based on census and survey findings in rain forests of Africa, Southeast Asia, and South America.

(After Struhsaker, 1981).

ably too high, because in censuses one is more likely to notice larger groups, and thus rather likely to find large mixed groups.

One clear conclusion from Table 8.4. is that such polyspecific grouping is more common in Africa than in the other two continents. This seems not to be explained by external factors such as population densities of the species taken separately.

Struhsaker, like Waser (1980, 1981) discounts the possibility that these primates are associating by chance, though on different mathematical assumptions. He attributes some of these associations to common food sources, in particular the rare, but highly preferred ones, usually figs. Several other authors have come to the same conclusion in different sites.

Sometimes the associations merely involve feeding in the same place. Sometimes one species actively benefits. Redtail guenons cannot bite through the tough rind of *Monodora* fruit, and sometimes "approach and even line up behind mangabeys as they open these soccer-ball sized fruits . . . when the mangabey departs, usually leaving (the remains of) the fruit attached to the tree, the redtail has access." However, mangabeys are more likely to join redtails than vice versa, so such observations of single food species do not explain everything.

Another feeding advantage often proposed is that a slower-moving monkey flushes insects, which a more agile species of monkey catch on the wing. Thorington (1967) first suggested that squirrel monkeys use cebus as "beaters," and the idea has been applied in many forests including Makokou and Kibale (Gautier-Hion and Gautier, 1974; Klein and Klein, 1973; Rudran, 1978; Waser, 1977). However, Struhsaker says he has never *seen* this happen, and that neither have the others, although insectivorous birds often catch insects scared up by monkeys. Terborgh (in press) also discounts insect feeding in the associations of the squirrel monkey and cebus.

In short, feeding probably underlies most of the associations in space, but rarely with active benefit of one species to another. Feeding, however, is not all; predation plays a major role.

In Africa, the crowned hawk-eagle can be seen or heard over the Kibale forest every day. They are apparently capable of killing ten kilogram adult male black-and-white colobus (Struhsaker and Leland, 1979). The commonest association of primate species is the small redtail with the larger red colobus. These are both likely to benefit, the redtail from the deterrent effect of a multimale troop of larger animals, and the red colobus from the redtails' greater vigilance. There are not quantitative data to prove the point, but the animals' own responses to each others' air raid alarms show they treat the hawk-eagle as a serious menace.

The incidence of such large birds of prey correlates with the frequency of polyspecific associations—highest in Africa, lower in South America where the harpy eagle is apparently rarer, and lowest in Asia, where the changeable hawk-eagle is too small to take monkeys past infancy (Struhsaker, 1982).

The species of monkeys that are small, vulnerable, and arboreal also seem most likely to form associations. Terrestrial primates like l'Hoest's guenon and baboons apparently have less predation and fewer associations (Struhsaker, 1982). Squirrel monkeys, in contrast, flock together with each other in huge but socially unstructured herds. It is, then, for protection that they follow along with the larger-bodied cebus, who add still further defense against birds of prey (Terborgh, in press).

Within these groups, the animals often interact surprisingly little. However, they do sometimes chase, slap, or sit together and groom. At Kibale the frequency of aggression between species in mixed groups correlates perfectly with the percentage of overlap in the two species' diets. The grooming is in inverse correlation, mainly from red colobus to redtail, perhaps attracting the redtails for their usefulness as hawk-watchers.

In Makokou, the different guenons similarly benefit by each other's reactions to predators. Spotnosed and moustached guenons are usually together. The spotnose is larger, noisier, and more evident to hunters; in a mixed band the spotnose male is most likely to be killed. A Gabonais proverb runs, "the little spotnose should take advice from the little moustache monkey" (Gautier and Gautier-Hion, 1969).

This flocking is an active process, with certain species more likely to join than others. Moustached guenons, following a talapoin band, may be drawn outside their normal range, because the scores of talapoins provide protection and let the moustaches enter open areas near villages. The grouping of spotnose and moustache is so frequent that the loud call of the spotnose male serves as interband spacing for both groups; the moustache male gives his call later or not at all. This may depend on a threshold difference: the loud call is given in alarm or to any loud noise, and the spotnose may be more easily alarmed (Struhsaker, personal comment). However, the spotnose call still functions for both groups together.

Finally, human predation, if anything, reduces the number of associations. Mixed groups were both more frequent and more complex in a high primary forest in the Makokou area than in more degraded areas near villages. This is also true at Kibale where primates flocked more frequently in uncut, mature forest. Humans see large groups preferentially when they are census taking—but also when they are hunting (Struhsaker, 1982).

SUMMARY

Different species may compete, cooperate, eat, or be indifferent to each other. The competitive exclusion principle states that too close competitors will not coexist.

The chapter describes the primates of five rain forests: Barro Colorado Island, Panama, with five primate species; Cocha Cashu, Peru, with at least 11; Kuala Lompat in the Krau Reserve of Malaysia, with seven; Kibale Forest in Uganda, and Makokou, Gabon with 12 primates each. In each forest there is the gamut from small, nocturnal, solitary or pair-living creatures to large-bodied ones in large multimale troops.

Two aspects seem crucial to understanding feeding niche. These are the content of diet and its location, particularly the size of patch where it occurs. Most species can be differentiated by one of these two dimensions. However, a full description must include timing and abundance of each food species; there may be no competition over an abundant resource, but scarce foods in a lean season are limiting. Although the differences of feeding and ranging behavior make sense within each primate community, there are few safe generalizations between forests.

African primates are more likely to form polyspecific associations than those of other continents. The species that feed from the same trees are most often found together. Predator pressure, especially from birds of prey, may explain more active associations. Within such polyspecific groups at Kibale, dietary overlap correlates with interspecies aggression, and inversely with interspecies grooming.

This chapter marks the end of the third of this book concerned with ecology. We have, to some extent, explained the evolution of society. We have partially explained mixed groups of different species. In each case food supply has allowed larger groupings or forced smaller ones; predators have favored large-group vigilance or small-group secrecy.

In each case, the environment seems to determine individual behavior, as, in Marxist economics, the means of production do. In each case, the individuals seem to be acting for their own benefit, as in classical capitalism, not for the good of their group or for their mixed-species flock.

Those with biological background say, "Of course." Perhaps there are students who are not so brainwashed who are asking, is this all? Don't primates ever *wish* to help each other? Don't they make decisions? Don't they, perhaps, *like* living in groups?

To answer those questions, we must turn the same story inside out. In the biologist's terms, the environmental pressures have selected, not just for social groupings, but for individuals with likes, dislikes, and desires.

PART TWO

SOCIETY

CHAPTER 9

GENE, ORGANISM, AND SOCIETY

SOCIOBIOLOGY

Social behavior, by definition, is the interaction of individual organisms. As individuals ourselves, we take a rather parochial view that such organisms are what matter most. This view is reinforced by the European intellectual tradition of the worth of the individual mind, body, and soul.

Individuals are one level of a hierarchy of organization. We are composed of cells, which are composed of genes and other chemicals. We, in turn, make up societies, which are loosely bound together in mammals and more tightly bound and interdependent in the social insects. Our conscious minds are at the individual level, so our logical and moral judgments start from the biased viewpoint of our own rank in the hierarchy. Cancer is called bad, when the anarchic power of a group of cells destroys the individual. (The viewpoint of a cancer-causing virus might differ.) Dictators who infringe upon individual human rights are also bad in the eyes of western democrats, even if they increase the power and wealth of their state, because we do not automatically believe the state has rights as a separate entity. There are times, however, when we lay down our lives willingly for a state; most people are ambivalent about how much they would sacrifice for their society.

Sociobiology is one current school of study of multiple levels in the hierarchy, particularly the levels of gene, individual, kinfolk, and wider society (Wilson, 1975). Among the various aspects of sociobiology is the question of when the entities at each level have conflicting interests and when their interests agree. Then there is the question of mechanisms—how do genes influence individual behavior, and individuals influence groups, or vice versa? Finally, there is the question of reproductive strategy. What patterns of reproduction lead to highest numbers of reproducing young, given the constraints of group life?

A great deal of controversy centers on the question of mechanism. There are many complicated, messy steps between a single gene and any bit of individual behavior, however stereotyped. Similarly, there are many steps and feedback loops between an individual's actions and his eventual effect on society. Some have argued that the intervening mechanisms are so complex that we cannot even try to compare levels. Furthermore, they argue we should not try, because simpleminded equation of genes and behavior can be invoked to justify group prejudice or the horror of genocide.

Those who focus, instead, on the conflict or concordance of interest find sociobiology illuminating. A given type of gene that occurs in an individual is likely to occur also in his close kin. It is fairly easy to calculate how many genes are shared, on average, with any class of relative. Hence, Haldanes' famous remark, "I would lay down my life for two brothers or eight cousins." (See Fig. 9.1A, B.) Therefore, the self-interest of a gene is not the fitness of the individual in which they are temporarily housed but the inclusive fitness of that individual and his kin, shading off rapidly within the confines of his species and even across species (Hamilton, 1970). Thus, if a plant or animal has some means for recognizing close kin and treating them better than others, the genes of that organism enjoy selective advantage over others' genes, no matter what the intervening mechanism. If the causes of such recognition are even partially genetic, the ability to treat kin differently will be selected from generation to generation. It should not be surprising that tadpoles from the same batch of spawn can "recognize" and refrain from eating each other, even if reared in isolation (Blaustein and O'Hara, 1981). It is much more surprising that pigtailed macaques reared with nonkin apparently prefer to look at and approach their own paternal half-siblings than an unrelated juvenile of matched age, sex, and activity (Wu et al., 1980). Perhaps this is done by some matching to the monkey's own phenotype (Holmes and Sherman, 1983).

Inclusive fitness helps explain the evolution of aid to family members. If the individuals habitually marry nonkin, inclusive fitness does not explain much wider altruism to the group or the species. The interests of self or close kin far outweigh the advantages to more distant relatives. However, if groups inbreed to a certain extent, or if close relatives remain together, the boundaries of the group may be the same as those of the family. Thus, loyalty to a small group could evolve out of simple family loyalty. Many, though not all, primate groups are closely related matrilines, which may nearly fulfill these conditions.

The only way that altruism beyond the immediate family could profit the individual is what is variously called "reciprocal altruism" (Trivers, 1971), or "you scratch my back and I'll scratch yours," or "cast thy bread upon the waters; it shall return to thee after many days." This can be advantageous behavior in the short term if the individuals concerned recognize each other, so the first altruist gets his investment paid back. Baboon males sometimes form coalitions: one mates while another guards his back against more dominant males. The process is repeated in reverse on later occasions, a case in which reciprocal altruism leads directly to reproductive success (Packer, 1977). If such a process is to go beyond individual tit-for-tat into group loyalty, the individual's interests must be very closely bound to the group's. For this we may look either at insect societies, or move down a level to treat the organism itself as a group, while the self-replicating "individuals" are genes.

FIGURE 9.1. AUNTY MAY'S CHOICE

Aunty May's brother and his wife died, leaving two little boys. May chose to bring up her nephews and care for her aged parents. She did not marry, though one son would have carried as many of her genes, on average, as both nephews, and further children would have multiplied her reproductive success.

There are at least four immediate morals:

1. Genetic kinship falls off rapidly with distance of relationship. Each individual has half his parents' alleles on average and passes half to his own children. Each further step decreases the shared alleles by half, so two nephews are "worth" only as much as one son. A large group is not close kin unless it is inbred.

2. The calculation of reproductive value is a function of age. If Aunty May's parents are too old to bear children, and if they do not help in some way toward bringing up her sons or nephews, they are "worth" nothing. On the other hand her nephews are worth slightly more than an un-born son, for Aunty May does not incur the risks of pregnancy or lost investment in a child that dies in infancy.

3. People do not always act in their apparent genetic interest. The implications of points 1 and 2 can be explored mathematically, but the content as I have written it was intuitively obvious to Aunty May. Statistics reflect average behavior in a population; they do not produce the decisions of a particular person, who happened to be an independent-minded spinster anyway.

4. People do not like such calculations, which imply that love can be quantified.

Imagine a virus that provokes cells to divide unchecked. The virus is only slightly contagious between animals. It can multiply within an animal as the cancerous growth spreads, and between animals either by direct contagion or if the infected host is a parent who lives long enough to reproduce. The longer the host lives, the more likely the virus is to spread to other animals. Therefore, viruses that slow down their multiplication of cancer cells within the host may have selective advantage over viruses that infect more cells but die with the organism. Selection might also go the other way—for more contagiousness even at the price of more virulence. (A virus that is highly contagious but not very virulent has the greatest reproductive success of all, that is, a common cold, rather than cancer.) In the first case, which is the interesting one, a group of virus genes' "self-interest" becomes increasingly similar to that of the host organism, until it becomes indistinguishable from other pieces of nonfunctional DNA—a neutral gene now, not an anarchist. If it mutated to produce some enzyme that aided the survival and reproduction of its host, it could even aid itself by contributing to the whole—a useful member of society at last.

Reciprocal altruism at this level is what E. G. Leigh calls the "parliament of the genes." If one gene so behaves as to destroy the balance of the whole, all are destroyed. In some but not all circumstances, the rest of the genome evolves to neutralize a truly selfish and deleterious gene (Leigh, 1971, 1977).

The relevant parameters include migration rate of units between groups, relative selection rates at unit and group level, and the number of groups or time span under consideration. For most vertebrate societies, migration rates are high and selection is much harsher for individuals than for groups. All this makes group selection in vertebrates an extremely weak force compared to selection at individual or family level (Maynard Smith, 1976).

"However, if the frame of reference is shifted down one level such that . . . genes . . . are thought of as "individuals" and single cells or organisms are thought of as groups . . . in most sexual reproducers literally millions of such groups go extinct for each group formed . . . . (A mammalian) egg came from one of nearly a million oocytes that developed in the first trimester of fetal life . . . fertilized by one of several hundred million sperm per conception. . . . What selection pressures could produce the fantastic ritual that goes on when so many sperm beat their heads against the vitelline layer of an egg until the microvilli around a single sperm extend to interdigitate around that sperm's head and draw it through the plasma membrane, thereby initiating a chain reaction and the release of charged ions that so change the plasma membrane as to make it henceforth impermeable? That seems like a lot of song and dance for a random result" (Hartung, 1981). Sperm have differential success at every stage of the long migration up the female reproductive tract (Quiatt and Everett, 1981). Genes do not die by "individual selection" unless they mutate. They cannot migrate as "individuals" unless they hitchhike by viral transmission. Hartung concludes that just as group selection is weak at the level of society versus organism, so is it strong at the level of organism versus gene.

Why put all this background in a book on primate behavior? Because primate societies, in evolutionary terms, are about at the stage of protogenes drifting about a mud puddle four billion years ago. Individual units can migrate and even exist for a while on their own, although they might function better in groups. Groups won in the end; cellular life evolved. We humans are aware of conflict of interest between ourselves, our family, and our group. Our great religions attempt to spread the family interest by exhortations to treat all mankind as brothers. Sociobiology has been attacked precisely because it insists we aren't all brothers (or sisters, either).

The thought experiments of Table 9.1 may serve as a litmus test for your own attitudes toward this kind of analysis. Both in these imaginary cases, and in the true story of Aunty May, I think it helps to underline the contrast between statistical predictions over evolutionary time and the urgent individual lives summed up by the statistics.

I think it also helps to point out the argument need not be qualitative; ('tis versus 'tisn't). It is a quantitative one; how much can units migrate? How much does survival depend on self or group? If we see that our own marvelous bodies are produced by a parliament of genes evolved to work in their common interest, it is easier to see that we organisms could also work together for society's survival and take less than another billion years to get around to doing so.

LEARNING AND INSTINCT

Learning and *instinct* are loaded words. They carry overtones of predestination against free will, racism against egalitarianism, vitalism against mechanism, although it is not always clear which vice or virtue will be attributed to which.

Whatever one's persuasion, one must admit that in some real sense primates learn more than other mammals. This book largely deals with the emergence of primate intelligence—the capacity to learn and to organize learned information. We must therefore consider some meanings of learning and instinct before seeing how primate social behavior and primate intelligence are both learned and innate.

It may help to go back to a much older controversy, which science has settled (or at least outgrown). In the eighteenth century "preformation" battled "epigenesis." Preformationists believed that human shape was immutable. They saw in the sperm a minute homunculus already in his ordained pattern, ready to inflate into adulthood, for how else could human shape arise? (See Fig. 9.2.) "Their ardor for their cause was not dampened even by the absurdity . . . of the encasement concept—the implication that each miniature must in turn enclose a miniature of the next generation" (Patten, 1958), so that all succeeding generations of humanity must have been preformed one inside the other in Adam's testicles.

This is the anatomical equivalent to the naive, extremist view of instinct, a view that is not held by any modern biologist, although it has been set up occasionally by environmentalists as a straw man to attack. The opposite naiveté that man is born with his mind a blank slate, on which the environment writes and *wholly* determines his subsequent behavior, has never penetrated biology. What might be the anatomical equivalent? That the germ plasm is neutral, de-

FIGURE 9.2. *The homunculus. Drawing of a spermatozoon with a preformed human in its head. (After Patten, 1958, from Hartsocker, Essay de Dioptrique, Paris, 1694.)*

veloping according to its embryological environment? That a sea urchin egg placed in the appropriate womb would grow into a man or a kangaroo, or a human ovum dropped into high tide might produce a sea urchin?

Kaspar Friedrick Wolff, in 1759, set forth his conception of *epigenesis*, development by differentiation, in which new shapes are organized out of those that exist already, but need not resemble them. This became the foundation of modern embryology. We take it for granted that genetic information is coded in the nucleus of the fertilized egg. At each stage—the fertilized egg, the dividing egg, the embryo with primitive cell layers, the later embryo sprouting limb buds—the parts of the embryo are formed by both heredity and environment. Which one determines that the limb bud shall cleave into fingers rather than toes? Both, for if the buds are grafted on other parts of the body at an early stage, organs will grow appropriate to the part of the body where they are transplanted, that is, appropriate to their environment. At a slightly later stage, the pattern of each cell lineage is fixed, so that a transplanted forelimb becomes a forelimb, even though it sticks out from the abdomen. Or is this a fair statement? Perhaps one should say that a forelimb bud left in its proper place learns from its environment to become a forelimb, and transplanted too late cannot give up the forelimb habit (Patten, 1958).

In embryology, it seems that the inherited codes set the most general pattern first: the cephalocaudal axis and division into neural tissue, muscle-bone tissue, and visceral tissue. Then progressively finer structures differentiate, rudiments of each organ, and at last the details within each organ differentiate. At each stage there is enormous redundancy of control, so that minor differences in growth do not upset the functional patterns. It is hardly fair to trace the coiling of each kidney tubule back directly to the action of genes in the egg. The tubules coil within the general pattern set by their "environment," the kidney.

The important question, then, is, "How much latitude for variation has a given organ at any stage in life?" At the earliest stages the cells are not even fixed by organ, and later on by organ but not by detail of organ. Throughout life they retain some responsiveness. Muscles strengthen with exercise, and viscera hypertrophy under demands for more adrenalin or increased urine flow. The digestive system lengthens under a high vegetable diet, shortens under a meat diet (Hladik, personal comment).

Inherited or *environmentally determined* are *relative* terms. They indicate the *degree of variation* that related organs, animals, and behavior patterns will show in different environments. If all the members of a species, or of an inbred strain, show a single characteristic in spite of very different environmental treatment, the characteristic is relatively innate. If, however, there is great variability under different treatments, one can say that the trait is relatively more determined by environment.

Marler and Hamilton (1966) stated, "When a geneticist speaks of an inherited trait he refers not to a characteristic of one individual but to the difference between two individuals or groups of individuals, or populations. . . . The method of exploring the relative contribution of these factors to variation in the population is to hold either genotype or environment constant, and observe the effect of systematically varying the other in experimental populations. . . . At no point is the inference drawn that a particular trait in a given individual is inherited; rather, a certain difference between the traits of the two individuals is shown to be inherited."

Thus, there are three fundamental points to defining environmental or innate control:

1. Anatomical and behavioral traits are both learned and innate: biological organisms interact with their environment at every level, at every moment of time.
2. Learning and instinct must be considered as unfolding epigenetically from stage to stage. An adult animal may urine-mark its territory, with a complex behavior sequence in some sense innate; yet this is a final flourish in the use of the excretory and nervous systems that are themselves a product of genetic control and environmental fields within and surrounding the embryo. Thus, to speak of learned or innate control, one must specify the level of detail and the stage of development concerned.
3. An individual does not just have innate traits. Rather, an innate trait appears the same in several genetically similar individuals reared in different environments, or it differs in several individuals of different genetic stocks raised in identical environments. The term *innate* is relative. There are always ways to change the environment enough to modify even highly innate behavior. The relative innateness of a trait thus depends on the range of environmental conditions sufficient and necessary for the behavior to appear.

DIFFICULTIES IN RIGOROUS MEASUREMENT

Two common procedures in studying learned and innate behavior are *deprivation experiments* and *teaching experiments*.

Deprivation experiments remove apparently relevant factors from the growing animal's environment. If the animal then responds like its normal conspecifics on first presentation of a stimulus, one can say that the response is relatively innate. For instance, male stickleback fish have red bellies when courting. Males that have never seen a red object or another fish will respond when they first see a crude, redbellied fish model. They attack it as they would attack a rival male, so both their perceptual recognition and their attacking motor patterns are innate, at least with respect to fish shapes and red things.

There are many methodological difficulties with the deprivation experiments, which Lorenz (1965) lists at the end of *Evolution and Modification of Behavior*.

First, only positive results count. If the fish does *not* attack the model rival, the experimenter may have simply used an inadequate experimental technique or general bad rearing that blocked other, more widespread systems than the particular system one was trying to test.

Second, the experimenter must be thoroughly familiar with the animal's normal behavior. This allows him to recognize disjointed components of larger behavior patterns. It also helps avoid general bad rearing.

Third, if a deprivation experiment is testing sensory recognition, it needs as many different animals as there are sensory cues to test. Animals are commonly biased toward learning biologically important information very rapidly, so if one tests a particular animal a second time, it may really respond to what it learned on the first test. On the motor side, however, the animal may show in a single test that it "knows" a whole series of action patterns.

Fourth, the test must offer an adequate set of stimuli to release the behavior. It is no use examining innate nest-building patterns in rats unless the rats have a familiar sheltered area to build in. Even wild-reared, experienced mother rats will not nest if you dump them into a strange, bare cage.

Fifth, the experimental and control animals must have genotypes as close as possible. Comparisons across species are likely to be nonsense.

Skinner, talking about the *teaching experiment*, again stresses difficulties of technique—the need for controlling and recording the exact sequence of stimuli and reinforcements. He shapes learning in the rat or pigeon in his Skinner box by rewarding successive approximations to the response desired, then trains by a controlled schedule of rewards. The results of even this highly simplified system can seem complex. When a pigeon responds to several intermittently flashing lights with an occasional series of pecks, interrupted by the random delivery of a pellet of grain, an intelligent lay observer might well be baffled by what was going on. Only the cumulative record of schedule, response, reward would illuminate the pigeon's (or the experimenter's) performance. How much more complicated then is human learning, with no Skinner box to cut off irrelevant input and no wires leading to a set of pen recorders that note exactly what was done to us and what we have done in reply.

It is paradoxical that all Lorenz' technical warnings concern keeping the environment rich enough for the experimental animal to behave normally. Lorenz' ideal is microdeprivation—to remove only the relevant experience being tested, leaving the animal normal in all other respects. Skinner, in contrast, prefers to deprive the animal of every experience except the particular item being taught; Skinner's "teaching experiments" involve far more deprivation than Lorenz' "deprivation experiments." This is leading perhaps into the metapsychology of scientific theories rather than the psychology of animals.

Some examples may indicate the complexity of methodological problems. One is the series of experiments begun by Harry F. Harlow on rhesus macaques that were raised from infancy without social contact. The early experiments showed that such macaques were sexually inept, did not know how to rear their own infants, and were aggressive (sometimes suicidally aggressive) to members of their own species. However, the bad rearing led to such massive disruption of response to other animals that one could not say which small cues or actions of sex or motherhood might still be relatively innate. In contrast, a positive result is convincing, however unexpected; Wu and colleagues (1980) raised pigtail macaques with other unrelated juveniles. They were then offered the choice of approaching one or two other monkeys or an empty cage. One of the monkeys was the experimental animals' own paternal half-sib, the other an unrelated animal matched with the half-sib for age, sex, and mobility. The experimental animals had never seen their siblings, and did not even share a uterine environment with them, just a quarter of their genes on average; yet, they preferred to watch and approach their own kin rather than unrelated strangers. There seems to be an inborn means of kin recognition and a bias toward approaching kin. In a normal social environment this bias would then be reinforced by all a primate's social learning of its own kin.

How would one determine whether a baby's smile is innate? A baby gives full social smiling, with eye-to-eye contact, sometime after six weeks of age. On the motor side, then, one must allow no practice in smiling for at least six weeks. It would not be enough to feed the baby intravenously to avoid windy smiles, for babies also smile at changes in the environment, including human voices. Much better to reversibly paralyze the facial nerve. Then, when the paralysis is lifted, if the baby smiles, the pattern is innate. However, if the baby does not smile, one is faced with Lorenz' first caution: negative results do not count. Perhaps the nerve was simply damaged; perhaps the child has learned that attempts to smile do not work (that is, bring no response from adults); perhaps one has a more subtle, generalized effect of bad rearing; a child subjected to such an experiment might not feel like smiling.

On the sensory side, the child should be deprived not only of seeing smiles but of seeing faces, which seem to be a cue for full social smiling. Reversible blindness for six weeks would seem to be the answer here, with the same queries if the child, afterward, would not smile. The alternative of never seeing a human being might imply no cuddling and no stimulation, which is certainly bad rearing for a small baby.

This is the sort of experiment that would give conclusive answers. Perhaps it suggests how brutal is much research on species for which we have less sympathy than human babies. It also shows how bad rearing, which shocks us sentimentally, may also destroy the conclusions scientifically.

Of course, there are ways to approach such certainty. Blind babies not only smile at voices but turn their eyes toward the voice—eye-to-eye contact independent of sensory feedback. Men of all races and cultures smile in the same way, which suggests that the smile is, in fact, highly innate. However, without the possibility of stringent proof, it may be that all normal humans also learn social smiling in early infancy, reinforced by their mothers' delight in the first windy grins, just as all normal mallard ducks learn in the first hours of life that a duck, not Konrad Lorenz, is their mother (Eibl-Eibesfeldt, 1972, 1973; Pitcairn and Eibl-Eibesfeldt, 1976).

CHARACTERISTICS OF LEARNED AND INNATE BEHAVIOR

Although it is only by experiment that one can *prove* a behavioral item depends, or does not depend, on a particular range of previous experience, it is much easier in practice to recognize behavior that is *probably* at the innate or learned end of the spectrum.

Highly innate behavior is usually: (1) Widespread or universal throughout the species. (2) Stereotyped in pattern on the motor side. (3) Responsive to highly simplified models containing only a few cues of the total normal situation, on the perceptual side.

If the behavioral item is a social recognition or response, both the motor act and the perceptual cues may be exaggerated or ritualized. That is, there may be selection on both sender and receiver for an unambiguous message, a situation strikingly different from the usual background, thus, bright colors or repetitive, showy courtship dances. (Ritualization will be discussed in Chapter 10, "Communication").

In contrast, largely learned behavior is characterized by *variability* (1) among individuals of a species; (2) in motor pattern; and (3) in types and possible multiplicity or complexity of perceptual cues.

Primates depend on learning more than do any other mammals; their specialty as an order is complex and variable behavior. On the other hand, ethological theory has largely concerned the more fixed responses of fish and birds. (But see Ewer, 1968, for discussion of innate behavior patterns in mammals.) Ethology has much to say about primate communication—the expression of the emotions. It says something about emotion itself, in that it assumes there is a biological basis for and selective processes at work on sexual behavior, threat behavior, and so forth. However, the social relations in a primate troop are largely learned, some inevitably learned from any mother of one's own species, and some uniquely learned from the particular individuals that surround each primate child.

A large part of the fascination in studying primates is to see how an order of animals has evolved such dependence on learning.

SUMMARY

Sociobiology is the study of conflict or concordance of interest between gene, organism, and society. Each level consists of units that organize to create larger groups. Natural selection can operate at any level of organization, but seems stronger at the organism level, not the levels of single gene or social group. This means that genes commonly have the greatest chance of survival by cooperating to produce viable organisms. Mammals' self-interest (and that of close kin) usually outweighs selection for survival of larger community groups, because we have more ability to migrate between communities than genes between genomes. Weaker selection exists for survival of both "selfish" DNA and social groups of organisms, which is one factor in our own ambivalence toward conflict of interest between ourselves, our family, and our community.

I find such analysis illuminating because I see no logical contradiction between reductionist explanations of behavior in terms of genome and prior experience and structuralist explanations in terms of consciousness and purpose. Many other authors believe that choosing one type of explanation must exclude the other. Embryologists have outgrown similar agony over the origin of anatomical structure and now agree that structure can be analyzed but still exists.

One mistake of reductionists (especially as caricatured by structuralists) is overemphasizing either the inherited or experiential causes of structure. Anatomical and behavioral traits are *both* learned and innate. The relative importance of learning or instinct depends on the range of environmental conditions sufficient and necessary for a behavioral trait to appear. If it appears in animals reared under a wide range of conditions, including deprivation of apparently relevant experiences, the trait is relatively innate; if it needs a particular, very narrow range of prior environmental conditions, it is relatively learned. Experimental proof is technically difficult. However, relatively innate traits are often (1) widespread throughout a species, (2) stereotyped in motor pattern, and (3) responsive to highly simplified perceptual models. Relatively learned traits tend to be variable (1) among individuals, (2) in motor pattern, and (3) in sensory cues.

THE BURNING NURSERIES

A thought experiment:

Arrange 300 nurseries to burn with equal ferocity. In each burning nursery lay a four-month-old baby. Test three matched groups of 100 mothers each.

A. Twenty year-olds with their first and only child.
B. Forty year-olds with their first and only child.
C. Forty year-olds with their fourth and youngest child.

Do you expect more mothers of Group A, B, or C to rush into the flames in an attempt to save their children?

A second thought experiment:

In your first experiment, suppose 60 group B mothers rushed into the flames, 50 group C mothers, and only 40 group A mothers. Equal proportions were killed, which suggests you adjusted your methodology (fires) equally. However, the group differences were not statistically significant. How do you slant your write-up of the conclusions?

a. The experiment should be repeated with a larger sample. This is the intellectually honest course.
b. The lack of significance suggests sociobiological calculations are irrelevant, unethical, and untrue.
c. The results vary in the expected direction. This suggests sociobiological considerations of reproductive value may be important for predicting behavior.

CHAPTER
10

COMMUNICATION

We rarely tell a person "I like you," and hardly ever, "You bore me." We rarely need to say it; a deep look in the eyes or a glance away, a step, or a shift of the shoulders is enough. Saying the phrase would suddenly raise the situation to a different level of intensity. At still higher intensity of love or anger, speech again becomes inadequate: we physically caress or hit each other (Argyle, 1967).

Most personal communication is nonverbal. Although we value love sonnets and love letters, courtship can do without them. In fact, the content of conversation may be quite irrelevant to the relationship between the speakers; it means only "Please pay attention to me for I am noticing you," which is what van Hooff (1967) calls "grooming talk."

There is much debate whether primate gestures can refer to external facts in the environment, as words do. This will be discussed at length in Chapter 20 on language. To leap to the conclusions, it seems clear (to me) that vervet alarm calls, chimpanzee directions for finding hidden food, and the pant-hoots of wild chimpanzees at a food source all direct conspecifics' actions toward their environment, in many ways like true language.

But nonverbal communication in humans retains at least three advantages over language. First, it is trustworthy. It is largely unconscious, and so we find it difficult to lie or be deceived with gesture—great actors are rare. Then, it is capable of infinite subtlety. It is an analog process, sliding smoothly from one form to another, usually not in discrete steps, so any nuance is possible. Finally, it is fast. The eyebrow flash or flicker of gaze is perceived and answered in kind before a word is said. We can attribute the same advantages to nonverbal communication in other primates; we use such nonverbal "intuition" to communicate with them. The external semantic meanings of vervet alarm calls are exceptional, in part because they refer to gross categories of danger that demand grossly contradictory responses for survival, so we can elucidate them by heroic experiment. The meaning of a mere social noise, in its context, might be, "I could grow interested if you give me a bit more encouragement and he keeps looking the other direction. . ." which we for some reason think is simpler than an alarm call meaning "run for cover."

SIGNAL, MOTIVATION, MEANING, AND FUNCTION

An act of communication has four aspects, which can be called *signal, motivation, meaning,* and *function* (W. J. Smith, 1968, 1977).

The *signal* is the form of the act: the squawk or stare or stink itself. The signal can be photographed or tape-recorded or gas-chromatographed. One may have difficulty in choosing the size of unit to treat as a single signal—what is a whole and what merely a component—and the observer's estimate of meaning will determine the size of his units. However, the signal can, in principle, be described in terms of form alone.

Motivation refers to the feelings of the animal sending out the signal. We ascribe such feelings according to unequivocal acts that sometimes occur with the message. If the animal has been known to shriek and run away, or to growl and bite, we may call the shriek fearful and the growl aggressive. Furthermore, one can extrapolate from a shriek to a squeal, or from a growl to a low growl, and say that quite probably signals of common form have similar motivations. Thus, the squeal may indicate some degree of fear, the low growl a degree of aggression. One can also extrapolate to some extent between closely related species. If a related species growls but never bites, quite possibly the growl originated in aggressive situations, although it may or may not still indicate aggressive feelings. Although internal states (or feelings) are presumed to cause the signals, empirical data come from the statistical correlations between a given display and the likelihood of unequivocal attack or fleeing. One can also measure relation to physiological state: the feeling of hunger can be equated to hours of deprivation of food or to blood sugar level. This is clearly easier for hunger than for such feelings as dislike or love.

Traditionally, the three big motivational categories for social signals are fear, aggression, and sex. However, these three axes correlate only very loosely with the uses of many signals. Alternatively, one can pick motivations that correspond to every usage of what seem to be unit signals. These motivations turn out to be very general, such as tendency to locomotion. Factor analysis can contribute here, but at least some of the difference seems to depend on the size of component chosen as the signal.

Meaning of a message one determines from the reactions of the other animal. This depends on the total context. Thus, a direct stare may often mean a threat, but the threatened animal will respond to the stare with a whole range of behavior, on the basis of its assessment of the sender's eyebrows, ears, nose, stance, age, sex, behavior in the previous minutes, lifetime history, and spatial position, as well as whether the receiver has hay fever that puts him in a foul temper anyhow. Messages out of context are often so vague as to be meaningless (Marler, 1965; W. J. Smith, 1977).

Finally, *function* is the evolutionary advantage of giving a particular signal with an associated range of meanings.

Many of our common descriptions of communication imply all four aspects together. For instance, alarm bark carries connotations of the sender's alarm, the receiver's alarm, and the evolutionary function of warning the kin group, as well as some indication cf the signal form. The difficulty is that the four are rarely in one-to-one relation. For instance, lemurs scream (*signal*) when seized, or when they see a flying hawk, or when another lemur screams; the *motivation* may be high-intensity shock, or sudden terror. However, the *meaning* is much more precise: "take cover." The *function*, finally, is as an air-raid alarm, saving close kin from attack by birds of prey.

EVOLUTION OF DISPLAYS

Some social communications are rigidly stereotyped; some vary through many subtle shades. It may be important to convey particular information by a clear, unambiguous message, even though the information is thus simplified. Other situations require, or allow, gradations and modifications of meaning.

A communicative display evolves through recognizable steps. It begins as a by-product or single aspect of otherwise useful behavior. Later, it may become ritualized if it is under selective pressure in its own right for more effective communication. Some ritualized behavior may then be used in new situations with a concurrent shift of meaning. Later still, it may be the fate of many displays to drop out of the repertoire (Fig. 10.1.), and be replaced by new displays that convey the meaning more strongly (Lorenz 1935, 1965; Tinbergen, 1951; Moynihan, 1970).

Moynihan (1970) observed that virtually all "classically trained" ethologists report that the repertoire of vertebrate species include 15 to 35 major displays. "Crude and approximate though these figures may be . . . these are comparatively narrow limits. No species has been found to have only two or three displays of this type. Nor has any species been found to have as many as 40 or 45."

Moynihan concludes that the number of stereotyped signals that an animal can usefully receive is limited, either by its brain capacity or by its surprise and delay in responding to a very rare signal. Marler (1976) demurs, "Our ignorance of the communicative significance of both within-category and between-category vocal variations . . . renders problematical all estimates of the size of the repertoire." One difference is that Moynihan is talking about major, stereotyped displays, Marler about the total continuum of communication, pointing out that it is always possible that animals lump or divide differently than their observers.

Why and how does a display become stereotyped? The origins of displays lie in events that occur at moments of high emotion. However, they cannot interfere with other urgent needs of the situation, such as speedy locomotion. There are several groups of such auxiliary gestures, which provide the origins of most displays: autonomic responses, intention movements, and displacement activities.

Autonomic responses are produced by the sympathetic and parasympathetic nervous systems. Sympathetic discharge readies the organism for fight or flight. It shunts blood from the skin and internal organs to the brain and muscles in

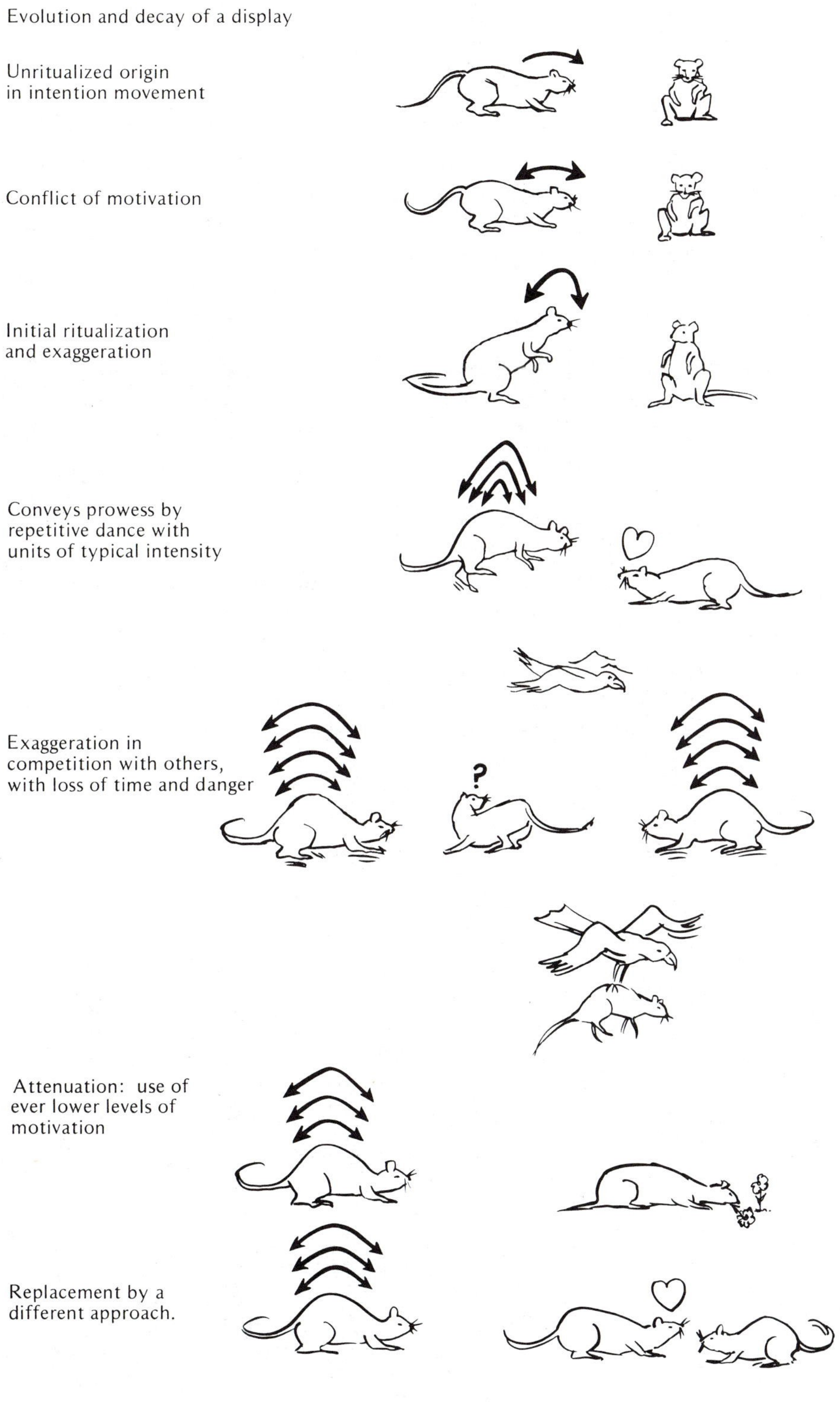

Evolution and decay of a display

Unritualized origin
in intention movement

Conflict of motivation

Initial ritualization
and exaggeration

Conveys prowess by
repetitive dance with
units of typical intensity

Exaggeration in
competition with others,
with loss of time and danger

Attenuation: use of
ever lower levels of
motivation

Replacement by a
different approach.

FIGURE 10.1. *Evolution and decay of a display.*

195   *Communication*

readiness for physical exertion. Thus, one may go white with fury, or its opposite, parasympathetic and pink, bristling haired or sweaty palmed. Darwin (1872) called this the "direct action of the nervous system."

A second common origin for displays is from *intention movements*. A slight start, a bobbing as though about to leap, may indicate the intention of movement. One of the most widespread mammalian and primate signs of dominance or of threat is erect, stiff-legged posture while staring at the adversary or subordinate. It is easy to see how this could have originated from the first movements of attack. These are what Darwin called "serviceable associated habits." Facial expressions made with vocalization, such as the O-shaped mouth with loud howling, might fall in the same class. One major set of movements is the protective responses, in which an animal draws back the mouth corners and shakes its head to dislodge noxious food or a bad smell. A whole battery of primate fear grins may be traced back to the protective responses. These are not so much the intention of movement as the movement itself, taken from a general context into a social one (Andrew, 1963b).

A third group of displays comes from *displacement movements*, apparently irrelevant activities that appear when an animal is caught in a situation of conflict. Thus, a blackheaded gull, confronting a strange female, may neither fight nor court her, but break off from the beginning of either to preen itself or pull up a few blades of grass. A student facing examination drums his fingers on the table or twists his hair. A lecturer confronting television cameras in his classroom rhythmically headflags between the staring lenses and the reassuring chart on the blackboard behind him. These displacement gestures can become the basis for communication to the blackheaded gullesse, the examiner, and the unseen audience.

Although pure intention movements, autonomic expressions, and so forth may communicate emotion, we do not call them *displays* until they have reached the second stage, that of ritualization (Huxley, 1914). Ritualization, in the biological sense, means that an item of behavior has been subject to selection that has increased its communicative value, in particular, that has made it less ambiguous and more easily interpreted.

Ritualized behavior can be recognized by just those properties that resemble human ritual. First, it is exaggerated. The original flushing becomes brighter; the slight bob to spring becomes flinging the head in a sharp jerk. Second, it is stereotyped and stylized. A gesture is less ambiguous if it is always performed in a single form, at a single level of intensity (Morris, 1957). Third, it may often be repeated. Repetition not only reinforces the message if it was not received the first time but also takes over the function of communicating intensity. In short, just as one would expect to recognize ritual if dropped into an unknown New Guinea tribe, so one would expect to recognize display in an unknown species of bird.

The selective pressures that lead to ritualization may be pressures either to reveal or to conceal motivation. Sometimes unambiguous information about motivation can favor the genes of both sender and receiver as in signals that mean, "Take cover, kinfolk," or "Help, Daddy!" or "Sex, handsome?" Sometimes, instead, an unambiguous signal form disguises ambiguity of motivation. If two males contest a female or a nest site, they would not gain by signaling their opponent, "If you hold out two more minutes, I'll give up." They gain by giving a constant threat signal until they give up at an unpredictable moment.

The familiar defensive threat—the snarl of the cornered cur—is used when an animal communicates its desire to run away but still serves notice it will bite if pursued too closely. The ritualized interanimal contests for a prize may not show this defensive component but consist instead of threat in standard form until one party gives up or escalates (Maynard Smith and Price, 1976; Dawkins and Krebs, 1978).

Many threats relate to the body size and fighting ability of the contestants. Red deer stags compete for harems by half-hour-long roaring duels. Roaring pitch and tempo correlate with size and stamina, but escalated contests are rare because of the risk of injury and because subordinates (called sneaks) steal matings during prolonged contests. The two males roar with gradually increasing tempo, paralleling each other's performance, until one suddenly gives up (Clutton-Brock et al., 1982; Dawkins and Krebs, 1978).

The catch in such rituals is that the signal may be faked. A gene that lets an animal grow or bush out its hair increases apparent size. Large toads give deeper croaks than small ones; a deep voice in most vertebrates means large reasonating chambers, and thus large size. Cold also lowers the toad's pitch, so small toads dive to the chilly bottom of the pond and emerge with their croaks artificially deepened. As Katherine Whitehorn says, many a female thinks she has found a large warm toad and finds only a small cold toad beside her on the pillow next morning. Escalation calls the bluff and checks faking. In the long term, then, such faked displays could slowly be replaced by new signals that are more reliable.

This brings us to the decay of displays. As individuals compete, a display might become rarer if it is replaced by others until it is too rare to be subject to selective pressure, as cave fishes' eyes disappear because there is no continuing selection for vision. A much commoner fate for a display in active use may be that it becomes exaggerated or overfrequent under the pressure of individual competition. A mutation that increases the hair bristling, or perhaps the size of laryngeal sacs, or the tempo of a roar exaggerates the display without increasing real fighting or sexual prowess. This may be selected in spite of escalation testing the process until counterpressures (especially from predators) eliminate the most exaggerated individuals and give advantage to some other display with the same function. Another form of exaggeration is to give the display more frequently, at lower levels of motivation. Then, a new and different means of making a strong statement might undercut such bragging (Moynihan, 1970; Smith, 1977).

The parallels with human, learned gestures and phrases are obvious, whether it is the latest scientific catchword or the popular forms of dancing. Many of the pressures are the same: to be noticed in the initial stages, then to communicate unambiguously, then to communicate more strongly, and finally, a communication so overstated and overused it loses meaning in competition with the next craze. These cycles are also well known in fine art and music and have similar parameters of formulation, formality, baroque exaggeration, and finally dissolution.

COMPONENTS, CONTEXTS, AND GRADATIONS

At first sight, the units of communication seem naturally given. Most vocalizations have a beginning and an end; most facial expressions appear as transitory changes, with a relaxed face before and after. We assume that choosing the unit is largely a technical problem, of making sure what the senses of the animal record, and recording likewise.

The technical problem exists, of course. Bushbabies apparently hear ultrasonics; cebus monkeys may be red-green color-blind, and nobody knows what primates do or do not smell.

However, this is a still more important logical problem. Almost any unit that one chooses has subcomponents that vary somewhat independently, and in turn is itself a component of larger units, in which it will contribute to different meanings. In this sense, speculations about the size of the communicative repertoire have no meaning. A primate may integrate its lifetime experiences in the response to any given act.

Earlier ethologists were saved from complexity by concentrating on rigidly stereotyped displays. Thus, one could begin with a behavioral repertoire of reasonably distinct signals, analyze the situations in which each display occurred, and then reach conclusions about the underlying motivation.

The striking aspect of primate displays is that few are clear-cut: most grade into each other. As Rowell (1962) pointed out, the agonistic noises of rhesus monkeys form a continuum. Although one picks out certain modal barks and grunts that are given most frequently, there are intermediates between each of the main types. Marler goes further to say that in wild chimpanzees *every* call is part of a continuous series. Table 10.1 shows the relative proportions of

TABLE 10.1. Intermediate and Modal Chimpanzee Vocalizations

Call Category	A (Intermediates)	B (Typical)	B/A (Variability)
1 Waa bark (most variable)	92	84	0.9
2 Scream	173	216	1.25
3 Bark	54	95	1.76
4 Grunt	25	43	1.79
5 Squeak	146	264	1.81
6 Wraaa	7	18	2.57
7 Whimper	54	175	3.24
8 Pant-grunt	64	285	4.45
9 Pant	13	81	6.23
10 Rough grunt	7	83	11.86
11 Pant-hoot	52	648	12.0
12 Cough	0	33	33
13 Laughter (least variable)	0	271	271
Totals	343($\times$2) = 686	2,313	6.7

Categories are listed in order of decreasing variability.

(After Marler, 1976.)

chimpanzee calls that Marler could assign to the main, modal types and those that were clearly intermediates (Marler, 1976).

Similarly, primate facial expressions are appalling to classify. Some major attempts are those of van Hooff (1967), Grant (1969), and Blurton-Jones (1972). Van Hooff (1967) speaks of compound expressions, such as the "bared-teeth face," with specifications of what eyes and eyebrows do at the same time. Grant categorizes mouth and eyes separately; Blurton-Jones takes it down to what each lip is doing.

The size of components in a continuum is crucial, both for studies of motivation and for comparing species.

On motivation, let me paraphrase Blurton-Jones (1972), who takes the example of brow raising in children to illustrate some differences of theory. Children raise their eyebrows in conversation (especially with adults), when they are surprised and when they are fleeing. One can classically interpret conversational and surprised brow raising as indicating a slight tendency to flee, or else say that they are evolved greeting displays that have lost their original fleeing motivation. Adults of many or all human cultures flash eyebrows in greeting (Eibl-Eibesfeldt, 1972). On the other hand, Andrew (1963a) looks for what is common to the various situations, and might in this case turn to Darwin's explanation that brow raising allows one to scan a wide visual environment. The Darwinian hypothesis is very attractive. However, if a larger unit had been taken—the wide-open mouth and raised eyebrows of the child running away— this would be a "fear face."

Both gestures and vocalizations can thus combine several parts, which stand alone in other situations. Marler makes the fundamental point that such combinations are phonological, not lexical, in known situations for both mammals and birds. That is, the combination has a new meaning that generally bears little or no relation to the meanings of its parts. For instance, "the 'snort,' introducing roaring of a male black-and-white colobus is used separately as an alarm call. The introductory notes of a pant-hooting chimpanzee are often similar to a 'whimper' call which, used separately, is a mild alarm signal" (Marler, 1977). When we try to teach an animal human language, we concentrate on combining signs to give new, combined meanings—"open key," "water bird." Nonverbal communication does not work by adding one gesture on to another in a digital string.

It works, however, by shading, or transitions from one gesture to another, as analogs for intermediate emotions. In comparing species, the importance of using small components lies in the fact that what one chooses as modal or related groups of displays for a particular species are often intermediates for another species. For example, we would probably pick "crying" or "cry face" as one of our modal human expressions. Chimpanzees make cry faces, but as an intermediate transitional expression when they are changing from their "whimper face" to their "bared-teeth scream face." Thus, our modal crying is their transitional form, whereas their modal whimper we would more likely consider transitional in ourselves. The same applies to the vocalizations associated with the faces, although the noises must be measured and analyzed separately, not just assumed to correlate with particular facial expressions.

A second example is the smile. In old-world monkeys, smiles or grimaces frequently grade into fearful screaming; in the chimpanzee they are related to bared-teeth screaming but also have a greeting and reassuring function, whereas in ourselves smiles more often grade into laughter or play faces. Thus, the same continuum would be grouped differently for man and monkey. Describing subcomponents in relation to overall context is crucial for what seems at first glance even such obvious groupings as smiling with laughter.

Compound or modal expressions are extremely useful. If the observer has overcome his technical problems and managed to record a reasonably natural sample of behavior, modal expressions can tell him the communicative biology of the species: what it is important to communicate clearly. This can only be determined by the reactions of members of the same species. For example, Japanese macaques give at least seven different variants of "coos," the soft macaque vocalization of contact and affiliation.

When Steven Green (1975, Green and Marler, 1976) recorded coos in the field, he believed them to be a single class of sound, all much alike to the human ear. On his return, he tried dividing the recorded calls into subtype, as shown on sound spectrographs, and correlating them with social situation (Table 10.2). There was no one-to-one correspondence, but different variants are usual in different situations, as the statistics show. Still later laboratory tests showed that Japanese macaques (even lab-reared, long-isolated ones) easily distinguish the coo variants, whereas stumptailed and rhesus macaques find them as baffling as humans do (Zoloth et al., 1979; Beecher et al., 1979). Still later tests have showed that Japanese macaques distinguish early and late peak coos categorically, not as a graded series. Petersen (1982) synthesized a series of coos by computer, with the peaks progressively shifted. At one point, Japanese macaques abruptly switched from classifying the coos as early to classifying them as late, whereas a rhesus and a stumptail were ambivalent and confused by the same task. This sort of categorical perception is what we also use for speech consonants: we (and several other mammals) perceive a stepwise jump from b to p in a computer-generated series which in fact moves smoothly from one consonant to the other. The Japanese macaques perform more accurately when the coo signals are played to right ear, not left. This suggests they are processing this fine distinction on the left side of their brain, as we process speech.

Other primate call series are being analyzed by this linguistic approach, particularly the trills of pygmy marmosets and the chirps of cotton-top tamarins. Again there are variants which are used in different situations, and by animals at different distances. Individuals may call in predictable sequence (Snowdon, 1982).

It seems that these units are stereotyped more by fine-tuned perceptual mechanisms than by the overt ritualization of classical displays. This gives us some clues to the origins of the mechanisms of listening to speech.

We still need intelligent empathy to draw up the catalog of behavior. Unless we have guessed appropriately what to quantify, quantifications don't help. If we start looking in the right direction, the quantifications give us a precise notion of what is happening, in analysis of the form of the message, in apparent motivation of emitter, and in meaning as judged from the recipient.

Then, armed with quantitative correlations, we can return to qualitative speculation. The first type of coo, the "double low," is given by solitary males struggling after the troop in the mating season. As the troop would disappear

TABLE 10.2

Code Type		Distinguishing criteria			Other feature–
	Name	Midpoint pitch	Position of highest peak	Duration	
⌒⌒	Double	≤510 Hz	N.A.	N.A.	Two overlapping harmonic series
⌒	Long low	≤510 Hz	N.A.	≥0.20 sec.	N.A.
⌃	Short low	≤590 Hz	≠ 1	≤0.19 sec.	N.A.
⌒	Smooth early high	≥520 Hz	<2/3	N.A.	No dip
⌒⌒	Dip early high	≥520 Hz	<2/3	N.A.	Dip
⌒⌒	Dip late high	≥520 Hz	≥2/3	N.A.	Dip
⌒	Smooth late high	≥520 Hz	≥2/3	N.A.	No dip

Type of coo vocalization

	Low			High — Early		High — Late		
Situation	Double	Long	Short	Smooth	Dip	Dip	Smooth	
Separated male	XXXX XXXXXXX XXXXXXX		XX	XX		XX	XX	26
Female minus infant	XXXX	XXX						7
Non–consorting female	XX	XXXXXX	X					9
Female at young		XXX XXXXXXX	X					11
Dominant at subordinate			XXX XXXXXXX			XX		12
Young alone				X XXXXXX XXXXXX	XXXXXXX	X		23
Dispersal				X XXXXXXX	XX	XXXXX	XX	17
Young to mother				XXXX	XXX XXXXXXX	XXXX	X XXXXXXX	26
Subordinate to dominant				XXX	XXXX XXXXXXX	XXX XXXXXXX XXXXXXX	XXXXX	43
Oestrous female					X	XXXX XXXXXXX XX XXXXX	XXXXX XXXXXXX XXXXXXX XXXXXXX XXXXXXX	52
	24	19	14	32	31	56	50	

201 *Communication*

from view, a male sat and called double lows after them, or twice when a trailing female briefly answered, he called to her alternating calling and masturbation. The other situation when Green recorded double lows was mothers vocalizing to the corpse of a dead infant. Both mothers and males moved lethargically and called at high intensity to unresponding partners. Does one then look for a common motivation? Or does one say the motivations (including the hormonal state of the caller) and the meanings in the troop context are so different that these are probably still two different kinds of cooing for a Japanese macaque?

OLFACTORY COMMUNICATION

Each of the different modes of communication has its own advantages. Perfumes spread in all directions (Fig. 10.2). If rubbed on a branch, they persist in space and time. "The scent mark in olfactory communication is what writing is to language; the message may be conveyed to the receiver long after its emission" (Schilling, 1979). Scents often reveal an animal's underlying physiological and sexual state; relatively few indicate transient changes of mood like the skunk's stink or the sweat of human fear. Odor is the groundswell of mammalian communication (Keverne, 1980).

An odor, like vocal and visual signals, may have many components. Mammals can usually distinguish species, sex, and individual by odor. This is true of brown lemurs and marmosets (Epple, 1975; Harrington, 1977, 1979). It is at least partly true of humans. Twenty-five German couples were each given nightshirts that they wore for seven consecutive nights. Then, in subgroups of

FIGURE 10.2. *Olfactory communication: the ringtailed lemur male scent marks his tail with glandular spurs on his wrists, then shivers it at opponents like a perfumed feather duster. (A. Jolly.)*

ten, they were asked to judge which shirt had their own odor, which their partner's, and which ones were male or female. They also rated the smelly shirts as pleasant or unpleasant.

About 80 per cent of the people could distinguish their own or their partner's odors. (Many mistakes went with picking out a "pleasant" odor as one's own or one's mate's.) Eleven of the women and five men accurately distinguished male from female shirts among the couples who had been asked to wash with special scentless soap and to use no perfumes or deodorants. In a second group of 25 couples who kept to their normal hygiene, the same number could identify their own and their partners' shirts, but only two women and one man could distinguish male from female. Although commercial perfume advertisements emphasize sexual attraction, it may be that in our crowded society we minimize the more personal cues that broadcast our sex to strangers (Schleidt, 1980).

Fascinating as this experiment may be, it also illustrates many of the problems that have beset experiments on scent communication—small number of individuals tested, highly variable responses, an unknown amount of different past experience, and a general conclusion that is a thought provided by the tests but hardly proved by them. These experimental problems combine with a more fundamental one: primate scents do not function like insect pheromones that switch others' behavior on and off. Primate scents, like other primate signals, seem to influence rather than control the receivers' actions.

The difference between such influence and control pervades the controversy over vaginal pheromones. Michael and his coworkers have shown that rhesus (and human) vaginal secretions change over the menstrual cycle. The fatty acids produced near ovulation attract the rhesus male, greatly increasing his responsiveness to female solicitation. Natural or synthetic mixtures of similar compounds applied directly to the female's sex skin have much the same effect (Michael and Keverne, 1968; Michael and Bonsall, 1977; Michael et al., 1977). Goldfoot and his co-workers, however, could not reproduce the critical factors that led Michael to call vaginal fatty acids a "primate pheromone." Anosmic males are attracted to females and copulate successfully. Successful mountings depend more on female initiation of the sequence than her particular state of smell. Sexually experienced males preferentially sniff at synthetic fatty acids on a block of wood, but virgin males did not. Finally, a synthetic smell of green peppers swabbed on a rhesus backside seemed as attractive as the "pheromone," presumably through sheer novelty (Goldfoot, 1981). Goldfoot concluded that although the cycles of smell undoubtedly play some role in sexual behavior, they are but one of many sex signals, not triggers that unleash unmodified lust. Rogel (1978) has reviewed the whole field, including attempts to test sexual odor communication in humans (eg Fox and Taub, 1980, Doty et al., 1975). She reached the same conclusion as Goldfoot: olfactory cycles exist, and undoubtedly play a role, but not an overriding one. She also adds that there is an interesting bias—several experiments suggest that women are more sensitive to musky odors than men, but so far (mainly male) scientists have concentrated on female primates' sexual odors, without asking whether male primate smells might play an even greater role in attracting females.

Another example of odor signals that may have varying meanings in different contexts appears at the other end of the primate scale—the urine wash-

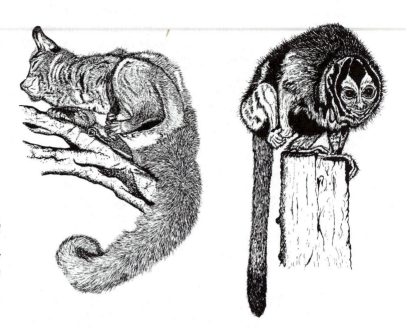

FIGURE 10.3. *Lorisoids and some New World monkeys urinate on a cupped hand, "wash" hind feet, then scrub their scent on the branches. Thick-tailed bushbaby and owl monkey. (From Dixson, 1983.)*

ing of lorisid prosimians and some new-world monkeys (Schilling, 1980). They urinate on a hand, then wipe the hind foot on the hand, usually repeat the process with hand and foot of the other side, and then go off leaving a trail of sticky footprints, sometimes scrubbing feet on the branch as they leave (Fig. 10.3). At its extreme, cebus monkeys progress to washing head, neck, and shoulders with urine (Robinson, 1979).

In squirrel monkeys, male urine washing increases sharply in the reproductive season, as does female sniffing at marks (Candland et al., 1980). Individuals recognize familiar from unfamiliar neighbors, and urine marks may play a role in troop cohesion. However, in other primates urine marking may be unconnected with location in the territory, season, or anything else except novelty of caging or situation (Andrew and Klopman, 1974), which also provokes other marking—ano genital in brown lemurs and nasal in squirrel monkeys (Schwartz et al., Hennessy et al., 1980). Urine marking has been suggested as serving many functions: trail marking in slow loris (Seitz, 1969), thermoregulation in cebus (Robinson, 1979), and increasing the stickiness of the grip in Senegal bushbabies (C. Harcourt, 1981). Harcourt points out that the only galago that does not urine mark is the needle-clawed bushbaby, which relies on its claws to climb thick trunks in search of gum, rather than grip on twigs. Candland and colleagues (1980) may be right in suggesting that urine is used opportunistically as a cue and in many different ways even by monkeys that do not have rigidly developed marking behavior or nasal structures.

Scent marking branches has a complex role in territorial communications just as vocalizations do. Sifaka in the rich gallery forest at Berenty scent-marked in a ring round their territories, an odorous fence (Mertl-Milhollen, 1979). Ringtailed lemurs in the same forest, with more fluid, overlapping ranges, scented along their trails, and respond differentially to individuals from near and distant troops (Mertl-Milhollen, in preparation). Olive cebus monkeys urine-washed at random throughout their fully overlapping ranges, but more frequently and violently during intertroop encounters (Robinson, 1979).

Thus, the use of scent marking can be highly context-dependent. This does not make it less important. Odor is implicated not only in primate sexual attractiveness and territoriality but in male-male aggression in ringtailed lemurs, owl monkeys, and saddlebacked tamarins (Mertl-Milhollen, 1977, 1979; Hunter and Dixson, 1980; Epple, 1980; Epple and Conroy, 1979), and in infants' recognition of the mother in squirrel monkeys and humans (Kaplan et al., 1979; Russell, 1976). Many primates, as well, have specialized scent structures—chest and throat glands, wrist spurs and pectoral glands, and frequently glandular ano-genital regions. Our own axillary hair with its sweat and sebaceous secretions, as well as much of our genital regions, are structures we would unhesitatingly call scent organs in any other primate species.

Finally, there are a few clear cases of control of other animals' physiology, real pheromones in those aspects mediated by odor. Sharp breeding seasons, as in ringtailed lemurs, may involve day-length changes and social odors acting together (van Horn and Eaton, 1979; Evans and Goy, 1968). Adults may delay maturation of their young ones of the same sex, particularly in monogamous species like marmosets and tamarins (Kleiman and Mack, 1980). Finally, human females (and rats) can entrain each other's menstrual cycles (McClintock, 1971, 1981; Graham and McGrew, 1980).

Menstrual synchrony in humans is presumably achieved through unconscious odor cues. (Odor alone is effective in rats, according to McClintock and Adler, 1978.) Most of the women studied were not aware of each other's cycles, let alone of the effect they might have on each other. The ancient core of the brain that links directly to emotion is also linked to the olfactory nerves. The social taboos and giggles we imposed on talk about smell may derive from dim knowledge that our chemical readouts are dangerous and our reactions sometimes overwhelming, if only in the vivid imagery that Darwin linked to half-forgotten odors, or the madeleine dipped in linden tea that wafted Poust into seven volumes of recreated memory.

TACTILE COMMUNICATION

Tactile communication plays a major part in primate life. Primates as an order are *contact animals,* in Hediger's (1950) term. Mothers carry the young for long periods on their bodies. Adults frequently sit or even sleep together in furry clumps. Above all, primates groom each other.

Old-world monkeys and apes part each other's fur with their hands, removing fine particles with fingers or lips (Fig. 10.4). One zoo chimpanzee even removed an object from another's eye while the second submitted for long minutes to the operation (Miles, 1963) Prosimians (except the tarsier) lick and scrape each other's fur with the toothcomb, a specialized structure composed of all the lower canines and incisors (Fig. 10.5). The toothcomb probably originated as an organ for chiseling gum off trees, but it is used now by all species to groom. There is even a built-in toothbrush, the sublingua, for cleaning hairs from between one's teeth (Rose et al., 1981). Bishop (1962, 1964) suggested that

FIGURE 10.4. *Vervet guenons groom using their hands, like other monkeys and apes. (Courtesy S. Bearder.)*

FIGURE 10.5. *Brown lemurs groom, using tongue and specialized toothcomb. (Courtesy J. Buettner-Janusch.)*

FIGURE 10.6. *Reassurance kiss of infant gorillas. (Courtesy P. Coffey and the Jersey Wildlife Preservation Trust.)*

among the prosimian ancestors of higher primates grooming fur may have also played a role in the evolution of fine control of the hand. Grooming helps to remove dirt and parasites, but it is much more than that; it is the social cement of primates from lemur to chimpanzee. Anthoney (1968) describes its ontogeny in the baboon from an infant's suckling the teat, to grasping and sucking fur, to the adult pattern. This accords well with the apparent emotional overtones of grooming and being groomed. As with all other gestures, though, there are differences even among related species: bonnet macaques sit in cuddlesome clumps, whereas pigtailed macaques maintain an arm's-length distance unless they are close kin (Rosenblum et al., 1966). Japanese macaques apparently groom kin rather than consorts, whereas baboons may groom consorts for far longer than the estrus period, over months of association (Baxter and Fedigan, 1979; Ransom and Rowell, 1982).

Moynihan (1967), surveying the new-world monkeys, concludes that adult mutual grooming was probably originally a precopulatory pattern only, which it still is in some small, hole-nesting forms, and that it took on general social functions independently in the various lines, and decreased in frequency in the howler and squirrel monkeys.

Finally, in addition to grooming, there is a huge repertoire of patting and nuzzling lumped as greeting behavior, as well as the agonistic contact of cuffs and bites and even kicks. Chimpanzees, particularly, pat each other's hands, faces, and groins, lay a hand on each other's backs in reassurance, and kiss in affection (Fig. 10.6).

VISUAL COMMUNICATION

Visual communication is perhaps what we think of first, if someone says "nonverbal communication." This salience arises because humans and higher primates look at what we pay attention to. A movement or detail may register in the corner of one's eye, like an odor in the air or a background noise, but then we swing round into active visual monitoring, much as a prosimian might move to actively sniffing a scent mark. Vision gives distance and direction cues for most messages, although commonly at closer distances (especially in foliage) than do smell or sound. It may convey both long-persisting messages about others' states, like scent marking, or transient happenings, like sound.

Visual communication involves facial and ear expression, hair erection, general posture, and tail position. Visual communication "must include the specific characteristic of the senders' morphology. It may be partly for this reason that the dynamic elements of visual signals are often similar in many different primates" (Marler, 1965). Species, size, sex, and all the individual characteristics including dominance and kinship status accompany every smile or scowl, and thus need no repetition. Properly, visual signals include these long-term markers. Most primates, like birds, are visual animals and can see in color, so groups of sympatric species may be distinguished by brilliant pelage or bizarre face masks, as well as by call or scent. The marmosets, lemurs, and guenons include some of the showiest of mammals.

Transient gestures have evolved in step with the permanent markers, as Lorenz showed long ago for the displays of ducks. Kingdon has elegantly analyzed the visual displays of guenons (1980). Hamlyn's guenon swings its head in jerks, which displays the discrete T-pattern of nose and brow. De Brazza's guenon has a white face and beard, bluish nose, and orange crescent brow, all framed in black on brow and upper arms. Its rear repeats the front view—blue scrotum, orange anal hair, white buttock fur, and black outline. The de Brazza gesture which corresponds to Hamlyn's discrete head jerks, swings the whole de Brazza monkey from side to side, displaying head and rump in turn. Sometimes gestures even predate anatomy; fur patterns may evolve faster than motor patterns.

The sexual swellings of females in estrus function as much shorter-term markers. Sexual swellings are oddly distributed among taxonomic groups. There is a rough correlation between swelling and the number of males in a troop. Where a female attracts several males from a few hundred yards' distance for multiple matings, it may pay her to advertise. If there is just one male, the close-range indications of behavior and smell may be as efficient. For widely dispersed animals in very thick forest, swellings are irrelevant. A female orang, if she feels receptive, must locate a mate by his distant calling in Borneo, or travel with him for some days or months keeping track of him in Sumatra, not expecting him to see and find her through the branches.

Bodily posture is one of the most consistent communicative gestures throughout mammals as a whole. Confident or threatening animals hold themselves straight, look big, and walk with stiff-legged swagger. Submissive ones hunch over, crouch, or lie down. Any marks of species or sex are thus displayed by the first and concealed by the second. There are also specialized postures such as the arch display of owl monkeys and tamarins, which exaggerate stiff-legged threat into a ritualized stance that looks rather inappropriate for launching real attack and functions as an ambivalent gesture that promises social contact (Moynihan, 1976; Rathbun, 1979). Tails high and tail lashing seem as though they should be an easy way to convey whole-body alertness, but do not correlate with dominance status in baboons. A few primate tail gestures are special signals like the odorous scent-shivering of ringtailed lemurs, but most

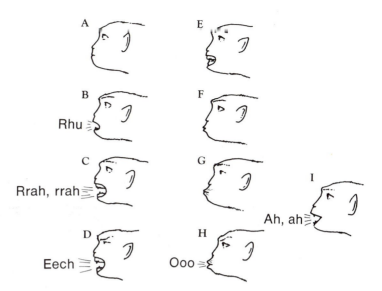

FIGURE 10.7. *Compound facial expressions in the macaque: A, Tense-mouth face. B, Staring open-mouth face. C, Staring bared-teeth screaming face. D, Frowning bared-teeth screaming face. E, Bared-teeth silent (lip chattering) face. F, Lip-smacking (teeth-chattering) face. G, Protruded-lips face. H, Pout face. I, Relaxed open-mouth face. (J. A. R. A. M. van Hooff, 1967.)*

tail postures seem more directly linked to locomotion (Hausfater, 1977; Mukherjee and Saha, 1978).

Facial expression plays a very special role in communication (Fig. 10.7.). We all have eyes front and turn our heads and eyes where attention is. The two eyespots are built into innately recognized schema from the mouse lemurs' fear of staring eyes (the probable predator) to human babies' reflex grinning at two black dots on a piece of paper (Coss, 1977; Spitz and Wolf, 1946). The eye shape and direction play a major role in all primate communication, especially the prosimians with their relatively immobile mouths and tethered upper lip (Pariente, 1980).

Recognition of some facial expressions, particularly threat faces, may be inbuilt in rhesus monkeys reared in isolation from their own kind (Sackett, 1966). However, recognition seems to mean only general activity. Isolated rhesus, reared apart from social contact, neither send nor react to facial signals appropriately (Miller et al., 1967; Redican, 1975). Distinguishing individuals faces is important in the lives of most social primates. Experimentally naive rhesus rapidly generalized from distinguishing two monkeys' full-face portraits to discriminating photos of the same individuals in three-quarter view, in reduced passport format, and in black-and-white (Rosenfeld and Van Hoeser, 1979). This rapid learning does not prove that rhesus conceive the pictures as other monkeys, but their startle, flight, and cowering at the back of the cage when monkey faces flash on the screen suggests that they do.

As the permanent facial markers changed with the evolution of each new species, the dynamic ones remained conservative. There are many primate homologs of human facial expression (Fig. 10.8A,B,C). The homology is quite straightforward in that the same muscles are used in our different-shaped faces and in that the evoking situations are recognizably similar for each face. However, the balance of expressions and the ones that commonly merge with each other are different in each species. Van Hooff's categorizations are usually taken as the basic ones for primates. Redican (1975) in an extensive review discusses

FIGURE 10.8. *Tense mouth faces, commonly given in confident attack. A, Squirrel monkey (drawn by J. Wojcik, courtesy of B. Marriott.)*
B, Gombe Stream chimpanzee in charging display. (Courtesy H. Van Lawick.) C, Heads of state are commonly photographed giving confident threat gestures. (Wide World Photos.)

A

B

C

which social expressions appear in which species. Many new-world monkeys lack some expressions of old-world monkeys, or even use totally different components; marmosets raise rather than lower the brows in threat. Even new-world monkeys' faces may closely correspond to van Hooff's original description.

One facial expression that may serve as an example is the smile. Some insectivores draw back the corners of their mouths and shake their heads when they bite unpleasant food or smell a strong smell. Many human babies do the same if offered unwanted food; so do juvenile lemurs that bite a rotten tamarind pod. Andrew (1963) proposes that this primitive rejection or protective response, which originally served to dislodge food from the mouth, evolved into a fear grin of monkeys and apes. If there was an intermediate stage when the animal grimaced at the mere smell of a conspecific, more particularly a threatening, dominant conspecific, the response could then be ritualized into a visual signal of fear and submission. From there, its emotion could change from real fear to appeasing greeting; old-world and new-world monkeys and apes all use the grin in both situations, at different intensities. When two animals meet, either or both may grin, but the subordinate grins more on the whole. There is a further refinement in both baboons and chimpanzees; the dominant animal may grin, or rather smile, reassuringly at a timid subordinate that fears to approach.

In chimpanzees, the smile has split into at least three distinguishable types— a horizontal version used by subordinates as a fear grin, a rapid vertical flick used by dominants to reassure subordinates, and an open-mouth form that is used during affiliative, relaxed behavior, like grooming (van Hooff, 1972). In chimp, as in man, smiling grades from the grimace of fear through the nervous grin at one's boss or the boss's smile back, to the happy smile.

The smile illustrates evolution of a single gesture from noncommunicative origin to ritualized display. It also illustrates how displays can change in mean-

FIGURE 10.9. *The smile, or silent bared-teeth face is a complex gesture. The smile probably originated with rejection of unpleasant food. In adult strepsirhines such as the ringtailed lemur, it is given in submissive defense with a "spat call." (Courtesy J. Buettner-Janusch.)*

FIGURE 10.10. *The smile begins in ontogeny with equivocal "windy grins" of digestive discomfort. (Courtesy P. Coffey and the Jersey Wildlife Preservation Trust.)*

FIGURE 10.11. *The squirrel monkey face given by a subordinate threatened by attack of a dominant. (Drawn by J. Wojcik, courtesy of B. Marriott.)*

FIGURE 10.12. *Stumptail macaque juvenile, nervous grin, greeting dominant. (Courtesy of M. Bertrand.)*

FIGURE 10.13. *Stumptail macaque in full grimace at approach of dominant. (Courtesy of I. Bernstein.)*

FIGURE 10.14. *Chimpanzee uses various grins in fear, begging, and greeting. (Courtesy P. Marler.)*

ing. The various bared-teeth faces of the old-world monkeys frequently shade into each other, and all bear a fairly clear relation to fear. In man, however, the grin of real fear is rare. With us, the bared-teeth expression of happy smiling is much more likely to grade into the high-intensity play face of laughter.

FIGURE 10.15. *Humans smile in greeting and reassurance, but the subordinate still grins more on the whole. (Wide World Photos.)*

VOCAL COMMUNICATION

The auditory channel can receive sounds from far and near and all sides at once. Sounds fade rapidly. Long, continued bellowing, or the Kloss gibbons' high-pitched trills are evolved ways to show off one's size and lung power, at considerable cost in energy, not a casually grown adornment like a moustache (Whitten, in press). The advantage of sound is to catch, not keep, attention, or to convey much information in a short time, not necessarily to keep the information in mind.

As an arbitrary division of current research on vocal communication, let us first discuss some constraints of the environment, then constraints and correlations with social structure. (The question of whether animal sounds can have semantic meaning will be left to Chapter 20.)

Low-pitched sounds carry farther through leaves than high-pitched sounds of the same initial volume (Marten et al., 1977; Wiley and Richards, 1978). There may also be sound "windows" at medium frequencies, which vary with temperature and time of day (Waser and Waser, 1977). The long-distance calls of ringtailed lemur, lepilemur, and sifaka are concentrated at such sound windows (McGeorge, 1978.) On the whole, sounds carry farther at dawn than other times of day, both in desert conditions (Henwood and Fabrick, 1979) and in the Kibale forest, when there is a sharp dawn temperature gradient between the canopy and the air warming above it (Waser, 1977).

Among major factors in the environment are other vocalizing species. In one Malagasy forest, McGeorge Durrell (in press) shows that the various bird, insect, and lemur species partition the airwaves much as radio stations do, broadcasting at either different frequency or different time of day or night, so there is little overlap of frequency. The one bird, a coua, that makes calls very similar to ringtailed lemurs may be an evolved lemur mimic, although the reasons are unknown (McGeorge, 1978). Similarly, the calls of Kloss gibbons are timed before and after, not competing, with the deafening dawn cicada chorus of Siberut (Marler and Tenaza, 1977; Whitten, in press).

Identifying a call does not just depend on its loudness. Pure frequency sounds are easier to identify than sounds in which the frequencies are blurred together by white noise. Pulsed calls and predictably changing calls are still easier, because they give a listener warning to alert his attention and then continue to confirm the message.

Thus, low-pitched bellowing carries farthest through the forest; the deep roars of howler monkeys may be one of the loudest of all animal noises at source, and can be heard for miles. Many other long-distance calls are higher-pitched, but make up in incisiveness for their relatively lower carrying power. Indri songs sound very much like the falling notes of massed air-raid sirens, for much the same purpose—they command attention against any auditory background. Still higher and more birdlike are the trills of titis and marmosets. Like birds, they warn off others unambiguously in the midst of a multispecies forest cacophony.

Localizing the source of a sound is easiest for noises with sharp onset and termination, like barks. Thin whistles with imperceptible start and end are much harder to place (Marler, 1967). Changing pitch can be more accurately located than a steady tone (Brown et al., 1978, 1979). Territorial or spacing calls are relatively easy to locate. There may even turn out to be consistent differences in locatability between true territorial animals that sing to keep others from a general area, and individual-space animals that sing "Keep away from us right here and now."

Intense alarm calls are often high-pitched screams or ventriloqual whistles without sharp beginning and end. Less dangerous predators can be mobbed with barks or chirps that draw attention to the predators' position. Many birds have two sorts of alarm calls for aerial and ground predators, as do ringtailed lemurs, white sifaka, and red colobus. However, the locatability is only one aspect of the differences, perhaps mattering most to very small birds and mammals who can hide from predators if their alarm calls do not give them away. The bellows of a sifaka troop or the earsplitting screams of lemurs might startle and confuse a stooping hawk, rather than fool it.

Thus, any animal call has some combination of carrying power, noticeability, and locatability built into its form. The whoop-gobble of grey-cheeked mangabeys is a long-distance call, which spaces troops (see Chapter 7). It is loud at its source (up to 57 decibels), but managabey loud grunts, staccato barks, and screams are as loud or louder. Of all the mangabey calls it suffers least attenuation in the Kibale forest. In group grunts (also about 60 decibels at the source) cannot be heard by a human more than 100 meters from the troop, loud alarm and agonistic noises carry 300 to 400 meters, but a whoop-gobble carries at least one kilometer. Its form makes it highly noticeable; the whoop alerts other animals (and primatologists) to start listening. The gobble then carries information about which individual is calling and from where (Waser, 1977).

The whoop-gobble is at one extreme of a continuum. The other extreme is the soft grunt of close contact, whose meaning depends on two animals already paying attention to each other and knowing the context of each gesture. We are much farther from understanding these soft, variable primate calls than the simpler messages that can be hollered throughout the woods.

Social structure has various regular correlations with vocal style. Pair-living primates—new world titis, indri, and, of course, gibbons—duet rather frequently (Robinson, 1979; Serpell, 1981). Gibbon duets may have evolved from a primitive pattern like the Kloss'. Kloss gibbon males sing alone, as do females. They climb to the tops of emergent trees to sing. It seems clear that in the Kloss singing is mainly directed to challenge others outside the group. Among the Moloch, Mueller's, black, lar, agile, pileated, hoolock, and the siamang, the male and female portios of the duet are more and more juxtaposed. Each partner's section necessarily complements the mate's or else the whole sequence aborts. The agile and pileated gibbons' great calls accelerate to a climax that so transfixes both partners that local hunters know this is the time to stalk them; the gibbons are oblivious to anything but their joint song (Tenaza, 1976; Marler and Tenaza, 1977; Whitten, 1982; Brockleman, in press; Haimoff et al. 1982; Deputte and Goustard, 1978).

Some species' repertoires consist of highly discrete and separate vocalizations. Other species have many intermediate, or graded forms, as Rowell and Hinde first pointed out for rhesus monkeys. Animals that live out of sight of each other, in leafy environments, have more clear-cut vocabularies, whereas those whose vocalizations go with visual contact may have subtle, intergraded vocal repertoires. Thus, the forest guenons of Kibale have small, discrete repertoires, and savannah vervets, larger ones. Social complexity may outweigh this forest savannah difference. Black-and-white colobus, in their small troops, have fewer things to say than red colobus in their large troops (Marler, 1970, 1971, 1972; Struhsaker, 1981). In primate society it is often necessary to know individuals: even squirrel monkey infants give their own distinctive "isolation peeps" (Symmes et al., 1979).

Chimpanzee vocalizations are at the other pole from gibbons—complex, intergrading, frequent, and with all ages and sexes giving all types. Again this correlates with the social structure, in which chimpanzees wander alone but identify others by the individual characteristics of their long-distance pant-hooting and run the gamut of expressions on reuniting in ever-changing groups. In contrast, gorillas remain together throughout the day. Their commonest noise is the soft belch of contact and feeding in thick vegetation. Three of their calls are given only by adult males, as though the silverback takes responsibility for decisions of the group (Marler, 1976; Marler and Tenaza, 1977; Marler and Hobbett, 1975). In short, analysis of vocalization, its function in the group, its comparative use in different species, and its ontogeny in the individual is a powerful means of analyzing social structure, because vocalization reveals what is important for the animal to communicate (Green, 1981).

SUMMARY

Any communicative act includes the *signal*, which is the transient form of the act, the sender's *motivation*, the *meaning* apparent to the receiver, and a long-term evolutionary *function*. These often do not relate in a simple one-to-one manner.

Communicative signals originated as by-products of other behavior, especially autonomic changes, intention movements, and displacement behavior. During evolution the signal may be ritualized to eliminate ambiguities by exaggeration, stereotyping, and repetition. Overexaggerated signals may be replaced by others, just as anatomical forms may be. Some signals change motivation and meaning in evolution, as in the series protective grimace → fear grin → smile.

It is very difficult to catalog other species' communication: they may split a graded series of calls or gestures more or less finely than their observers do, or take account of different amounts of context in reacting to a signal. Meaning, as in deciphering an unknown language, can only be guessed from reactions of native speakers.

Different communicative channels have different advantages. Scent is diffuse and persistent, visual display is directional and almost always associated with the displaying individual; sound is transient, so allows much information in a short time and can seize attention at any moment, in any direction, and at short or long distance. Long-distance courtship, or territorial announcement is usually olfactory or vocal; predator alarm almost always vocal.

Human nonverbal communication is rich and varied. Many or most interpersonal signals are nonverbal, conveying emotion rather than facts about the external world.

CHAPTER 11

DEMOGRAPHY

DEMOGRAPHY, CAUSATION, AND THE SOCIAL MILIEU

Demography is the study of births, deaths, and migrations, the processes that give a population its structure. The power of demography is that it can describe these processes mathematically, by summing the lives of many individuals (Fig. 11.1.). It cannot predict the life or actions of any one individual, only indicate probabilities.

Demographical accounts are descriptions, not causes. The sun rose every morning during human history. We may infer that it will rise tomorrow. The cause of its rising is not past risings; the causes are centrifugal force and gravity, or if you prefer, a kink in space-time. Similarly, a high infant mortality rate may lead one to infer that a given infant has a 40 per cent chance of dying in the first year, but it dies from measles, not demography. The human population of the earth will reach more than six billion by the year 2000, barring nuclear war, just from the children of children already born. The causes are not mathematics but human hope and desire.

Obvious? Perhaps, but all too easy to muddle, when combined with considering group influence on the individual, and group interest versus self-interest.

We are used to the idea that individual births, deaths, and migrations give rise to population structure and dynamics, but it requires more thought to see how the statistical outcome of demographic processes affect the individual. The following exposition is from Altmann and Altmann (1979):

> The number and distribution, according to age, sex, and kinship, of potential social partners are produced by demographic processes. . . . Changes in the size and composition of a group are brought about by just seven processes: birth, death,

emigration, immigration, maturation, group fusion, and group fission, each of which is susceptible to small-sample fluctuations (demographic 'drift'). . . . A formal model for predicting group size and composition from four of these processes, that is, birth, immigration, death, and emigration (hence, the BIDE model), was developed by Cohen (1971). According to the BIDE model, a population in which the per capita annual rates of birth, death, and emigration are 0.177, 0.173, and 0.55, respectively, and in which the immigration rate is 0.548 individuals per group—all reasonable values for Amboseli baboons *(Papio cynocephalus)*—will have an average group size of about 50. However, in such a population about 9 per cent of all individuals will be in groups smaller than 30, and 11 per cent will be in groups larger than 140. Therefore, any effects on behavior and social relations of such very large and very small groups will be recurring phenomena, a point to which we shall return. The literature on primate social groups contains numerous speculations about the adaptive significance of group size. . . . In the BIDE model, group size is a consequence of autonomous birth, death, and migration processes. If the BIDE model assumptions are correct, natural selection must act on the birth, death, and migration rate, rather than on group size per se, and thus it is these rates, not group size, for which it is appropriate to seek evolutionary explanations . . .

A few rough estimates, calculated under simplifying assumptions, will serve to illustrate the great potential for large differences in social milieu. We look first at the kin composition of a social group. The expected numbers of any class of relatives available in a group depends on that group's recent demographic history. In a population in which births greatly exceed deaths and dispersal from natal groups is low, each individual will grow up surrounded by relatives. That is exactly what happens in rapidly expanding primate populations. By contrast, consider a group of primates at or near a stationary condition, Alto's baboon group in Amboseli as of 1975. In such a group, what is the chance that a liveborn neonatal infant has a living next-older sibling? From our data (J. Altmann et al., 1977 and in preparation) on mean interbirth interval and female life expectancy, we estimate that on the average the number of offspring in an adult female's lifetime is eight. Since one out of eight infants therefore will be the offspring of a primiparous female, only seven out of eight individuals have any older sibling, living or dead. The probability that such a sibling will survive from conception to age 22 months (the mean interbirth interval with a surviving infant) is 0.46. Thus the probability that a liveborn infant will have a living next-older sibling is L (0.46) M 0.40. In such a stationary primate population, most adult females will not survive long enough to be grandmothers. Thus, an infant's available playmates usually will not include either siblings, nieces, or nephews. Cousins (probably offspring of half-sibs) are more likely. However, even this likelihood will be reduced because two reproductively mature sibs may not produce offspring sufficiently close in time.

Grandparents, especially of first- or second-born infants, and older siblings, especially of later-born infants, may play an important role in an infant's life if they survive. Moreover, the occasional cases of many surviving close kin may be quite dramatic in the impact on the social group as a whole, as well as on the individuals involved. Our aim is to point out that numerous surviving close kin will be uncommon, not that they will be unimportant.

The preceding examples were based on mean values, but in small social groups chance deviations from mean values are likely to be very large. To illustrate this, we consider gender of playmates rather than kinship. What is the chance that an infant will have in its group a potential playmate of the opposite sex that is within three months of its age, that is, another infant born during the six-month period centered on a given infant's birth date? Suppose that the infant lives in a group of 50 baboons, a size that would seem to be large enough to be buffered against small-

Bold lines and solid symbols denote males; thin lines and open symbols females; horizontal lines show parentage and birth date. Gray dashed lines show travel outside the 4 groups. Death from intraspecies killing indicated by I; from poachers, P; from natural causes, N.

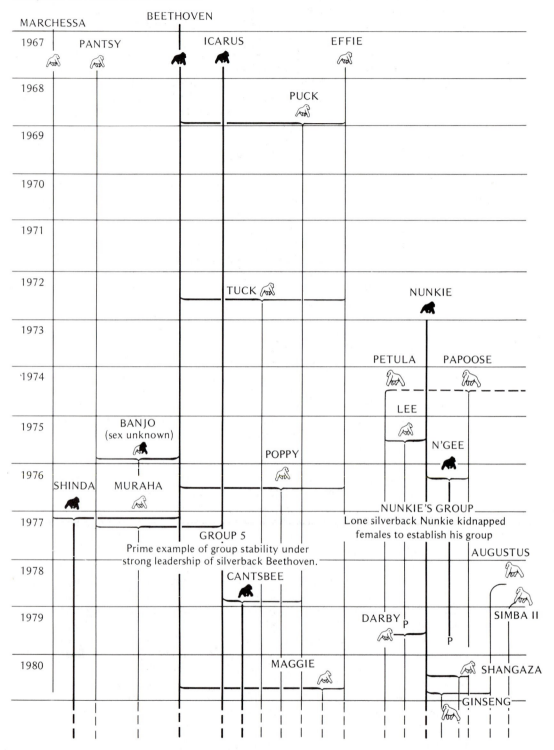

FIGURE 11.1. *The changes in four groups of gorillas in the Virunga Volcanos over thirteen years. Groups' size and composition changes with births, deaths, immigration, emigration. Only*

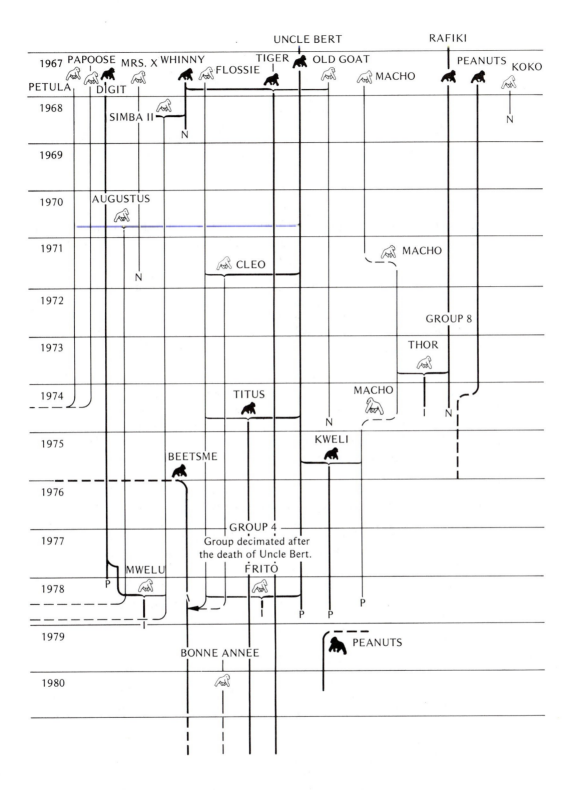

such devoted long-term studies reveal the history of individual relations. Because most primates live in very small groups, their companions depend on behavior or accidents at the individual level. However, long-term evolutionary trends result from some demographic summing of individual lives. (Courtesy D. Fossey and the National Geographic Society.)

sample effects and that is about average size for stationary baboon populations. Alto's group is a group of that size. During 1975 the 15 adult females of this group gave birth to five infants per six-month period, none stillborn. Assuming equal sex ratios at birth, the probability that five out of five infants would be of the same sex is about 0.06, so that even if all infants survived, about six infants out of every 100 in social groups of this size would not have any available playmate of the opposite sex within three months of their own age. Furthermore, the probability that exactly four out of five infants will be of the same sex is about 0.30, which means that in groups of this size, almost a third of all half-year cohorts will include an individual with no same-sex associate. If some of these infants do not survive the first year of life, the chance that, at the time these individuals enter the juvenile play groups, some will have no choice in the sex of their playmates becomes even greater; mortality during the first year of life among liveborn baboons in Ambelosi has been 29 per cent (J. Altmann et al., 1977). Beyond that, lack of a sharply defined breeding season would further increase the chance that some infants will be born at a time of year in which few (or many) others are born, thereby exaggerating variability attributable to small-sample effects.

In our discussion of the BIDE model we pointed out that even if life history parameters are uniform throughout a population, some individuals will, by chance, find themselves in a much smaller group than will others. They will therefore be more susceptible to effects of small-sample fluctuations in number, gender, and kin relatedness of available companions.

Such fluctuations may reverberate over a surprisingly long period of time. Dunbar points out that in groups the size of most primates', and the size of most primate study areas, it is dangerous ever to assume equilibrium. In simulations of a founder population of four-year-old females, with parameters taken from Dunbar's own study of gelada baboons, the simulated populations took about 75 years to reach stable age structure. Similarly, assuming a year with no infants, female maturation at four years, and male maturation at six, perturbations in adult sex ratio could still be seen after a simulated 22 years. In a real population, such perturbations would reflect back directly on behavior. Gelada males obtain harems either by joining an established harem as follower, or by taking over a harem, fighting and defeating its leader. Takeovers mainly work for large harems, where peripheral females support the new male. Large harems tend to form in the first place where the female-male ratio is high. Thus, a disaster that killed the infants of one year could seven years later appear as a female-biased sex ratio, with large harems and fighting among the males (Dunbar, 1979).

MORTALITY AND SURVIVORSHIP

G. E. Hutchinson (1978) writes, "If we want to think intelligently about how population growth is controlled by natality, we find paradoxically that it is best to start by thinking about death."

Picture a *cohort* of individuals in a population, all born at about the same time. If 1,000 are born, some die in early infancy and some each month or year thereafter. A *life table* is a list of the proportion of our cohort that survives to a series of times, x. If you plot the proportion surviving against age, you draw a *survivorship* curve. You can use the same data to plot qx, the age-specific mortality, or the fraction λx of those alive at any age x that are likely to die between age x and x + 1.

The calculation of survivorship and death rate from field data is beautifully laid out in *Techniques for the Study of Primate Population Ecology* (National Research Council, 1981). This is a handbook that every working primatologist should own.

We may picture geometric forms of survivorship curves, as shown in Fig. 11.2. The convex curve shows the survivorship of organisms that live to a definite age and then all die at once. Experimentally maintained populations of some invertebrates actually do have convex survivorship curves. Such a curve for humans is the unspoken goal of modern medicine: that no humans should die before they reach four score and ten.

The central shallowly concave curve indicates a constant death rate, in which the same fraction of the initial cohort dies at any age. Some wild populations of passerine birds and lizards seem to follow this pattern. For them, predation and hardship take a constant toll.

Finally, the deeply concave curve represents enormous infant mortality but little loss of adults.

Wild mammal survivorship curves are more complicated than these ideal simplifications. For all, there is a period of high infant mortality, followed by juvenile and adult patterns that differ for each species. If we compare three well-studied species, the Dall sheep, Danish roebuck, and grey seal each loses half the cohort in the first year. Then, Dall sheep adults have little mortality until about age six, and Danish roebuck, high, constant adult mortality. Which sex is more likely to live longer depends on the relative rigors of maternity and paternity in the respective breeding system: opposite directions for the sheep and the roebuck. The grey seal shows extreme divergence between male and female. With the seals' system of ferociously defended harems, there is enormous mortality of males after they reach breeding age. For humans, in developed countries in the twentieth century, we have enormously reduced infant mortality, by dealing with the dirt, cold, and malnutrition that turn mild diseases into killers, and by immunization against the most common dangerous diseases. We have not, however, eliminated infant mortality. "We may be faced with the possibility that the evolutionary expense of providing really adequate infantile care is just too great" (Hutchinson, 1978).

The relative life span of men and women varies with the number of children and amount of food a woman has. The men of one New Guinea group remark, "Of course, men need many wives: women wear out so fast" (Grey, personal communication). In all industrial countries, in contrast, women outlive men.

In spite of all our group counts of primates, we have little quantitative primate demography. This is because cross-sectional age counts do not really tell you the survivorship of any one cohort. You do not know which cohorts went through good years or famines or are recovering from the onslaughts of poachers. You must extrapolate from recent history or the state of the environment to tell if the population is changing at a constant rate or in uneven jerks. Only recently have a few studies produced the kind of long-term data that tell us life history processes (Harcourt et al., 1981; Dunbar 1980; Southwick, 1980; Teas et al., 1981).

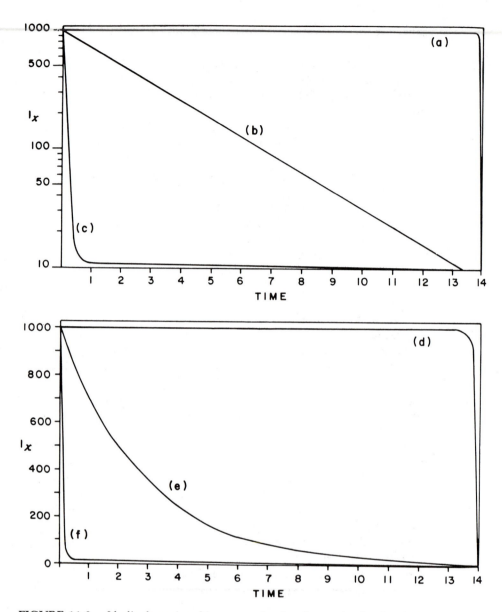

FIGURE 11.2. *Idealized survivorship curves, showing the number lx of an initial cohort of 1,000 organisms which survive to a given age. The upper graph presents the data on a logarithmic scale; the lower graph has same data on an arithmetic scale. Convex curves a and d represent a population with almost all mortality in old age. Concave curves c and f show populations with almost all mortality in infancy. The central curves, b and e result from constant likelihood of mortality at any age. (After Hutchinson, 1979.)*

Some of the best data are on the toque macaques of Polonnaruwa in Sri Lanka. Dittus (1975, 1977, 1979), in a five-year study of a population of 370 monkeys, produced an age-specific mortality graph shown in Fig. 11.3. In this population the death rate of female juveniles and infants from zero to two years

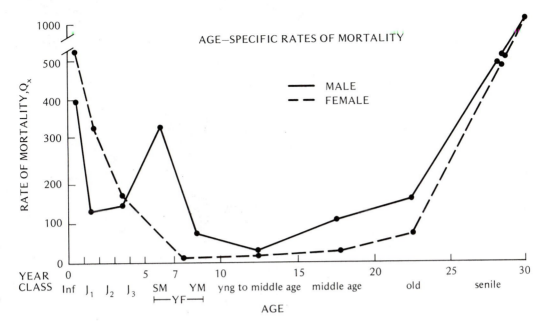

FIGURE 11.3. *Rates of mortality (qx) per 1,000 entering each age class, in toque macaques at Polonnarua, Sri Lanka. Young adult females (YF) span the subadult male (SM) and young adult male (YM) age classes. Infant and juvenile females had much higher death rates than infant and juvenile males, but after five years of age males were more likely to die than females. (After Dittus, 1975.)*

is much higher than that of male juveniles and infants. Then, as females mature to childbearing age, their death rate falls to near zero. Subadult males, however, change troop, and fight their way into new hierarchies, or hang about on the periphery as outcasts. Their mortality peaks sharply. They continue to die off as adults, so there are eventually more old female survivors.

These toque macaques live in a forest surrounding ancient temples. They have no predators and are essentially food-limited. The female children die because they are the least dominant members of the troop. They spend most of their time finding food, and it is then most likely to be taken away from them (Fig. 11.4.). Even subadult males do not generally die from the wounds of fights, but from inanition as they fall too low in the hierarchy to claim food for themselves. In a famine year, the number of deaths increased, and the same pattern of differential mortality grew even sharper (Dittus, 1979).

Dittus (1980) has reviewed several other studies of old- and new-world monkeys. Although these are cross-sectional censuses, not followed cohorts, the pattern of high infant female mortality and high subadult male mortality seems very similar. Three macaque populations that are abundantly fed by man still seem to have more male than female juveniles, though the imbalance is not so great (see Table 11.1.). Before jumping to the conclusion that this is normal for all primates (and gruesomely familiar to us), let us consider another life table, for the howlers of Barro Colorado Island (Froelich et al., 1981). This one is less firmly based, being derived from a cross-sectional study of weights and toothwear, and a census of a total of 200. Froelich and colleagues produced a

FIGURE 11.4. *A young male toque macaque, who has fallen in the male hierarchy and is in emaciated condition, removes food from a juvenile female's cheek pouches. (Courtesy of W. P. J. Dittus.)*

TABLE 11.1. Cross-sectional Counts of Age and Sex Ratio for Old-World Monkeys

Species and Location of Study	Total no. Animals, Sexed	Infant	Sex Ratios, ♂:♀ Juvenile	Subadult and Adult
Wild Populations				
Toque macaque Sri Lanka	417	1.1	1.5	0.6
Rhesus macaque North India	68	1.3	2.1	0.5
Bonnet macaque South India	156	?	1.4	0.8
Savannah baboon Kenya	179	0.8	1.7	0.8
Savannah baboon Uganda	62	?	2.1	1.0
Savannah baboon Kenya	179	1.0	1.4	0.6
Vervet Uganda	205	?	1.2	0.7
Populations Provisioned with Food by Man				
Japanese macaque Takasakiyama, 1962	724	1.0	1.3	0.9
Japanese macaque Takasakiyama, 1965	1,020	0.9	1.2	1.0
Rhesus macaque Cayo Santiago, 1967	607	?	1.1	0.6

(Dittus, 1980).

life table (Fig. 11.5.), assuming population growth at a constant rate and ages as found in the longitudinal studies of Glander (1980) and Jones (1978) in Costa Rica.

The resulting survival and mortality curves, like those for the macaques, have a peak of subadult male mortality. Unlike the unfed macaques, there are more surviving juvenile females than males, and more old males than females. The difference in infant and juvenile mortality may be because these howlers are an expanding population, not sharply food-limited. It may also be a trait of the species. In the group that Glander (1980) studied for more than ten years, nine of nine infant females survived to one year and six to two years, whereas only 8 of 15 males reached one year and four, two years. The male-female difference is statistically significant.

Among the chimpanzees of the Gombe Stream, too few infants died to reveal early sex difference in mortality, or else the chimpanzee mothers manage to give equal care (or lack of it) to sons and daughters. Young adult males, as usual, are more vulnerable (Fig. 11.6.). The high mortality of adult males in the Kahama community was the result of intergroup attacks (see Chapter 4). Even after 15 years, the numbers are too small for generalizations, which in itself is instructive, for this is one of the first and most intensively studied of primate populations.

AGE AT MENARCHE

The number of children born by a female can be described by the age at which she begins reproduction, the age at which she ceases reproduction, and the interbirth interval.

Primate females begin reproduction anywhere between a year and 13 years of age, considering the gamut from mouse lemur to chimpanzee. The significant fact is that they are slower to mature than other mammals of equal body weight. The majority of monkeys, and even the larger lemurs, do not bear offspring before three years of age.

Timing of the onset of puberty depends on environmental factors, especially nutrition. There is often a period of adolescent sterility, when a female shows the external signs of estrus or menstruation, but does not yet ovulate. This also varies in length with nutrition and other environmental factors. The reproductive stages of female baboons are shown in Fig. 11.7.

Zoos are commonly finding that young males and females mature at faster rates than their wild counterparts and conceive before their zookeepers are prepared for the idea. In rhesus monkeys, in well-fed captive groups in California, the birthrate for three- to five-year-old mothers is significantly higher than for the same age group on Cayo Santiago (Smith, 1982; Sade et al., 1977).

This is in accordance with the far more extensive human data. There is controversy as to how much chronic malnutrition reduces fertility. However, the data unequivocally indicate that well-nourished populations of girls may reach menarche at 11 to 13 years, as in present-day Europe and America. In nineteenth-century Europe and various populations in developing countries, the age at menarche ranges upward to 18.8 years, as among the Bundi of New Guinea. Whether this affects eventual fertility depends on age at marriage or

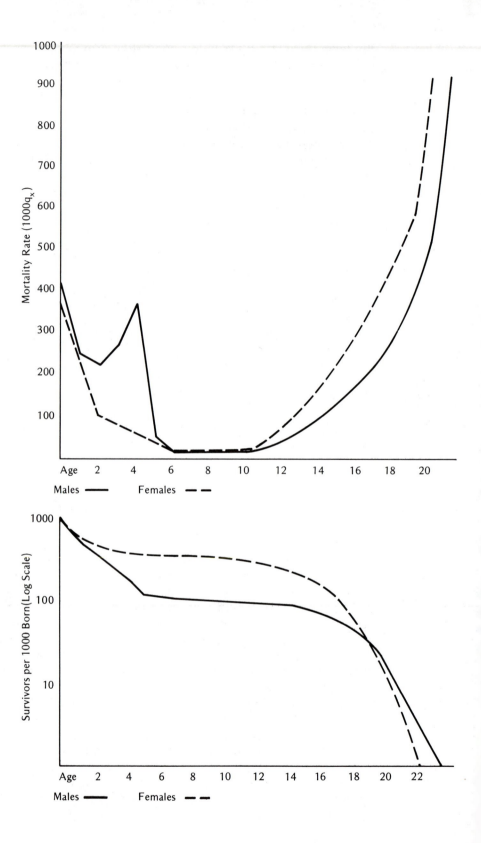

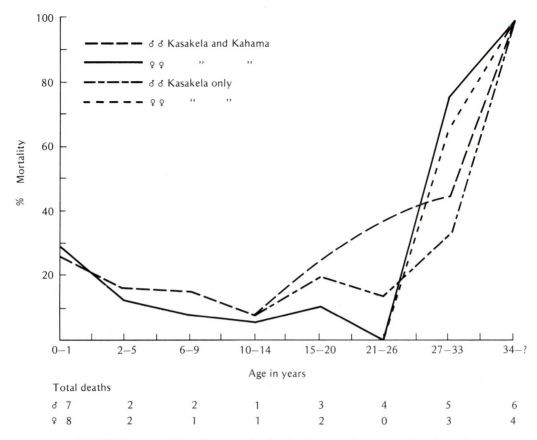

FIGURE 11.6. *Mortality rates for Gombe Stream chimpanzees. Females had very low mortality up to old age—a highly convex survivorship curve in terms of Figure 11.2. Males of the Kasakela community died of disease and accident, but Kahama males also died from assaults by the Kasakela community. (After Goodall, 1982.)*

regular sexual liaison, as well as contraceptive practice, and the addition of an extra year or so to human reproductive span has only a small effect. Although obviously complex, the relationship between menarche and nutrition seems undisputed (Bongaarts, 1980; Tanner, 1962; Short, 1976).

In primates, the addition of an extra year to the breeding span has a relatively greater effect. Besides differences between populations, there are important differences within populations. At Cayo Santiago, three-year-old daughters of high-ranking females are more likely to conceive than are low-ranking three-year-olds (Drickamer, 1974). Because status is usually passed on from

FIGURE 11.5. *(Opposite page) Demography of Barro Colorado mantled howlers. A shows age and sex-specific mortality rates and B, survivorship. Mortality and survivorship curves are different ways of plotting the same data, but one emphasizes the individual's chances of dying at a given age, whereas the other emphasizes the number, sex, and age of an individual's surviving companions in the population. (Froelich et al., 1981.)*

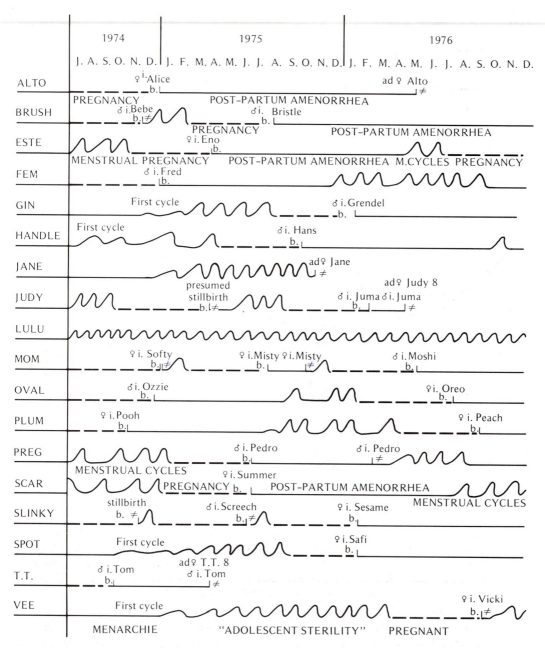

FIGURE 11.7. *Reproductive stages of adult females in Alto's baboon group, Amboseli, Kenya, July 1974 through December 1976. On the line for each female, births are indicated b, deaths ≠, cycling by a wave roughly corresponding to changes in the sex skin, and pregnancy by a dotted line. (After J. Altmann, 1980.)*

mother to daughter, this means that high-ranking genealogies grow faster as a whole.

Koshima Island is the site of one of the famous primate troops. Between 1964 and 1971 the Japanese macaques of Koshima Island invented and transmitted new behaviors such as washing sweet potatoes and "placer-mining" wheat grains thrown amongst the beach sand. (This "precultural" behavior will be discussed more fully in chapter 21.) However, they outgrew their environment and the scientists' capacity to provide for them. In 1972 the supply of sweet potatoes was stopped, except for particular testing. During the pre-1964, semi-wild period, the population grew at 9 per cent per year, during the period of artificial feeding at 15 per cent, and during the period of famine it fell by 4 per cent per year. As population fell, so mean age at first birth rose from 6.2 to 6.8 years. (It was apparently still lower at the start of the study, but this is probably a consequence of knowing only the youngest females' ages.) More to the point, age at menarche was related to weight, and weight growth was related to status. The only two young females heavy enough to enter breeding condition in their fourth and fifth year were A-lineage females. Of infants born during the famine period, only those of A and B lineage survived to be weighed (Mori, 1979).

A second troop of Japanese macaques, at Ryozen, has gone through similar changes. Again, there were different body weights and the two years' difference in primiparous age (Sugiyama and Ohsawa, 1979). The correlation between weight and age at first breeding appears in captive Japanese macaques, as well (Wolfe, 1979).

Various other social factors, besides nutrition, may affect physical maturation. In saddleback tamarins, young animals with older mates conceived, or successfully inseminated their partners, at younger ages than pairs of young animals caged together. Whether this is a purely behavioral or a behavioral-endocrine effect is not known (Epple and Katz, 1980).

INTERBIRTH INTERVAL

The major factor in total natality is the interbirth interval. (See Figs. 11.8., 11.9.) This interval is set by the interplay of physiological times of gestation and lactation with the environmental timing of birth season. Many primates give birth only during highly restricted seasons. Others, including humans, have more or less sharp peaks of frequent birth, but some births throughout the year. Still others seem to give birth with no seasonal bias, although we might find slight bias given more data. The calculation of birthrates from census data is, of course, different for ideal birth pulse or birth flow population models—the one with all births ascribed to the same day, the other with births spread through the year (National Research Council, 1981).

One primate group, the callitrichids, have a postpartum estrus (Epple, 1970). Their females are thus pregnant and lactating at the same time. They give birth to twins at about six-month intervals, in the wild as well as in captivity (Soini, 1982; Lunn and McNeilly, 1982). Owl monkeys do not have postpartum estrus, but can apparently produce an infant every eight to nine months (Hunter et al., 1979). Both owl monkey and marmoset males carry the infants, which lightens the females' physical and physiological burden.

Cheirogaleids and Galagines sometimes have postpartum estrus, but rarely become pregnant. In the wild, both Malagasy and African forms tend to highly seasonal breeding, so infants conceived at postpartum estrus have little chance of survival. One exception is the mouse lemur, whose gestation period is so

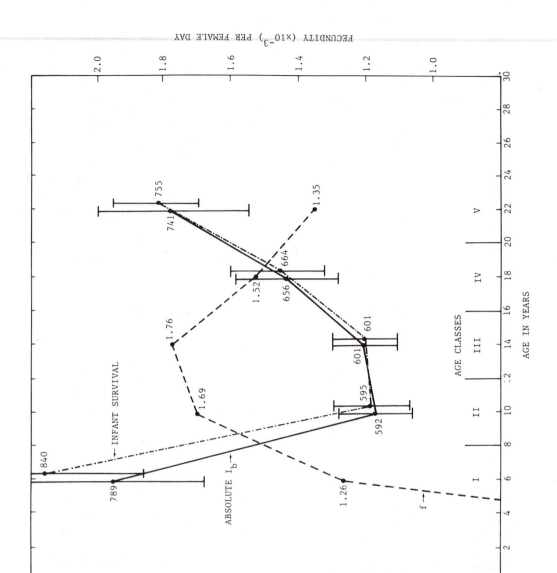

FIGURE 11.8. The relationship between female age, absolute interbirth interval (—), interbirth interval for mothers with surviving infants (—·—·—), and female fecundity (– – –), in the baboons at Gilgil, Kenya. (After Strum and Western, 1982.)

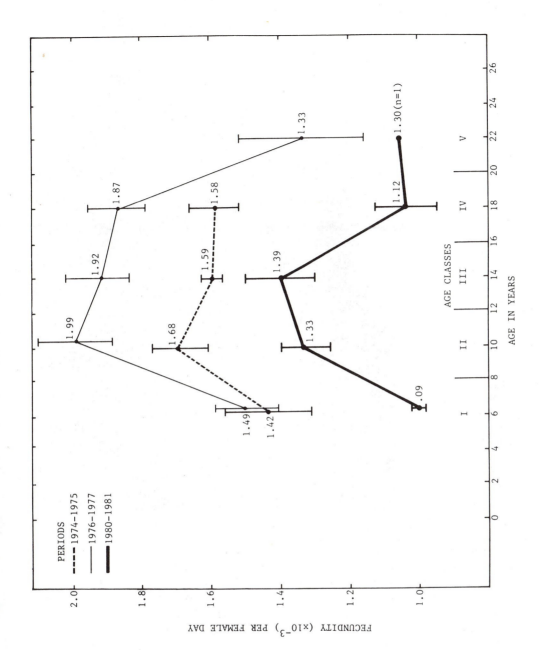

FIGURE 11.9. *The influence of nutrition on age-specific fecundity for baboons at Gilgil, Kenya. During the periods of lower fecundity there was heavy competition for food from grazing ungulates: cattle, sheep, impala, and Thompson's gazelle. (After Strum and Western, 1982.)*

FIGURE 11.10. *Interbirth interval in captive orangutans and other great apes may be two years or less if the infant is removed; interbirth interval in wild Bornean orangs is around eight years. (Courtesy P. Coffey and the Jersey Wildlife Preservation Trust.)*

short it can produce two litters during one rainy season (Foerg, 1982a; Charles-Dominique, and Martin, 1972; Martin, Horn, and Eaton, 1979).

A typical birth interval for macaques or baboons or langurs is in the one- to two-year range. For the chimpanzees of the Gombe stream, if the previous offspring survives, the second is born an average of five years later (range 4.5 to 7.5 years). For orangutans in Borneo, eight year intervals seem normal (Fig. 11.10.) (Goodall, 1983; Galdikas, 1979, 1981).

For the larger new-world monkeys, old-world monkeys, and apes, the chief determinants of birth spacing are the season (or lack of it) and the length of lactational amenorhea. Lactation, in turn, depends on the previous infant's survival. Thus, births relate to the pattern of infant deaths.

The period of amenorhea after giving birth is shorter in many humans than in many other primates, relative to body weight. We do not know if this has to do with earlier food supplement for the baby, nutrition to the mother, or underlying biological factors.

Lactational amenorhea is, however, much longer in some human popula-

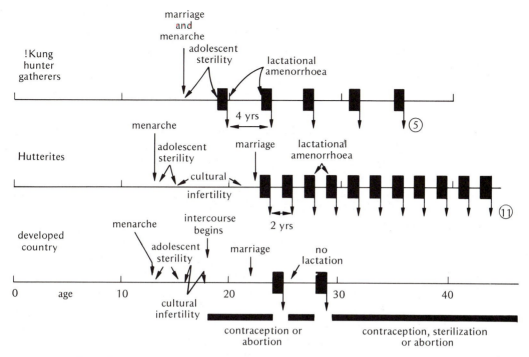

FIGURE 11.11. *The changing patterns of human fertility. Women in developed countries spend far more time cycling and less in either pregnancy or lactational amenorhea than do !Kung hunter gatherers. The difference between nomadic !Kung, with their four- to five-year birth interval, and the Hutterites, well-fed North Americans practicing no contraception, resembles the difference between wild and captive great apes. (R. V. Short, 1976.)*

tions, such as Bangladeshi women and nomadic !Kung bushwomen, than in others (Fig. 11.11). It is not clear how much this has to do with chronic under-nutrition, particularly as to fats. When !Kung settled as agriculturalists, their interbirth interval shortened from five to two years (Howell, 1974; Lee, 1980). If this is not a simple nutritional effect, it may relate to suckling frequency and intensity, and the feedback processes on female physiology (Huffman et al., 1978; Konner and Worthman, 1980).

One thing does seem clear: females buffer the effects of mild environmental deprivation, losing weight themselves during lactation but still providing milk for the child. There may be a relatively abrupt biological transition to real famine conditions when women stop cycling and conceiving. Further, if a lactating mother, human or Japanese macaque, is starving, her child is likely to die first, thus leaving somewhat more chance that she will live (Mori, 1979).

If the infant dies, the mother begins cycling and conceives again in a much shorter time. This is complicated by seasonal effects, because many primates give birth only at restricted seasons, others throughout the year. The speeding up of reproduction can in some circumstances favor infanticide by rival males (see Chapter 12).

Birth season interests us because it is so little apparent in the contemporary world. In human populations there is usually a low peak of births. This is culturally determined, with a peak of conception in spring, and at major festivals such as harvest or Christmas. Climate may well influence the birth season indirectly, but we do not respond automatically to external triggers such as day length. This is shown by racial difference in birth season between different shades of South Africans and by Puerto Rico's recent change to U.S. timing (Dyson and Crook, 1980; Cowgill, 1964).

Birth peaks and seasons in primates differ between populations of the same or closely related species. In general, an infant makes the heaviest physiological demands on its mother in the middle of the lactation period, when she must feed and carry a large, growing, heavy child, using a far less efficient organ than the placenta (Altmann, 1980). We shall discuss this further in Chapter 13. What matters for the moment is that there may be an irreducible minimum birth interval for many species because a baby born in the wrong season would not survive. Glander (1980) points out a subtler effect. More howler babies in his troop survived if born in the same months as other babies than if born alone although the groups of births occurred in different seasons. It is not clear how the timing of conception or the fact of survival happened in these clustered births.

MENOPAUSE

The final factor in natality is the time of menopause, or death if that happens first. Among western women the average age of menopause has risen from 47 to 50 years since the nineteenth century (Tanner, 1962; Short, 1976). This presumably reflects a nutritional change; menopause occurs much earlier in some developing countries.

Primates rarely live till menopause and may not even have one in most species. The conception rate declines with advancing age, at least in macaques. There are a few isolated reports of wild postreproductive females (Waser, 1978; Tilford, 1981). Rhesus seem to be the only primates with well-documented menopause in captivity (van Wagenen, 1972; Hodgen et al., 1977).

Chimpanzees also decline in fertility with age. The oldest laboratory female who produced a living infant was 38 years old. However, cycling does not seem to stop; the oldest female chimpanzee of known age is still cycling at 48. Graham (1979, 1981) suggests that menopause in women and lab rhesus is a consequence of our artificially prolonged life span, rather than something that would have occurred in our wild ancestors, let alone been of selective advantage. Others argue the usefulness of grandmothers to hunting-gathering hominids, particularly as children's dependence lasted longer and demanded more adult care. If a woman's energy and life expectancy declined to the point where she could not expect to raise another offspring of her own, better to stop, and put the remaining fraction of energy into helping out her daughters' reproduction. The difference between female and male patterns would then reflect female's greater investment in care-giving during our Pleistocene past (Gaulin, 1980). In particular, if carrying infants is the major constraint on birth spacing for nomadic women, then a grandmother who helped carry toddlers could allow her daughters to raise more children (Mayer, 1982).

REPRODUCTIVE VALUE AND
REPRODUCTIVE POTENTIAL

The *reproductive value* is the estimated number of female offspring that remain to be born to a female of a given age, in a stable population (Wilson, 1975; National Research Council, 1981). It is calculated from the age-specific fecundity rates and the expected survivorship of females of a given age. Thus, reproductive value is low for infant and juvenile females, who may not survive to reproduce. It is also low for aged females who have finished reproduction. Reproductive value is of evolutionary interest as a quantitative index of the force of selection acting on traits that differ by age. We have already introduced it as an intuitive concept in the thought experiments of the burning nurseries.

The *reproductive potential* of a female is the total number of offspring she can raise. Thus, for a Gombe Stream chimpanzee, if she has her first infant at 13 and then gives birth every five years, she could produce a family of five by the time she is 35. Flo actually did seem to give birth to five offspring, but the fourth and fifth died in her old age, probably the result of her own age and death. So far, only three of the Gombe females are thought to have successfully reared as many as three offspring, although they might have an older, unsuspected offspring and thus a family of four. On the whole, it seems that for primates under natural conditions family size is surprisingly small.

However, it is also clear that the maximum reproductive potential is surprisingly large. Chimpanzee mothers whose babies die, or are forcibly removed in the laboratory, may breed again after two years. For an animal that first gives birth at ten, which is quite possible in captive conditions, and lives to 40, the family size climbs toward 15. Increased nutrition similarly shortens the interval. It seems that when conditions are favorable, primates can take advantage of the situation.

Ripley (1980) argues that some weedy species of primates are facultatively *r*-selected. These species—rhesus macaques, Hanuman Langurs, and ourselves—can move into new habitats and can shift gear in breeding when the habitat allows. There are many aspects of primate behavior that slow population growth, including infanticide. These may operate such that the population is kept in partial balance with the environment, even though the ultimate selective pressures were those on an individual to increase the representation of his genes. Then, in the rare cases when a new environment opens, the checks are lifted and we see, for a while, exponential growth.

GROWING, DECLINING, AND
STATIONARY POPULATIONS

Populations are rarely if ever stationary, that is, in a state of zero growth or decline. Much of the theory of equilibrium age structure assumes a population that is stable, not stationary. A stable population is one in which the age distribution is constant. A stable population may grow or decline at a constant rate. In a stationary population, the constant rate is zero, still with stable age distribution.

Even the assumption of a stable population is not really likely, for there will always be fluctuations from year to year. The life table for Barro Colorado mantled howlers reveals such fluctuations. There is a significant underrepresentation of the cohort who were seven years old in 1976. They were one-year-olds, just being weaned, during the disastrous year of 1970, when it rained during the dry season flowering period and the fruit crop failed. Disparities among male cohorts at age nine, 13, and 15 to 16 years may date back to other hard years, and reflect the relative vulnerability of male two- to three-year-old juveniles who are subordinate to females and who face the rigors of social exclusion from optimal foods but cannot yet tolerate much leaf toxin (Froelich et al., 1981).

MATRILINES AND PATRILINES

Chapter 6 discussed the fact that most primate groups are clusters of matrilineal kin, whereas males migrate between troops. Some are not. Both males and females may then change group, or, in chimpanzees apparently only females.

How closely related are these kin? Much of current theorizing describes where altruistic behavior ought to occur, but we are only now achieving the genetic studies that confirm animals' relatedness (for review of kin selection studies, see Kurland, 1980).

Among the solitary and monogamous primates, the offspring generally leave to form a new group. There we have examples of a female potto who abandoned her range to her daughter (Charles-Dominique, 1977), female orangutans whose daughters remain in adjacent or overlapping ranges (Rodman, 1969), and a gibbon group whose maturing son temporarily took over his mother as mate as well as his aged father's territory (Chivers and Raemaekers, 1980). There is close conflict of interest here within and between groups; the parents balance each offspring's success in finding a mate and a home range against the interests of younger offspring. This is the extreme case of the rhesus' discrimination against elder daughters. In monogamous species the eldest of either sex is peripheralized from the group.

In the common case of neighboring groups being more distant relatives, successful displacement of the next group may mean success for one's own close kin. On Cayo Santiago, dominant groups like dominant matrilines outbreed more subordinate ones (Sade, 1980).

The most detailed genetic studies have been done at Cayo Santiago, an island off Puerto Rico, which Carpenter stocked with rhesus macaques in 1938 (Carpenter, 1942). After initial aggression, the rhesus formed troops where each individual is now numbered, named, and pedigreed. They form one of the longest-studied, free-ranging populations.

As the rhesus groups enlarged, several fissioned into new smaller groups, with consistent spatial separation, stable female membership, and aggressive intertroop encounters in which the monkeys attacked former friends and relatives. One such split took two years to complete from the time the first signs of division appeared (Chepko-Sade and Sade, 1979). Sixty-eight matrilines have been involved in four such troop divisions. Only nine of the matrilines split; the other 59 went as a block to one daughter troop or the other. Of the ones that split, the usual pattern was that an eldest daughter (the lowest-ranking rhesus in a matriline) went to one daughter troop with her offspring, escaping the dominance of her mother and sisters in the other troop.

Whole social groups fissioned when the average degree of relatedness fell below that of second cousins. Geneologies fissioned when the degree of relatedness fell below that of first cousins. The upshot was generally that members of new troops after the fission were more closely related than members of the original troops that split. However, the males still carried complex patterns of relationship from troop to troop (Cheverud et al., 1978; Oliver et al., 1981).

This is probably true for many wild primates as well. When the groups split, they become rivals, much more closely related within than between groups. However, the Cayo Santiago and well-known Japanese macaque troops are *growing* populations, favored by provisioning. The effect of provisioning on the Ryozen troop is shown in Table 11.2. In the period of provisioning, the dominant lineages greatly increased their reproduction. It is not difficult to understand—they monopolized the feeding ground where there was a concentration of peanuts and soybeans. When feeding stopped and animals foraged widely through the habitat, breeding slowed and no longer differed by class (Sugiyama and Ohsawa, 1982). Thus, a stationary or shrinking population may have smaller matrilines and fewer troop splits. (Stopping artificial feeding here equalized the lineages, in contrast to Koshima island where it increased the difference between lineages.)

The genetic linkages in wild baboon and macaque troops may be even stronger through the male line than the female. If one male is particularly dominant or successful at mating in a single baboon troop, a cohort of like-aged youngsters may all be his progeny (Altmann, 1979). Of course, this generally is true in one-male harem groups such as most langurs'. In Himalayan rhesus monkeys, with small groups, high male mobility, and high infant mortality, an

---

TABLE 11.2. Reproductive Success in Ryozen Troops of Japanese Macaques by Feeding Regime and Social Class

Social Class	Total Females	Infants Born	Infants Died	Birthrate (%)	Infant Death Rate (%)	Total Surviving (Two yrs.)
1969–1973 Artificial Feeding						
A (central)	15	12	0	80	0	80
B	28	17	3	64	18	50
C	23	9	3	39	33	26
D (peripheral)	15	10	1	60	10	60
1974–1980 Natural Diet						
A (central)	40	15	4	38	24	28
B	20	4	1	20	20	15
C	41	16	4	39	27	39
D (peripheral)	33	10	4	30	40	18

(Sugiyama and Ohsawa, 1982).

individual may have more paternal siblings in another troop than maternal siblings in its own, so there may be close genetic relationship *between* troops (Melnick, in press).

Finally, within a large, multimale group, particular males may associate and mate, especially with females of a single matriline. This makes sense, because these females are associating with each other, and a male who befriends one finds himself surrounded by the whole sisterhood. This, of course, still further increases the genetic relationship of the matriline in the next generation (McMillan and Duggleby, 1981; Grewal, 1980).

There are, of course, caveats. We have often assumed mother-daughter relations in wild troops on the basis of grooming and proximity. Without prior knowledge, it is easy to include some adoptive relations (Walters, 1981, and Chapter 13). Furthermore, there is occasional female migration reported for both baboons and macaques (Rasmussen, 1981; Burton and Fukuda, 1981). The actual relationships in any one case may turn out to be far more complex than suggested by models of typical species' behavior.

The degree of genetic relation between troop members returns us to the discussion of sociobiology and the conflicts of interest between unit and group. If the group as a whole is closely related, it might increase an individual's inclusive fitness to sacrifice some of his personal fitness for the good of the group. Both the theoretical considerations of the effect of migration and the empirical studies of wild troops indicate that primate groups are nowhere near this level of relatedness. At most, the Cayo Santiago groups approach such a level, but even they would require much increased inbreeding for group selection to play a major role.

It is still conceivable that group advantage played a major role in our own ancestors. If proto-warfare over hunting territories strengthened some groups and crippled others, there could have been great differences in group success. However, if the groups maintained the level of outbreeding usual for wild primate troops and almost all human villages or hunting units, the "success" would be difficult to select genetically. One group's success would usually be at the expense of their kin in the next group.

If you suppose some larger scale tribal loyalty on the scale of a few hundred *inbreeding* individuals, you might propose a selective advantage to tribal scale competition. Or you might suppose that offspring of successful groups *selectively* outbred, choosing mates from other similarly successful groups. Again, by juggling parameters, one could argue for enhanced selection for the more competitive genes, as well as for the altruism to compete at some personal risk for the good of the group. We are back at the uneasy speculations that bridge between genetic evolution and mind. Did the capacity to recognize and remember allied groups and enemy groups ever become a selective force strong enough to favor genetic predisposition to group combat and to patriotism? (Eibl Eibesfeldt, 1982).

INBREEDING AND INCEST

Incest taboos have occasionally been offered as one of the Rubicons between human and animal. "If social organization had a beginning, this could only have consisted in the incest prohibition, . . . (which) is, in fact a kind of remodeling of the biological conditions of mating and procreation (which know no rule, as can be seen from observing animal life) (Levi-Strauss, 1956, quoted by Demarest, 1977).

Actually observing animal life leads to the conclusion that incest avoidance is the normal rule. In most primates, out-migration by subadult males or females or both removes them from their natal troop, and thus from the possibility of close incest, although there may be an optimal level of partial inbreeding.

Besides this general propensity to migrate, adult sons seem to specifically avoid mating with their own mothers. During the son's early years this may be because of the mother's dominance, but it commonly persists to adulthood even if they remain in the same group (Goodall, 1971; Itani, 1972; Demarest, 1977). Even when caged together for three successive breeding seasons, an adult Japanese macaque only single-mounted his mother, although each performed full mating sequences when tested with other partners (Itoigawa et al., 1981). In 15 years of research at the Gombe, only one instance of mother-son copulation has been observed (Tutin and McGinnis, 1981).

Brother-sister mating is also rare in chimpanzees, and two adolescent females consistently avoided the courtship of their brothers. In the harem groups of gorillas an infant was born from a father-daughter mating. After its death from infanticide, the female changed to mating with her half-brother (Fossey, in press). However, the small confined population of the Virunga gorillas may force more inbreeding than normal.

Why do animals go to such lengths to avoid incest, facing the hazards of migration and of making one's way in a different group? Ralls and Ballou (1982) found that 15 of 16 primate colonies in ten different institutions have higher mortality of inbred young over those whose parents are not known to be related. These young do not die of breeding. They die because they are genetically somewhat more predisposed to catching whatever disease is about, or because they move more feebly and thus stimulate less mothering, or because they do not assimilate their food so well and thus are more vulnerable to disease, abuse, and accident. These results are in line with those from other mammals (Frankel and Soulé, 1981; Ralls et al., 1979), and from humans.

One study has attempted to measure the cost of inbreeding in wild primates. Packer (1979) recorded infant survival among Gombe Stream baboon troops, where he knew the life history of each of their fathers. In fact, he found only one male fathering infants on close female kin—a male whose troop had fissioned during his childhood. He changed troop like all other males, but happened into the other half of his own natal group. All four of his offspring died within a month of birth, but we do not know whether this is a likely outcome of any baboon inbreeding, or bad luck of that particular male and his female cousins.

Inbreeding depression is not uniform among mammals. Domestic animals are selected not just for particular traits like speed or milk production, but for ability to deal with the inbreeding that such selection entails.

Some wild species, for example, bonnet macaques, may normally have a higher rate of inbreeding than their congeners (Wade, 1979). However, the general lesson is that inbreeding a wild species almost always has demonstrably ill effects. Sexual reproduction with all its effort to mix genes seems, in primates, to be worth migrating to an interestingly different genetic assortment.

SMALL GROUP EFFECTS AND NONOPTIMAL SOLUTIONS

Demographic calculations should be on the basis of a sample of several hundred animals over several years, to smooth out the vagaries of random differences between groups and years. However, an individual lives in just one group at one time. The inevitable variations from any statistical average raise the question of *nonoptimal solutions.*

If we assume that there is a single reproductive or behavioral strategy whose genetic basis has had a selective advantage, we look for the strategy that would have succeeded in the average conditions faced by many generations of the species over evolutionary time. If these strategies seem unreasonably complex and fine-tuned, opponents of sociobiology accuse its upholders of inventing "Just So Stories"—theories that purport to explain the origin of any kind of behavior without actually proving an evolutionary link.

Janetos and Cole (1981) consider some of the differences between optimal strategies for foraging and mate choice, and some simpler strategies. For instance, they consider possible difference in mate fitness if the female can freely remember and choose the fittest of all the males she meets (optimal strategy), if she is constrained not to go back to males she has left, and so at each step she decides whether to mate or not with a falling threshold of acceptibility until she mates with the last male regardless of quality, or finally a much simpler version of the female having a fixed threshold. In the simple version, she mates with the first male who exceeds her set expectations or else the last-comer, regardless of quality. Under the assumptions of this model there is very little selective advantage for the optimal over the simpler-minded solution. The particular assumptions do not matter so much here as the general point that simpleminded, rule of thumb, common sense solutions may serve almost as well in *variable* situations as an attempt to figure all the angles. There might even be an advantage to having some play in the system, some deliberate nonoptimality, to allow for changing situations. We have seen that this seems true in foraging behavior. Great tits persistently keep up low-level sampling of nonoptimal foods or patches, even after they have apparently mastered the reward contingencies of the situation. This is, of course, adaptive in the longer term, because nature is rarely so mechanical as human theories suggest.

This is leading dangerously close to Part 3, and the idea that mind developed to provide rule of thumb generalizations which were simple enough to follow, but subject to random breakdown and variation: eg. I'm bored with this, let's try something new.

SUMMARY

Demography is the quantification of life processes within populations. It does not predict the behavior of an individual, but the range of conditions in its population. Individual behavior is influenced by demographic processes in the short run, as individuals find themselves in groups with high or low mortality and fertility, and in growing or declining populations. It is also influenced in the long run by selection for life history strategies.

Mortality rate is usually high in infancy and old age. Subadult males may have high mortality in most primate populations. In one and possibly several macaque populations, infant females are more likely to die than infant males, as a result of food deprivation, but this is not true for all groups or species. The number of infants born to females depends on age at beginning and end of reproduction and interbirth interval. Age at beginning reproduction depends in part on females' nutritional status. Interbirth interval depends on whether the previous infant survives, because most primates except Callitrichids have a period of lactational amenorhea. Thus, total number of births may be a poor indicator of females' fitness; one needs to know total surviving offspring. Interbirth interval also depends on whether the species or population is tied to one birth season and, in part, on females' nutritional status. Few wild primates reach a menopause, although the conception rate decreases with age. Reproductive value is an index of the number of offspring a female of a specified age can still expect to survive to bear. It is highest at the start of reproduction. Reproductive potential is the total number of offspring a female can raise. This is surprisingly low for most primates: eight for Amboseli baboon females, and perhaps four or five for Gombe chimpanzees.

In growing populations an individual may have many surviving relatives. In growing matrilineal groups an individual may have many of its mother's clan in the troop, but in stationary populations with high migration rates the closest relatives may be paternal sibs, perhaps in other troops. Migration between troops helps avoid inbreeding depression, as do personal inhibitions against incest between closely related animals. All such calculations, however, are subject to wide random variation in small groups such as primate troops. Calculations of optimal strategies or unidirectional natural selection should be tempered by the consideration that any one group is unlikely to represent the average for the population. An individual needs the flexibility to deal with the nonaverage, nonoptimal group where he or she happens to live.

CHAPTER
12

COMPETITION

DEFINITIONS

Competition, anger, aggression, territoriality, dominance, and *status roles* are often wrongly lumped together. They are not even the same kind of word. *Anger* is an emotion; *aggression,* a behavior; *territoriality, dominance,* and *status* are social relations. Evolutionary competition underlies all the rest.

An ecologist would define *competition* as "the active demand by two or more individuals . . . for a common resource or requirement that is actually or potentially limiting" (Wilson, 1975). Individuals may compete within or between species for resources in the environment, such as food, water, or nest sites. They compete within species for mates of the opposite sex. Darwin described the evolutionary differences between resource competition and competition for mates when he considered the rules of natural selection and sexual selection.

Anger, and its overt expression, aggression, occur in only some competition. Competition may be a simple "scramble," in which the first-comer appropriates as much as possible of the resource, without apparent interaction with other individuals. Aggression appears when there is some actual contest between the individuals (Wilson, 1975).

Aggression may involve fighting, killing, and even cannibalism of the other animal. Most animals have modified their fighting techniques to include some means of escape or surrender, which lets them survive to fight again another day. It usually also pays the victor to desist, saving himself time and energy and avoiding further risk of injury by a desperate foe. It is not true, however, that only man kills his own kind. As Wilson (1975) points out, lethal violence commonly is discovered in mammals somewhere past 1,000 hours of observation. If a Martian zoologist matched George Schaller's 2,900 hours of Serengeti lion watching by spending 2,900 hours observing aggression in a randomly chosen human population, he would probably see only juvenile play fights and a few angry shouting matches among adults. Humans by this test are "a relatively pacific species."

Aggression may be defused or ritualized in each encounter. It may also be regularized over a long period as territoriality or dominance. We have already considered territoriality in Chapter 7. Dominance relations are a similar long-term imposition or acceptance of one's position and one's rights of access to resources, in a spatially overlapping society.

The idea of dominance has a checkered history. Schjelderup-Ebbe, who discovered the pecking-order of hens, enlarged his findings to a Teutonic theory of despotism in the structure of the universe. For instance, water eroding stone was "dominant" (Gartlan, 1968b). Schjelderup-Ebbe called animals' ranking "dominance," and many workers, with an "aha," recognized dominance hierarchies in many vertebrate groups.

Among the primates, dominance soon seemed even more baroque. Baboons and macaques mount each other in a gesture of assertion or present their rumps to be mounted as a gesture of submission (Maslow, 1936). Rarer variants are the stronger animal backing up to the weaker and forcing the weaker to mount, which happens among bonnet macaques (Simonds, 1965), or else juvenile males (playfully?) mounting adult males, as in hanuman langurs (Jay, 1965). This sexual gesture, transferred into threat-submission situations, combined with Zuckerman's (1932) description of zoo hamadryas baboons fighting to the death over females and the general influence of Freudian theory, all led to a view of aggressive dominance as a universal primate ranking, inseparably bound to sexual priorities. Of course, Carpenter's (1934 in 1964) howlers rarely threatened each other at all and mated promiscuously within the group, but people then said new-world primates might best be ignored.

At last, DeVore and Washburn (1960) pointed out that in a savannah baboon troop a recognized hierarchy stabilizes the society. Far from continuous vicious pecking down the line, the hierarchy makes it possible to avoid fights in most situations, because each animal knows the other's strength and respects his rights.

With more recent analysis, dominance has been sorted out into component parts. First, there are *priorities to essential resources,* such as food or water, and *priority for mates.* Second, there is the *threat hierarchy,* which may be complicated by coalitions among animals and not be a linear peck-order. The threat hierarchy may or may not be closely correlated with priorities for any given species or environment. It may also be much more clearly shown by subordinates' avoidance behavior than by dominants' overt threat. In this case, it is more useful to describe the hierarchy by *approach-avoidance* (Rowell, 1966).

Various other attributes have joined the two basic notions of priority and of threat hierarchy. In some species, such as white-faced cebus and gorillas, one big male challenges strangers and other threats to the troop. He also suppresses fighting within the troop. These are two aspects of what Bernstein calls the *control role* (Bernstein, 1966; Hoff et al., 1982). In other species, one animal sets the direction for troop movement. This leadership function may coincide with other aspects of dominance; in gorillas it is the silverback who decides troop movement as well as virtually everything else. It may be quite separate;

in hamadryas harems, the day's course is often set by an old male who has relinquished other privileges including mating rights. Many other behaviors such as grooming may also correlate more or less loosely with the threat hierarchy.

Is there still a reason for speaking of dominance, or is the word too contaminated by its multiple meanings for further use? In "Dominance, the baby and the bathwater," Bernstein (1981) argues that dominance relations between particular individuals are a clear part of daily life in most primate troops. Relative tendency to dominate others may even be heritable and thus subject to natural selection. We can recognize this fact, as the monkeys do, without committing ourselves to any single listing of rank orders, that are, at best, a partial description of that particular troop, day, and measure chosen, and at worst are an artifact of our ability to count.

FREQUENCY AND DEGREE OF AGGRESSION

Why do so many animals have conventional forms of competition? It would seem that if an animal can maim or kill his competitor by a sneak attack, he could usually gain by doing so. Then why don't they? If you regard a contest in terms of game theory, the best strategies depend on the potential payoffs. If there is a great deal to gain and little to lose by escalating a contest, one should expect violence. This would be the case for elephant seals, for instance. Their winners obtain huge harems; losers will probably not mate at all and might as well be dead in terms of prolonging their genes. In the more usual primate case, winners obtain a few matings, but they usually will have other chances later, while the risk of serious injury in fighting is a effective deterrent.

Imagine a population of "hawks" who always escalate fighting until the opponent gives up, and "doves" who offer ritualized combat but retreat if threatened with violence. A dove randomly wins or loses the prize if he disputes it with another dove, but always loses to a hawk, with a small chance of injury while retreating. Hawks win over doves, and sometimes over other hawks, but often get hurt. Neither strategy is best in all situations: in a population of doves, it would pay to be the only hawk, but in a population of mainly hawks, it would be best to hide as a dove for a while until the hawks killed each other off. The evolutionary stable mix of doves and hawks depends on the sizes of payoff and penalties (Maynard Smith, 1974).

		Meeting				Meeting		
		D	H			D	H	NFU
payoff	D	½+	−	payoff	D	½+	−	½+
to	H	+	±	to	H	+	±	±
					NFU	½+	±	½+

A third strategy, called retaliator or no first use, may be better in most cases than either hawk or dove. The no first user escalates only in response to escalation, and otherwise fights conventionally. It only risks injury if it confronts a hawk, and it wins more often than doves do. Changing the payoffs can nullify even this strategy, though. If the negative payoff is not only serious injury but mutual assured destruction of oneself and all one's kin, a gene that is still here to reproduce has never taken that gamble and lost.

It is fairly easy to add a term for time and energy expenditure during the ritualized contest but the moral is the same. The degree of violence that is profitable depends on the relative payoffs and the frequency of different strategies in the population (Maynard Smith, 1974, Maynard Smith and Parker, 1976).

This sort of calculation offers some hope of formalizing the vast differences in aggressive behavior within and between species. The overall conclusions for primates seem to be that (1) primate competition generally exists, either as mutual exclusion or a threat hierarchy; (2) competition is usually expressed in infrequent ritualized confrontations; and (3) many or all species seem capable of violence, even though it is rare. The gamut from the mildest threats to full-scale battle seems to be generally present, but the commonest form is low-cost or ritualized.

DIFFERENCES BETWEEN MALES AND FEMALES

Males, by definition, are the sex with smaller gametes. Females produce fewer larger ova: females are committed to investing more reproductive effort in each gamete than do males. The differences probably arose in our earliest sexually reproducing ancestors, as it was more efficient for some sex cells to actively seek out more sedentary partners, while the sedentary partner contributed more cytoplasm and yolky food to nourish their joint product. This initial difference in investment can be reversed in a few species as the offspring grows: male sea horses and cichlid fish and anis (a tropical bird) raise the brood while the female goes off and mates again. In mammals, with the addition of placental nourishment of the embryo and then lactation, there is almost no species in which the energy the father devotes to each infant regularly exceeds that of the mother. In fact, the only individual cases that come to mind are some human families.

The mother's total reproductive potential is limited by the amount of energy she has to devote to the care of her infants, after providing for her own survival in order to care for them. Her immediate competitive goals then are for food for herself and her young and whatever indirectly helps ensure this food, particularly territorial or social status. These ambitions are limited but continuous. Most wild primates do not seem to have all the food they would like. Extra food leads to bigger, healthier offspring, and to rapid population growth; in most wild, unprovisioned populations, females are usually in competition with each other. Their ambitions rise beyond day-to-day feeding success when they can pass either physical strength or social advantage in reproduction to the next generation by assuring a daughter's food supply, or raise a successful son who may many times multiply his mother's genes.

Males, in contrast, are playing for higher stakes within their own lifetime. They may fertilize many females and thus have the potential to leave many more offspring than a female could. In some species this high payoff makes it worth considerable risk to be successful, including that of serious injury. For the unsuccessful ones, the serious injury, or even death, may materialize or they may just produce no offspring. Thus, males (or, to be more precise, the sex that invests less in each offspring) are under more pressure to compete for mates.

This competition again has two aspects. The competition may be within members of the same sex, by prowess, in Darwin's term, with the female then accepting the winner. Males, on the other hand, may compete by the exercise of charm directly to the female, who then chooses which mate she prefers. Either form of sexual selection may run unconstrained, with ever more dazzling or ornate adornments being chosen by the females, or ever more imposing intra-specific weapons determining the winner of conflict among males. Darwin proposed that such sexual selection would be checked only by natural selection, particularly in the form of predators. The interaction has recently been illustrated in the long-tailed widowbird of the Kenyan savannahs. When the trailing tail feathers were cut from one group of birds and glued onto others, more females chose to nest in the territories of males with double-length tails even than with normal controls, or than with the same males before their cosmetic operation (Andersson, 1982). The females apparently preferred supernormal stimuli: even more extreme versions than exist in nature. Female preference constantly acts to increase tail length, whereas predators and the limitations of energy counterselect for shorter tails. The actual length results from the balance of two opposite forces.

BODY WEIGHT, CANINE SIZE, AND TESTIS SIZE

Males and females are different sizes in many species. The argument of the preceding section suggests that greater male size results from sexual selection. This is the common assumption, since Darwin. However, in many mammals and several primates it is the females who are larger. There seems to be no single, simple explanation for large females; there are a whole range of different possibilities, different for each taxonomic group (Ralls, 1976). Primates with larger females include the potto, the blackhanded spider monkey, the common marmoset, the Sunda Island, banded, and whitefronted langurs, and the moloch gibbon. I can think of nothing that these species share except large female size, certainly nothing that differentiates them from their closest relatives.

Excluding these species, degree of sexual dimorphism correlates with body size, breeding system, habitat, and diet. (Fig 12.1A, B). That is, relatively larger males occur in heavier species, in polygamous rather than monogamous species, in terrestrial rather than arboreal species, and in frugivores rather than folivores (Clutton-Brock et al., 1977; Ralls, 1977; Leutenegger, 1978). There is considerable controversy about which effect matters most. Recently, Leutenegger (1982) has pointed out that some 80 per cent of the variance in sexual dimorphism depends on body weight. He argues that the primary selective pressure is natural selection for large body size in both males and females, whether for predator protection, wide range, or coarse diet. If males are genetically more variable than females (there is a little evidence this may be so), then selection for large bodies would proceed faster in the males.

A second explanation for sexual dimorphism is divergence of feeding niche.

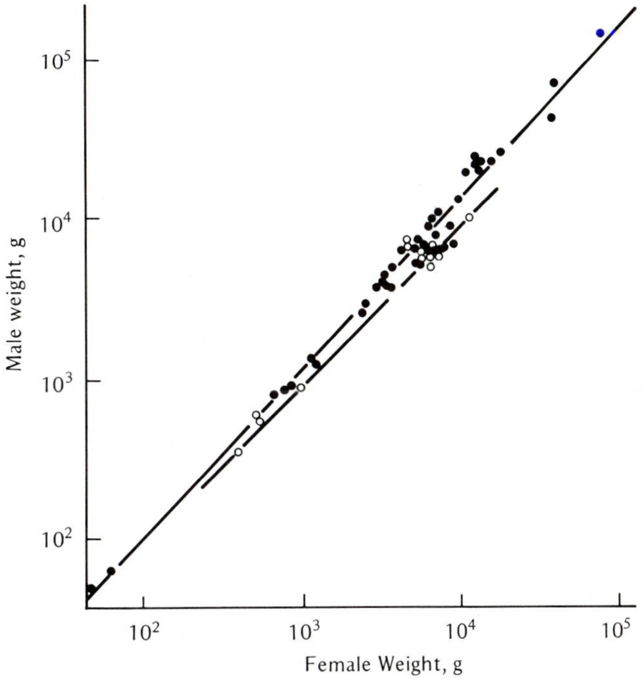

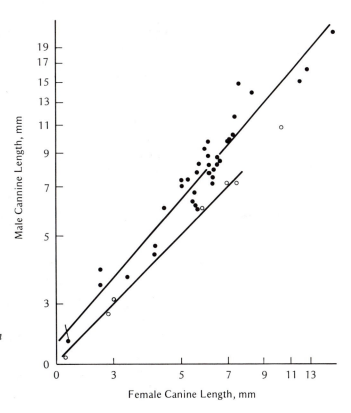

FIGURE 12.1A. *Relationship between male and female body weights. The upper line (P) represents the regression for 42 polygynous primate species (closed circles). The lower line (M) is the regression for 11 monogamous species (open circles). Polygynous species are slightly more dimorphic than monogamous ones. They also include larger animals that tend to be more dimorphic. (Leutenegger, 1982.)*

FIGURE 12.1B. *Relationship between male and female maxillary canine length. The upper line (P) represents the regression for 37 polygynous species (solid circles). The lower line (M) is the regression for 5 monogamous species (open circles). Polygynous males have significantly longer canines than monogamous ones. (Leutenegger, 1982.)*

The male orangutan shambles along the ground, and the females are more arboreal. He is a "phlegamatic red titan (who) consumers vast quantities of unripe fruits and mature leaves, the junk foods left by the more discriminating females" (Hrdy, 1981; MacKinnon, 1977). Food may be a constraint without being different, which may in turn be the primary basis of size difference (Martin, 1980). Female blue monkeys consumed more calories than the larger males, just to support pregnancy and lactation (Coelho, 1974). Males may then be freer to put more calories into personal growth.

In Leutenegger's calculations only some 7 per cent of sexual dimorphism is attributable to mating system. A slightly earlier paper, however, shows that monogamous primates have roughly equal body size, but polygamous primates have larger males. Monogamous primates are, on the whole, not very big. One could turn the argument around and credit large size as well as large dimorphism to sexual selection on the males (Leutenegger, 1982).

A related set of arguments concerns canine dimorphism (Fig. 12.2.). In a few species—the monogamous common marmoset, Geoffrey's tamarin, and moloch gibbon—females have larger canines. Taking the residue of species,

FIGURE 12.2. *A Celebes macaque male displays his size and his canines in a tension yawn. (Courtesy P. Coffey and the Jersey Wildlife Preservation Trust.)*

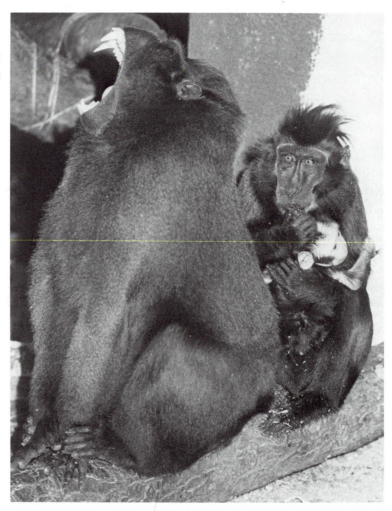

canine dimorphism increases with canine size, with polygyny, and with terrestrialism, but has no relation to diet. Again, does one argue that total size by natural selection matters most, or that sexual selection, as reflected in the mating system, matters most? (Harvey et al., 1978a, 1978b; Leutenegger and Kelly, 1977; Leutenegger, 1982).

Even if a relatively small part of the variance relates to the mating system, it is one of the most interesting parts. Three of the Oligocene primates of the Fayum, the probable ancestors of modern monkeys and apes, are sexually dimorphic in size and canines like modern polygynous species (Fleagle et al., 1980; Martin, 1980). We presume that these earliest old-world monkeys were living in polygynous societies.

The weight differences between human males and females fall into the "mildly polygynous" class (Alexander et al., 1979). It seems that the human basal stock was certainly not like the obligate monogamous mammals. There are also racial differences in dimorphism. "Amerindians are most dimorphic, followed by Asians and Europeans, with inhabitants of Africa and New Guinea more or less alike and least dimorphic" (Eveleth and Tanner, 1976, quoted by Alexander et al., 1979). This seems directly opposite to recent marriage patterns, where Africans and people of New Guinea are most polygamous. Alexander and colleages believe, however, that they can distinguish a group of "ecologically monogamous" people who live in marginal habitats, as opposed to culturally monogamous ones who do not. The ecologically monogamous people (mainly hunter-gatherers) have little dimorphism. This may be the result of low nutrition stunting the males; if better food conditions led to more male growth, it would imply a polygynous evolutionary background.

All of these bodily size and canine size calculations relate to female:male sex ratios. They do not relate to the number of males in a group. The size of testes relative to body size, in contrast, indicates most clearly the difference between single and multimale social groups (Fig. 12.3). It seems that where more than one male may mate with the same female, there is sperm competition. That is, the male can increase his chances of fertilizing the egg by producing

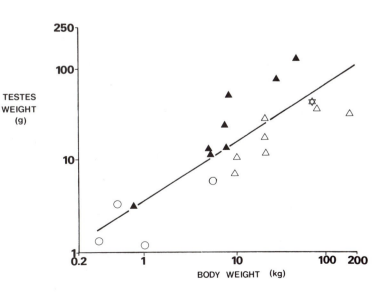

FIGURE 12.3. *Relationship between combined testes weight and body size for primate genera with different social systems:* ○ *monogamous,* Δ *single-male,* ▲ *multimale,* ☆*human (After Harcourt et al., 1981.)*

more sperm. On this measure, man falls (just) on the side of single-male groups (Short, 1981; Harcourt, 1981; Harcourt et al., 1981).

MALE OR FEMALE PRIORITY

There are three main patterns of priority for food between the sexes as well as for rights of way on branches and associated threat. In obligate, monogamous species, male and female seem to have about equal rights, along with equal body size and canine size. In most polygamous monkeys, all the apes and the solitary prosimians, males take precedence over females. This follows naturally from the various pressures that favored large male body size, weapons, and aggressive temperament (preceding section; see, also, Hrdy, 1981). Even among "normal" male-dominated primates there is a spectrum of behavior. Female Japanese macaques may unite to drive out immigrant males and even bite the males. Female savannah baboons apparently never dare bite or oppose their much larger males, though female support can make a great difference to a male's status within the troop (Packer and Pusey, 1979; Strum, 1981).

What can we conclude about the few species where females actually dominate males? First, all four social lemurs that have been well studied have female food priority. This is true for the monogamous Indri (Pollock, 1979), the small, group-living white sifaka (Richard, 1978), the brown lemur (Jones, 1980), and the ringtailed lemur in its larger troops (Jolly, 1966). Females of these species insouciantly remove food from in front of their males, occasionally cuffing the male over nose or ear. Indri females regularly feed higher in the trees, among the richer foliage (Pollock, 1979). All this would be inconceivable to a baboon or macaque male.

The males are not less aggressive than the females. Ringtails and sifaka maintain a year-round hierarchy of threats. Ringtails stink-fight, running their tails through the scented spurs of their forearms, then shivering the scent forward at an opponent, who may give a spat call and withdraw, or may reply in kind. More subtly, males watch each other, retreating as others advance, or redirecting a threat to yet another animal. Females have a loose hierarchy of their own, but are not concerned with the male game of jockeying for position. The mating season is brief and bloody. Males attack each other in jump-fights round an estrus female, slashing downward with the canines. In several seasons, the winner of observed fights has been a subordinate male, not the top of the usual threat hierarchy. The ringtails seem to be an extreme case of non-correlated measures of status, with the threat hierarchy having little to do in the short term with priority for either food or sex. Sifaka may be as baffling: males may have status-determined access to their group females, but physically fight intruders during the short mating season. We know nothing of the mating season in other wild lemurs or Indri, except that they too mate at a highly restricted period of the year—for Indri, perhaps only once in three years (Jolly, 1966, Richard, 1978, Pollock, 1975).

Hrdy (1981) thinks this seasonality may be the clue to the evolution of female dominance. Two monkeys resemble ringtailed lemurs in their female priority to food: the squirrel monkey in the New World and the talapoin in the old. These are both small-bodied, greenish, swamp-living monkeys. They congregate in huge troops—scores or perhaps hundreds of animals together. Casual counts tend to be too high. They baffle observers as well as predators with an impression of multitudinous small bodies flickering through the trees. If disturbed while sleeping, talapoins simply drop into the river and swim away.

Squirrel and talapoin males tend to keep separate from their females, out of the breeding season. Talapoin females gang up on males in cages, take their food, and may even kill them. The males may be as active in maintaining separation as the females. In one experiment there were no conflicts over scattered food, but two squirrel males managed to hold off eight females to gain access to a bottle of fruit drink (Leger et al., 1982). As usual, the exact conditions of the competition have much to do with the outcome.

In a seminatural environment, at any rate, squirrel monkey males traveled separately from the females, except during the mating season. Then they fattened up "like football players with padded shoulders" (Hrdy, 1981); they fought each other and wounded each other (Fig. 12.4.). Individuals changed rank as a result of each fight, as unpredictably as ringtailed lemurs (Baldwin, 1968; Scollay and Judge, 1981).

Perhaps female dominance has something to do with seasonal breeding in small-bodied primates. Hrdy suggests that the males conserve energy, laying low until they can make their great challenge during the mating season. Jones replies that this strategy makes sense only if it somehow benefits the male to lay low. If, for instance, it gives his offspring a greater chance of survival, he

FIGURE 12.4. *A squirrel monkey male, Scar Eye, who intermittently led his troop, (A) in fatted condition during the mating season, and (B) thin, six months later. (Courtesy F. V. Dumond and Monkey Jungle.)*

A   B

253   *Competition*

FIGURE 12.5. *A patas resident male in Kenya chases a would-be challenger from his one-male group. (Courtesy D. Olson.)*

might relinquish food to them. This would explain the self-sacrifice of the monogamous Indri, but does not apply to the others more than any other primate.

There is one more case of a female-dominated primate that does not fit any rules. Patas males look like typical harem overlords. They are twice as large as the females, with red and turquoise genitals and long canines. Young males are not tolerated in the harem, but roam as bachelor bands. Sexual selection seems on the rampage (Fig. 12.5.).

However, females as a group take precedence over males both in captivity and in the wild (Hall, 1964; Rowell and Hartwell, 1978). The male hangs back as the troop goes down to drink, not as a chivalrous guard against predators, but because he is not allowed down.

Patas, like the others, are seasonal breeders. There may be yet another clue here. Patas forage in highly dispersed manner. If there is danger, the male may bounce on a bush in a stylized distraction display. Females scatter, diving for cover, scooping up their own or any other babies as they go (Hall, 1965; Zucker and Kaplan, 1980). This behavior led Chance and C. Jolly (1976) to describe them as "acentric" animals. Unlike macaques, baboons, and many others whose males at least sometimes confront a danger and defend the troop, while the females put themselves behind the males, patas females' attention is directed away from the male if there is a threat. All the same, Struhsaker and Gartlan (1970) watched three males chase down a jackal and rescue an infant from it, so patas are capable of defense.

What, if anything, can we conclude? At least, as Hrdy says, primates are not wholly locked into male dominance, although male dominance is clearly the norm. We can conclude that female food priority is not confined to any particular body size, or difference of body sizes, or social structure. It seems to go with seasonal breeding, although many seasonal primates also have male dominance. It often seems to go with males spatially foraging separately, though many male-dominated primates also have spatially separate males, such as orangutans and chimpanzees. The separateness may be imposed on the males by the females; in ringtails, it is only subordinate males who lag behind the troop as a Drone's Club. Finally, it goes with minimal male defense against predators. Here, the lemurs are an equivocal case. The best one can say is that lemur males and females are equally active in mobbing ground threats and equally hasty in diving away from hawks. It could be that the lack of large carnivores on Madagascar has played a part in allowing female dominance (A. M. Jolly, personal communication.)

I do not think the riddle of female dominance is solved. Ralls (1976) reviewed the mammals in which females are larger-bodied than males. She decided there is no simple explanation. Perhaps female dominance is as complicated.

MALE DOMINANCE HIERARCHIES: BABOONS

Let us turn from competition between the sexes to competition within each sex. There are a variety of ways to compete. We have already considered territorial exclusion in Part I. We have also considered confining most overt aggression to a brief breeding season in the description of ringtailed lemurs and squirrel and talapoin monkeys. Let us now describe three species in which males have fairly continuous dealing with other males and potential rivals: baboons, hanuman langurs, and chimpanzees.

Baboons were the original model for competitive males exerting their rights over females. Zuckerman (1932) watched chacma baboons at the Cape of Good Hope, and a colony of hamadryas in the London Zoo. The hamadryas fought over estrus females, sometimes so violently the keepers could not intervene for two days to remove the female's body. It was not until the studies of Kummer (1967) and his successors that we learned wild hamadryas form small harems. The male disciplines his females into following him. He bites any female that strays too far on the back of the neck. She screams loudly and returns to the group. Experimental release of captured females shows that it is the males who run the system. A hamadryas female released among savannah baboons learns that she can wander unmolested through the troop, whereas a savannah female released among hamadryas is bitten and chased until she learns to follow one leader—only, she is eventually likely to run right away. There are a few hybrid troops on the species' boundary with intermediate behavior.

Even in this extremely possessive social system, there is some male tolerance. Males do not try to steal each other's adult females, even if the apparent bond is only half an hour old. They assess the female's preference for her partner: if she seems bonded, there is no challenge, irrespective of the female's attractiveness to either male (Backmann and Kummer, 1980). They can form new harems only by adopting a juvenile female and waiting until she matures, or by tagging along as a subadult with an established group. The group leader tolerates the hanger-on, who gradually matures and eventually takes over mating rights within the group. At this stage the young male tolerates his aging leader, who retains "respect" and the leadership of troop progressions.

Hamadryas sleep in vast numbers on the few safe cliffs and forage as single unit groups through the day. Once in a while a combat flares that ally the males of several unit groups against those of several others. This reveals a "band" structure between the level of harem and that of sleeping aggregation. There is, thus, a level of organization comparable to the multimale troop of savannah baboons.

That multimale troop has shaped our most common picture of primate dominance. Zuckerman (1932) saw the males as in competition for the permanently receptive females. Maris (1969) watched baboon males cooperate in attacking a leopard, and saw them as protectors of the group. Hall and DeVore (1965) wrote of the stabilizing effect of known hierarchy. They checked the threat hierarchy by throwing pieces of food between animals. Dominants had priority to the food. However, in one troop a coalition of two individually subordinate males could dominate the male above them. All these, with Maslow's (1936) study of dominance gave an impression of correlation between the various aspects of dominance. That is, it seemed that sexual success went with food priority, threat, and still other roles like predator defense, and the protection of mothers with young infants against harassment by troop members (Fig. 12.– 6 A,B,C,D).

Then this view began to disintegrate, as one looked at individual character. In Stoltz and Saayman's (1970) baboon troop, Barker was young, aggressive, and individual head of the threat hierarchy. Toothless old Yogg, instead, was groomed by most females and did most copulating. In Ransom's (1981) study of the Gombe baboons, two adult males joined the Beach Troop. "David and Grinner differed markedly in the tactics they used . . . David chose the slow process of establishing dependent and supportive relationships with resident males, a process that seemed to have little disruptive effect on the status quo. Grinner went it alone, avoided new relationships, and aimed for the top. He harassed and ultimately replaced Moon as highest status male . . . His arrival in B troop shattered the status quo and began a long period of social instability which culminated in a troop-split."

There have been two views of the relation between threat hierarchy and access to resources, particularly reproductive females. Rowell (1972) believed that her Ugandan forest-edge baboons changed troop so often that every male occupied every rank during his life, and both ranks and reproductive success equaled out. Altmann (1972) drew up a priority-of-access model in which dominants had first choice of female; the more females that came into estrus at once, the more males could mate. Hausfater (1975) attempted elegant quantitative analysis of rank and reproduction. He found that there was indeed some correlation, but the alpha-ranked male in his troop mated less often than expected, and the low-ranking males, more. Again, baboons shifted rank frequently so that one would need lifetime data to see whether a short time near the top would really lead to more offspring than steady career near the middle. Similar data emerged from Packer's (1979) and Seyfarth's (1978a,b) work—some correlation with rank, but nowhere near perfect. The deciding factors are, rather, which males will join in coalitions, and which males in consortship, with female as well as male choice of partner important.

The relationship of rank to food resources is equally equivocal. Amboseli baboons of all ages, sexes, and ranks fed for approximately equal lengths of time. Low-ranking animals were more likely to be interrupted by other animals, which may impose a social cost in shorter bout length. Animals of adjacent ranks tended to eat similar food: differently ranked animals ate more divergent diets. All the rank-related effects were fairly obscure, though. On the whole, rank differences in feeding needed to be teased out statistically (Post et al., 1980).

FIGURE 12.6. *Adult male baboon aggression in baboons is common, but is mainly bluff. A. charging run, and B., sudden stop with stiff-armed threat, eyebrow raising, and pilo-erection. C and D., when challenged, males resist, fights with chases, slapping, and canine slashes can break out. There is still a low incidence of injury to antagonists. (Courtesy T. W. Ransom.)*

Strum (1982) most recently clarified baboon tactics. Her troop, the Pump House Gang of Gilgil, had a fairly linear threat hierarchy at any one time, though with many cases of reverse threat up the line. Mating success did not correlate with rank, but with tenure. The lead four animals in the threat hierarchy were three newcomers and one young male maturing to adulthood. The three males who had been resident more than a year in the troop had many female friends and did most of the consorting. Strum suggests that the newcomers are using aggressive tactics to place themselves among the older males and to gain membership in the troop. Meanwhile, the older resident males use much less costly affiliative strategies and a minimum of aggression directly related to reward.

A

B

C

D

FIGURE 12.7. *The dance of the adult males. The baboons' split-second timing in echoing each other's actions shows their opponents that these two are a united coalition. Such coalitions are to both individuals' advantage, which may explain evolved capacities for friendship. (Courtesy T. W. Ransom.)*

This is the Barker-Yogg and Grinner-David difference translated into differing strategies at different ages or levels of experience. (Note that the population at Gilgil is expanding, with many estrus females and high infant survival. This may be only one situation along a continuum (Hrdy, personal communication). Males may even monitor other troops before changing. Are they assessing the nature of potential mates and rivals?

The baboon case is worth spelling out in such detail because it has been the central model in every stage of the study of primate dominance. If you read almost any popular account of the subject, the author is usually talking about savannah baboons. He may describe a monolithic system in which the most aggressive animal claims rank, food, and females, or he may describe a variable

strategy changing through life, depending as much on coalitions as cooperation, and on the individuals within each troop (Fig. 12.7A,B,C,D). It is all the
same baboons in different decades of study.

MALE INFANTICIDE, ESPECIALLY IN LANGURS

When Phyllis Jay went to study hanuman langurs in North India, she watched
a multimale troop that seemed to offer a welcome contrast to the savannah baboons. Under normal conditions when the hierarchy was not changing, males
scarcely ever threatened each other. There was a rank change during her study,
accomplished with no more fuss than the erstwhile beta male whooping and
bouncing on trees at the end of the mango grove, and thereafter walking proudly
before the alpha. But then Sugiyama studied hanumans in a crowded population at Dharwar in South India. There they lived mainly in single or age-graded
male troops, with roving bachelor bands of excess males. Sugiyama watched
four troop leader changes, one provoked by his experimental trapping of the
harem male. Bachelor males came in, fought the resident male, and among each
other. The winner ousted all other males, adult or juvenile, from the troop. He
also attacked small infants, slashing them with his canines. The mothers eventually abandoned their dead or dying offspring. The females came into estrus,
having lost their offspring or even before, during the confusion of the takeover, and they mated with the new male. It seemed langurs were not so gentle
after all (Jay, 1965; Sugiyama, 1965a,b).

Sugiyama's report and Mohnot's (1971) that he had watched males killing
three infants near Jodhpur challenged Sarah Hrdy to study hanumans. She
originally planned to study the effects of crowding on behavior, for she thought
infanticide must be the result of social stress. She watched at Abu, a sacred
mountain where langurs are tolerated by townspeople and fed by pilgrims.
During nine male troop membership changes, 17 infants disappeared. Hrdy saw
males repeatedly stalk, attack, and wound infants, but did not actually see one
kill, although local residents reported killings. (Fig. 12.8). Far from being in-
adaptive for the infanticidal males, it seemed that the new males eliminated their
predecessor's offspring, speeded the females' return to breeding condition, and
thus greatly increased the chances of their own infants reaching independence
before their own tenure ended and they in turn were replaced by another infanticidal male. This is only one of several situations where infanticide occurs
in mammals. Other situations have different evolutionary backgrounds. In par-

FIGURE 12.8. *Male langur
makes off with an infant, whose
tail projects to the left of his
mouth. Two old females attack
him and temporarily rescue the
infant. (Courtesy S. Hrdy and
Anthro-Photo.)*

ticular, human infanticide is usually performed by the mother or a trusted "wise woman," not by stepfathers (Hrdy, 1979; Ripley, 1980).

Hrdy concluded that male and female reproductive advantages were so different that the two sexes were virtually at war over infants' survival. Phyllis Jay Dolhinow (1977) and several of her students (Curtin and Dolhinow 1978; Boggess, 1979, 1980, in press) reacted strongly, saying that over most of their geographical range langurs do not commit infanticide, and that this must be a pathological response to human crowding and interference. It could not be a normal, evolutionarily adaptive part of the species' repertoire.

Chapman and Hausfater (1980) then codified the conditions that would lead to infanticide. The relation between male tenure, interbirth intervals between surviving infants, and the degree to which mothers can be hurried into estrus determine the rewards to infanticide. It is intuitively obvious that if mothers have a very short period of lactational amenorhea so that losing the baby makes little difference, or if they breed seasonally so that mating cannot be hurried, then there is little or no advantage in infanticide. However, given year-round breeding and long lactational amenorhea, then it is also obvious that an infanticidal male in a population of noninfanticidal males will always have more breeding opportunities than his competitors. He begins breeding earlier, or at least as soon as would a noninfanticidal male, and his infants survive after his replacement. In other words a mutation for infanticide should spread rapidly in such a population.

What if the trait has already spread, so the replacement male has a good probability of being infanticidal? Then, if a male's tenure ends when his own infants are not weaned, his offspring will be killed. Thus, there are lengths of tenure for which the noninfanticidal strategy works better. It may be better to start breeding later, leaving the next generation behind as robust juveniles and small embryos. (Fig. 12.9.). For different lengths of tenure it should be possible to predict the proportion of infanticidal males you would expect in a population at equilibrium (Hausfater et al., 1982). Note, however, that they predict both longer periods and higher amplitude of breeding advantage to infanticidal males than to noninfanticidal ones. On average, if a male cannot predict his tenure, he would do better as a killer.

If the theory is so strongly in favor of infanticide, why don't we see it more often? One reason is that often the conditions favoring infanticide are not met. Another is that parents may have evolved counterstrategies that removed the

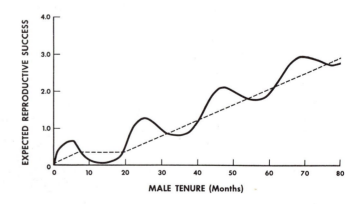

FIGURE 12.9. *Expected reproductive success of infanticidal and noninfanticidal male langurs as a function of male tenure (see text). (After Chapman and Hausfater, 1979.)*

conditions. The third is that we were not looking. The strongest evidence for evolved infanticide comes from rodents. The Bruce Effect, which is abortion or resorbtion of embryos by female mice exposed to the urine of strange males, seems to be evolved to preempt infanticide, saving the female the extra effort of bringing doomed young to full term. Infanticide in lions of the Serengeti was known well before the langur case (Shaller, 1972; Bertram, 1975).

Male infanticide has been observed among many other primate species. It is strongly suspected in purple-faced langurs (Rudran, 1973), silvered langurs (Wolf and Fleagle, 1977), black-and-white colobus, and savannah and hamadryas baboons. It has been observed and documented in red-tailed guenons, blue guenons, red colobus, red howlers of the new world, chimpanzees, and gorillas. What seemed at first to be pathological abnormality of a few overcrowded hanumans now seems to be an evolved, adaptive behavior that crops up in many primate lines as well as in other mammals (Marsh, 1979; Struhsaker, 1977; Butynski, 1982; Crockett and Sekulik, in press).

Table 12.1. (after Fossey, in press) lists observed and inferred killings of gorilla infants. It should be said that all of those took place during a time of great disturbance when poachers were killing adult males; few or none have happened more recently now that male leadership has stabilized. Five of six presumed killers were males, none the infants' fathers. In two cases the female transferred to the killer's group and the killer sired the next infant, and in two more the killer copulated with the mother or another group female. In four cases where the intervals to the next birth are known, they are much shorter than the normal 39-month interval if the infant survives. This fits well with the arguments for adaptive male competition by infanticide. All the killings occurred during violent physical confrontations, either within or between groups, and in all witnessed killings the male deliberately charged, snatched, and bit the infant, usually through the skull.

The one case in which a female was the presumed killer was also the only case of cannibalism; 133 fragments of infant bone, teeth, and hair were found in the dung of dominant female, Effie, and her young adult daughter, Puck. Effie gave birth two to five days after the killing. This will be reconsidered in the section on female competition, along with the cannibalistic female chimpanzee. Even this gorilla case resulted in a shift of male-female bonding; the mother transferred her allegiance from the aging silverback, her own father, to a recently matured half-brother.

Similar conclusions emerge from the red howlers of Hato Masaqueral in Venezuela. One infanticide and five cases of males stalking infants not their own, countered by maternal aggression or defense by the father, were observed. (See Fig. 12.10.) The age and number of infants who disappeared during adult male change is highly significant (Crockett and Sekulik, in press). Combined with Rudrans' (1973) prior account of male infanticide in the same howler population, the Venezuela study leaves small room for doubt.

Infanticide may also happen in multimale troops. Savannah baboons occasionally attack infants; in this case, it seems to be immigrants who were not members of the troop when the infant was conceived (Busse, in press). One such immigrant also attacked pregnant mothers, perhaps causing three abortions (Pereira, in press). Chimpanzees are the most peculiar (Bygott 1972). Five of nine known cases were attacks by a group of troop males on "stranger" fe-

TABLE 12.1. Infanticide Among Gorillas, Virunga Park

Victim	Age, Months	Sex	Mother	Sire	Killing Seen	Killer	Circumstances	Killer Cop With Female	Female Transfer to Killer's Group	Interval to Next Birth	Sire Next Infant
Thor	11 mo.	F	Macho	Rafiki	+	Uncle Bert	Rafiki's death		+	25 mo.	Uncle Bert
Frito	3 mo.	F	Flossi	Uncle Bert	+	Beetsme	Bert's death	+	−	14 mo.	JP
Muelu	8 mo.	F	Simba	Digit	−	Nunkie	Bert's death?		+	40	Nunkie
Curry	9 mo.	M	Bravado	Beethoven	−	Lone silverback?			?		
Banjo	5 mo.	?	Pantsy	Beethoven	−	Effie?	Female cannibalism?		?	11	Icarus
Phocas	1 d	M	NoName	Paddy	+	lone silverback	Male cop with another group female		−	—Female died of wounds	

Three more dead infants found after hoot series and outbreaks, in lesser known troops. Two with deep (canine) puncture wounds, especially in skull. One 30-month female attacked by group when introduction attempted.

(After Fossey, in press).

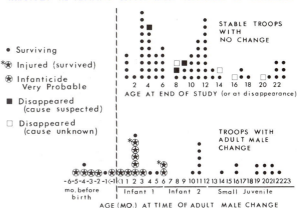

• Surviving

*⊛ Injured (survived)

⊛ Infanticide
   Very Probable

■ Disappeared
   (cause suspected)

□ Disappeared
   (cause unknown)

STABLE TROOPS
WITH
NO CHANGE

AGE AT END OF STUDY (or at disappearance)

TROOPS WITH
ADULT MALE
CHANGE

mo. before birth   Infant 1   Infant 2   Small Juvenile

AGE (MO.) AT TIME OF ADULT MALE CHANGE

FIGURE 12.10. *Survival of red howler infants, Venezuela. The number of infants disappearing strongly suggests infanticide, in addition to the cases actually seen. (After Crockett and Sekulik, in press.)*

males from outside the troop. In the course of three observed attacks the infants were snatched as their mothers tried to defend them. Instead of biting the infants fatally, as gorillas do, the males have displayed with the still-living body, flailing it as they run, eat it, and, most gruesome of all, groom its remains. For chimpanzees infanticide is not the uncomplicated murder it seems to be for gorillas or langurs. Neither is it the normal hunt for prey. Goodall (1977) has argued it is a by-product of aggression toward the strange female. The males may guard, consume, and groom the body for hours after its mother's disappearance, a behavior unlike any other. Of eleven cases of chimpanzee infanticide summarized by Fossey (in press), only one was followed by the mother's joining her attacker's community.

Meanwhile, the controversy among langurologists goes on, as summed up by Boggess (in press) and Hrdy's reply (in press). This brings us to the point that many circumstances do not favor infanticide. Even if it is more common than once thought, the payoff depends on the particular situation. If the mother's period of amenorhea is short, it would be advantageous to kill only very young infants. If the period of takeover is long and uncertain with several males competing, there would be little point in killing infants for yet another male's advantage. If females mate with several males, or if a male takes, loses, and retakes a troop, an infanticidal male must somehow guard against killing his own offspring. Intensive studies of several hanuman populations have followed many male changes, with sexual competition and even fighting between males, but no evidence of infant killing or suspicious disappearance (Boggess, 1979, 1980, in press; Curtin, 1981, 1982). This does not seem to be a function of lack of stress, population density, or freedom from human interference so much as reflecting seasonal breeding and the slowness of the takeover process. In areas where male bands mate sometimes with troop members, or where a male immigrant belongs to the troop for many months before achieving dominance, there is no infanticide. It may be that in seasonally breeding populations a male can afford tolerance, or that the risk of killing his own offspring would be too great.

What counterstrategies are there? The chief one is the langur female's pseudo estrus. Many primates mate when pregnant, both in captivity and in the wild.

Pregnant langurs may be particularly prone to mate with new harem lords, which may fool the male into accepting the subsequent offspring as his own. Females attempt to avoid new leaders, but this rarely seems successful. They may also counterattack directly. Hrdy saw only older females attack the male, which may mean that those with the least to lose defended their grandchildren. However, there is little record of females joining to overwhelm an infanticidal male by their superior numbers. It may be that evolved female competition prevents them from doing so. There is no evidence that females refuse to mate with their infant's killer. This would mean postponing their own reproduction, at great cost to themselves, as well as denying their offspring the genes of a "successful" father. There seems no way that individual selection could evolve such self-restraint.

Males, on the other hand, defend their own offspring and females. Male baboons have been seen to unite to mob an infanticidal immigrant. Two effective counterstrategies seems to lie with the males—either to triumph by force or, more subtly, to live with enough tolerance so that potential successors just *might* be fathers of the current infants, and so have little reason for infanticide.

CHIMPANZEES: BAND OF BROTHERS

After considering potentially lethal competition between males in single-male harems, we may turn to a strikingly different social system: the chimpanzees. Female chimpanzees change groups, but males rarely if ever do so. Thus, the males are likely to be closely related. Furthermore, they depend on each other throughout life for political support.

The gorillas provide a transitional case. Although each group usually has only one silverbacked male, younger blackbacks may mature within the group. In Visoke Groups 4 and 5, two young males grew up as "crown princes"— Icarus, probable son of Beethoven, and Tiger, who was sired by the previous silverback of Group 4 but grew up in close proximity to the leader, Uncle Bert. Uncle Bert sired Tiger's mother's succeeding children. Two other young males in Group 4 spent less time in proximity with Uncle Bert: Digit and Beetsme who immigrated as an adolescent. It seemed that males who had belonged to the group since early infancy might remain to take over, whereas others emigrated. Icarus has indeed remained, and Beethoven has accepted him as a second breeding silverback, in spite of his killing one of his father's aging consorts. We do not know what would have happened in Group 4, because Digit and Uncle Bert were both killed by poachers (Harcourt and Stewart, 1981; Fossey, 1981). Before Digit's death, however, he sired a daughter, Mwelu, within Uncle Bert's group, in spite of Tiger's closer relationship.

Thus, gorillas, like hanuman langurs, apparently may have groups with more than one breeding male. Because we have longer-term individual knowledge of gorillas, we can see how tolerance, or kinship, or both between males can lead to younger males being admitted to full status. This means that a male lineage may form the core of a continuing group.

In chimpanzees, this tolerance among males frequently involves promiscuous mating with estrous females. Two to 12 males may copulate with the same female, in succession. Males near a copulating pair commonly show erections, and then mate in turn, particularly if the male's brother is mating. Usually there is no apparent aggression between the males. A male may reserve a female for himself by leading her away from the others as a "consort"; his success in this maneuver largely depends on the female's willingness to follow unobtrusively. Only the alpha male seems able to keep females to himself in an open situation, either by overtly charging other males, or passively remaining close to the estrous female. The sharing of mates, with sperm competition but relatively little behavioral competition, binds the chimpanzee group together. For females, it helps assure that all the males tolerate or protect the infants, for almost any male might be a father (Tutin and McGinnis, 1981).

The gang attacks by group males on their neighbors were described in Chapter 7. The affiliative bonds between chimpanzees will be further considered in the next chapter. For the moment, we may ask how male chimps achieve a particular dominance status.

Their dominance is complex. Noë and colleagues (1980) factored out the various possible indicators of dominance in a captive group, and found three principal factors that gave three different orders of ranking. Straight agonism—most threats and appeasement—fell into one group, but bluff, that is, displaying with raised fur and stamping, was common to three young males and a female, not just the top three males in the threat hierarchy. Finally, competition for places to sit or social companions seemed settled in the favor of high-ranking females, whether this was the result of the females enforcing their rights or of the males chivalrously allowing them precedence.

The interactions of captive chimpanzees are highly complex, and often involve many animals at once. Recipients of aggression commonly recruit, or attempt to recruit aid. In one sequence of behavior from this colony, the second-ranking male Luit tested the alpha, Yeroen, by attacking a female named Spin who was sitting near Yeroen. Spin fled. Spin's friend, Dandy, an adolescent male, moved between the two, giving submissive oh-oh barks. Luit desisted, but again charged Spin. Luit and Spin actually fought, while Dandy walked round Yeroen uttering shrill barks in the direction of the fighting pair. The sequence continued with intermissions, involving a female who did not like Spin and joined the attack, Luit kissing yet another female on the mouth, a third male who supported Luit, and Yeroen charging a fourth female who fled, vomiting. At one point Spin sought aid from Yeroen by embracing and kissing him on his shoulder, while pointing and shrill barking at Luit. Yeroen merely stepped aside when Luit lunged at Spin (de Waal and van Hooff, 1981).

Such complex interactions made it clear that Luit was taking over the role of alpha male, not through direct challenge, but through shifting power plays and allegiances within the whole colony. Between challenges, the males consoled each other with hugs and grooming (de Waal and Roosmalen, 1979). This colony, in a two-acre enclosure in Arnhem Zoo, has more freedom to interact, or ignore each other, than most captive colonies, but has been studied in more detail than almost any wild group. De Waal (1982) recounts the changing coalitions with a rare combination of quantitative observation and frank admission that the chimpanzees have a good deal of foresight, and even social calculation, in their bids for power. (See Figs. 12.11., 12.12., and 12.13.) The elder male, Yeroen, was alpha for the first three years of the colony. Luit, a younger, stronger animal, took over leadership during several months of challenges and

FIGURE 12.11. *In Arnhem Zoo, Luit, displaying on trunk with hair erect, separates the aging alpha Yeroen from a female companion Krom. (Photo by F. de Waal, from de Waal, 1978.)*

separations of Yeroen from his social companions. When Luit eventually became alpha, he was supported by a maturing male, Nikkie, and by the females. During his rise to power, Luit attacked the females, or joined in fights on the winner's side, chastising the loser. When Luit became alpha, however, like Yeroen before him, he intervened to help losing females or broke up fights evenhandedly, while females submissively greeted him. Meanwhile, the wily old Yeroen formed a coalition with young Nikkie. After many permutations, Nikkie rose to alpha status, including Luit's submission, and sexual access to females. Yeroen, however, retained the control role, arbitrating fights and receiving females' greetings (Figs. 12.14 and 12.15A,B).

In the Gombe reserve, four alpha male chimpanzees have succeeded each other since Jane Goodall's studies began. Goliath, the first, "was not a large animal, but was in good physical shape, had a quick temper, and a fast, im-

FIGURE 12.12. *Yeroen has chased Luit into a tree, and holds out a hand with a grin of reconciliation. The whole group's attention is on Luit, except two who eat leaves Luit threw from the tree. Later, Luit threw down enough for everyone, which defused the tension. (Photo by F. de Waal, from de Waal and Roosmalen, 1979.)*

FIGURE 12.13. *After Luit becomes alpha, he is challenged in turn by Nikkie. Here, a bout of intensive reconciliation grooming between Nikkie and Luit. (Photo by F. de Waal, in* Chimpanzee Politics, *1982.)*

FIGURE 12.14. *Yeroen and Nikkie groom each other, ostracizing Luit. Such subtle interactions have as much to do with chimpanzee dominance as the rare fights that finally reverse status. (Photo by F. de Waal, in* Chimpanzee Politics, *1982.)*

FIGURE 12.15A. *Chimpanzee females individually are no match for males, but males do not use their canines on females. Here Luit slaps Spin so hard she rolls in the sand, indirectly challenging Yeroen's power to protect her. (Photo by F. de Waal, in* Chimpanzee Politics, *1982.)*

FIGURE 12.15B. *The females in coalition play a major role. On the left, Yeroen and three females take on the maturing male Dandy, who displays, right. (Photo by F. de Waal, in* Chimpanzee Politics, *1982.)*

pressive charging display." Mike took over the alpha position in the early 1960s without ever fighting. He intimidated the others by charging with clattering empty kerosene cans from Jane Goodall's camp. Mike ruled for more than six years, but in 1971 he was physically attacked and beaten by Humphrey. Humphrey was a large, brutally aggressive animal who made unprovoked attacks on many members of the community. Unlike his predecessors, he had no known or suspected brother supporting him. He kept his rank for only 20 months until his decisive defeat by Figan, supported by Figan's older brother, Faben.

Figan rose to alpha rank over about three years. Faben, who although older has been subordinate since his arm was paralyzed in the polio epidemic of 1966, and Figan combined to attack Evered, Figan's agemate and rival. Between 1970 and 1972 Figan and Faben traveled more and more together, and a series of challenges culminated in a full-scale fight between Figan and Humphrey. Faben was present, but did not help, while Figan not only attacked Humphrey but also attacked and displayed at the old male Hugo and the adolescent Jomeo. After this Figan was clearly dominant over Humphrey. Several months later Figan and Faben together treed and attacked Evered, and from then on Figan was clearly alpha. The brothers continued to spend a good deal of time together communicating by "tonal grunts" unlike other males.

Figan was a relatively small-sized male at the time of his takeover (39 kilograms, as compared to Humphrey at 45 kilograms), but he was highly motivated and highly intelligent. As a mere adolescent, he was the one who used to lead other males away from the banana pile. When rising to power he could be seen surveying the group before he entered, as though planning the time and direction of his charging display for maximum effect (Riss and Goodall, 1977).

Dominance in chimpanzees apparently confers some advantage of possessing estrous females in an open situation (see Chapter 13). Dominance, or for that matter subordinance, means juggling the individual combinations of animals, and the likelihood of different individual reactions (Simpson, 1974). Affiliation is just as important as competitive prowess: three of four Gombe alphas have had the support of a brother within the group, whereas between groups the support of other males can mean life or death. This is still dominance, but a long way from the original conception of a linear peck order. Instead, among chimpanzees we see the evolution of political intelligence.

FEMALE-FEMALE COMPETITION

As we have said in previous chapters, a female's chief need is food. In the normal course of primate lives, more food means earlier maturation, bigger babies, more milk, and thus more surviving babies. If food is the chief need, why should female primates be sociable at all, and tolerate competitors foraging beside them? One answer is that many do not. Monogamous females viciously attack rivals. They may even inhibit their daughters' maturation. Captive lion tamarins have killed maturing daughters caged with their mothers too long (Kleimann, 1979).

We have seen Wrangham's answer in Chapter 6 that bands of related females could exclude less well-organized competitors from fruit trees and other clumped food sources. We have also seen that numbers of animals provide predator defense, particularly if large, well-armed males confront the predators. The preceding sections indicate that males are particularly useful to keep off other males, who may even be infanticides.

Given that females of many species have good reasons to put up with group life, competition continues within the group. This is usually competition for food; it may at times become physical abuse of other females or their infants.

Among many prosimians, chimpanzees, and orangutans, females usually forage alone, mediating their competition by spacing. In all group primates that have been well studied, there is a dominance hierarchy among the females. This is not expressed in such frequent or severe agression as among the males, but is continuously there nevertheless.

As Sarah Blaffer Hrdy writes in "The Woman Who Never Evolved," this competition is no less grim than the males'. The main difference is that females may have longer-term goals. It is females, in most primates, who pass on social status to their offspring. I shall speed through this section in hopes that students will read Hrdy's Chapter 6 as well.

Female competition begins with one's own nutritional status. The macaques of Koshima Island, as we saw in Chapter 11 on demography, died if they could not maintain their own weight while pregnant and lactating. The lowest two lineages did not produce any surviving young (Mori, 1979). The dominant lineages of macaques of Ryozen flourished during artificial provisioning when they dominated the feeding ground. When the food stopped, their reproduction fell to that of the subordinates (Sugiyama and Ohsawa, 1982).

It continues through mating. Gelada baboon females harass each other when in estrus or mating. The stress seems to lower the chance of conception by subordinate females (Dunbar, 1980; Dunbar and Sharman, 1977). The stress effect was very clear for mouse lemur females kept in social groups, not solitary as they would be in the wild. They had fewer estrous, more abortions, and higher infant mortality (Perret, 1980). In the free-ranging rhesus of La Parguea, dominant females bred younger, earlier in the season, and left more surviving young (Drickamer, 1974).

Predation may cancel out the effects that appear in food-limited populations. Among Amboseli baboons, there was no difference in survival of infants of high- and low-ranking mothers (Altmann, 1980) although there was a very strong difference in two troops of Botswana baboons (Busse, 1982). In fact, for the little vervets of Amboseli, infants of low-ranking mothers died of nutrition-related causes, but high-ranking mothers, who tended to travel in the front of the groups, lost an equal proportion of infants to predators (Cheney and Seyfarth, 1981).

Infants commonly take a "dependent rank" from their mother while they are with her (Kawai, 1958). There is no mystery to this: the spoiled offspring of a highly dominant female learns that it can safely threaten others from its mother's shadow. If a subordinate female's infant tries the same thing, it may be chastised, or its mother grabs it, cringes, grimaces, and runs. How long the infant enjoys this dependent status depends on its sex and species. Dominant young males may grow up with a certain bravado, or even not transfer from their natal troop, but in the normal course of events they will transfer. Then their mother can bequeath them only sound physical condition and confident character. Daughters, on the other hand, generally remain near their mothers. The mother galago or potto may cede a portion of her range to a growing daughter (Clarke, 1978; Charles Dominique, 1977). A mother macaque supports her daughters in confronting each other. In a growing, well-fed troop, this can lead to a highly predictable social structure. A mother "caps" her daughters' rise in rank and backs a newly matured, four- to five-year-old daughter against any older sisters. The female hierarchy then starts with Matriarch A, followed by her four- to five-year-old daughter, say A-d$_3$, followed by A's six- to seven-year-old daughter, say A-d$_2$, and then A's oldest daughter, A-d$_1$. The series goes on with Matriarch B and her daughters, then the C lineage, and so forth. If the troop has not divided or fused, A, B, and C may themselves be sisters of increasing age. The mothers' behavior would make sense if they were calculating that their youngest mature daughter has the highest reproductive potential but also reserving their own status to support still other, future young. It also may be that a mother's help does most for her four- to five-year-old daughter because a primiparous six-year-old has a low probability of having her infant survive; helping the youngest daughter may result in most grandchildren (Koford, 1965; Chapais and Schulman, 1980; Schulman and Chapais, 1980).

Such predictable hierarchies have actually appeared in the provisioned macaques of Japan and of Cayo Santiago. However, more normal troops that face drought and predation deal with longer birth intervals and aged or dying mothers. In Alto's group, a baboon troop that was studied for more than ten years at Amboseli, the majority of daughters have retained their mothers ranks. However, where the age gap between daughters was greater than two years, the older might remain dominant over the younger. If a mother was aged, her daughters rose above her. More intriguing, daughters whose mothers fell in rank or died after the daughter was two have tended to reclaim their "proper" status—the one their mothers held during their infancy (Hausfater et al., 1982).

The lineages, as well as individuals, sometimes change rank. A medium-ranked female in one Japanese macaque troop, aided by the alpha male, successfully challenged the alpha female. Almost all the members of the two lineages also shifted (Gouzoules, 1980, 1981).

The tightness of baboon female hierarchies appears in their day-to-day interactions. Adult females attempt to groom females of higher rank. High-ranking females have less tolerance of upstarts who do not know their place, so in practice the baboons groom females only one or two steps up in rank (Seyfarth, 1956). This ranking extends to immatures. Juvenile females prefer to groom their superiors, either juvenile or adult, complicated by the attraction of small infants, such that adult females with infants are most attractive of all. Young males prefer to associate with adult females in estrus, not lactation. Cheney (1978) speculates that this could result in looser bonds among the lower-ranked matrilines. If the high-ranked juveniles almost always associated with their own kin group, but the low-ranked ones commonly abandoned their kin in favor of social climbing, the low-ranked ones would have more divided affections.

Langur females also aid each other, but they decline more in reproductive capacity with age, and apparently decline as well in status: young langur females are much higher ranking than their elders (Hrdy and Hrdy, 1976;). Computer simulations can turn a macaquelike hierarchy into a langur-like one either by increasing the rate at which young females win status fights, or else by increasing the death rate, which means that elderly mothers do not "cap" their daughters' rise (Hausfater et al., 1981). Langur female relations are much less tense than baboons and macaques. Perhaps their leaf diet means they are not in such strict competition (McKenna, 1979).

The mantled howler monkeys, like langurs, have a reversed age hierarchy. It seems that there, young females force themselves into new troops, entering at the top of the hierarchy (Jones, 1980). In one well-known troop, such young females never succeeded in bearing an infant that survived. However, once replaced and pushed down to second and then third rank, they reproduced most successfully. For a howler, the advantage of being boss may be that you eventually reach second place (Glander, 1980).

Females then compete by discriminating against each other's infants. This is very equivocal; much allomothering seems to help, freeing the mother to rest or feed, teaching the allomother the skills of motherhood, and seeming to please both animals. Females, as we have seen, may also rescue each other's infants from danger, including infanticidal males. The darker side is that many allomothers abuse infants, particularly the multiparas. Infants have died when kidnapped by females of an alien troop (in langurs, Hrdy, 1977) or by a more dominant female of the same troop (in baboons, Altmann, 1981).

The worst case is certainly the Gombé chimp Passion, who, with her daughter Pom and son Prof, has snatched, killed, and eaten at least three infants of her own community (Fig. 12.16.). One hopes that at least this case is really pathological, but the gorilla Effie who ate another's infant raises the fear that it is standard behavior. Goodall (1977) points out that Passion was achieving all-too-effective elimination of rival lines.

There is some indication that females can influence the sex of their offspring, perhaps in response to female-female competition, and can detect the sex of others' offspring. Thick-tailed galagos in the wild, dwarf lemurs and mouselemurs in captivity seem to have highly skewed sex ratios. Far more males are born than females. This would be reasonable from the mother's point of view if she would have to share very limited habitat with a daughter, whereas a son might disperse (Clarke, 1977; Perret, 1982). Among the Amboseli baboons, high-ranking females gave birth to more daughters. Again, this is reasonable if high-ranking females could assure their daughters high status in the troop. For low-rankers, a son might be lucky in a neighboring troop, whereas a daughter would get only the leavings at home (Altmann, 1980). In captive macaques, females have tended to alternate, giving birth mainly to the sex that was under-represented the year before (Silk, et al., 1981). It seems that female rhesus can respond to the sex of each other's unborn young. Females pregnant with daughters are more likely to be bitten, chased, and attacked than females carrying sons (Simpson et al., 1981; Sackett et al., 1975). This also makes sense if daughters will be lifetime competitors of other females.

Hrdy concludes that the feminists' lament that unrelated women do not bond like unrelated men goes back to an origin rather older than the human species—either in maternally linked competitive lineages like the baboons and

FIGURE 12.16. *Passion (upper right) with her son Prof and daughter Pom eating the three-week-old infant of Gilka, which they have snatched from its mother and killed. (Photo by Tsolo do Fisoo, courtesy of Jane Goodall and the Gombe Stream Research Centre.)*

langurs, or in individual competition, using male support to fend off other females, as in chimpanzees. But, as she says, "No scientist has yet trained a systematic eye on women competing in spheres that really matter to them. . . . Ethologists who have to contend with malarial swamps and subjects that hide in dense foliage 40 feet above them confront a task simple by comparison. How do you attach a number to calumny? How do you measure the sweetly worded put-down? Until we are able to solve such problems, evidence for this hypothesized competitive component in the nature of women remains . . . intuitively sensed but not confirmed by science" (Hrdy, 1981).

SUMMARY

Competition for resources such as food or mates underlies conflict between animals. Anger, which is an emotion, and aggression, which is overt interaction between conspecifics, may or may not occur during competition. Dominance, or status, is a long-term relationship between individuals. It may be measured in several ways: by priority to resources, by direction of threats or approach avoidance, and by other aspects such as defense against predators or stopping fights within the group. In some species and groups all these measures of dominance correlate. Often they do not.

Most aggressive encounters are low-key and ritualized, varying from the frequent aggression of baboons to almost none in sifaka outside the breeding season. All species seem capable of wounding or even killing if watched long enough. Game theory, by specifying risks and payoffs, offers a theoretical approach to the frequency of violence.

In most primate species males dominate females on all measures. Monogamous primates have relatively equal status. Among all studied lemurs, patas, and talapoin monkeys in the old world and squirrel monkeys in the new, females have priority to resources. These primates are all seasonal breeders (like many others), but seem to have little else in common.

Male dominance hierarchies may be uncorrelated to food priority (lemurs and others), or be highly correlated (macaques and baboons). In multimale groups alpha males do more mating than others, but others mate and father some infants. In single-male groups one male may father all the infants during his tenure. Occasional male infanticide has now been reported in langurs and at least six other unrelated species when a new male takes over the breeding group, and (in baboons) when a male immigrates and could not have fathered any group infants. Within multimale groups, male threat dominance and mating success are complex, depending on enlisting the aid of others. This is particularly true in chimpanzees, in which males and their own brothers remain in the natal group.

Female competition is subtler and less violent than males'. However, females preempt each other's food, and harass or sometimes kidnap, and, in at least one case, eat each other's babies. Females pass on dominance rank to offspring who remain in the group, and nutritional status to all offspring. Thus, females have a long-term stake in successful competition, which extends to differential treatment of each other's sons and daughters in utero.

CHAPTER
13

SEX

"A female brown capuchin changes behavior radically at the beginning of estrus. First, she becomes very skittish . . . Second, she usually wears a grimace . . . Third, the estrous female produces a distinctive vocalization, a more or less continuous soft whistle, which at high intensity grades into a strongly pulsed hoarse whine. She may give this call for over three hours without pause. Fourth, she follows the dominant male of the group for long periods . . . approaching him very closely and touching or pushing him on the rump, or shaking a branch near him, then running away . . . The dominant male appears to react with disinterest and mild intolerance . . . and frequently chases the female away." (Janson, in preparation). Her estrus lasts four to six days. The dominant male only consents to mate an average of one and one half times during this period, in spite of the female's persistent importunity. Finally, on the last day of estrus, she turns her attention to the group's subordinates. As many as four different males may mount and thrust in ten minutes. The dominant chases the subordinates for the first half of the day, and then loses interest again, while the others go on mating.

The brown capuchin is an extreme example of female *proceptivity*, or attempts to initiate sexual behavior. She is *receptive* during estrus to advances only by the males of her current choice. As far as the dominant male is concerned, she does not seem very *attractive*, although the subordinates react otherwise.

These three different aspects of sexual behavior coincide with the actions of different hormones. A macaque female can be rendered attractive by swabbing appropriate smelling fatty acids onto her vagina (Chapter 10). Males, as well as females, have different degrees of proceptivity, receptivity, and attractiveness, although the hormonal basis is not so well known. Successful sexual encounters result from coordinating these aspects in both male and female (Beach, 1976). (Fig. 13.1., A, B, C, D).

Even if we know some of the hormonal bases of sex, there remain the ev-

274

FIGURE 13.1. *A hanuman langur in estrus solicits a male. B. He mounts. C. Another adult female rushes up to harass the pair. D. The male dismounts to chase her off. (Courtesy D. Hrdy and Anthro-Photo.)*

olutionary questions. Most primate males attempt to guard estrous females from potential rivals. How can the dominant brown capuchin afford to sit back and count on his females, during the peak of their estrus, besottedly soliciting only him?

The answer seems to be ecological. Brown capuchins, like many primates, undergo food shortage during the dry season. Unlike their congeners, the brown capuchin's main food sources at this time are especially small trees, with close-set, ripe fruit. The dominant male can easily control access to all the fruit. He wards off subordinate males and keeps them from supplanting his favored females and juveniles. In a couple of instances, a male did not defend two juveniles that were unlikely, for different reasons, to be his own. It seems, then, that the female brown capuchin is making an instinctive "marriage de raison." She mates at peak estrus with the animal who will keep her children fed (Janson, in preparation).

This chapter begins with the immediate aspects of mating: receptivity, proceptivity, cycles, and seasons. It will then digress to other affiliative behavior, such as grooming and cuddling. Then it will go on to courtship, sex and individual affiliation in monogamous species, in macaques and baboons, and in the great apes, as influenced by their ecology. Eventually, we will deal with humans and the hypertrophy of human sexuality.

COPULATION

Mammalian intromission may last a few seconds, in the Ugandan kob, or 11 hours or more in small marsupials. The act of mating, like other aspects of life, seems to relate to many others. Timing is in part a question of reproductive physiology, in part of the likelihood of being interrupted by rivals, and in part the danger from predators. Likewise, the roles of male and female interact with the rest of their behavior and anatomy: it seems that spiny anteater and hedgehog males persistently follow the female, but do not attempt contact until the female rolls over and invites it (Ewer, 1968).

The timing and sequences of primate copulation likewise vary. Copulation patterns may differ even in species of the same genus. The liontail macaque and the crab-eating macaque are multiple mounters, with a series of about ten mounts preceding ejaculation. The toque macaque mounts once and then thrusts about 15 to 20 times before ejaculation. Stumptail macaques are single mounters, but thrust on average 60 to 65 times before ejaculation (Fooden, 1980).

The primary function of all these patterns is to transfer sperm to the female in such a way as to reach the ovum. A part, at least, of the performance involves stimulating the female to produce the appropriate vaginal fluids to aid in sperm transport. In many mammals, although probably not primates, sexual stimulation is necessary to induce ovulation. Kleiman (quoted by Eisenberg, 1981) points out that the presence of spines on the penis of caviomorph rodents correlates with brief and few intromissions and may stimulate the female in minimum time. Some primates, such as the bushbabies, have penises with backward pointing spines. The greater bushbaby and the stumptailed macaque apparently form a true copulatory tie, like dogs. It is not clear whether either spines or tie induce maximum stimulation, or merely prevent indifferent females from leaving, or both.

Female orgasm, with its tightening of the uterine and vaginal muscles, presumably aids in stimulating ejaculation in the male, and in conserving and transporting the sperm. Uterine or vaginal contractions have been recorded in chimpanzees, and in homosexually mounting stumptail macaques (Allen and

Lemmon, 1981; Goldfoot et al., 1980), and clitoral tumescence in rhesus (Burton, 1971), and pygmy chimpanzees (Savage Rumbaugh, and Wilkerson, 1978). Female primates indicate sexual excitement and sexual climax in various other ways—reaching back in a stereotyped manner to clutch at the male (Zumpe and Michael, 1968), changes in bodily tension, facial expression, respiration, and vocalization (summarized in Allen and Lemmon, 1981; Hrdy, 1981).

It has been categorically claimed that only human females experience orgasm, and as firmly claimed that it is common to a number of other primates. The human-only school follows Desmond Morris (1967) in linking it to human females' nearly continuous receptivity throughout the menstrual cycle. The others point out both that the immediate response may increase the chances of fertilization at a particular mating, and that sexual pleasure has to do with far more than the consequences of any single mating. Oddly, the controversies all concern the presence or absence of orgasm; all authors assume that if the female has any discernible reaction, it is evolved, at least in part, for pleasure. Females may, on the whole, be able to find mates of some sort for the actual act of fatherhood. On the other hand, few females raise all the offspring of which they are capable. Sexual proceptivity, receptivity, and attractiveness may all contribute to bonding with a particular male or males, which then improves the female's longer-term chances of raising her offspring. The two schools of thought converge as soon as one admits that for female primates as well as humans copulation is only one part of a sexual relation, and that in other primates as well as humans it often occurs when the female is not ovulating.

The importance of the coordination of the two partners appears at an extreme in orangutans. Wild orangs, particularly subadults, are the only wild primates known to rape unwilling females (MacKinnon, 1979; Galdikas, 1981). In zoos, this takes the form of females submitting to males caged with them. Females caged beside a male with a door too small for him to get through could time their own entrances, and in these circumstances mate mainly at mid-cycle, in relaxed fashion, and, in the preliminary tests, with a high rate of conception (Nadler, 1982). Similarly, in the wild a female in full estrus seeks out a fully adult male by following his long calls through the forest. "In view of the small size of the orangutan penis and the inherent difficulties of suspensory arboreal copulation, it seems that only with female cooperation can intromission be successfuly achieved. Such cooperation is usual only within an established . . . consortship" (MacKinnon, 1979). Conception mostly or perhaps only takes place within such mutually attracted consort pairs (Galdikas, 1981).

CYCLES, SWELLINGS, AND SEASONS

All primate females alternate a brief fertile period at ovulation and a longer period of infertility while the hormonal cycle is either preparing for the possibility of pregnancy or replaying to produce another ovum. The new-world monkeys have much shorter cycles than old-world monkeys, often of the order of 7 to 15 days, and a somewhat different hormonal mechanism. However, the primates tested seem to have in common a mid-cycle surge of estrogen just before the egg is released, followed by secretion of progesterone, which prepares the uterine wall for implantation. If pregnancy does not occur, the uterine wall is shed. A little menstrual bleeding occurs in new-world cebus and howlers, as well as the apes and many or most old-world monkeys.

The external genitalia may also change through the cycle (Fig. 13.2.). Lemur genitalia flush pink during estrus, and so do bushbabies' and tarsiers', which may close and seal over during inactive periods. Most of the new-world monkeys, except howlers, have prominent swellings. Within the old-world monkeys, apes, and even humans, it is common to have slight swelling and flushing of the labia at estrus. In a few species, such swelling can engorge labia, anal area, and much of the skin around. Baboons, talapoins, and stumptail macaques have swellings; rhesus macaques do not. Chimpanzees carry swelling like grapefruit or even small watermelons; gorillas swell just enough to be perceptible to favored zookeepers, but orangs not at all. These swellings are affected by circumstance even in individuals of the same species; isolate, captive baboons grew larger swellings than normal wild ones (Rowell, 1967).

The distribution of sexual swellings seems to correlate with life in multimale troops. The females are visibly advertising their condition. Monogamous and even harem primates, whose males presumably keep track of their females from short range, have no need to hang out a flag that can be seen for many yards through the trees. In multimale troops, in contrast, females commonly mate with more than one partner. As already said in the section on infancti-

FIGURE 13.2. *The female celada has a bare chest patch. During estrus both her chest and rump redden and develop an outline of white tubercles. (Courtesy R. Dunbar.)*

cide, if there is a possibility that each male fathers the infant, each may refrain from harming it, or even aid the infant and its mother.

Mating takes place primarily at mid-cycle, particularly in the species that have marked external sexual swellings. Much has been made of Zuckerman's original distinction between such "estrus cycles," and the "menstrual cycles" of such species as rhesus and humans. In estrus cycles the female mates only during a sharply limited period around ovulation. In menstrual cycles the most obvious external changes come during menstruation, not estrus. Mating takes place fairly continuously throughout the cycle, although often with a peak at probable ovulation and a sharp fall at menstruation. These two types of cycles, however, are ends of a continuum. They seem to have the same hormonal bases and could just as well have the same name. Caging conditions can transform behavior: rhesus, for instance, are less choosy when in close confinement (Gordon, 1981).

Many aspects of behavior, besides copulation, change over female cycles. There are the odor signals emitted by the vagina (see Chapter 10), and many other inviting or repulsing behavior patterns (Cochran, 1979). There may be changes in female aggressiveness and general tension, which is well known for humans (Michael and Zumpe, 1970; Mallow, 1981). But overriding the internal changes, there is usually much longer-term change: pregnancy generally occurs after few cycles, and then months or years of lactation prevents further ovulation. Many primates mate during pregnancy, but few during the lactation period.

Even if pregnancy does not occur, many or most primates mate at only one season, so cycling stops after a few months (Lancaster and Lee, 1965; Michael and Zumpe, 1976).

When Sir Solly Zuckerman initiated research on primate sexuality, he thought that continuous readiness to mate underlay primate social structure (Zuckerman, 1932). He argued that prosimians were solitary and seasonal, whereas higher primates like baboons and humans had permanent social bonds based on quasipermanent sexual excitement. The wealth of data that have appeared since he first wrote have proved him wrong. Ringtailed lemurs live in troops that mate during two weeks a year. Gibbon, siamang and indri live in faithful monogamy, apparently mating during a few cycles every three years. Social life obviously can continue with only sporadic sex.

Seasonality is highly complex and labile from population to population. Many primates' breeding seasons are triggered by changes in day length (Van Horn, 1980) and a few may directly reflect changing food supply (Rowell, 1979). Populations of the same species in different climates may breed at different seasons, and bringing primates into captivity may greatly prolong the season. Also, some species may breed seasonally, but related ones in the same habitat do not (Charles-Dominique, 1977; Rowell and Richards, 1979; Rudran, 1973; Cross and Martin, 1981; Ewing, 1982).

The unifying explanation seems to be time of weaning (Petter-Rousseau, 1968; Coe and Rosenblum, 1978). Lactation is the period of maximum stress for a mother, when she needs most energy to support herself and her young. Weaning is the period of maximum stress for the infant, when it is most vulnerable and can eat only the tenderest and least toxic of adult foods. When there is a sharp mating and birth season, they may fall at an apparently arbitrary time of year, which can be understood only by looking ahead to the time of weaning (see Chapter 14).

A further complication is the social synchrony of female cycles. This is likely in prosimians (Jolly, 1965; Boskoff, 1978), occurs between members of rhesus matrilines, and is best documented in people (see end of this chapter and McMillan, 1981). Introducing rhesus females who have been stimulated with estrogen into a free-ranging troop can bring other females, as well as males, into season (Vandenbergh and Drickamer, 1974; Gouzoules, et al., 1981). This communication of breeding state may serve to sharpen the response to seasonal environmental changes such as light. It may have rather different functions, as in the mantled howlers who gave birth at various seasons of the year, but significantly often in synchrony. Similarly, adjacent harems of gelada baboons may be synchronized within, but not between groups (Dunbar and Dunbar, 1976).

Females may suppress as well as enhance each other's sexual cycles. This may not be surprising in mouse lemurs, which usually live much of their lives alone (Perret, 1982). One would expect group caging in captivity to be particularly stressful for such animals. Captive savannah baboons have longer estrus cycles than normal in the wild, and wild subordinate gelada baboons conceive less frequently than the dominant members of their harem (Rowell, 1967a, 1967b; Dunbar, 1980). There is probably an interaction between hormonal interference, active female harassment of mating females, and even food supply.

Male hormones and activity, as well, can influence female cycling. This is well known in mice, in which the urine of a strange male brings females into estrus and may even cause them to abort their litters. It is now becoming known in primates as the "strange male effect." Chimpanzee females caged next to a new male may come into estrus overnight. As we saw in the section on infanticide, pregnant female langurs (Hrdy, 1977) and patas (Loy, 1981) may show full estrous behavior during pregnancy if a new male takes over the harem.

MIXED FEELINGS: GROOMING, MOUNTING, AND HAMADRYAS ADOPTION

Much primate behavior cannot be neatly boxed into categories like aggression or sex. (Fig. 13.3 A, B, C.) Grooming is one such behavior. It clearly originated from mother-infant contact. All mammalian mothers lick their newborn infants. Primates have developed grooming to a fine art, and enlarged it from maternal behavior to many other contexts, including courtship (Anthoney, 1968).

Primates part each other's fur with their fingers and extract grains of scurf with fingers, lips, and teeth. They have evolved special adaptations for grooming. The toothcomb of lemurs and lorisoids is formed of all their lower canines and incisors, flattened, forward pointing, and arranged for scraping. It may have originated for chiseling tree gum, but it is now a social organ. It is even possible that the saltiness of our own sweat evolved to reward the diligent groomer with crumbs of salt.

Levels and uses of grooming vary enormously between species. Many new-world monkeys remain as a pair or social group with little or no grooming except as a prelude to copulation (Moynihan, 1976). In many old-world monkeys

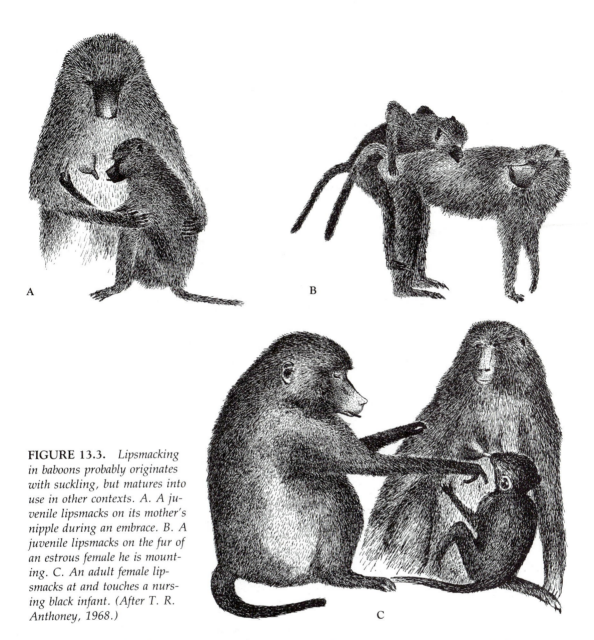

FIGURE 13.3. *Lipsmacking in baboons probably originates with suckling, but matures into use in other contexts. A. A juvenile lipsmacks on its mother's nipple during an embrace. B. A juvenile lipsmacks on the fur of an estrous female he is mounting. C. An adult female lipsmacks at and touches a nursing black infant. (After T. R. Anthoney, 1968.)*

and the chimpanzees among the apes, bouts of grooming reflect daily relationships. Male chimpanzees groom according to patterns of friendship and appeasement (Simpson, 1974). Baboon and rhesus males loll in front of a female, who is expected to groom them back. Females groom up the hierarchy, each female currying favor with her immediate superior (Cheney, 1978; Seyfarth, 1976). Among bonnet macaques, males and females groom each other reciprocally, without reference to the ill-marked dominance status (Simonds, 1965).

Estrus, not surprisingly, changes the amount of grooming. Males tend to groom females far more attentively when the female is in estrus. But sex is not all. Michael and Zumpe (1976) gave eight female rhesus an operant test, in which the female could work for access to one of a pair of males, with the males presented in pairs. Each female developed a particular preference for one of the five males—not the same male for each female. The most favored partner for five of eight females was the male who mated and ejaculated most with that female. All eight females, though, settled on the male who groomed them the most.

Grooming is not the only such equivocal gesture. Wickler (1967) has reviewed many cases of the change in meaning of gestures that evolved in one context and came to be used in another—in particular, sexual signs that have taken on a different meaning in the social context. (Fig. 13.4.) The classic case is dominance mounting, in which an animal of either sex asserts his authority by mounting his subordinate, sometimes even with thrusting. Mounting of this sort may be a kind of punishment, as when a subordinate has been fighting or threatening near a dominant, and the dominant breaks up the fight and mounts one of the participants. It may also involve a kind of permission. When a subordinate wishes to take a piece of food, he may "present" his rump to the dominant, who briefly mounts and then allows him to take the food. Bonnet macaques mount in any dominance order, and even in other monkeys mounting is not strictly tied to the direction of dominance. Excited estrous females are quite likely to mount as well. The presenting and mounting, although clearly sexual in origin, are now just as clearly ritual signs of excitement, self-assertion, appeasement, or even—for bonnets—friendship, in a wide social context. (See Fig. 13.5A, B, C, D.)

However, Wickler points out that the underlying motivation may well be as complex as the gesture itself. Presenting, after all, appeases *because* it is the gesture of a sexually receptive female. Dominance mounting should divert an animal from aggressive, potentially injurious behavior to sexual behavior. It is an advantage both to "fool" and be "fooled" in this situation. Given the possibility of using sex in potentially aggressive situations, the device works best if it stays fairly close in feeling and form to real sex. Being anthropomorphic, if we pity the gibbon's nearly sexless monogamy, the macaque might well pity our dull set of status symbols. Board meetings would be so much more amusing with vice-presidents mounting each other down the line . . .

The initial pair formation of hamadryas baboons is a completely different case of a set of gestures evolving away from their original function while probably retaining much of the original emotion. (See Fig. 13.6.) Kummer has shown

FIGURE 13.4. *Erection in several species is both sexual invitation and inter-animal threat. Here, Ugandan black-and-white colobus threatens the observer. (Courtesy M. Rose.)*

FIGURE 13.5. *Males in many species mount each other as dominance, not sexual relations. Females may also mount, especially when in estrus, and the patterns are much like those of copulation. A. Male Celebes macaque positions estrous female. B. Heterosexual mount. C. Lateral embrace by adult females. D. Female-female mount. (Courtesy A. Dixson, 1977.)*

A

B

C

D

FIGURE 13.6. *Juvenile female hamadryas baboon cuddles against her subadult male leader during a fight. (Courtesy of H. Kummer.)*

that hamadryas respect each other's priority to wives. When one captured hamadryas male is introduced first to a strange female, other stronger males do not attempt to take her away if her own behavior shows that she is bonded (Bachmann and Kummer, 1980). A maturing male, then, cannot steal an adult wife. Instead, the young male "adopts" a female juvenile, hugging and carrying her, protecting her from outside aggression for a year or so until she grows up enough to mate. The male gives the gestures of maternal care to his child bride, and throughout a female's life she depends on another stronger animal—teasing her family from that animal's side, enlisting her defender's aid in feuds, and expecting punishment if she strays too far. She shifts all this juvenile dependence directly from her mother to her mate (Kummer, 1968).

We know all these complexities of behavior in ourselves, of course. We know how many marriages also contain a parental-infantile relation, how mothers may love their sons with different emotions from those reserved for daughters. All of psychoanalytic theory is based on the transformation of child-parent love into sexual love as the normal pattern of development of our species. We know, as well, how even a single gesture combines motivations. The macaque presenting to be allowed a piece of food is much like the secretary batting her eyelashes to be allowed typing mistakes—not just sex involved, but a measured proportion of sexiness, calculated to species and situation, and even to changing fashions in women's emancipation.

FINDING AND KEEPING PARTNERS

We have given two examples of the wider context of primate sexual relations: the marriage de raison of the brown capuchin and the child bride of the hamadryas. There are a great variety of other patterns of affiliation, or, in the jargon, "reproductive strategies."

For monogamous primates, it is vitally important to find a mate who is healthy, active, and compatible, for each partner's lifetime reproductive success depends on the other. Young male gibbons sing alone, without defending a territory. In the one gibbon pair that has been known since a strange female, Sylvia, first appeared to add her great call to the notes of the lone male Stanley, neighboring troops that had paid no attention to Stanley alone converged to battle with the duetting pair (MacKinnon, 1978; Chivers and Raemaekers, 1980). Gibbon species differ in their gender roles, as well as in their songs. Kloss' gibbons, on Siberut Island, sing at separate times. There, the female defends the territory against other females, males against males. The lars, in contrast, seem to function much more as a pair (Whitten, 1982; Tenaza and Hamilton, 1977; and Tilson, 1981). Widowed and widower gibbons are known in several species, singing alone for months or even years without finding new mates (Whitten, 1982; Chivers and Raemaekers, 1981; Brockleman, personal communication).

Gibbon pairs mostly sing to form their bond; marmosets actually mate. Newly formed laboratory pairs of common marmosets mate far more frequently than pairs that have been established for several months (Woodcock, 1982). Because marmosets commonly give birth to two pairs of twins per year, much of their social interaction involves transfer of infants from the male to the female for nursing.

Most harem-forming primates seem to be a simpler and more straightforward case. Females by and large remain with the group; males challenge the incumbent male and fight their way in. After such a change, the females come into estrus, or, if pregnant, behave as if in estrus, and mate with the new leader. The situation is more complex in such animals as howlers, gorillas, and red colobus, whose females migrate. No one is clear how either such males or such females choose which groups to join.

Within multimale troops of baboons and macques, we know much more about individual affiliations. Particular baboon males are friends of particular females (Ransom and Rowell, 1972; Ransom, 1981). They may associate with each other when the female is not cycling, and the male may carry the female's infant. This is a far more precise relationship than the general male role as defender of the troop. In the Gombe Beach troop, one old male, Harry, spent much of his time near or grooming a high-status female, Myrna, and her offspring, Loy. Although he preferred Myrna as a consort partner during her estrus, he was commonly displaced by another male, and so may not have been Loy's father. "Occasionally Harry even exhibited some indecision when faced with a choice of associating with Myrna or his (different) consort partner of the moment" (Ransom, 1981).

Frequently, savannah baboon males attempt to isolate a female for themselves as a consort throughout her estrus; others harass the pair and try to divert the female to themselves only at the peak of estrus (Figs. 13.7A, B). The activity round the female, and fights or even dominance reversals among savannah baboons contrast with the more relaxed attitude of gelada males. In their one-male harems, the fighting occurs when the male takes over the harem, and a female's estrus hardly changes the day's activity. The pairs do not even mate so frequently, as there is unlikely to be competition from another male's sperm (Dunbar, 1979).

FIGURE 13.7A. *A female baboon increases tension by approaching another male. (Courtesy T. W. Ransom.)*

FIGURE 13.7B. *The consorting male appeases the other by presenting. (Courtesy T. W. Ransom.)*

The interaction and sometimes conflict between particular females' attractiveness as long-term social partners and their temporary attractiveness when in estrus have been found in many other savannah baboon troops (Strum, 1982; Seyfarth, 1978). Low-ranking males may initiate personal relations before a female is in estrus, sometimes with juveniles, in a close parallel to the hamadryas pattern. Meanwhile, high-ranking males also make such bonds with females. Later, as the male ages, the female still seeks him out, which accounts for some of the mating success of old male baboons (Seyfarth, 1978a, b; Ransom, 1981).

The macaques have a similar variety of patterns—consortship, conflict, and affiliation in different degrees. The Barbary macaques of North Africa offer one extreme. Males seem to respect each other's possession of estrous females, but only so long as the female is in close proximity, and it is the female's choice whether to stay or go. Females consorted with three to ten different males per day in a troop with only eleven males to choose from. Mounting and ejaculation in these circumstances is "straightforward and brief." This multiple chance of fatherhood may relate to Barbary males' particular fondness for the troop's infants (Taub, 1980, and see next chapter).

Rhesus have both multiple mating strategy and consortships. In a large outdoor corral the multiple mating system was more common, but high-ranking animals could maintain consortships (Wilson et al., 1982). All authors agree that females of any rank engage in sexual activity, but there is some evidence that on Cayo Santiago high-ranking females mate more often with high-ranked than with low-ranked males. Macaques are not so commonly reported as baboons to have "friends," but this may be a result of being watched more in semi-captivity than in the wild. Male Japanese macaques do have friends, whose babies they care for, but they may in fact be the mother's kin rather than sexual partners, and mate rather *less* with her than with unbonded animals (Enomoto, 1978). Female Japanese macaques, like the other macaques, play an active role in initiating sexual behavior and choice of partner, including homosexual mounting of each other (Wolfe, L., 1979).

The only general summary about the choice of partner in these multimale troops is that it varies immensely. The only pervasive conclusion is that this is not just a passing relationship during estrus, but that it forms part of much longer term associations between individuals (for example, in savannah baboons) or a female and all the troop males (such as in Barbary macaques).

How do our nearer relatives, the great apes, choose or compete for partners? The highly different social structures of the four species involve different sexual tactics.

Gorillas control access to their females at all times. The dominant silverback may tolerate younger adult males in the group, but the silverback accomplishes nearly all mating. He challenges and violently fights strange male intruders. This leads to bodily dimorphism; a male is at least twice the female's weight. However, he has small penis and testicles for body size, and little purely sexual assertiveness. Nearly all copulations are initiated by females during their three-day estrus. Mating lasts 1.5 minutes: no one interrupts a silverback gorilla. Orangutans have a rather similar one-male system, although it works by male territorial defense, not male shepherding of the harem. Orangs again have large male size, small male genitalia, adult male intolerance, long (11-minute) copulation, and an active female role in soliciting an adult male during the two days of estrus, although she is sometimes "raped" by subadults at other times. In complete contrast, chimpanzees live in multimale communities. Defense does not depend on individual size, but on cooperation by bands attacking and even killing neighbors who are found alone. Thus, there is little bodily sexual dimorphism, but large genitalia as sperm competition partly determines reproductive success. The pink female estrous swelling is visible from hillside to hillside, attracting any male in sight, for ten days during which the female often mates with many partners, perhaps enlisting all their support later when her infant is born. Copulation is initiated by either sex and is finished in a tenth of the time gorillas take, or a hundredth that of orangs. Other males in the queue rarely interrupt but do not have long to wait. Chimpanzee males use a variety of reproductive strategies—possessiveness, which only works for the dominant males; consortship, which requires both partners to cooperate in sneaking silently away from the group; or mating last with most sperm within the group (Harcourt, 1981; Short, 1981; Tutin and McGinnis, 1981). Among the three dominant males of the Arnhem Zoo, with their shifting power coalitions, there were clear cases of "bargaining," when males groomed, greeted, and even made begging gestures to dominants, and then were allowed to mate (de Waal, 1982). (See Fig. 13.8.)

The final great ape is the pygmy chimpanzee, the most intriguing of all (Fig. 13.9A, B). In the wild, pygmy chimpanzees seem to live in fairly large, flexible communities. They are far more often in mixed traveling groups than the common chimp, whose grouping is mainly males alone, or males with an estrous female. In laboratory and zoo they are the sexiest of primates, mating nearly every day, merely with the excitement of sharing food. This happens in the wild as well, for instance, when sharing hunted meat. Estrus in the wild seems to last 15 days, as opposed to six in common chimps, and female swellings never fully detumesce. Estrus, writes Kano, is the normal condition of a female pygmy chimp. All the apes mate in a variety of positions, but ventroventral, standing, and still other positions seem far more frequent in the pgymy pair. Furthermore, they gaze into each other's eyes, both as a prelude to mat-

FIGURE 13.8. *In Arnhem Zoo, a female chimpanzee awaits the outcome of "bargaining" between the three adult males. (Courtesy F. de Waal.)*

A   B

FIGURE 13.9.   *Pygmy chimpanzees at Lomako, Zaire. Males spend more time traveling with females than do common chimpanzees, females are receptive longer, and both preludes to copulation and copulation itself are complex and variable. (A. Courtesy of R. Malenky. B. Courtesy of N. Badrian; Lomako Forest Pygmy Chimpanzee Project.)*

ing and during copulation, apparently monitoring the partner's state, and changing behavior accordingly. Varied and responsive sexual behavior may be one important part of a general capacity for social communication and intelligence (Kano, 1979, 1980; Savage-Rumbaugh, 1978; Badrian and Badrian, 1977; Badrian et al., 1981).

HUMANS

Humans are certainly the most sexy primates. We can mate throughout the menstrual cycle (though many cultures have strong taboos against doing so), throughout pregnancy, throughout the seasons. Some cultures have strictures against mating during some periods of a woman's life, but if these cut out several months or years while she is pregnant or lactating, the culture allows the husband other wives, so that he can go right on mating.

Zuckerman (1932) claimed year-round sex as the prerogative of higher primates. Morris (1967) got it right: it is the privilege of the naked ape, or perhaps we should now include pygmy chimpanzees. It is all the more impressive if one considers Short's 1976 conclusion that twentieth-century western women are unusual in the length of time they spend going round the cycles of ovulation and menstruation (see Chapter 11). A Kalahari bushwoman, with later menarche, earlier menopause, and most of the time between in either lactation or pregnancy, might spend no more than two years of her life actively cycling; yet for much of the time she is sexually active.

The complex of behavior that goes with our semi-continuous sex may have involved division of labor between the sexes (Morris, 1967; Lovejoy, 1981). If men were commonly off hunting, and women more sedentarily gathering berries or termites, behavior that led them to take an interest in each other and share food could be highly adaptive. If infants were beginning to need longer and longer care and to be more helpless in the early years, the women would need more help from the men, and bonding that led particular men to help particular women might favor the offspring of those men and women. This can be caricatured as a male chauvinist "hunting hypothesis" or a female chauvinist "infant-care and food-gathering hypothesis," but in practice both elements come into both versions. Both versions lead on to the evolutionary pressures for quick thinking, social responsiveness, cooperation, manipulation, and memory and foresight for events in the past and the future. It is true that the more male-oriented versions stress cooperation between men in the hunt, and the more female-oriented versions stress the interdependence between men and women. It is still open to argument whether the relative lack of bonds between unrelated women springs from the relentless competition of our female primate ancestors (Hrdy, 1981), or the relentless cultural imposition of male superiority.

Women have at least, a strong physiological effect on each other. Women who associate closely are likely to menstruate at the same time (McClintock, 1971; Graham and McGrew, 1980). This is particularly true if they have little contact with men. One could suggest all sorts of explanations: sexual competition for the mate's attentions, shared care of eventual babies, or mere byproduct of the scent of each others' hormones. It is sobering to realize that we have so much influence on each other; yet, until recently, the effects were considered coincidence or old wives' tales.

One point not in dispute is that we have deep personal bonding between people of opposite sexes. We have evolved sexual advertisements, including the permanently protruding female breasts and the male's (for body size) enormous penis. These are not indiscriminate advertisements. Virtually every human culture has a rite of marriage and rights of marriage (Stephens, 1963). The bonding depends in part on mating itself. Women's ovulation is "concealed"; few women experience any sign that they are ovulating, and a woman cannot reveal the precise moment to the man even by telltale gestures. When a man returned from the hunt, he might mate on any day; if he mated often enough, he would father his wife's child. Lionesses seem to go through similar performances; few of their protracted consortships produce cubs.

One interesting suggestion is Nancy Burley's (1979) argument that concealed ovulation frustrated early female attempts at birth control. Burley cites the desire of women in many cultures to bear fewer children than their husband wants. There is good evidence that infanticide is widespread among hunter-gatherers, and it may have been practiced far back in the Paleolithic to keep down family size. Perhaps, says Burley, concealed ovulation was selected to "fool" women into bearing more babies than they planned. However, this supposes that women were consciously or unconsciously refusing to mate during estrus. It seems easier to believe the Zuckerman / Desmond Morris argument that concealed ovulation "fools" both male and female into enjoying unproductive sex, increasing their commitment to each other, and to their eventual family.

None of this decides whether the bonds are monogamous or polygamous. As we have seen, we have the sexual dimorphism of species in which there are typically fewer males than females (Chapter 12), and the testis/body size of a (just) single-male group primate. More telling, we do not have the extreme adaptations of obligate monogamy. It is easy to picture the original human family as being either monogamous or polygamous, but closely bonded over a period of years, much like human families today. They might also have lived within a multimale community that split and joined, like chimpanzees or hamadryas baboons today. The band could help provide defense against predators and share food and aid across a wider group. The families within the band would, however, remain bonded, unlike the promiscuous chimpanzees, in part through their sexual pleasure.

SUMMARY

Sexual behavior involves coordinating the proceptivity, receptivity, and attractiveness of both partners. Copulation sequences differ in timing and in initiatives by either partner between species, with females in some species taking an active role. Female orgasm has been documented in chimpanzees and stump-tailed macaques.

All female primates have an ovulatory cycle, which is much shorter in new-world than in old-world primates. Most females are in estrous for very little of their lives, being usually pregnant or lactating. Many only breed at one season of the year. External swellings at estrus are common in species that form multimale troops. Females of some species, including humans, may synchronize estrous cycles with each other.

Many other behaviors have a sexual component, including grooming, dominance mounting, and hamadryas male adoption of juvenile females.

Strategies for acquiring mates include hamadryas adoption, sung courtship in monogamous gibbons, fighting the resident harem leader in langurs, and "friendships" in savannah baboons. Dominant gorilla males control harems, orang males control ranges, and in both these apes the females seek out the dominant male when in estrus. Chimpanzee males may either attempt to mate last and with most sperm, or lead a willing female discreetly away as a consort. Only the dominant male can openly keep a female to himself. Pygmy chimpanzees mate more frequently than common chimpanzees, for more of the estrous cycle, and in situations of general, not just sexual, excitement.

Many primates mate sometimes when pregnant and when not strictly in estrus; the species in which this seems commonest are the pygmy chimpanzee and the human. In humans it forms part of the personal bond between particular males and females who also depend on each other for food and infant care. Virtually all human societies have a rite and rights of marriage. We do not know what such mating means in the lives of pygmy chimpanzees.

CHAPTER 14

MOTHERS AND INFANTS

AGE STRUCTURE

The life span of primates is long. It is by no means uniquely long. Elephants probably live as long as men, and the Galapagos tortoises may live 200 years. Absolute length in any case matters only in relation to other species with whom the animals come in contact and to environmental fluctuations in or between generations.

Lengthened life span is a progressive trend throughout the order. The relative length of different phases of life is very different from that of other mammals. The immature phases form a relatively large proportion of the primate's total life, as shown in Fig. 14.1. This is important for the individual. He has a long period of dependency on his elders and a long period in which to learn the appropriate behavior for an adult of his species. It also is important from the point of view of the troop and its social composition. In troops of most species there are many infants and juveniles who must be cared for. The infants and juveniles also contribute to the life of the troop, forming a center of attraction that may strengthen the social bonds among adults. Juveniles usually make the few innovations in behavior that can be incorporated into the culture of the troop. All these points will be discussed in later chapters, but all relate back to the physical stages of growth.

Primates as an order have a very long pregnancy. Even in mouse lemurs pregnancy lasts three and one half months and in true lemurs, four and one half months. (Compare this with three weeks in mouse-lemur-sized mice and three months in lemur-sized cats.) Pregnancy length in higher apes is about the same as our own. However, our own newborn are much less advanced than monkey or ape newborns, either in motor control or in percentage of growth. The length of human pregnancy seems to be the result of two opposing tendencies. The human brain is much larger in proportion to the body than an

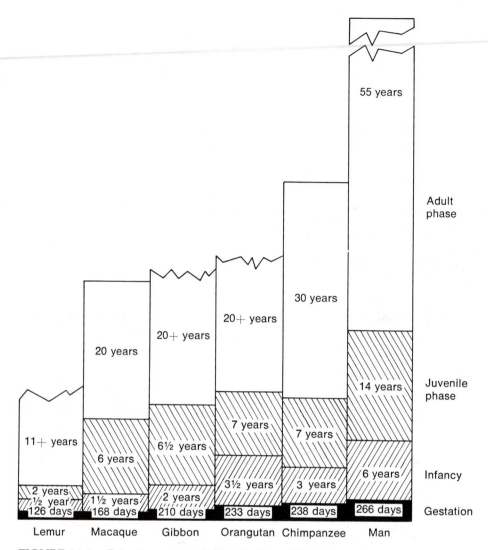

FIGURE 14.1. *Primate age spans. (After Napier and Napier, 1967.)*

ape's, and a large part of the growth of the brain takes place early in life. However, the child must be delivered through the mother's pelvis, which in turn is constrained by the demands of bipedal walking. Therefore, much of the brain growth must take place after birth: "The gestation period is terminated in man and other primates when the size of the head is consonant with a safe delivery" (Napier and Napier, 1967). (See Fig. 14.2.)

The infantile period, during which a child is physically and emotionally dependent on the mother, also grows progressively longer as one ascends the primate order. Unlike gestation, it doubles in length from ape to man. In part, this relates to the immaturity of our newborn. The first year of our infancy is really what Ashley Montagu calls *exterogestation*, a nearly helpless stage in which the infant is still transported and nourished like a fetus, although it can see and hear and feel the outer surface of its mother. The more active primate baby,

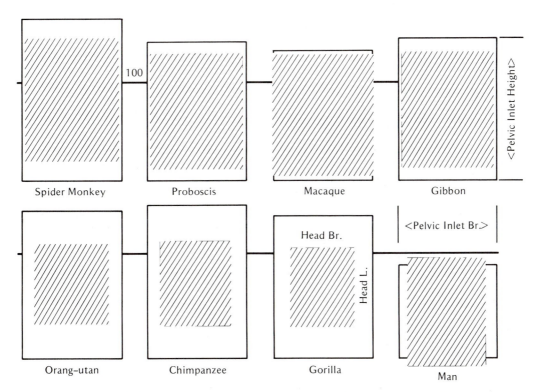

FIGURE 14.2. *The relation between the average diameters of the birth canal of adult females and average head length and breadth of newborns of the same species, all diagrams reduced to the same birth canal breadth. (After A. H. Schultz, 1969.)*

like our preschool children, is still emotionally bound to its mother. Much of the initial social learning takes place in this prolonged period of dependency.

The juvenile phase, lasting until puberty, is still more prolonged. In this phase, as in infancy, much must be learned by the growing primate, and there has to be time to learn it. Puberty and menarche come just before or during a spurt in physical growth. Relatively little is known about just what the juveniles in wild troops are doing during this long period. The Freudian theories of human development lay great stress on our "latency period," during which we are supposed to be becoming human. Wild juvenile primates are obviously playful and obviously initiate many interactions among themselves, but few that influence adults of the troop. They seem to have none of the inhibitions of sexual play that are imposed on our youngsters, and thus no Freudian "latency."

Human subadults may have a period of adolescent sterility when conception is unlikely to occur, even though sexual relations take place. The same stage is obvious in many primate species whose females may not conceive during their first few estrous periods. In some species, such as the savanna baboon, young males are much slower to reach full growth than females, which leads to an apparent imbalance in the adult sex ratio.

The longest absolute increase in a phase of life from primate to man is the adult phase. The zoo-cosseted chimpanzee or gorilla may spend 30 years as an adult, whereas a man can look forward to about 55 years of adult life. Preservation or elaboration of a body of culture by adults over several generations of growing young is essential to perpetuate tradition as we do.

One of the most influential overall theories of human development points out that adult men retain many characteristics of juvenile or even fetal primates. Man anatomically resembles an orangutan or chimpanzee child. His face does not have brow ridges and his skull is set vertically on top of his spinal column without flexing forward as in quadrupedal mammals. Behaviorally, he retains the curiosity and inventive play common to other mammals' obstreperous young. This theory was elaborated by Bolk (1926, in Napier and Napier, 1967). Aldous Huxley (1939) took it to the logical conclusion by imagining that an enlightened savant of the eighteenth century found the secret of immortality. After 200 years, he, alone among men, had lived long enough to grow up—into a gorilloid, glowering from under his massive brow ridges at a world of no interest to him. On the whole, although "fetalization" may have been one of the mechanisms of human evolution, it seems that each of the traits has a clear selective function, not least of which is the contribution of human curiosity to human society. We, like other animals, have a mosaic of adaptations, of which many may result from prolonging growth beyond where other primate adults stop (Gould, 1968).

GROWTH RATES

Mammals, by definition, depend on their mothers. However differently young mammals grow up, all begin by needing their mother's milk, her licking, and grooming as they enter the world.

However, there is a vast difference in the preparation for life. Some species have *precocial* young: the young are born with eyes open and are fully furred, perhaps able to walk or run from the first day. Some have *altricial* young, naked and helpless. The terms, of course, are relative—the hare is more precocial than the rabbit, but the acouchi, which runs on the first day of life and nibbles food on the second, is more precocial still. Further, not all the aspects need go together. The naked, helpless young of the tree shrew are left alone in their nest from birth and manage their own thermoregulation. The kangaroo, an inch-long, acephalic embryo, one month after conception crawls unaided the long eight inches from the womb to its mother's pouch, where it can settle for eight months more to finish gestation. In one animal the temperature regulation and in the other control of the forelimbs have developed out of all proportion to the rest (Bourlière, 1964).

It might seem that precocial young should grow up more quickly. After all, they are more like adults than naked, hairless, altricial young are. The opposite is true: mammals that keep the young in utero until a fairly advanced stage apparently also can afford to let them grow up relatively slowly. If one plots the growth rate of the young during gestation for a wide variety of mammals against the average weight of adult females, it is clear that precocial young grow more slowly than altricial young, relative to their mother's weight; that is, their advanced stage at birth may reflect long gestation times, rather than a rapid, energetically demanding input of energy during gestation. (See Fig. 14.3.) It is also clear that very large mammals (this means mainly large ungulates) have

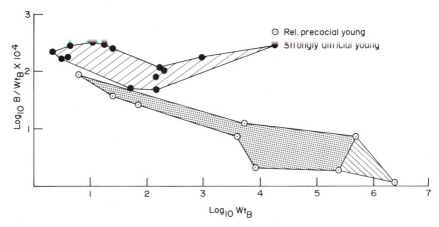

FIGURE 14.3. *Weight gains per unit body weight of females during gestation plotted against mean body weight of adult females. Species producing precocial young invest less in growth per day per unit body weight than do those species bearing altricial young. (After Eisenberg, 1981.)*

precocial young, whereas the smallest mammals have altricial young (Eisenberg, 1981).

The next step is lactation. The rate of growth continues very much in step with the rate of growth during pregnancy (Fig. 14.4.). That is, the precocial group does not suddenly speed up or slow down at birth, relative to the others. The precocial ones continue putting on weight faster in absolute terms, as so many are large mammals, but if growth during lactation were redrawn in terms of gain relative to mother's body weight, it would look much like the graph of growth during gestation. (See Fig. 14.5.) Even during lactation, the precocial ones make less of a daily energy demand on their mothers, although they may stay around longer. Note the very important fact that all mammals grow more quickly during the nursing than gestation periods. This is the time of maximum demand on the mother (Eisenberg, 1981).

It may turn out that the growth of the infant's brain is even more interesting and important than the growth of the infant's body in terms of demand on the mother's physiology. The brain is an organ of high metabolic demand and is expensive in energy for its weight. Martin (in press) suggests that the mother's metabolic turnover constrains fetal brain size, as well as the growth of the infant's brain during lactation. Precocial mammals in his sample typically had brains some 4.5 times larger than altricial ones at birth and 45 per cent larger as adults for similar body weights. Precocity (and lengthened gestation and growth) thus goes with greater capacity for intelligence as measured by brain weight. Primates are an extreme: in primate evolution both fetus and newborn have a brain weight approximately twice the size for body weight than all other mammals taken together, except the toothed whales, which are intermediate between the primates and the others (Sacher, 1982, quoted by Martin, in press).

The difference between human and the great apes' brain growth rates is also instructive. Human baby weights and human newborn's brain weights are about double those of newborn great apes, with roughly similar mother's body weight and gestation length. Thus, human mothers devote relatively more, even

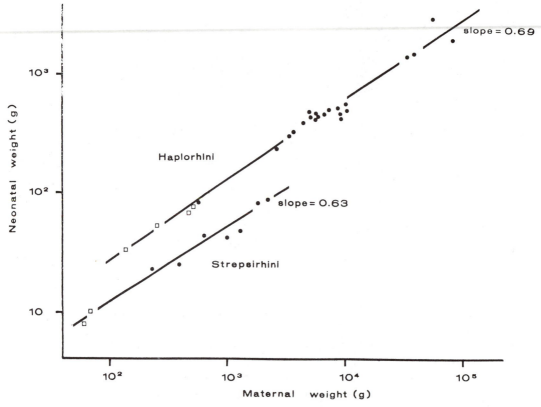

FIGURE 14.4. *Among primates, haplorhini have heavier newborns and longer gestations than strepsirhines; that is, they are more precocial on the whole. (Courtesy of W. Leutenegger.)*

around twice as much, energy over a standard time to fetal development. At birth the relative proportions of brain and body are much the same as in apes.

The great divergence between humans' and apes' brain growth comes after birth. We maintain the rapid fetal pattern of brain growth until at least 12 months after birth. This suggests that the particular amino acid composition of human milk allows for our peculiar postnatal brain development (Gaul, 1979; Martin, in press). This also means that the energy demands of lactation, great in any primate, may be particularly great in humans.

How can we imagine that mammalian mother care evolved? Eisenberg starts with a reptilian ancestor, which laid relatively small-yolked eggs in a nest, and guarded and incubated the eggs past hatching. When it developed milky secretions from its sebaceous glands that could nourish its young, it was officially a mammal. One stock, the monotremes, remained roughly at this stage. The marsupial-eutherian stock, however, developed teats and live birth, and so could either go on building nests or else carry the young around clinging to the teats. Marsupials of several lineages developed pouches to carry their extremely altricial newborns. Portmann (1965) argues that the first eutherian mammals had young that were intermediate—not quite so embryonic as modern rats or mice, but still helpless nest dwellers. The young of truly precocial mammals seem to

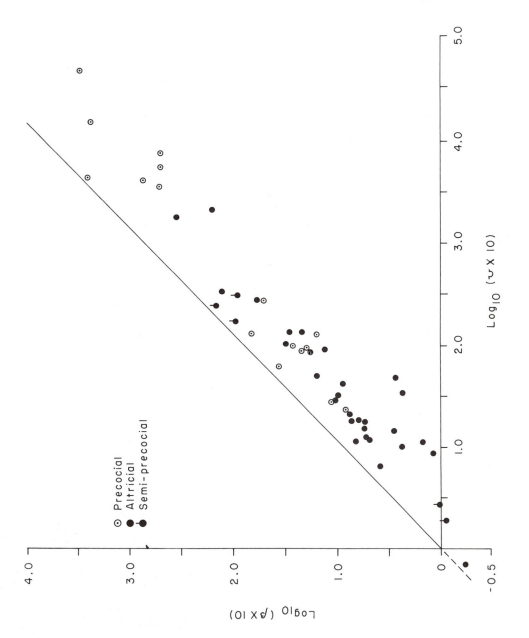

FIGURE 14.5. *Mammalian litter weight gains per day during
gestation (y-axis) plotted against
weight gain per day during lactation (x-axis). Weight gains per
day for all species are greater
during lactation. Precocial and
altricial forms have the same
trends. (After Eisenberg, 1981.)*

have prolonged gestation to include the nest phase. Their embryos start with open eyes, then close them in utero, then reopen them before birth.

It may be that these early mammals were arboreal (Martin, 1968b). They could have built tree nests, like many present day prosimians, and the older juveniles might have clung to the parent's body like arboreal mammals today (primates, sloths, bats, opossums, colugos, pangolins). One of the crucial bits of evidence here is that the hind foot of many mammalian embryos has an opposed big toe, although this grasping toe closes again to make running or a walking foot in ground-living mammals.

Some primitive primates, the nest-building prosimians, then may retain ancient patterns of maternal care (Figs. 14.6. and 14.7.). The higher primates have babies that cling to the fur from birth (Fig. 14.8.). This seems to go with relatively slow development, a slow energetic input over a long time. Again, we are in the realm of K selection, in which the fluctuations of the environment are likely to fall within a infant's life span. This raises the important question of what happens if motherhood goes wrong?

If an embryo is malformed or the mother is stressed at a very early stage, she may resorb the embryos. Altricial mammals such as mice apparently have more capacity to do this than precocial mammals. At a later point, the fetus is aborted or stillborn. Many mammals then cannibalize their offspring, as they also do if disturbed during parturition, which likewise recycles the offspring's

FIGURE 14.6. *Lorisoids commonly build nests and carry the young by mouth like kittens when they must be moved. Note the hunched up posture of the infant Senegal bushbaby. (Courtesy S. Bearder.)*

FIGURE 14.7. *Senegal bush-baby twins are "parked" while their mother forages. (Courtesy S. Bearder.)*

FIGURE 14.8. *The rarely photographed woolly lemur carries its infant on the fur, like higher primates. (Courtesy R. D. Martin.)*

TABLE 14.1. Divergent Systems of Mothering

	Gestation Length	Infant Rides on Mother's Back	Infant off Mother 30% of Time	Infant Travels and Feeds Alone
Ringtailed lemur	135d	1 week	6–7 weeks	2½–3 months
Brown lemur	120d	4 weeks	11–12 weeks	3½–4 months
Ruffed lemur	90–100d	never	0 weeks	1½–2 months

energy. This is again more likely for altricial than precocial mammals, particularly not the large precocial herbivores. Still later a mother may abandon her young. Desertion costs depend on whether she may produce future offspring and thus on what proportion that infant or litter is of her future lifetime's expectation. On the whole, the slow-growing precocial baby is likely to be a greater investment—again, K-selected.

This discussion has dealt with primates in a lump as slow growing, fairly precocious, large-brained and K-selected, compared to other mammals. We now need a cautionary tale to show how systems of mothering can diverge even among closely related animals (see Table 14.1.).

The brown lemur's young develop slowly (Fig. 14.9A.). They are born fully furred and cling transversely under their mother's loins like a money belt. They stay put there for about the first month, although other group members groom the bits they can reach. Infant sharing is rare after this. Ringtailed lemurs, in contrast, ride lengthwise under their mother's belly like young monkeys and apes. They transfer briefly to other troop members within the first days of life, and ride on their mothers' backs at a week (Fig. 14.9B.). Further infant development is equally fast; both ringtails and brown lemurs reach sexual maturity at the same age, two and one half years. The differences seem to relate to ecology. The ringtails are more terrestrial and do not always have so far to fall. Far more important, the ringtails may live in drier, more seasonal scrub, whereas browns are confined to lush closed-canopy woodlands. The ringtails must reach weaning quickly before the summer food supply runs out (Klopfer and Boskoff, 1982; Sussman, 1977).

A third lemur, the ruffed or variegated lemur, was for a long time classified in the same genus as the other two. However, ruffs have almost hairless, helpless offspring. They usually give birth to twins, but in captivity may produce quintuplets. The mother either leaves them in a tuft of lianas or an epiphytic bracket fern, or else she builds a nest of leaves and, in captivity at least, may line it with her own hair (Fig. 14.9C.). When she needs to transport her young, she carries them in her mouth, like kittens. They develop extremely rapidly, like other altricial young. Probably they have little or no social contact outside their mother: females are generally intolerant of others in captivity although one mother and daughter occasionally tolerated and even suckled each other's offspring. The young travel and feed by themselves even earlier than true lemurs (Kress et al., 1978; Wolff, personal comment; Petter-Rousseau, 1962). It is against expectation that this rain forest form should be relatively r-selected, but it certainly is; someday a field study may tell us why.

FIGURE 14.9A. *Brown lemurs ride transversely for two to three months. Here, rare twins. (Courtesy P. Coffey, and the Jersey Wildlife Preservation Trust.)*

FIGURE 14.9B. *A ringtailed lemur newborn clings longitudinally to its mother's belly, then rides on its mother's back within a week. (Courtesy P. Coffey and the Jersey Wildlife Preservation Trust.)*

FIGURE 14.9C. *Altricial infants of the variegated lemurs are usually left in a nest. This mother pulled the fur from her thigh to line the nest. (Courtesy J. J. Petter.)*

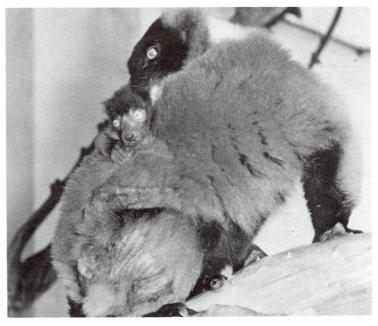

PREGNANCY AND BIRTH

Most primates give few external signs that they are pregnant. There is a familiar though by no means universal human sequence of slight nausea in the first trimester, placid well-being in the second, and lethargy in the third. It is often not easy to see signs of external swelling in captive primates, let alone shifts of mood (Fig. 14.10.) At most, keepers notice increased restlessness for a day or so preceding birth, which may correspond to the human tendency to pick that day to start painting the house. Our relative insensitivity may not be shared by other primates: pregnant females may be differently treated by troop members (Noyes, 1981).

The hormonal changes, in contrast, are quite reliable indicators of pregnancy in humans and other primates, although as usual there are differences between species and between individuals.

The anatomy of the placenta is significantly different between strepsithines and the higher primates, including Tarsier. Strepsithines have epitheliochorial placentas. That is, the mother's blood remains spearated from the embryonic placenta, by the lining of the walls of the maternal blood vessels, the uterine lining, and the connective tissue between these two. This is presumably fairly inefficient in transferring oxygen and food to the embryos. Tarsier and the higher primates have hemochorial placentas. Some of the maternal uterine tissue breaks down, so that the embryo's placenta is in direct contact with the maternal blood, although a layer of embryonic tissue prevents maternal and fetal blood from actually mixing. This more efficient system perhaps allows for the greater birh weights of higher primates. Of course, the distinction is not absolute. Galago has a disc of more closely linked tissue within its epitheliochorial placenta, and the new-world monkeys only achieve full hemochorial status late in gestation.

FIGURE 14.10. *Pseudopregnant gorilla. Pseudopregnancy occurs in many mammals, from mice to human beings. Because a baby is never born, pseudopregnancy can be taken to extremes. (Courtesy D. Sorby.)*

Still, placentation remains one of the chief distinctions between higher and lower primates (Hill, 1932; Le Gros Clark, 1960; Leutenegger, 1980).

Birth takes place in three stages. In the first stage, horizontal contractions of the uterus enlarge the pelvic canal. This may or may not begin with breaking of the waters, the amniotic sac of fluids that surrounds the baby. In the second stage, longitudinal contractions expel the infant. It is only at this point that the mother can aid the birth process by voluntary bearing down. The third stage is expulsion of the placenta or afterbirth.

Primate labor differs from ours in being shorter, commonly about two hours. This is not surprising, given the relative size of fetal head and maternal birth canal. However, it is probably underestimated because we may not notice the mild contractions of stage 1 (Lindberg, 1982). It also takes place far more often during the night (for diurnal primates) or day (for nocturnal ones). There are always exceptions: patas seem to give birth by day as do orangs. In the patas, this minimizes predation during the night in an animal that relies on hiding, not group defense (Chism et al., 1978, 1983). Statistics for a few hundred normal human births also show a clear nocturnal peak at about 2 or 3 A.M. Normal human labor, for both primiparas and multiparas, averages two hours shorter for births that end at night than in the daytime (Jolly, 1972, 1973).

Following the third stage, the mother usually eats the placenta (Fig. 14.11.).

FIGURE 14.11. *Eating the placenta. (Courtesy P. Coffey and the Jersey Wildlife Preservation Trust.)*

This may contribute to her nutrition at a crucial time, allowing her to pay full attention to the baby, and may remove an odorous object that would attract predators to the vulnerable infant. It may even serve fundamental aspects of hormonal transfer or mother-infant bonding (Poirier, 1977). Hutchinson once suggested defining *Homo sapiens* as the primate that does not normally eat the placenta.

Treatment of the placenta aside, birth seems to be a remarkably conservative process throughout the order (Brandt and Mitchell, 1971). This leads to two speculations. First, this is a highly innate and conscious process in humans unless anesthetized. One is aware of, but surprised by, the transition from first to second stage labor contractions, which is triggered by processes out of one's own control. The psychologists or anthropologists who claim humans have no innate behavior, or who imagine innate behavior as somehow divorced from conscious awareness, probably haven't had a baby.

The second set of speculations concerns "natural" childbirth. All the complications of human labor have been occasionally observed among wild or captive primates: still birth, breech presentation, placenta previa, sometimes leading to death of the mother. The one birth observed in a wild chimpanzee was apparently difficult or even painful, as judged by its unusual length, and the mother's grimacing, trembling, and grunting (Goodall and Athumani, 1980). Natural primate childbirth is not always safe or easy.

However, there are repeated suggestions that feelings of relaxation or comfort during birth can ease and speed labor. This relates to shorter human labors at night and to the apparent reluctance of primates to be watched during childbirth (Naaktgeboren and Bontekoe, quoted by Lindberg, 1981). Van Waagenen (1972) admits "rhesus in labor have been observed to abruptly subdue contractions when they focus their attention on the incoming food truck . . . (or) to voice an opinion on a dispute . . . and it is difficult to give up the belief that some animals being watched for the imminent birth of an infant can quietly, in turn, watch the personnel and retain the fetus until after the 5 o'clock locking of the animal quarters."

A study which I will forbear to cite as a reference gives fascinating data on the diurnal cycles of amniotic pressure and uterine contractions, and how these change one or two days prior to the beginning of labor in different patterns for day and night deliveries. This could obviously be important to know for human obstetrics. The 17 rhesus that were tested, however, were apparently physically restrained in some way for five to eight days between surgery and delivery, or they would have pulled out the catheters inserted in uterus and femoral artery. They were kept unanaesthetized in an upright sitting position in a restraining chair throughout labor and delivery. This position is unnatural for a rhesus at the best of times, let alone during delivery. One cannot help thinking that in a study of the time course of labor, more sympathy for the animals might have produced better science.

RECOGNIZING A BABY

How does a mother recognize a new baby? Among primates, as among people, experience must play a major role. Most troop-living monkeys have met or even

FIGURE 14.12. *Black-and-white Colobus group. Infant colobines have distinct natal coats, which may relate to their pronounced infant-sharing. (Courtesy P. Coffey and the Jersey Wildlife Preservation Trust.)*

held and cuddled other infants before they have their own. (The attraction of juvenile females to new infants will be discussed in the next chapter.) Infants of many species are clearly marked; in many the natal coat is a different color than the adult's. Black-and-white colobus have white babies (Fig. 14.12.), and dusky langurs have orange ones (Alley, 1980). The baby rhesus monkey has a dull brown coat like its parents, but a pink parting in its hair seems to be particularly attractive to the females. Our own children are of course a markedly different shape than adults. Lorenz has made a beautiful diagram of the so-called cute response (Fig. 14.13.). Children or animals or cartoon characters with very large heads and short legs in proportion to their bodies look cute to us, the more so if they have large eyes in their large heads. This applies to baby fish, baby birds, baby dogs, artificially babified dogs such as the Pekinese, and babies. Gardner and Wallach (1965) experimented with silhouettes of real babies' heads and "super babies" that had even larger foreheads and smaller chins, as well as more adult faces with receding foreheads and protuberant chins. They presented a series of these silhouettes to college students, both male and female, and asked which looked the most babyish. The students, particularly the female students, consistently picked a super baby rather than the real, normal baby head. Like a herring gull chick pecking at a larger, redder beak than would be conceivable in its own parent, we respond to an exaggerated cartoon baby as being more real than the real thing.

Primate mothers lick off their babies and groom them into fluffiness after birth. In other mammals smell plays a major role in infant recognition, and in later recognition of individual infants. This is probably true for primates as well.

There may be still other cues. Caged slender loris apparently accepted any infant as their own, including ones of very different age and galagos did not discriminate their own young for several days after birth. Since loris usually park their young on a branch during the night and galagos have nests, it may be a position rather than an individual cue that attracts the mother, as for so much else in lorisoid behavior (Swayamprabha and Kadam, 1980, Klopfer, 1970).

Finally, lost and distressed babies make noises (Fig. 14.14.) If they are really

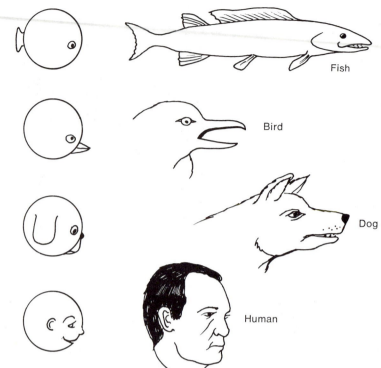

FIGURE 14.13. *Babies have large rounded heads, large eyes, small noses, chins, and legs. Adults are elongated and angular. The similarity of young vertebrates may result originally from the cephalocaudal gradient of embryonic development. Many vertebrates have probably evolved recognition of this generalized baby shape, the "cute response." (After Lorenz, 1950, in 1971.)*

Fish

Bird

Dog

Human

FIGURE 14.14. *Newborn gorilla. Infants' expressions and vocalizations, as well as general shape, distinguish them. (Courtesy P. Coffey and the Jersey Wildlife Preservation Trust.)*

abandoned, it will not matter if a predator finds them; they will die anyhow. The only hope is to make their mother hear first.

Recognition of the baby would seem a likely thing to need no prior experience. The primiparous mother must be assured of treating her first young as an infant, not a stranger. An interesting sidelight on this is that human mothers, after a normal birth when they have not had too much anesthetic, say that they are overcome by a feeling of joy and tranquility after the baby if born. Obviously, culture tells them they ought to feel so, but the physical wash of emotion, rather like that after completed sexual intercourse, may have much deeper sources. Such a physiologically determined feeling of joy and tranquility in primates, who have not been nattering and knitting for the baby for the previous nine months, may allow them to accept the infant without the resentment that would normally greet a total stranger squirming about their bodies.

In summary, it is not really known how mother primates recognize their young. There are plenty of physical markers of the newborn and every likelihood that unlearned processes are deeply involved, but experience certainly plays a considerable role as well. The qualitative aspects of a mother monkey's care are very similar in most species. A baby monkey does not survive unless his mother treats him as a baby, not a stranger, allows him to climb on her and suckle, cleans him off, and to some extent cuddles and supports him. However, the quantitative differences among species, individuals, or mothers in different situations are enormous.

REFLEXES OF THE NEWBORN

The newborn infant is born with a repertoire of reflexes: grasping, crawling, rooting, sucking. In many primates these are fully functional at birth (Fig. 14.15.). A squirrel monkey may aid its own delivery by crawling upward with its forelimbs before its hindlimbs are delivered. Even humans can often support their own weight by their hands at birth. One suspects, however, the idea that Spartans tested their infants by hanging them by their hands off the edge of a roof

FIGURE 14.15. *Primate and human infant reflex movements are closely similar. (Courtesy P. Coffey and the Jersey Wildlife Preservation Trust.)*

was a story put about by the Athenians. The Moro Reflex, the convulsive flap of arms toward the midline if you slap a baby's mattress or let its head drop back sharply, is the ancient primate gesture of seizing the mother's fur as she jerks into motion. Other primate babies, like ourselves, may not be able to cope completely. A mother gorilla or chimpanzee must usually support her infant with one hand whenever she moves for the first two or three months of life. Many other primate mothers compensate for an infant's disability, and crippled young may survive for months, even in wild troops.

We now know that human infants in the first days or hours of life are capable of learning scent recognition, and will suck in longer or shorter bursts to produce their own mother's voice, rather than another mother's (Decasper and Fifer, 1980). At this early stage young primates have some recognition of a mother as warmth and pleasure, although it may not be of mother as any sort of defined object. Harlow's (1965) experiments with motherless monkeys disproved the then-current theory that primates learn maternal attachment as an adjunct of "cupboard-love"—that is, by association with a source of food. He gave rhesus infants a choice of bare wire pseudo mothers and terrycloth pseudo mothers. They clung semi-permanently to the soft mother surrogates, whichever one had the milk.

These early reflexes at once begin to be elaborated and restricted, as mother and infant begin to create a new, personal bond, out of the raw material of their nervous systems. Grasping and rooting may be as mechanical as a knee jerk, but unlike a knee jerk, they evoke emotion in the one who receives and, quite possibly, in the one who gives. Helen Blauvelt, who has studied infant behavior in both animal and man, perceptively called the neonate's repertoire "the reflexes of love."

LACTATION

The extra nutritional needs of a lactating mother are lower per unit of time in slower growing, precocial mammals. The energy requirements of voles may be 150 per cent more than normal when lactating, whereas humans increase by only about 30 per cent. There are needs for specific nutrients as well, such as iron and calcium, which may limit the infants' growth as much as calories themselves (reviewed in Daly, 1979).

Milk composition differs between every species of mammal. Mammals of different styles of mothering have different proportions of protein, fat, and sugar. Infants which are left behind in nests are given relatively high fat content, for warmth and energy to tide over the intervals between feeds. Highly precocial mammals drink milk with both high fat and sugar levels, which gives immediate energy to run after the mother. Primate milk, including our own, supplies more protein for growth, since so many of the energy needs are supplied by the mother's own body heat and carrying. As mentioned in the section on growth rates, human milk possibly allows for rapid post-natal brain growth (Martin, in press).

Suckling frequency also varies immensely between mammals. At one extreme, the tree shrew squats over her young squirting milk into their mouths for about 10 minutes every two days. Primates are near the other extreme, with babies suckling as much as every 20 minutes, and often clinging to the teat as a comforter in between times.

The duration of lactation has been discussed in Chapter 11. It may be only a few months, as in the lemurs, or the marmosets which give birth twice a year, or it may last three or even five years in chimpanzees and orangutans. In most primates other than the Callitrichids, lactation effectively blocks ovulation.

The lactation period is the time of greatest demand on the mother's energy. She is attempting to carry and feed a large infant, using a far less efficient organ than the placenta. Many women lose weight during this period, and in food shortage, this is the time when women and female primates are most vulnerable.

ATTACHMENT AND IMPRINTING

Mother-infant interaction is a continuous process between two individuals, modified from day to day, in terms of the biological state of mother and child, and in terms of all that they have done and felt for each other. However, it is possible to distinguish some qualitative shifts of emphasis, particularly in human infancy.

The most potent theoretical model comes from the work on imprinting in birds. Konrad Lorenz (cf. 1970) first described imprinting. He also practiced it—hopping through the spring grass, quacking, with a train of baby mallards behind him to the bewilderment of the neighbors (Lorenz, 1952). The mallards accepted him as their mother as long as he remained of a reasonable size, crouched not more than three feet high, and made an intermittent noise, such as quacking. From this beginning came Lorenz's categorization of *imprinting:* (1) It was a process that occurred in early youth, the so-called *critical period,* and after that stage of life could no longer occur; (2) it was a process in which the young duck or bird learned the characteristics of its parents' species, not so much the characteristics of mother as an individual, but the generalities that would later allow the young bird to identify its own appropriate sexual partner; and (3) further, the process seemed at the time to be irreversible (Cf. Lorenz, 1935 in 1970).

Since Lorenz's original work, the definition has somewhat softened and broadened. It is now realized that nearly any kind of learning occurs more easily at one stage of life than at another. The learning of species' characteristics may be reversible; it depends on the circumstances. It also depends on the species; even fairly closely related animals may differ in the degree to which their adult sexual preferences are unlearned behavior, or learned during a kind of imprinting, or learned fairly broadly throughout the period of growth. Further, as in most important biological processes, there are redundancies of control, so that a bird is far more likely to learn right than to learn wrong.

However, with this wider definition it is possible to see a common process occurring in many young birds. When the bird is newborn, it has a tendency to approach, follow, and cuddle against any conspicuous object, particularly if the object is moving or flashing and making an intermittent noise. This sort of object has the properties of a reinforcer: the young bird will work to turn on a flashing light. It seems to reduce distress when the bird can cuddle up to an object, or even see one; it makes quieter calls, stops distress calling, and slows down activity unless the object is moving away, in which case it follows the object. After the first unspecific period when almost any conspicuous object will elicit approach, the young bird learns very rapidly the characteristics of its own particular object and will no longer respond to other objects. Following this comes a period of rising fear when the chick actively avoids strange objects, running to its own "mother" if present. If the chick reaches the age of fear responses with no imprinting object, it will no longer imprint but will flee from all novel stimuli. In this case, it may in fact be imprinted on the gray walls of its cage for want of a more adequate mother. In the wild this system normally leads to learning one's own mother's characteristics; a duckling that hatches and sees a fox as its first imprinting object might as well follow the fox, as it is lost anyway. By and large, the system works fairly well. The duck is likely to learn that a duck is its mother.

Hinde (1966) reminds us that birds and mammals have been phylogenetically separate since they were derived from their reptilian ancestors. As few reptiles have maternal care, maternal behavior in birds and mammals may have been separately developed, and any similarities can be no more than analogous. However, mammals as well seem commonly to go through a period of approach to any conspicuous object, in the likelihood that it will be their mother, followed by a period of learning ever more particular characteristics of their own mother and then of rising fear of novel situations and stimuli and strangers. Whether or not this early learning channels sexual responses in the adult mammal is still being studied. In many species this is clearly not so, and animals may be raised in total isolation from conspecifics, yet still mate normally as adults (Scott, 1968).

It was John Bowlby (1969) who forcefully pointed out the importance of imprinting concepts in human development. Children gradually learn to recognize and cling to or follow human beings in general, then their particular caretakers. Loss of the loved ones, as when a child is removed to hospital, sometimes leads to deep depression, and in extreme cases to an emotional shallowness that prevents further commitment. Bowlby put this in evolutionary terms: the child that knows and stays with its own mother, fearing strange people and places, is partly protected from predators of its own or other species. In other words, the primeval human toddler lived in an environment that selected for active love.

Rajeki and colleagues (1978) have recently reviewed Bowlby's theory and several other psychological theories of infantile attachment. They raise objections to all the versions. Their main doubt about Bowlby's is that infant primates may be bonded to biologically inappropriate objects, but so may chicks and ducklings. The advantage of such early learning over an innate recognition is precisely that it allows more flexibility or richness or detail in the choice of imprinting objects.

A second series of objections concerns the usefulness of speaking of stages of growth, when growth is a continuous process for mammals (Brainer, 1978; Kagan, 1978). We don't pupate and emerge as butterflies, even if adolescents sometimes wish they could. It is clear, however, that various groups of abilities appear together, so there is a sense in which stages of attachment can be useful concepts.

ATTACHING TO AN INDIVIDUAL Age groups

The first stage of infancy is the "reflex stage," which we have already discussed. It is now clear that even in this early, helpless period, human babies may recognize their own mother's voice and smell. Four- to five-week-old squirrel monkeys recognized their mothers by odor, even if she was anaesthetized, and a hood over her head, but not if she had also been washed (Kaplan, 1977). It takes us about six to nine weeks until we seem to recognize a mother's face. The same period may last only one to three weeks in a rhesus (Rowell, 1963b). It could be even shorter in colobines, who regularly lend their babies to other females. A colobus or langur baby might thrive better if it could quickly recognize its own mother (McKenna, 1979, 1981).

The second stage begins when the infant begins attachment, or, if you like, imprinting behavior. This is the period when the baby learns to recognize his mother, her species, and herself. In the human baby, this recognition shows clearly as social smiling. Smiling is at first directed to very general stimuli; two black eye spots on a sheet of paper are enough to evoke it, and so are many simple changes in the environment. But it becomes narrowed first from general changes, to eye spots, to something resembling a human face, although grossly distorted faces and masks will evoke smiling for a time, to at last the mother and familiar people alone. This is what seems to the theorists to resemble imprinting in the duckling, the narrowing from any conspicuous change to species-specific stimuli to recognition of the mother herself.

Many people have studied attachment behavior in human babies: the smiling response (Ambrose, 1963); young babies sent to hospital (Schaffer and Emerson, 1964); development in Ugandan children (Ainsworth, 1967); differences in attachment between babies in normal families and those in institutions (Spitz and Woolf, 1946). Schaffer's work is particularly clear on babies sent to hospital; under two months, infants seem to show no effect of either separation from their mothers or reunion with them; from two to seven months, a baby separated from its mother lies in the hospital neither crying nor vocalizing very much, just quietly scanning the environment, and when sent home is also markedly unresponsive, even limp at first (from 20 minutes to four days after reunion with the mother), then rather suddenly picks up again. Still older babies, who fear strangers, may protest and scream at separation.

It is in this attachment or imprinting period that the child becomes cute. A newborn Caucasian infant is pink if not red, wet and wrinkled and scrawny, in fact incredibly attractive to the mother, but not the sort of round-faced, eye-to-eye contact smiley baby who sells soap flakes to the troop at large. (African newborns are often also pinkish, but their dark eyes give an early impression of eye-to-eye contact.) The round smooth face, the large head, in short the "cute" gestalt, indicate really not a new baby but an attaching baby.

In primates there has been little direct work on the early period of attachment. The infants are still clinging to their mother most of the time as they have since birth, and are still physiologically dependent on her, so there is not much overt behavior that could be compared with the human smiling response.

The nearest parallel to the human social smile is the chimpanzee's biting and play-face, which become obvious around week six. Before this, the mother pays fairly little visual attention to her baby's face. When he starts to bite everything, and to make the play face that might, at this point, be called "ready-to-bite" face, the mothers start to pay much more attention, and to stroke their babies' faces and put their index fingers in the babies' mouths. This starts one of the first reciprocal games, where the mothers "stroking and poking alternates with the baby's playface and biting" (Plooij, 1979). Such games and interactions in which the adult and baby take turns stimulating each other, each sensitive to the other's cues, are probably a major precursor of later social awareness.

FEAR OF STRANGERS

The third stage is that of rising fear of novelty and specificity to the mother. Perhaps too much has been made of fear of strangers in human infants because this might fit so neatly with ducklings' imprinting if it were true of all babies. A few babies, at least, in the second half of the first year, begin anxiously to reject strangers. Others do not cry if a stranger picks them up, but look repeatedly back and forth from the face of the stranger to the face of the mother, obviously aware of and comparing the differences. Schaffer (1966) shows that the child can in fact recognize the differences as early as three or four months: it is not until eight months or so that he becomes fearful. In rhesus monkeys the comparable period perhaps appears at about nine to ten weeks in a laboratory group situation, when a group of new calls appears: lip smacking, which is a social attraction signal among monkeys, pleasure noises, social grooming, and the beginning of the fear grin and the fear call, the geckering screech that goes with the grin. By this age the monkey baby has typically lost its distinguishing infant marks: the hair parting of the rhesus baby or the black coat of the young baboon.

As usual, fear of strangers differs in different species, sexes, and environments. Pigtailed macaque infants, reared by their mothers in a laboratory group, showed fairly little tendency to move away from a cage containing a stranger. Pigtail mothers are highly restrictive, rarely allowing their infants to contact non-kin. Bonnet macaque female infants retreated from strangers in the same test, grimacing and screeching, but males reacted like the pigtails. Bonnet mothers give their infants more independence, so their infants have more responsibility in maintaining the bond, and females are particularly subject to mistreatment by non-kin. Experience certainly plays a part: bonnet infants raised alone with their mothers approached strangers at any age (Rosenblum and Alpert, 1977).

PARENTAL INVESTMENT AND INDEPENDENCE

At some point, the infant becomes independent of its mother for food and transport although they may continue to associate with each other for much of their lives. Trivers (1972, 1974) reasoned that the mother's and infant's interests may well conflict at this point. The infant is bent on its own survival. The mother, however, may prefer to hold back some resources toward her own survival and the production of future offspring, or toward the care of the infant's older siblings. Maternal efforts at weaning and young primates' weaning tantrums would then be the expression of resource competition within the family.

There is clearly a shift in the restrictiveness of mothers as infants mature. At first, the mother is responsible for initiating and maintaining most contact, while the infant leaves her. Later, it is the infant who is chiefly responsible for joining the mother. However, as Altmann (1980) points out, each infant is a relatively large proportion of a primate mother's output. Further, an infant that has already survived to weaning has passed a vulnerable period: future off-spring should be discounted by the likelihood they will die. An infant has an extremely high stake in its mother's health and survival: wild baboons or-phaned at weaning rarely, if ever, survive. Finally, the infant of a rather old mother who is herself at risk might fare better if it becomes self-sufficient fairly early. In short, the infant's interests in its own independence and, failing that, its mother's well-being, coincide closely with the mother's—far more closely than one would judge by merely counting genes.

MOTHERLESS INFANTS

One of the most famous series of primate experiments has been Harry Har-low's work with motherless monkeys (Harlow and Harlow, 1965). (See Fig. 14.16.) It began with an attempt to find out the stimuli of motherhood. Harlow

FIGURE 14.16. *A baby rhe-sus keeps contact with his soft pseudomother while reaching to drink from the hard pseudo-mother. (Courtesy Wisconsin Primate Laboratory.)*

thought at first that he had cracked the code of adequate mothering: all you need is terry cloth.

But then the babies grew up. They turned out to be strikingly abnormal in their behavior. One might perhaps have suspected as much from their totally abnormal upbringing and the fact that while growing up they sat and rocked in the corners, a stereotyped rocking typical of human mental defectives as well as isolate monkeys. The motherless monkeys could not mate. Males mounted with impossible orientation, obviously with sexual excitement but little hope of reproductive success. Females would not accept mounting by males. Harlow placed several of these females in a "rape rack" accessible to experienced males, so they became pregnant and give birth. They mistreated their offspring; in fact, they treated their own newborn babies much as they would treat a rat that entered their cage, hitting it away, stepping on it, and grinding its face into the cage floor. Rhesus babies have a fair degree of motor control, and some of these infants succeeded in nursing by their own efforts, although their mothers continued to reject or ignore them (Fig. 14.17.). More of the first babies eventually had to be removed. However, motherless monkeys were considerably better with their second and succeeding babies, so that even they learned from experience.

FIGURE 14.17. *A motherless mother rejects her own child. (Courtesy of Wisconsin Primate Laboratory.)*

In the course of mistreating their infants, they added still more to the understanding of attachment. Infants will attach even to such punishing mothers. Cloth surrogate mothers that blasted air at the infants, or suddenly extruded brass spikes through the ventral surface, or even hurled the infants to the floor failed to drive off the infants, who spent even more time clinging to these monsters than controls with terry cloth surrogates. It would be interesting to know if these infants became even more abusive as parents themselves (Rajeki and Lamb, 1978).

At this point I should note some changing attitudes in science. Harlow's early work was greeted as a blow for motherhood—liberation from the simple-minded "cupboard-love" learning theory that babies were simply conditioned to "love" whatever fed them. His later revelations of the psychotic behavior of isolates have been equally received as fascinating insight into the importance of motherhood. Separation and isolation experiments were widely repeated with different species, and for both shorter and longer times of isolation. The drastic reaction of pigtail macaques to separation from their mother became a "model" for depression and despair in children (Kaufman and Rosenblum, 1967). Hinde (1977), and others, showed that for a few individual infants the effects of even short-term separation may persist for years. His results have affected pediatric hospital practice and parent's visiting rights. Isolate-reared chimpanzees' exaggerated fear of novel objects resembles human autistic children's (Menzel et al., 1963), and might give clues to that bizarre condition. Environment affects the degree of abnormality (Roy, 1981; Kaplan, 1981; Bajpai, 1980).

One of the most interesting studies reveals the different effect of almost total sensory and social isolation for the first six or seven months on infant rhesus, pigtail, and crabeating macaques. The rhesus were profoundly affected on all measures. Pigtails retained some capacity for exploratory play with objects; crabeaters some capacity for social play. The detrimental effects of isolation are not monolithic, but differ between species (Sackett et al., 1982).

Finally, it became clear from Harlow's own work that contact with an infant peer group could partially compensate for maternal deprivation (Fig. 14.18.), and younger peers could even act as "therapists" for isolates, coaxing them

FIGURE 14.18. *Contact with other juveniles can largely make up for lack of mothering. "Together-together" monkeys. (Courtesy of Wisconsin Primate Laboratory.)*

back toward normal social behavior (Suomi et al., 1974). In recent years motherless females have received all sorts of therapy in the Wisconsin colony—rearing with peer groups, demonstrations of mothering by experienced females, housing in stable social groups during pregnancy and birth, and being left with their first child for at least two days or even until weaning, if they accept it. Their success as mothers has greatly increased.

Meanwhile, there is a new direction of attention, away from simple environmental conditions and toward the animal's own reactions. It seems that females who react to short-term separations by depressive behavior as adolescents are far more likely to become abusive mothers than females who underwent the same separations, but seemed more resilient at the time (Suomi and Ripp, 1983).

The real questions for the historian of science are why we had to look at monkeys to find all this out? Why were we ever surprised at the results? And once the first dozen rhesus had been reared in isolation and the first surprise was over, did the later results justify the misery knowingly inflicted? It would be too simple to accuse one person or laboratory. Rather, consider the scientific climate that will not let one look at a baby monkey hunched in a corner, endlessly rocking or chewing its own extremities, and then conclude it is unhappy. Science, in its original meaning, was simply knowledge. It is frightening that we can so easily impose amnesia on our own, long-evolved, primate social knowledge, in the name of experimental rigor.

ORPHANS AND ADOPTIONS

How much of the distress of separated infants appears in more natural settings? How much is, in fact, irrelevant in evolutionary terms, because such an infant quickly dies?

As usual, the answer varies with species and individual, and with social setting. Dolhinow (1980) describes short-term infant separation in hanuman langurs, the species in which she first revealed that some primates have extensive baby-sharing among adult and juvenile females (Jay, 1963). In experiments with a large caged group, mothers were removed for two weeks when the infants were six to eight months old, that is, largely eating solid food, and out of contact with their mothers to the point where some infants did not notice her absence for several hours. Like macaque infants, once they did notice, they cried, ceased playing, and began to search actively. Of eleven infants, two died within three days, refusing to eat, two managed more or less on their own, and seven were "adopted" by other adult females, entirely at the infants' own initiative and choice of foster parent (Fig. 14.19.). When the mothers returned after two weeks, five of the infants remained with their adoptive caretaker. Most mothers set about reestablishing relations with the other adults, chasing and pulling hair, with no particular attention to their own infants. In this same colony, three infants transferred to a foster-mother at weaning even with their mothers present, and one when its own mother died (Dolhinow 1982).

Dolhinow concludes that the chief factors affecting the infants' adoption are (1) that this is an infant-sharing species; (2) the restrictiveness of the mother; the more independence she allowed her infant beforehand the more it was able

FIGURE 14.19. *An eight-month-old male langur successfully adopted a foster mother, although she had an eight-week infant of her own. Here both attempt to gain contact. (Courtesy P. Dolhinow.)*

to impose its own adoption; (3) the availability of potential foster mothers and whether they were lactating at the time; and (4) the presence of other infants who competed for the other females, and also tired out the abandoned infant with their invitations to play.

The largest study of wild orphans to date are the 40 known over a four-year span in a troop of Japanese macaques, where adults, including mothers, have been frequently shot by local farmers for raiding corps (Hasegawa and Hiraiwa, 1980). Weaning among these northern climate macaques begins at about ten months, and juveniles of 12 months still cannot cross difficult passages without aid. Two-year-olds are physically, if not emotionally, independent.

In this troop, infants were not isolated after their mother's death. Most weaned orphans survived. Among Dolhinow's langurs, it is usually adult females who adopt orphans in the wild, but in these Japanese macaques it was most often the female siblings, or in one case a cousin, that adopted the young. When young males were the chief companions, they usually adopted males more as playmates than babes. There were two cases in which adult males took over close care of female infants, both carrying and grooming them. This is a well-known phenomenon not only among Japanese macaques but among savannah baboons, where males protect orphans against predators and other baboons, and carry them on the long daily trek (see Tables 14.2. and 14.3.).

317   *Mothers and Infants*

TABLE 14.2. Survival, Foster Parents, and Orphans' Age in a Troop of Japanese Macaques

| | Orphans | | Chief Caretaker | | | | |
| | | | Adult | | | Immature | |
Age	Number	Died or Disappeared	Male	Female	Sib	Relative	Unrelated
0–1	9	7	1				1
1.1–2	8	3		1	2	2	2
1.1–3	6	1	1	1	2		1
3.1–4	7	0		1	3	1	1
4.1–5	4	1			3		1

(*After Hasegawa and Hiraiwa, 1980*).

TABLE 14.3. Foster Parents and Sex of Adoptee in a Troop of Japanese Macaques

Orphan

| | | Male Adopters | | | Female Adoptors | | | |
Sex	Adult	Immature Sib or Relative	Immature Unrelated	Adult	Immature Sib or Rel.	Immature Unrelated	Orphans Died	Total Orphans
M	0	0	3	1	1	2	3	10
F	2	0	0	2	11	2	6	23

(*After Hasegawa and Hiraiwa, 1980*).

Thus, the deep depression of mother-separated infants is not so apparent in the wild if the infant is physically able to survive. However, Hiraiwa (1981) has documented much longer-term effects of orphaning, in this same troop. Orphans, when they became mothers themselves, were more aggressive to their own first infants than other mothers were. Half their infants died, accounting for most of the troop's infant mortality. These mothers had been raised by their own mothers for the first year and had as many opportunities as other juveniles to observe other babies of the troop. They have, however, missed any chance to interact with a next-younger sibling. If this makes a difference, one would expect to find different maternal behavior among last-born females in a sibship. It may be that there are long-term and subtle effects of mothering itself, whether or not an infant has a younger sib.

Jane Goodall's (1968, 1971) account of orphan chimpanzees stimulated others to appreciate the possible devastating effects of orphanhood. An unweaned one-year-old and a three-year-old with no siblings died. Another three-year-old was adopted by her nearly-adult sister. After a period of lethargy and depression she apparently returned to normal by her sixth year. The best known of the orphans was Merlin. Orphaned at three, he was cared for by his six-year-old sister Miff, although she was too small to carry him about or defend him in crises. Merlin "regressed" to infantile behavior. He became oblivious to the signs of impending displays by adult males, so he was often picked up and dragged along when they charged. His termite fishing remained at three-year-old level, in spite of hours spent with Miff as she probed termite hills; only his attention span, not his technique, matured. Perhaps he was brain-damaged by illness or starvation just after his mother's death, or perhaps he was psychologically crippled by his mother's loss and Miff's inability to replace her. At six, Merlin seemed to recover mentally enough to begin playing again, but he was so emaciated that he would have probably died during the rains even if he had not then succumbed to polio.

The strangest of the chimpanzee orphans was Flint, the first baby that Jane Goodall knew well. Flint's mother was an aging female called Flo, whose family figures large in all films and books of the Gombe Stream chimpanzees. Flo attempted to wean Flint at the usual age of three, but he resisted, persisted, beat her, and threw tantrums to such effect that he was still nursing at five. She came into estrus, and bore an infant, Flame. Flint went on suckling and rode on Flo's aged back even while she carried the infant on her belly. Flame died, perhaps of influenza, at about three months, when Flo was also ill, but Flo managed to recover. Flo and Flint continued their close bond, even nesting together, until Flint was eight and one half, by now clearly an abnormally strong attachment. Flo again caught influenza, this time fatally.

Flint remained sitting by his mother's body for many hours. Over the next six days he grew more and more lethargic. His older sister Fifi and brother Figan visited him, but could not coax him to activity. He developed gastric enteritis and peritonitis. Three weeks after Flo's death, Flint died.

REDUNDANCY

At this point, one should redress the balance of reporting. Although lack of mothering can apparently have long-term effects, and gross deprivation generally does, most biological systems have a great deal of redundancy, such that a slip at one stage can usually be compensated at many other periods. The social growth of infants is no exception. The fact that wild orphan macaques survive and reproduce at all, that captive langurs may voluntarily change "parents" at weaning, that peer play can largely compensate in the laboratory, with laboratory measures for maternal care, all testify to the robustness of the system. Even using the far subtler measures of human social response, we know what a variety of maternal care results in adequate human beings (Dunn, 1976). One wise article on the dangers of overstressing primate maternal deprivation deals with the deplorable record of captive gorillas. The majority of zoo gorilla babies are rejected by their mothers (Fig. 14.20.). Most of these females suffered early isolation—captured as infants, their own mothers shot, and then a more or less traumatic journey to their childhood home in a cage. However, their eventual success as mothers seems to relate more to whether they are living as adults with a male (usually the infant's father), and with other familiar social companions than to their own past history. If a gorilla mother is isolated with her baby or if she is socially stressed by changing companions, she often

FIGURE 14.20. *A gorilla treats her infant as an object. Captive great apes have a record of abusive motherhood. (Courtesy P. Coffey and the Jersey Wildlife Preservation Trust.)*

ends by rejecting or abusing the infant. (Remember the Bruce effect in mice—there is no use rearing an infant if a new male would probably kill it.) In a stable social setting, with a known male, such apes may be normal mothers. Like motherless rhesus, these gorillas often do learn; later infants may be lovingly reared, when the first ones were rejected (Nadler, 1980). (See Fig. 14.21.)

Successful motherhood, then, is far more complex than the all-or-none childhood isolation experiments would suggest, or even isolation at particular

FIGURE 14.21. *The same gorilla successfully mothered later infants. (Courtesy P. Coffey and the Jersey Wildlife Preservation Trust.)*

stages of growth. The normal system is rarely stressed to breaking, and the mother and infant establish a bond that is perpetuated from generation to generation.

INFANTS' SEX AND PARITY

Primates, like people, treat infants of different sex and parity differently. Caged baboons and gorillas are fascinated by infants' genitalia, particularly male infants' (Rowell, 1967; Hess, 1973). However, most studies cannot distinguish the mother's own age and experience from the influence of siblings, or the undoubted individual differences of the infants themsleves.

Male and female rhesus infants do behave differently, the males being rougher and initiating more play and more social threats. Rhesus mothers withdraw more from male infants, play with them more, and punish them more. They restrain and retrieve and keep contact more with females. These are the "reciprocal" behaviors that both answer and reinforce the original bias of the infants, already set by the presence or absence of prenatal androgen (Mitchell, 1979).

Hooley and Simpson (1981) take on the problem of interacting influences in a study of 42 mother-infant rhesus macaque dyads, living in small social groups in Robert Hinde's laboratory at Cambridge. The primiparous mothers were, naturally, younger than multiparous ones. This goes with the primiparas being less dominant, less confident, more excitable, and receiving more aggression. (Remember these are matrilineal, highly competitive monkeys, confined together.) The primiparas restricted and protected their daughters more than their sons. Low-ranking mothers in several species, including baboons and this same rhesus colony, restrict their infants more in the presence of dominant, potentially kidnapping "aunts." (Rowell et al., 1964; Altmann, 1980). Among bonnet macaques the dominant females are particularly aggressive to female infants (Silk et al., 1981). In rhesus, this differential aggression starts before birth: mothers pregnant with daughters are more likely to be attacked or wounded (Simpson et al., 1981; Sackett, 1981). Thus, the primiparas' restriction of their daughters is probably not a simple reaction to a first-born female infant, but rather guarding the baby female against the group.

In contrast, multiparae restricted their daughters less than their sons, particularly if the sons had siblings two or more years older. The older siblings were likely to play roughly or hit the infant, and to carry the infant off. Male infants were particularly attractive to the older sibs. The higher-ranking, confident multiparous mothers, then, did not need to protect their daughters from other females, but they did protect their sons from being swept off into boisterous play. This ties in with the "adoption" of some male orphans in the Japanese macaque troop by other juvenile male playmates.

The different treatment of infants leads on to a far more surprising finding. The sex ratio of macaque and baboon infants apparently varies with mothers' rank, the lower-ranking animals having more sons, and the higher-ranking ones more daughters. The adaptive explanation may be that in these matrilineal monkeys, daughters remain with their mothers, bound by their mothers' rank and home range, but sons emigrate. A high-ranking female bequeaths her rank and reproductive success to her daughter. In Silk's colony, the chief cause of female infant mortality was aggression by dominant females: someone else's daughter is your own daughter's lifelong competitor. Low-ranking females, then, might as well have sons who will emigrate and try their luck elsewhere.

The strength and direction of the bias in sex ratio obviously depends on social system. It was even stronger in Clark's (1978) bushbabies. Clark suggested resource competition as the basis. A bushbaby's daughters share the limited rich environment of the streamsides; males were relegated to drier upland habitat. When Trivers (1972, 1974) originally suggested such apportionment of sex ratio and resources, he was thinking of deer and caribou, where a dominant, well-fed mother might invest in strong sons who would win harems. In fact, red deer invest more lactation time in sons than in daughters (Clutton-Brock et al., 1982). One would expect similar bias towards sons for dominant females in primates like chimpanzees where the females migrate, or like langurs where a successful male may lead a large harem.

The mechanism is still unknown, although it seems to be something that can change from year to year in proportion to the survivorship of the young of the previous year (Silk, et al., 1981).

THE ECOLOGY OF MOTHERHOOD

So far, this chapter has largely discussed maternal care (or lack of it) in laboratory settings. Of course, female primates must eat, and generally, travel with their troop. For much of their lives they are eating for, and transporting, two.

There are many studies of motherhood in the wild, and a few generalizations that emerge. Arboreal primates, for instance, are slower to become independent than terrestrial ones (Chalmers, 1972; Sussman, 1977). The probable evolutionary reason is that falling off a tree can be fatal.

The most comprehensive work to date is Jeanne Altmann's *Baboon Mothers and Infants* (1981). Altmann set herself the problem of relating energy demands on the baboons to their styles of mothering, and these in turn, to the harsh statistics of demography. The baboon population of Amboseli has been steadily declining as the area goes through a long cycle of salinization, which kills the fever trees. Baboon infants there grow slowly, five to six grams a day, like laboratory animals whose mothers were kept on low protein diets. Half the infants die by two years of age, mainly from predation and disease.

Baboon mothers' time is spent in walking, feeding, resting, and social life. The amount of time feeding is greatest during the infant's fifth and sixth months of life, at the expense of time spent walking, resting, and socializing. Because the mother must keep up with the group, which does not appreciably shorten its daily route even on the day of a birth, the shorter walking time means cutting out walking between food patches and not walking after the infant to retrieve it. Socializing is high for mothers with young infants, because other troop members approach to groom the infant, but falls off somewhat without the mother's actively avoiding it.

The feeding data can be compared to a model based on assumptions of the mothers' daily needs. Starting from the observed time budget, where mothers spent about 17 per cent of their time resting and 23 per cent walking, Altmann

FIGURE 14.22A. *Mother vervet threatens her growing juvenile after it attempts to suckle. (Courtesy M. Rose.)*

FIGURE 14.22B. *Five-month-old baboon throws a tantrum as her mother walks on, ignoring her. (From J. Altmann, 1980, by permission.)*

calculated that mothers would need to feed 43 per cent of the time to maintain their own body weight. This is part of an extensive group of studies in the same area, so feeding rates, protein and caloric content of food, and limiting toxins are as well understood here as in any primate site (Post et al. 1980; S. Altmann, in preparation). On two different assumptions of the extra demands on the mother to wholly provide for her infant's needs, these demands would cut out all social time, and intrude into resting time sometime between the sixth and ninth month.

Altmann suggests that there must be strong selective pressure for the infant to provide some of its own food as early as possible. (See Figs. 14.22A. and 14.22B.) At least some of the mothers probably did not maintain their own body weight in early lactation. Mori's (1979) data showed that for the Koshima Island Japanese macaques, the mothers that did not maintain their weight were most at risk of losing their own infants and dying themselves. Thus, an infant that helped nourish itself could also improve its mother's health and its own survival. The decrease in mothers' observed feeding time after the sixth month does not show that the mothers cannot cope; it shows that the infants are helping their mothers to cope.

The infants find it easier if they are weaned in a season when there are suitable "weaning foods." At one extreme are the umbrella tree (*Acacia tortilis*) flowers—soft, sweet, and easy to pluck. One sick and crippled infant survived for two months on umbrella tree blossoms shaken down to the ground as the others fed. At the other extreme, grass corms can only be dug by older animals, and these are one of the few dry season foods. Baboons have not developed seasonal breeding, but like most human groups they probably have highly seasonal patterns of infant mortality.

Individuals vary considerably in individual styles of mothering. Altmann found no difference in infant survival by maternal rank, for the 21 infants in her study. However, there were distinct differences in the restrictiveness of mothering. Restrictive mothers restrained their infants more, initiated contact with their infants more often, and left them less often, and less often ignored or punished their infants than the other group whom Altmann terms "laissez-faire." "Laissez-faire," of course, includes a range of behavior from easy-going to downright rejecting, just as restrictive includes a range from protective to imprisoning. Dividing the 12 mothers into seven restrictive and five laissez-faire, she found that high-ranking mothers tended to be laissez-faire, and low-ranking ones restrictive, which confirms many cage studies. Maternal "nervousness," as measured by frequency of glancing about at others, was even more highly correlated.

There was a slightly higher rate of disease and death among the infants with laissez-faire mothering. Altmann points out that it would be too simple to stop there. A more independent infant of a "laissez-faire" mother is better equipped to survive if the mother should die. It also may help prevent her death, or improve her long-term health by its own earlier weaning. Thus, restrictiveness may be a result in a short-term gain, but at the cost of long-term loss. In particular, older mothers may do better to be "laissez-faire" because they are more likely to die before the infant matures.

This leads into a far more general discussion. Much of sociobiological theory has focused on the grimmer sides of life. What happens in food shortage? What happens when mother or infant dies? When resources are scarce, do the mother and her child conflict over its weaning?

Altmann argues that many of these speculations lead to cooperative, not conflict, solutions. It is usually in both mother and child's interest that neither shall die. It is in both their interest for the child to be weaned in reasonable time, with immense variations in what is reasonable time depending on external factors from the number of lions to what month the umbrella trees flower. Mother and child continue to depend on each other for aid long after weaning, as we shall discuss in the next chapter, so their solutions are in the framework of years of mutual amity. We can gain insights into the evolution of motherhood when it is tested to breaking. What we find, when we do gain the in-

FIGURE 14.23A. *Panamanian spider monkey makes a bridge for her child over a tree gap. The growth of independence in the wild is slow and complex, unlike the simplified laboratory situations. (Courtesy D. J. Chivers.)*

FIGURE 14.23B. *Hamadryas baboon mother makes a bridge for her child on an Ethiopian cliff face. (Courtesy H. Kummer.)*

sights, is that it does not usually break. There is an extraordinarily strong, resilient, and flexible bond, which we might as well call love. (See Figs. 14.23A and 14.23B.)

SUMMARY

Primates, as K-selected animals, have lengthed all phases of life. They are typically "semi-precocial" with slow-growing young born furred, with open eyes, and capable of supporting their own weight from birth. Primate brain: body weight at birth is twice that of most other mammals. Human young are secondarily altricial, more helpless than closely related primates, and with continuing brain growth in the year after birth. Placental anatomy distinguishes strepsirhines from higher primates. Hormonal changes during pregnancy are basically similar in higher primates. Birth goes through the same stages. Old- and new-world monkeys generally give birth at night. Humans have a slight statistical bias toward nocturnal birth. Lactation is the time of greatest energy demand on any primate mother, particularly humans who are supporting the rapid infant brain growth.

Many primate babies have a distinct natal coat. All share the "cute" gestalt: big eyes in a big round head. Newborns have a similar repertoire of grasping, clambering, rooting, and suckling reflexes. Mothers may have a hormonal recognition or acceptance of the baby at first. Human newborns become "cuter" after several weeks, at a period when they have more social contact with others of the troop, and are "attaching" to their mother as an individual. After six to eight months this attachment may express itself as fear of strangers and active protest when the mother leaves. These stages resemble imprinting in birds, but may not occur or may not be so clear in other primates.

Primate mothers and infants have a particularly high stake in each other's well-being. Each infant represents a large proportion of the mother's total reproductive potential. If the mother dies, the infant or even juvenile is also likely to die. Even a drop in her rank from ill health may affect the infant's prospects. Thus, although they conflict, for instance in preferred weaning time or degree of independence allowed the infant, the theoretical conflict is less than one would suppose from simply counting genes. Infant macaques and chimpanzees reared in social isolation are grossly impaired in sexual, maternal, and other social behavior. If past weaning, orphans in the wild sometimes survive by being adopted by other troop members, but they may still suffer social trauma. Even short-term separation from the mother produces long-lasting changes in some individual rhesus babies. However, barring death of one or the other, it is much more common to suffer food shortage than separation in the wild. The period just before weaning is one of greatest demand, and the stress depends on seasonal food supplies, the mother's status, and many other extrinsic factors besides her treatment of the infant.

Primate mothers range from restrictive to laissez-faire, with each species' norm at a different point on the continuum. In observed baboons and macaques, younger, and lower-status mothers seem more protective, and older, more dominant ones more laissez-faire. Protection guards from predation as well as other troop members, but laissez-faire infants become independent sooner, which gives them an advantage if the mother dies or conceives again. In two captive macaque groups, mothers pregnant with female infants were attacked more than mothers pregnant with males, and female infants were more harassed. Subordinate females in several macaque and baboon studies gave birth to significantly more sons; dominant females, more daughters.

Studying natural selection leads one to emphasize mortality and failure of bonding, to see what is selected against. The results are less often underlined. What is selected *for* is a strong, redundant, personal bond between mother and child, which we might as well call love.

CHAPTER 15

GROWING UP IN A TROOP

FATHERHOOD

It is a wise child that knows its own father. Put the other way, it is a wise male that is sure of his own offspring. Monogamous primates such as siamang or marmosets can be, presumably, almost certain. In multimale troops, males that consort with particular females may have a high probability of knowing their own young. The only absolute certainty is that of a male immigrant; he can be sure that none of the infants are his own. As we discussed in Chapter 12, he may react by infanticide.

The degree of male care reflects paternity in that obligate monogamous pairs have a high degree of male investment. Whatever the reason that males confine themselves to a single mate—whether because the food supply is sparse, and mates widely dispersed, or because the female cannot raise offspring alone—the male's best strategy then is to do what he can to help his one mate's young survive. If the species is obligatorily monogamous, of course, the selective pressures for male care are more constant over generations than if the pairs are only such because of temporary ecological constraint (Kleiman and Malcolm, 1981, for further discussion).

Within monogamous species, the gibbon male contributes mainly or only by defending the pair territory by song, and acrobatic display. Indri males sing and display, but also defer to their mates over the best feeding spots—a trait they share with other, nonmonogamous lemurs. Siamang males carry the infant for much of its second year of life.

In the smaller new-world monkeys males do by far the largest share of infant carrying. Father marmosets, tamarins, owl monkeys, and titis carry the young (Fig. 15.1.), transferring them to the female for feeding. As usual, the exact amount varies by species and also by individual group. In a few groups the mother may carry most, but in many carrying is almost exclusively male

FIGURE 15.1. *Among small monogamous new-world monkeys, the father or other male family member does most of the infant carrying. Silvery marmoset with his twins. (Courtesy P. Coffey and the Jersey Wildlife Preservation Trust.)*

after the first week or so of infant life (Hoage, 1977; Box, 1975; Epple, 1975; Vogt et al., 1979; Dixson and Fleming, 1981; Ingram, 1977).

One important adjunct to these findings is that young males also carry infants even more frequently than young females. In the laboratory these are almost always older siblings. If group composition is more fluid in the field, it may be that unrelated young males are also helping. Perhaps this resembles the monogamous wrens whose younger males may immigrate and queue up for the dominant breeding position, but also help with current broods (Dawson, 1978; Wiley, 1981).

Daly (1979) in a provocative article titled "Why don't male mammals lactate?" argues that the greatest demands on the lactating mother occur in the fast-growing altricial mammals like the bank vole, which may more than double its basal food intake while feeding its young. Daly thinks that milk supply in more K-selected, slow-growing, monogamous primates may not be a limiting factor. However, for a female marmoset or tamarin who may be lactating for twins and pregnant with two more twins at the same time it seems likely that the male contribution is crucial to the offsprings' survival in the wild. If it were physiologically easier for the male to lactate, he might do that too.

Among multimale groups of primates, we have already mentioned that hamadryas males "adopt" female juveniles. Savannah baboons may pay particular attention to infants of female friends. (See Fig. 15.2A,B,C.) They may or may not have consorted with these females, and thus may or may not be the infant's father. These males defend the infant as well as the mother in troop squabbles and have been known to adopt orphan infants (Ransom and Rowell, 1972; Altmann, 1980).

Perhaps the most instructive examples are the variations of macaque male behavior. In free-ranging rhesus there is very little interaction between males and infants. Confining them together may, however, promote strong ties of

A

B

FIGURE 15.2. *Relationships among adult males, infants, and their mothers in a Gombe Stream baboon troop varied according to age, status, and sex of the individuals. A. Female infants of low-status mothers were sought out by young adult males, recalling the hamadryas adoption of "child brides." B. A high-status male hangs on to an infant that he is using to inhibit the aggression of another male sitting behind. C. Young males usually dealt carefully with infants. (All courtesy T. W. Ransom.)*

C

affection. Japanese macaques, in contrast, often have very strong one-to-one relations between males and one- or two-year-old youngsters. These ties form especially during the birth season, then are dropped during the mating season, and may or may not be renewed later. The amount of such paternalistic behavior varies widely between Japanese macaque troops, and thus may be partly "cultural." Bonnet macaques have a much looser male hierarchy than the other two, with much more tolerance between adults, or adults and subadults. The male bonnets are quite likely to play with infants in general, not the Japanese special relationship of a single male with a single infant.

Finally, Barbary macaque infants may spend even more time with males than females. The males use babies for "agonistic buffering." A male carrying a baby is unlikely to be attacked. He may approach a more dominant male and "present" the baby. The infant is briefly inspected or groomed, and then ignored as the two males settle to grooming each other (Deag and Crook, 1970). Again, adult male attention to infants goes with more infant play and carrying by maturing males than young females (Burton, 1972). Ali, (personal communication) suggests that this behavior derives from bonnet and Barbary macaque females, rather than males, changing group. The males, then, would be kin to each other, so any infant is some relation. The extreme and "calculated" promiscuity of Barbary females (see Chapter 13) also ensures that any infant is a potential offspring.

It is intriguing to ask how much of this is true paternal behavior. It was long believed that in multimale troops there would be no way for animals to identify their fathers. Wu and colleagues (1980) presented extraordinary results. Pigtail macaque infants were reared alone but with some contact with non-kin peers to prevent strong isolation effects. These infants were offered a choice between unfamiliar, paternal half-siblings and unfamiliar non-kin matched for age, sex, and activity level. The experimental animals looked at their half-siblings more, and approached them more than the non-kin. This needs much more confirmation from different kinds of studies. One confirmation comes from the observation that in a captive rhesus group, paternal half-siblings grabbed at infants as much as unrelated juveniles, but the mothers resisted the half-siblings less than the unrelated juveniles (Small and Smith, 1981). The adult females in the group were not related, so there was not the round-about effect of a male mating with one matriline.

In savannah baboons, Busse and Hamilton (1981) think much of male infant carrying is actually protection of offspring against infanticidal immigrants. Packer (1980) points out that even within a troop, males carry two- to eight-month-old infants "against" each other in fights. Five infants in his Gombe Stream troop were wounded and killed this way in the course of three years, when the usual agonistic buffering did not work. The countervailing benefits to carried infants in protection from other troop adults or from predators may have been substantial, but of course don't show—a wounded infant is obvious, but an infant saved from wounding is not. Packer notes much of the contact resulted from the infants' own initiatives. The infants distinguished clearly, and avoided, immigrants who arrived since their birth, who thus could not have been their fathers.

Taub (1980) takes this further, because in his Barbary macaque group agonistic buffering was not random presenting of just any infant, but reflected specific relations of certain males with particular infants. Finally, in a captive rhesus group, infants preferentially sought out their own fathers, perhaps taking their cues from their mothers' current associations (Berenstain et al., 1979).

Among the great apes, we have already mentioned that gorillas discriminate in favor of their own offspring. They not only do not kill them as infants but allow the "crown prince" to spend more time near them as a youngster and sometimes eventual breeding rights as an adult (Harcourt and Stewart, 1981; Veit, 1982). The other great ape males pay little attention to infants in the wild, but in captivity are remarkably tolerant of their boisterous games.

Apparently, John Miller, in 1771, first speculated that patrilineal inheritance of wealth in humans correlates with confidence of paternity. In societies in which a man is less certain of his relations through the male than the female line, he more likely leaves his wealth to surviving brothers, and thence to his sisters' sons (Hartung, 1981; Gaulin and Schlegel, 1980; Alexander, 1974). Hartung (1981) points out that within one generation there would have to be a phenomenal amount of sexual license to make it more likely that the sisters' sons are more closely related than the wife's sons, but, as one multiplies uncertainties over several generations, likelihood of genetic relation through the female line remains constant, whereas the uncertainties keep growing with every transfer in the presumed male line. With humans, of course, everything is complicated by the accumulation of surplus wealth. Pastoralists essentially accumulate capital that bears interest every year. Pastoralism is significantly associated with paternal inheritance and male jealousy. Fishing societies, on the other hand, have little way to continue accumulating individual capital once a family owns its nets or boats, unless they begin to dominate other families. Primitive fishing societies tend to maternal inheritance—and to extramarital sex.

AUNTS AND ALLOMOTHERS

When Thelma Rowell first saw rhesus females carry and cuddle each others' babies, she called it "Aunt behaviour" (Rowell, 1963; Hinde et al., 1964). Later, the term seemed to imply not only too much genetic relationship but too much kindness. We now call such behavior "allomothering" (Hrdy, 1977). (See Fig. 15.3.) With the change in terms has come a change in viewpoint. We now look not only at the babysitting that frees a mother to rest and feed but at the competition from females with offspring, or future offspring of their own.

Relevant variables include the closeness of relationship of mother and allomother, their ages, and relative status. They also include a broad evolutionary picture—do these females expect their own offspring to compete with this baby for resources?

In the monogamous primates by definition there are no other adult females in the group, so allomothering would only be by siblings.

Squirrel monkeys take an active interest in each other's infants. In the lab-

FIGURE 15.3. *Allomothering is common among primates, but its degree varies widely. A woolly monkey allows her oldest friend to touch the baby, seven days after birth. (Courtesy L. Williams.)*

FIGURE 15.4. *Hanuman langur infants are forcibly passed round the troop from the first day onwards. (Courtesy S. Hrdy and Anthro-Photo).*

oratory this sometimes turns into kidnapping of young infants, or so interferes with the mother that she then rejects the infant (Hopf, 1981). More usually it is just transfer or grooming of older infants.

The old-world monkeys differ sharply in degree of allomothering. As a generalization, colobines, including langurs, do; cercopithecines don't so much. That is, langur mothers commonly allow other females to take away their babies, even on the first day of life (Fig. 15.4.), but cercopithecus and baboons and macaques are far less likely to allow it. This broad phylogenetic difference may relate to the fundamental dietary difference. Colobines eat mature leaves, which are both abundant and widely dispersed on a tree or grove. Female colobines may not be in such immediate competition for food or feeding sites as are female cercopithecines. Thus, the dominance hierarchy is less important or rigid, so in the immediate present a female is less likely to confront a dominant that will not return the baby. In the future, when the babies are grown, they too will be part of a fluid hierarchy, ranked according to age rather than nepotistically by their mother's status. Thus the mother can allow her infant to be taken, because other females have little reason to abuse it and are likely to give it back (McKenna, 1979a,b, 1981).

Ironically, it is among the langurs that we see allomothering so indifferent as to amount to abuse. The baboons and macaques would not let it happen in the first place, unless the infant were kidnapped. (Fig. 15.5.)

This said as a generalization, of course, there is instant need to soften it. Baboons and macaques, as well as langurs, are much attracted to young infants, and within each group there is much variation.

Red colobus, unlike langurs or black-and-white colobus, allow no other females to contact their infants. Red colobus have female migration, so there is little or no kinship between troop females. Even juveniles who try to groom just get slapped (Struhsaker, 1975).

One of the few cercopithecine species in which there is a suggestion of female migration are the bonnet macaques (Ali, personal communication). They are at times the most likely to share infants in the lab, perhaps precisely because they are used to association with unfamiliar females. In contrast, pigtail macaques restrict their offspring, sharing only with matrilineal kin (Rosenblum and Kaufman, 1967). In at least one lab bonnet colony, however, kidnapping

FIGURE 15.5. *Infants are often threatened and abused by other females—here, rhesus in Nepal. (Courtesy J. Teas.)*

escalated to the point where it was the chief cause of female infant mortality and seemed to be a form of competition or spite on the part of the "allo-mother" (Silk, 1980).

Patas monkeys have a very high level of allomothering. Instead of stressing the competitive side here, Zucker and Kaplan (1981) turn to the special ecology of the species. Patas forage in widely dispersed groups. Where there is danger, it may be highly adaptive for any passing female to scoop up an infant and run, rather than leaving the infant to locate its own mother.

The same reasoning may apply in comparing more terrestrial vervet guenons and ringtailed lemurs with their more arboreal guenon and lemur cousins. The terrestrial forms allow their infants far more contact with other troop members. It may also tie in with the dangers of life on the ground, as Altmann (1981) points out for laissez-faire baboon mothers. A more independent infant may risk abuse from other troop members, but it could also be in a better position to promote its own adoption if its mother died.

Adoption could also be a factor in langur allomothering. Hausfater and colleagues' (1981) simulation of langur female age-graded hierarchies could be achieved by raising the death rate over baboon simulation of nepotistic hierarchies. If colobines tend to high maternal death rate, then their young, as in Dolhinow's (1980) experiments, may need to be able to organize adoption.

We thus find various arguments for allomothering: helping kin, competing with non-kin, and if necessary, adopting. One final argument is that young females may learn the techniques of mothering by caring for babies (Lancaster, 1971). This brings us to a consideration of the growth of juvenile behavior.

JUVENILES AND SIBLINGS

Little by little the infant turns its attention from mother to peer group. It has been played with at times and carried by older siblings, or males, or other fe-

FIGURE 15.6. *Flo gently stops her daughter Fifi from touching infant Flint. (Courtesy H. van Lawick.)*

males since early infancy (Figs. 15.6. and 15.7.). As its locomotor skills mature, it then spends more and more time in play.

Infants of many species have no like-aged peers. The gibbon infant at most has a sibling three or four years older to play with. In very small groups of any species there may be only one infant at any time. Mothering is constrained by being mammals and has some aspects in common throughout the primates, but juvenile behavior varies immensely between species, groups, and individuals.

The individual's own motor control develops slowly. Some groups of behavior appear together. In a troop of olive baboons, the behaviors of riding dorsally on the mother's back, bipedal standing, mounting, and "mouth-and-wrestle play" all appeared about the same time in infants 4 to 12 weeks old. The same coordination of posture and hind limbs presumably underlies them all and, when this is more or less organized, then the various social behaviors appear as a group. However, they then follow their own time course: dorsal

FIGURE 15.7. *Flo and her family. She grooms Fagen, her near-adult son (about 11 years old), while Figan (about seven years old) sprawls in the foreground. Four-year-old Fifi plays with Flint, the three-month-old baby.*

riding peaks in frequency at 12 weeks, wrestling at two and three years. Various other behaviors have apparently independent control. Presenting does not appear until long after mounting, although it seems even less demanding of general postural control (Chalmers, 1980).

The environment also affects the juveniles. We have already mentioned the importance of arboreal versus terrestrial life in increasing risks when an infant sorties unsteadily off the mother. If weaning foods are available at one season and not another, some infants can grow and begin to provide for themselves more quickly (Altmann, 1980). In Amboseli, juvenile vervets play more and show more aggression during the wet season; they have more energy to spare (Lee, 1981). There are often some foods that are too difficult for young animals to reach or to open. The young may beg them from the adults in primates from marmosets to chimpanzees (Silk, 1978, 1979). Again, the importance of these foods in the diet and thus the juveniles' independence will vary with season. In the laboratory or zoo there is a definite trade-off between social and object play; those animals with fewer or no playmates compensate by attention to objects (Fagen, 1981; Chalmers, 1980; Chalmers and Locke-Haydon, 1981). Changing to a more interesting environment can decrease the amount of stereotyped or self-directed activity (Bowen, 1981). Conversely, caging with conspecifics can increase the amount of social play to the point where even orangutans can seem like a fully social species (Edwards and Snowdon, 1980). Presumably, these effects of captivity are extreme versions of more subtle effects of variations in wild environments. Two orangutan mothers who live in nearby ranges may give their offspring the chance for an occasional bout of joyous wrestling (MacKinnon, 1978). Rhesus who came from cities where they thieved in the bazaars and slept on the rooftops were far more active than forest rhesus and more ready to contact and manipulate objects. Juveniles brought up in a rich and demanding environment may, thus, be more innovative than their country cousins (Singh, 1965, 1966).

The general picture throughout the primates is that infants become gradually more independent of their mothers and spend more and more time in the company of peers, if any (Fig. 15.8.). The key word is *gradually*. Most juveniles are still closely associated with their mothers when the next sibling is born. The interbirth interval depends on the older sibling's lactation period. Chimpanzee juveniles, especially female infants, attempt to interfere with their mother's copulations. It is not clear whether this may be a misinterpretation of the male as aggressing against the mother, although in real aggression infants keep well out of the way. Male juveniles are likely to interfere even in copulations of unrelated females and are tolerated by the adult males, so there may be a form of early sexuality involved. It is easy to interpret interfering with the mother as an unconscious attempt to postpone her next conception, although there is no evidence that the infant succeeds (Tutin, 1979).

When a new infant is born, the approaches of older siblings and juveniles may largely shape the mother's behavior. She is restrictive if the infant is of an age and sex to be carried off as a playmate (Simpson et al., 1981; French, 1981; Hamden and Rodman, 1980).

Youngsters retain special links with their mothers until late in life. If they do not change troops, the links may last throughout life, expressed in deference when they meet, or grooming beyond average (Goodall, 1971).

FIGURE 15.8. *Even normally solitary infants seek play and comfort when available. (Courtesy P. Coffey and the Jersey Wildlife Preservation Trust).*

The links sometimes reappear more strongly in time of need. "One male juvenile, WJ, received extensive wounds about the head and shoulders from unknown causes. . . . WJ's slow and awkward movements . . . attracted the attention not only of myself and his age mates but also of an adult female, Gina, whom I had long suspected of being his mother on the basis of marked physical resemblance (. . . a decidedly hooked nose). Gina stayed near WJ, groomed him, and actively protected him from his inquisitive playmates (and me) for several days until his condition had considerably improved. Following this incident, however, Gina and WJ rarely interacted" (Ransom, 1981).

Male and female juveniles differ in many traits. In general, males are more aggressive, more independent, and engage in much more rough-and-tumble play. Females remain more with their mothers and do far more play-mothering of babies. It seems highly likely that juvenile females learn many of the components of baby care, even though this is surely redundant with instinctive bias. As Richard Andrew is fond of pointing out, learning that is adaptive probably reinforces instinctive biases, and vice versa, in any species. The tendency of little females to play with babies is just such a system and a good candidate for instinctively slanted learning. The fact that people (and monkeys) may be good mothers with little prior exposure to babies shows that this is a highly redundant complex of learning and instinct, like many other biologically important systems (Lancaster 1971). In baboons the attraction to infants also leads immature females to associate with the infants' mothers, strengthening bonds within matrilines, or to females of high rank (Cheney, 1978).

Sex appears very early in primate play. Young males mount and thrust and females present to all sorts of inappropriately sized partners. Sex play continues into and gradually becomes true sexual behavior (Mitchell, 1979).

The sex differences in juvenile behavior are not absolute in any species. Males play with babies; females mount and wrestle. Any individual probably goes through the full range of its species' behavior at some time in its career to the degree that it is physiologically possible. There are, however, definite quantitative biases. Even these differ between species. Female golden lion tamarins are more aggressive as juveniles and more lethal as adults than their mates. Male juvenile Barbary macaques are more likely to carry babies than the females, just as they do when adults (Mitchell, 1979; Burton, 1972).

The troop provides a frame for juveniles' behavior, and juveniles are responsive to the temper of the troop. Baboons walk with infants and young juveniles generally in the center of the troop, protected from predators (DeVore and Hall, 1965; Rhine et al., 1980). This means they are also subject to the vagaries of troop social life. In one experimental study mother baboons were rotated out as their infants reached 90 days, so a group of perpetually shifting females was caged together. The high inter-mother aggression was reflected in high infant-infant aggression. Another group was left stable, with a single lead male. The infants showed less aggression, more exploration, and more rough-and-tumble play. When the male fell sick for a day and was taken out, the females bickered and one attacked a black infant. Such fortuitous changes in wild troops presumably set the various tones of infant and juvenile life (Young and Hankins, 1979).

There was even one wild "gang" of five juvenile rhesus monkeys, four two-year-olds released from captivity and one who joined them from a nearby wild troop. They maintained their identity for more than a year, led by a strong-willed youngster who played the role of dominant male. In the end, a solitary male took them over, consorting and mating with one of the maturing females (Makwana and Pirta, 1978).

SUBADULTS

Subadults are adolescents, beginning to show true sexual behavior, but not yet conceiving or fathering young.

There is far more literature on this period in males than females. For one thing, it lasts much longer in males. Female adolescence is characterized by a few estrus cycles in which the female does not conceive and does not seem very attractive as judged by the lack of attention from lead males.

In monogamous species this is a period of increasing conflict between mother and daughter. Lion tamarins and other callitrichids suppress their daughter's maturation hormonally. Removing a young female to the company of a breeding male can greatly speed her maturation (Kleimann and Epple, 1981). No one has seen a gibbon female actually thrown out of her natal group, but this may be because people have happened to watch groups with maturing sons. Young female red howlers were driven out of their natal troops by rival females, not their mothers, a quite different social situation (Crockett, in press).

Although species with female migration interest us greatly because of their resemblance to humans, there seem to be few accounts of why and how a young female chimpanzee or gorilla picks herself up and goes off to mate in another group, or why she sometimes then comes back.

Young males have a triple problem—rising successfully in the dominance hierarchy, maturing sexually, and, in most species, migrating to a new troop. There is often a lag of several years between puberty and social maturity.

To take three species for comparison, in male owl monkeys puberty (as measured by blood testosterone) begins about 313 days (range 211 to 337, for six animals). This is extremely early. In common marmosets, which weigh half as much, it starts at 250 days. The owl monkeys' caudal scent gland, which matures under the influence of testosterone, began to take on adult appearance at the same time. They did not achieve full body weight until about 500 days. Although this is a monogamous species, young males caged with adults were not retarded in their development (Dixson et al., 1980). Wild Japanese macaques do not begin to develop in testis size until they are four years old. Then they rapidly develop adult-sized genitalia between October and the mating season in January or February. They are not fully mature until the age of six. Here the rate of maturation depends in part on external factors, with the testis enlarging before the mating season and shrinking afterward throughout adult life (Nigi, et al., 1980). Orangutan males develop massive fatty cheek flanges when they are socially mature. If two males are caged together, one may suppress the other's development for years. Kingsley (1979) found males developed flanges anywhere from 11 to 18 years old, depending on their social companions. Before this they were sexually competent, but did not look like adults. In captivity, males as young as eight have fathered infants, and subadults in the wild attempt to "rape" females at ten. Fully adult wild males seem about 15 and upward (MacKinnon, 1979; Galdikas, 1981).

Juvenile and subadult males are "peripheralized" in several species, notably Japanese macaques; they form a halo round the troop when it is stationary. A mild form of the same process may result in the fact that subadult male baboons are often on the fringes of a baboon troop and are first to give the alert. As a juvenile male baboon matures, he becomes dominant to smaller juveniles and gradually moves from the position of infant among females to his own position as subadult (Lee and Oliver, 1979). When some baboon troops move, the subadult males are in the vanguard, the most vulnerable position. This may not be altruism on their part, but a combination of exuberance and exclusion. The result is that they serve as sentinels against predators, an advantage to the troop as a whole (Rhine et al., 1979).

The peripheralization of subadult males for many primates culminates in male migration between troops. Males may wander as solitaries for a time. In a few species they form all-male bands. Hanuman langurs and gelada baboons have such bands, which roam far more widely than the established harem groups. Boggess (1982) has documented relations between immature and adult male hanumans. In one troop at two different periods there were fairly relaxed intermale contacts when there was a single harem leader, but earlier, when the troop had several adult males, there was great tension between them all, including the immatures. There are few all-male bands in this area, and Boggess argues that there is fairly fluid male intertroop migration. A troop may become multimale by the incursion of a group of males, not by the slow maturing of the leader's sons. Moore (in preparation), studying all-male bands at Mount Abu, found that there were alliances within the bands. These do not seem to be brothers or close kin, which is what one would first suppose.

Cercopithecines, in the male as well as in the female line, seem to have closer links between kin. Rhesus macaques at Cayo Santiago may migrate preferentially to a troop where an older brother has migrated (Meikle and Vessey, 1981). In larger, nonisland colonies, these effects are probably much attenuated. One would expect long-term bonds between male kin only among such animals as gorillas and chimpanzees, among which some or all of the kin remain within the band.

THE ELDERLY

There is now a large amount of medical literature on aging primates. The field literature is still largely anecdotal (Fig. 15.9A, B). Some of the accounts of individuals include the aging, wounded rhesus macaque whose whole troop reduced its range to stay with him (Lindburgh, 1981). There are several accounts of "respect" for old hamadryas baboons, with worn teeth and frail physique, who still sit on their favored sunning rock or even lead the troop progressions.

FIGURE 15.9. *Relatively little is known about wild primate old age. Here, Sri Lankan toque macaque male Stumpy in middle age, 1971, and old age, 1980. (Courtesy W. J. Dittus.)*

A few females survive to die of old age, or disease associated with general weakness of old age, like Alto, the baboon, or Flo, the chimpanzee. The general status of these animals reflects their own species systems—Alto retained high rank into old age, but aging langur females fall to the bottom of the female hierarchy (see Chapter 12).

There has been a little speculation on the role of the elderly in the troop. At least one dyed-in-the-wool conservative Japanese macaque saved his whole troop from being trapped by not letting them approach the new object, although it seemed likely neither he nor the others had any experience of traps before (Miyadi, 1967). One old female mangabey led her group to water in a time of drought (Waser, 1978; see Chapter 5). Aging female langurs may attempt to save infants from infanticidal males, taking risks perhaps for their own grandchildren or great nieces and nephews (Hardy, 1977). It is not clear, however, how much of the aspects of postreproductive life is selected *for*, and how much it is a result of pushing the expression of deleterious genes to as late as possible in the life cycle.

THE TROOP, THE KIN GROUP, AND THE INDIVIDUAL

Part two, on society, opened with a discussion of the conflict or concordance of interest between levels of a hierarchy. We asked how much a gene is bound to the same interests as its organism, or an individual to his or her society. The furor over sociobiological analysis is an emotional reaction to intellectually analyzing the individual's behavior in terms of the interests of gene, kin group, and society.

Has the description of the spectrum of primate social behavior illuminated the lives of individuals in these terms?

We have at least seen several types of society in terms of functioning wholes. A monogamous indri-gibbon-marmoset-titi society correlates with the subadults of both sexes being prepared to leave their parents. It correlates with male care of young and females sharing territorial aggression—and a bond between the mated pair which we can only feebly describe when we record their morning duets or their evening tail-twining.

We have seen the "nepotistic" female hierarchies of the cercopithecines, in which a mother's rank almost forever impinges on her daughters', and contrasted them with the langur hierarchies where females' rank declines with age, as Sarah Hrdy says, like Hollywood. One system can be intellectually transformed into the other just by changing the parameters of life span. To live in one or the other, though, means for a baboon or rhesus female that she checks throughout the day on the whereabouts of female kin, females who might steal her baby, and any male she may count on for support. The langur female can, and does, let her baby be handed round through the troop while she insouciantly browses. The fact that we can '"explain" watchfulness or insouciance in terms of demographic parameters and inclusive fitness does not make the female's behavior somehow dwindle into a data point on a graph. She and her baby and her troop are still there in front of us.

The reductionist approach of sociobiology has coincided with an emphasis on the selfish and cruel aspects of primate life. In part, this is fortuitous and a result of the perpetual scientific search for novelty, not just truth. When western primatology began to get under way in the late 1950s and early 1960s, we, and particularly Jane Goodall, were hugely impressed by the fact that mothers care for infants so long and so tenderly. We were impressed by the fact that such a small proportion of baboons' swaggering and threats actually result in wounding or rank changes (Hall and Devore, 1965). We were just plain impressed by troops—coherent social units whose members fed and groomed and slept together. Small wonder that many of us explained much of individual behavior as being "for the good of the troop."

As we have watched longer, we have gradually seen the rarer behaviors of cruelty. Infanticide happens, but it is infrequent and brief compared to the constancy of motherhood. Death in baboon dominance fights and gang killings between chimpanzee groups also happen. When we had grown a little bored reporting goodness, murder stories livened up the primate conferences. And, at the same time we understood that the genetic interests of an individual or his close kin predominate. They may or may not turn out to be good for more distantly related members of the troop.

It seems that we are now about ready for another swing of fashion. Again, this relates both to theory and to prevailing intellectual climate. Many of the innovative ideas in primatology now are coming from ecologists, not students of social behavior. The students of social behavior are also discouraged by their own lack of job prospects and by the divisions within social anthropology, where, in a rational world, they would look for like-minded colleagues. The current greater activity is among ecologists and active conservationists, who talk in terms of genes in habitat—minimum effective breeding populations in sustainable reserves. The findings on social behavior become, or may become, the bases of species management plans, rephrased in terms of survivorship and gene flow.

The mother, guarding her baby, eating for two to produce enough milk for that baby to grow, is still the vehicle of survivorship. The subadult peering out from the safety of his or her natal group, raising the courage to launch out into the society of strangers, is still the vehicle of gene flow. We describe and analyze and explain. We must not think we have explained them away.

SUMMARY

Some male primates actively carry and play with infants, particularly in monogamous pairs. In some multimale troops males also carry and even adopt infants, some, but not all, the male's own sons. Adult females of many species are attracted to other's infants. This allomothering is particularly common in colobines and may correlate with their relaxed dominance hierarchy and widely scattered food supply. Allomothering frees the mother to feed unhindered, but especially in cercopithecines, may merge with infant kidnapping and abuse.

Juveniles play, males more actively, and with more rough and tumble. Play with babies follows species' typical patterns of adult care. In most species young females play more with infants; in species in which adult males care for infants, young males play with infants. Siblings may largely determine the degree of a younger sibs' freedom. Sexual and dominance play starts young in both sexes and continues until they become the adult version.

Subadults are sexually active, but not yet parents. Females have a period of adolescent sterility. Males have, usually, a longer period of socially suppressed reproductive behavior. At this time either sex or both may be physically peripheral to the parents' group, in the course of starting migration to their eventual breeding group.

Analysis of individual fitness, inclusive fitness, and genetic relations between troops have provided powerful tools for understanding the relations between behavior and troop structure. Analysis does not, however, make individual behavior disappear.

PART THREE

INTELLIGENCE

CHAPTER
16

PRIMATE PSYCHOLOGY

Professed psychologists traditionally attempt to simplify psychological processes. The history of many fields of science can be seen as the struggle of two opposing desires: the intellectual delight in reducing complex facts to simple, crystalline theories as against the more visceral delight in complex reality itself. This is particularly true of the study of the primate mind. People choose to work with primates because they are complex animals, mysterious, and quasihuman (Fig. 16.1). On the other hand, the same people create out of the primate's behavior a single theory or dichotomy to satisfy the simple mind of the psychologist.

GESTALT PSYCHOLOGY

Nowhere is this clearer than in the work of the founder of primate experimentation, Wolfgang Köhler (1921, 1927), although his simplifications seem complex by later standards. Köhler spent World War I sequestered on the island of Tenerife with a colony of juvenile chimpanzees and was thus among the original twentieth-century primatologists. He was attempting to prove the Gestalt theory of learning. Gestalt theory argues that there are qualitative leaps of understanding, when a whole idea or a whole pattern is perceived. This whole, or "Gestalt," is somehow greater than the sum of the individual parts—a new level of understanding, not just a collection of components. Köhler's famous experiments, teaching chimpanzees to pile boxes on each other or to fit two sticks together to reach a banana, stemmed from his interest in the pure theory of understanding. He minutely observed the chimpanzee's hesitations, its trials, its errors, and then the sudden swift solution. To him, sudden "insightful" solutions proved that the chimpanzee mind, like that of man, works by qualitative jumps when a new whole becomes meaningful. And yet, unlike later more

345

FIGURE 16.1. *Jambo, first gorilla born and raised to adulthood in captivity. (Courtesy P. Coffey and the Jersey Wildlife Preservation Trust.)*

abstract psychologists, he was delighted and fascinated by his animals. Köhler's long appendix on the social behavior of his colony remains one of the most complete and amusing descriptions of captive chimpanzees.

BEHAVIORIST THEORY

Köhler's school of Gestalt psychology was answered from the other side of the Atlantic by the American Behaviorist school. Its work stemmed from that of Watson and Thorndike. Behaviorists tried to eliminate qualitative jumps of understanding from their theory. They reduced all of behavior, human included, to minute accretions, as new acts were learned through the reinforcement of primary drives. Behaviorism was an atomistic discipline desperately eager to explain mind by a single mechanism. The experimental situations were designed to exclude every aspect of the animals' makeup except for what was relevant to the particular learning problem under consideration.

A review of the methodology of discrimination learning by Meyer and colleagues (1965) pointed out the annoying fact that monkeys most easily discriminate three-dimensional objects that can be handled, not two-dimensional stimuli. This is precisely the sort of qualitative effect that annoyed the early learning theorists. Worst of all, primates seem to learn social relationships, dominance orders, and predator signals associated with alarm calls much more readily than any of the psychologists' carefully constructed tests with "neutral" objects.

Qualitative differences in the kinds of things an animal will learn, qualitative jumps in learning, and innate bias in learning are all aspects that cannot be handled in classical learning theory, except by the roundabout means of saying that there is a stronger drive in some sense satisfied by the learning in question. The proponents of the classical theory postulated (and preferred) animals, or people, to be born blank slates on which to write the sum of learned behavior. This has led to obvious insensitivity about the nature of the animals and to many articles published as work on "the" monkey, which is an Indian rhesus, or even "the" animal, which is an inbred albino rat. (Ironically, Margaret Mead and other social anthropologists who argued that human behavior begins with a nearly blank slate used this as a reason to delight in the diversity of human cultures. The same premises do not always lead to the same conclusions).

COGNITIVE ETHOLOGY

The behaviorist school of thought is now on the wane, as people turn to the complexities of social behavior, tool use in the wild, and ape language learning. The most courageous step is that of D. R. Griffin, a cognitive ethologist, who reopens "The Question of Animal Awareness."

If one believes in the continuity of evolution, it seems we must attribute consciousness to animals. Our neural mechanisms are just like those of other mammals; our brains are bigger (except for whales') but little different qualitatively (Griffin, 1984; Passingham, 1982). "Unless one denies the reality of human mental experiences, it is actually [a] parsimonious [hypothesis] to assume that mental experiences are as similar from species to species as are the neurophysiological processes to which they are held to be identical. . . . (The) belief that mental experiences are a unique attribute of a single species of animal is not only unparsimonious, it is conceited" (Griffin, 1981).

B. F. Skinner (1981) replies that people who need to drag in consciousness as an explanation have not understood "selection by consequences." Just as natural selection of the fittest creates an illusion of evolutionary purpose, so repeating rewarded action creates an illusion of conscious purpose.

You may feel like eating a piece of chocolate cake. You probably do not think you want it in order to survive, or even to raise your blood sugar level. However, primates who ate ripe fruit that raised their blood sugar survived to become our ancestors. As a consequence, sugar is reinforcing to us. Perhaps they also left, as a legacy, our susceptibility to the addictive components of cocoa beans and the related cola nut. The fact that neither one may be adaptive in this generation does not stop them and their chemical mimics from being reinforcers. Similarly, a craving for chocolate cake relates to past experience with the stuff—previous taste and reward, not future. If your background includes no chocolate cake, this paragraph may misfire unless you read it in terms of presidential jelly beans or almond agar cubes in sweet soup.

If you now drop the book and go out in search of chocolate or almond agar, is that a purpose—a future goal or event, which in some sense causes your present behavior? In one sense, no, for our usual picture of time is that nothing in the future can *cause* an effect in the past. In another, important sense, yes. Our minds have evolved to such complexity that the combination of past evolution and experience can produce an idea of chocolate cake so strong you go out and pursue it. The two descriptions of causality are different ways of saying the same thing—events in the past can affect the present by means of an idea of the future. Such hierarchical feedback circuits are well within the domain of computer programming (Hofstadter, 1979).

We find consciousness difficult to understand because of its chief evolutionary advantage. An idea is a way of summarizing a situation. If we had to remember every previous experience with chocolate cake, re-abstract it from the glorious return from the camping trip or that disastrous birthday party, and re-calculate the chemical content, we would never have time to learn anything. An idea can jump all those steps to provide a basis for action (Bateson, 1973). Consciousness is a useful way of summing up the consequences and policy implications of past experience. Purpose is hindsight applied to an image of the future.

Thus, when B. F. Skinner declares that a lecture he has just given was no-more than a chain of conditioned reflexes, he may in a sense be right. (Should I call this a personal communication or a personal observation?) His lecture could be described at an even lower level by specifying the sequence of neurons that fired in his brain as he talked. However, Griffin would be equally right that this is neither the most revealing nor the most parsimonious way to account for the lecture, and that Skinner's self-awareness does not disappear while his neurones are firing.

Part three of this book is really a sketch for some future text on the evolution of logic. Whenever we bring in a psychological theory, we will find a tendency to *reify*—to imagine that a concept is a thing. Whatever the concept, however often we keep reminding ourselves that "tool use" is some part of the gradient from kicking a stone to designing a computer, or a "stage 4 baby" is an infant with a loosely related bunch of skills, we still can't help thinking that if the concept can be named, it is somehow real.

What I propose is that intelligence, like any other ability, evolved because it was useful. Reified protoconcepts might have been something like "that animal's anger" or "this nut-cracking stone." They might have led to further, useful action. They need not have been "right" in modern terms if they led to adaptive actions. The beginnings of attributing emotion to others could include both "The lion wants to eat me" and "The river wants to drown me."

This brings us back to the question of qualitative differences. Intelligence is like locomotor agility. For what? Scuttling on twigs or leaping between vertical trunks? Some attributes, like grasping hands, help with both; others do not. Our human logic is an immensely powerful tool for dealing with our environment. We may conclude that it is in some way congruent with the "real world." However, it is also a product of primate evolution. It may be biased in ways we shall never guess unless or until we consider the points of view of beings whose consciousness differs from our own.

Having declared my bias, let us turn in the next four chapters to an account of manipulation and tools, then to the ontogeny of cognition, the impor-

A

B

C

D

FIGURE 16.2. *Wolfgang Kohler anticipated many later tests of ape intelligence. A. Sultan joins two sticks, in the paradigm of insightful tool use (Chapter 17). B. Involvement of the social group (Chapter 21). C. Stacking boxes, an insightful or ''Stage 6'' combining of means and end (Chapters 17 and 18). D. Using a pole as a ladder.*

349   *Primate Psychology*

tance of play, and then to language learning in the great apes. The final chapter of this book puts all these together with social learning to try to show how intelligence evolved as biology's greatest advance.

SUMMARY

An age-old emotional conflict exists between the desire to understand the world by simplifying it and the desire to convey its riches in qualitative terms. Reduction and synthesis are too often treated as logical opposites instead of alternative views of the same processes. Primate psychological studies have been largely dominated by the atomistic simplifications of learning theory. Both early Gestalt psychologists and current cognitive ethologists are more interested in hierarchies of qualitatively different levels of understanding, including animal consciousness. This section (and this book) is biased toward cognitive ethology. It attempts to set the stage for considering human logic as a product of primate evolution. (See Fig. 16.2A, B, C, D.)

CHAPTER 17

MANIPULATION AND TOOLS

Our recent takeover of our environment and the environment of most other species is based on technology. Man has been defined as the toolmaking animal. Darwin speculated that our ancestors made do with small canines because they used sticks and stones as weapons instead of slashing teeth. Small wonder that primatologists snap to attention whenever their animals manipulate objects.

When Jane Goodall saw wild chimpanzees fashioning sticks into termite fishing probes and crumpling leaves into sponges for sopping up water, L. S. B. Leakey remarked that we would either have to redefine tools or redefine man. Benjamin Beck (1980) has reviewed almost all known cases of wild animal tool use, and finds it remarkably difficult even to define a tool. This chapter is drawn largely from Beck's account. Beck not only reviews the literature; he comes to a conclusion that I find wholly convincing. He argues that tool use is just one means of showing and using general intelligence (or cognitive structures, if you prefer that phrase). Similar complex behavior appears in social spheres, in cooperative hunting, and in feeding situations such as sea gulls' dropping of shellfish on hard surfaces or orangutans' calculating efficient routes between fruit trees. Tool use is a special case because of its importance in human, not primate, evolution.

This chapter, then, like the last, is a preamble to the discussion of primate intelligence. It summarizes primate's tool-using achievements in the wild and in the laboratory to show what behavior individual primates can manage and what we are trying to explain. It is only in the next chapter, on cognition, that tools are set in the context of more general intelligence. Then in Chapter 19 the ontogeny of tool use is subsumed under play. At last, in Chapter 21 we reach division of approach between the sexes and the differentiation of local cultures, including the chimps' rudimentary tool culture of hammerstones and dipping sticks.

HANDS

The primate hand, with fingernails and divergent thumb, originally evolved as an adaptation to arboreal locomotion. Elliot Smith (1927) saw that this was preadaptation for higher primate intelligence. Creatures with three-dimensional binocular vision and hands that could hold onto branches could also see, hold, and eventually think about tools and objects.

Not all primates have the "primate hand." Marmosets climb like squirrels with clawed digits. Spider monkeys in the new world and colobus in the old have lost their thumbs altogether and habitually dangle from branches with four hooklike fingers.

The actual movements of the hand are more phylogenetically conservative than hand shape. Prosimians have a wide variety of hand shapes (Fig. 17.1). In the lemuriform group these are as different as the little mouse lemur's generalized clinging hand with its starfish of fingers and the heavy grasping hand of the sifaka, which must catch hold of a trunk at the end of a five-meter leap. The lorisiform series goes from the Senegal bushbaby's, much like a mouse lemur's, to the three-fingered chameleonlike pincer of the potto. Each of these taxonomic groups has just one prehensive grip. Lemuriforms reach with parallel fingers, touching branch or food first with the finger ends. Lorisiforms reach with splayed fingers, touching branch or food first with the distal part of the palm. Then the closing hand molds to the object it encounters (Bishop, 1962, 1964).

Both new-world and old-world monkeys have gone beyond this stage, to differentiate "precision grip" from "power grip" (Napier, 1960). They have not only a motor pattern for locomotion but some separate control of the index finger and thumb, which are used for fine manipulation of bits of food or grooming tiny particles from each other's fur. In the new-world monkeys, the digits lie and flex in parallel. Their common precision grip is a sideways "scissors" action of thumb against index finger or index finger against third finger. Old-world monkeys, on the other hand, bring the thumb end against some part of the index finger. They may partially oppose the fingertips of thumb and index finger.

There seems to be some correlation with ecology among the monkeys. The ground-living macaque-baboon group is more dextrous, with more highly differentiated precision grip, than the guenon species. The gelada baboon has the most highly developed precision grip of all, because it largely feeds by plucking grass blades (Etter, 1973). This supports C. Jolly's (1970) view of early hominids as small-object feeders like geladas.

The apes have very marked precision grips in which the thumb is partially rotated and pressed to the side or tip of the index finger (Napier, 1961). Apes, like humans, can control one finger at a time, pushing the index through narrow openings to retrieve a grape (Welles, 1975). (See Fig. 17.2) Humans, though, are the only primate that can play the piano with all ten fingers (even if the others wanted to).

Humans are also the only primates that have consistent, heritable, right- or left-hand preference, and a species bias toward one side. Other primates may develop individual preferences toward using one hand on given problems. This commonly results in about a third of individuals being right-handed, a third

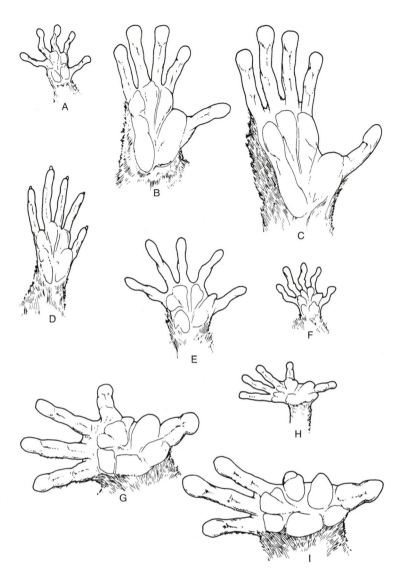

FIGURE 17.1. *Strepsirhine hands. A, mouse lemur; B, brown lemur; C, white sifaka; D, hapalemur; E, thick-tailed bushbaby; F, Senegal bushbaby; G, slow loris; H, slender loris; I, potto. (Drawn from handprints of walking animals by A. M. Kingsbury, from A. Bishop, 1962.)*

FIGURE 17.2. *Chimpanzee and human precision grips. (Courtesy J. Napier.)*

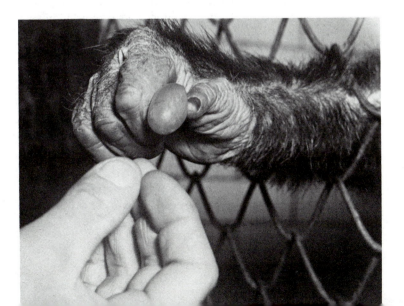

left-handed, and a third ambidextrous. Tests after a year or two show that many individuals have switched preference (Passingham, 1982). The degree of preference may depend on the task. Marmosets are likely to use a preferred hand for catching insects, which may speed up reaction time (Box, 1977).

Humans are predominantly right-handed, with few of us ambidextrous. The side of the brain that controls the dominant hand is usually the side that has the centers of speech. The dominant hand is not "better" in a simple sense than the other. Usually, there is division of labor between the two. In bimanual tasks, one hand holds or steadies an object while the other fiddles with it, in an analogy of the division between precision and power grips. Often the hands exert force in opposite directions (Bruner, 1968).

Our evolved fine control of the hand and our differentiation of the two hands link tool use with brain complexity, and this in turn to language, at least in the human lineage.

THE TABOO POLE: SPECIES DIFFERENCES IN CURIOSITY

Species differ not only in manipulative ability but in willingness to handle new objects. Harlow (1951) raised serious problems for the "punishment and reward" school of behaviorists when he found that rhesus monkeys opened locks from sheer manipulative curiosity.

One of the clearest illustrations comes from Bernstein's attempt to study social learning (1974). He built a redwood structure with six vertical poles. One was to be equipped with a meter-wide grid of electric wire. He reasoned that some monkeys would receive electric shocks, but others would observe and avoid the shocks by a kind of protocultural learning. His experiments instead demonstrate species differences in the approach to objects.

> The geladas displayed an amazing appetite for redwood and reduced the lumber portions of our apparatus to ground-level stumps. . . . Further testing with them was suspended . . .
> The sooty mangabeys ran and jumped rather than climbing . . . Vertical poles leading to nowhere had little appeal, and no animal approached the newly introduced grid during training, thus producing perfect avoidance learning in zero trials . . . Two years later we were still waiting to shock our first mangabey. . . .
> For all four macaque groups ((pigtail, stumptail, crabeater, Celebes black ape)) during the shock phase there was a dramatic *increase* in . . . use of the taboo pole.

Adult female macaques were rarely or never shocked. The alpha males reacted variably. The pigtail alpha received one shock and never climbed again, while the stumptail took a shock, chased his troop indoors, and then hung out the doorway threatening the offending pole. All the same, each macaque group, especially its young males, managed to destroy the grid; the pigtails broke the pole twice and the wires nine times. Groups of juvenile black macaques and crabeaters continued playing with live sparking wires they pulled loose. Then, in the postshock test period, everyone lost interest in the "taboo." As Bernstein ruefully concludes, "We failed to institute a protocultural tradition, but learned something about the power and function of play."

TOOL USE: DEFINITION AND CLASSIFICATIONS

What is tool use? Beck arrived at a definition: "The external employment of an unattached environmental object to alter more efficiently the form, position, or condition of another object, another organism, or the user itself when the user holds or carries the tool just prior to use and is responsible for the proper and effective orientation of the tool." In the next sentence he admits this definition is "tortuous." However, it takes such complexity to distinguish, for instance, a chimpanzee climbing a tree to reach food from the chimp positioning a stick as a ladder to reach food, and using a stick to scratch oneself from scratching oneself on a branch still attached to the tree. In short, tool using is hard to define sharply, and may or may not indicate intelligence. The hermit crab's shell is a tool by Beck's definition. We tend to think tool use is an obvious and important category because it is important to us, but, like other behavior, its origins were fairly simple and perhaps simpleminded.

Beck points out that most animal tool use serves one of four functions. Tools may extend the animal's reach: towers of boxes to climb on, sticks for knocking and hitting, probes to pry in narrow crevices. Tools may amplify mechanical force in hammering or clubbing or jabbing or digging. Tools may amplify gesture in display, like the oridinary chimps' brandished branches or Mike's clattering kerosene cans, which propelled him to dominance in the Gombe. Finally, tools used for sponging, wiping, and containing can increase the control of liquids.

WEAPONS

A more traditional way to classify tool use is by situation: tools as weapons, as feeding aids, or in bodily care. Aggressive and defensive weapons are tools. This could have begun when a primate either by chance or out of frustration dislodged a branch that fell near his enemy (Hall, 1963). New-world howlers, spiders, squirrel and capuchin monkeys drop branches as well as urine and feces. Woolly monkeys defecate into their hands and throw the feces at pursuing humans. Patas, guenons, macaques, baboons, and gibbons drop stones or branches. Capuchins and baboons not only drop but actively detach their missiles and throw them with some accuracy. Southwest African baboons chose stones six times larger than the average size of clifftop stones, weighing an average 583 grams, and moved the stones to a spot on the rocky canyon walls above the observers before releasing them (Beck, 1980; Hamilton et al., 1975). Hall (1963) pointed out that one could move by very small steps from occasionally dislodging a stone to this sort of apparently purposeful behavior.

One incident seems to confirm that baboons can foresee the consequences of such action. A male hamadryas loosened a stone on a clifftop, directly above a ledge where several juveniles were playing. He caught the stone in his hand and held on until the juveniles dispersed. (Baboons rarely, if ever, simply sit

FIGURE 17.3. *Gorilla displays with and eventually throws an object at the photographer. (Courtesy P. Coffey and the Jersey Wildlife Preservation Trust.)*

with a stone in one hand in other circumstances.) Then he let go, and the stone crashed down onto the deserted playground (Kummer, 1981).

Chimpanzees, orangutans, and gorillas aim missiles at intruders (Fig. 17.3), both in the wild and in captivity (commonly feces in captivity). They also brandish and flail with sticks and branches at intruders (chimpanzees clubbing the stuffed leopard, Chapter 4).

Most such behavior is directed at other species, usually predators. Chimpanzees brandish branches in aggressive display, and displaying gorillas throw foliage about, but it is mainly in captivity that a few of them use weapons to attack their own kind.

FEEDING AIDS

The second main context of tool use is feeding. Chapter 3 discussed Cebus hammering hard nuts on a bamboo node. Cebus monkeys and crab-eating macaques hammered open oysters with stones, and baboons opened sausage-tree fruit the same way, as well as bashing scorpions (Hladik, 1973; Marais, 1969). Chimpanzees transport appropriate hammer-and-anvil stones to beneath the palm nut trees (Chapter 3).

Fishing for termites with fine probes, or ant dipping with long stout sticks have only been seen in wild chimpanzees, as has sponging up water with a crumpled leaf. There is one account of a wild baboon using a stick to pry pebbles from a clay matrix and then swallowing them (we don't know why). Another account describes baboons using sticks to widen the entrance to subterranean insect nests (Kortland and Kooij, 1963; Oyen, 1979). These contrast with the repeated observations at the Gombe of chimpanzees' termite fishing while baboons sit watching, apparently eager for dropped termites, with no notion that they might start fishing themselves. Chimpanzees use sticks to hit or hook down out-of-reach branches, or as ladders to climb (see Chapter 21). They also use sticks and probes as olfactory aids in nonfeeding contexts, such as to investigate the vagina of an estrous female.

Various other tools are used to reach or clean food. Chimpanzees crumple leaves as a sponge to sop out water, to wipe out a strychnos fruit or the brain tissue from a colobus skull. Baboons and macaques regularly rub grass between their hands to rid it of grit, and crab-eating macaques use leaves in similar fashion to wipe off food.

Parker and Gibson (1979) argue that the search for embedded food has been a major force in the evolution of primate intelligence. Looking for the seed in the nutshell or the termite in its mound may involve some ability to imagine what is out of sight. (It does not *necessarily* demand chimpanzee imagination— a termite-parasitizing wasp or even an anteater may not have much imagery or might have a quite different kind.) Adding a second object, such as a fishing stick or hammerstone may mean that the chimp is coping with a complex, combined idea. Bringing the tool from a distance, from out of sight of the food, is still more demanding, because it means holding the idea over time. Again, though, the tool use is not the sole criterion; we may not wish to attribute the same mental mechanism to a Sphex wasp that brings a pebble to tamp down the entrance to her burrow.

Toolmaking involves yet another step backward from the goal. Many primates actively break off the branches they drop. Chimps strip the leaves from their termite probes and bite off broken ends. The Mount Assirik chimpanzees (but not those in East Africa) peel all the bark from their termiting twigs. It is the mental leap to such two-step foresight that 25 years ago seemed unique to man.

SELF-WIPING AND SELF-ADORNMENT

A third context of tool use is objects applied to the body. Cebus monkeys wipe odorous substances on their body. Besides their own urine, they anoint themselves with "ants, onions, alcohol, perfumes, ammonia, orange peels . . . and snuff" (Beck, 1980). This is similar to "anting" in birds and the self-anointing of hedgehogs and, dare one suggest, human use of scents to attract attention.

The opposite, wiping to clean off a sticky substance, also appears in cebus and wild baboons (Lawick-Goodall, et al., 1973; Beck, 1980). Chimps and wild orangutans use leaves to wipe off sticky fruit, mud, dung (especially diarrhea), ejaculate, and blood.

This brings us to a recurrent legend. Lemurs, howler monkeys, and other unlikely primates are often said by local people to apply medicinal plants to

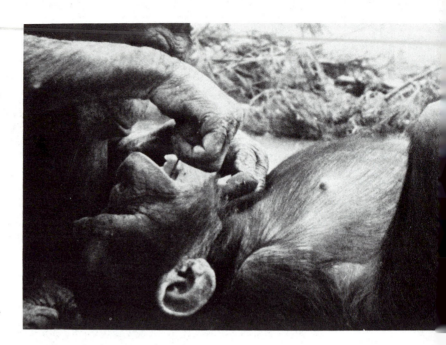

FIGURE 17.4. *Belle grooms Bandit's milk teeth with a stick. (Courtesy C. Tutin.)*

wounds. This is just the sort of tale people like to believe, and it is so far unconfirmed by primatologists, unless you count baboons and chimpanzees wiping off blood. It is so widely quoted and believed in different continents, though, that there just may be some basis for the story.

There is no doubt about apes' self-adornment. Wild and captive apes drape branches, leaves, and straw on heads and shoulders. Orangs especially drape themselves with leaves and vines in rain, hot sun, or when apparently "hiding" from humans (MacKinnon, 1974).

Finally, McGrew and Tutin (1973) saw a captive female chimp use sticks, twigs, and even a piece of cloth like dental floss to take out one of the teeth of a young male who was shedding his milk teeth (Fig. 17.4.).

NEST BUILDING

Nest building is somewhat neglected by primate psychologists. If birds and bushbabies do it, it need not require advanced mental processes. Or does it? Chimpanzees, orangutans, and gorillas build sleeping nests. They usually construct a new one every night, and sometimes rougher day shelters as well (Fig. 17.5.). Gorillas mainly nest on the ground, in a crudely arranged circle of vegetation. Chimpanzees and orangutans make elaborate woven structures in trees, with first large branches and then small ones broken across and interlacing. Jane Goodall (1962) reported a transient nesting fashion in the Gombe, when the chimpanzees began nesting in palm trees, which demanded new techniques. The nests of chimpanzees at Mount Asserik, where there are large ground predators, are higher in the trees than elsewhere (Baldwin et al., 1981).

Like other manipulative behavior, nest making improves with age. The young ape has time to learn because the child sleeps with its mother until three or four years of age. Bernstein (1962) gave nesting materials to the caged apes at Yerkes. Wild-born chimpanzees made good nests without tutoring. Captive-born ones did not, but two captives originally raised in human homes made circular nests reminiscent of real ones. This suggests both partially innate building patterns and the permanent damage that early isolation can do to complex manipulative ability (Menzel et al., 1970).

There seems no reason why human ancestors should not have built nests as well. One of the early Olduvai remains is a circle of stones that was perhaps a shelter. Prosimians commonly build nests. If prosimians, great apes, and one of the Olduvai hominids constructed such things, one cannot at any rate claim confidently that earlier protohominids did not.

One question is how terrestrial life might have influenced nesting. Gorilla ground nests seem degenerate structures compared with the tree beds of chimpanzees. However, one juvenile chimpanzee nest, one gorilla nest, and a large number of orang nests actually had roofs, so it is possible to imagine transitions from simple bed to real shelter (Goodall, 1968b).

More important is the transition from bed to home. Prosimian nests are breeding as well as sleeping nests, where altricial young can be left while the mother forages. Young apes are never left behind in the nest to sleep. However, on four occasions when chimpanzees had bad colds they went to bed at 4 to 4:30 P.M. and slept late, until about 9 the next morning, and two made elaborate daynests where they slept for three or four hours the next morning.

One mature male, who lost the use of both legs, probably from polio, slept for more than seven hours a day in the nest. He made two nests in one tree and three in another, and dragged himself from the ground to the lowest nest

FIGURE 17.5.   *A chimpanzee lolls in his day-nest at the Gombe Stream. (Courtesy H. van Lawick.)*

for the day, then into a higher one for the night, during seven days. On the last night before he had to be destroyed, he had dislocated an arm and was too weak to climb into the trees, but when his observers "broke off a pile of leafy branches he managed to work them under him, using one arm and his mouth, to form a crude ground nest" (Goodall, 1968, p. 199).

Finally, an orangutan built herself a nest during the day in which to give birth. In these apes one can see nest becoming home. Perhaps early humans built nest beds for other humans who were in need—small infants or women in childbirth—and thus eased the transition to settled family life.

ASSOCIATION AND INSIGHT

Köhler began it, with his famous chimpanzee Sultan. Let me quote him:

> Sultan is the subject of experiment. His sticks are two hollow, but firm bamboo rods such as the animals often use for pulling along fruit. The one is so much smaller than the other that it can be pushed in at either end of the other quite easily. Beyond the bars lies the objective, just so far away that the animal cannot reach it with either . . . he takes great pains to try to reach it with one stick or the other, even pushing his right shoulder through the bars. When everything proves futile, Sultan commits a bad error or, more clearly, a great stupidity such as he made on other occasions. He pulls a box from the back of the room toward the bars. True, he pushes it away at once as it is useless . . . then a good error, he pushes one of the sticks out as far as it will go and takes the second and with it pokes the first one cautiously toward the objective, pushing it carefully from the nearer end and thus slowly urging it toward the fruit. This does not always succeed, but if he has got pretty close in this way, he takes even greater precaution, he pushes very gently, watches the movements of the stick that is lying on the ground and actually touches the objective with its tip . . . the procedure is repeated . . . he puts the stick in his hand, exactly to the opening of the stick on the ground, and, although one might think that doing so would suggest the possibility of pushing one stick into the other, there is no indication whatever of . . . a solution. Finally, the observer gives the animal some help by putting one finger into the opening of the stick . . . this has no effect . . . the experiment has lasted over an hour and is stopped for the present as it seems hopeless.
>
> The keeper is left there to watch him. . . . Keeper's report: 'Sultan first of all squats indifferently on the box which has been left standing a little back from the railings, then he gets up, picks up the two sticks, sits down again on the box and plays carelessly with them. While doing this, it happens that he finds himself holding one rod in either hand in such a way that they lie in a straight line. He pushes the thinner one a little way into the opening of the thicker, jumps and is already on the run towards the railings, to which he has now half turned his back and begins to draw the banana toward him with the double stick. I call the master. Meanwhile one of the animal's rods has fallen out of the other as he has pushed one of them only a little way into the other, whereupon he connects them again. (Köhler, 1927: 113–115).

This, says Köhler, was *insight* into understanding the connection between food and the two sticks when Sultan suddenly saw the disconnected parts that he had held in his hands for an hour as a single "gestalt," a coherent whole.

Köhler's experiments were repeated by Birch (1945), who showed that previous experience with sticks contributes to the animal's eventual solution of problems using sticks. Then Schiller (1957) gave naive chimpanzees the same problems, but in play, with no rewards visible or offered, and Schiller's chimpanzees made all the same manipulations, stacking three boxes one on top of another, connecting sticks, and weaving string in and out of the cage wire. They arrived at these solutions even more quickly in play, when not distracted by a banana. Köhler's own account shows that Sultan did not achieve his first solution of the problem until he had more or less given up the banana, had withdrawn, and was fiddling about with the sticks. Then, having aligned the two sticks, he saw their connection with the problem he had been offered and ran forward to the bars again.

The current manipulative genius among chimpanzees must be a chimpanzee called Julia, who has been studied, trained, and otherwise fostered by Rensch and Döhl (Rensch and Döhl, 1967, 1968; Döhl, 1966, 1968). Julia was first taught to open boxes locked with 14 different kinds of fastening. She applied her training to opening several different fastenings to get into the same box. Eventually she could solve six-step series of boxes, where she had to choose her strategy by visually working backward from the goal (Figs. 17.6. and 17.7.). She generalized to opening unfamiliar boxes with miniature keys in miniature padlocks,

FIGURE 17.6. *A double series of locked boxes. The chimpanzee must choose only one of two tools she sees through the plastic lid of the initial double box. Each tool opens one other box, containing one tool to open yet another box. One series of five tools leads to one box with a banana, the other leads to an empty box. The only reliable way to reach the banana is to trace backward step by step from the goal. (After J. Döhl, 1968.)*

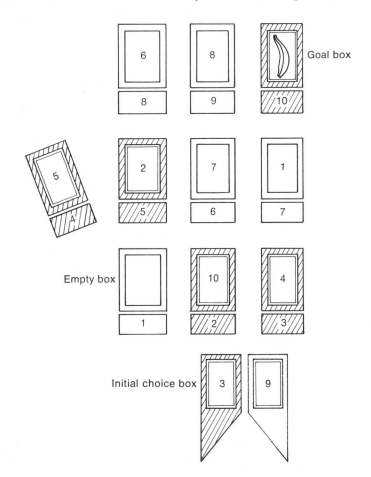

FIGURE 17.7. *Julia studies the series of locked boxes in Figure 17.6. (Courtesy B. Rensch and J. Döhl.)*

tiny screwdrivers that she had to guide with the opposite thumbnail into the minute screws, or even entirely new and unfamiliar fastenings. The only fear is that Rensch and Döhl will train Julia to become an expert safecracker and that some limit will have to be put on her activity.

Julia can also solve complex mazes. She guides an iron ring along the paths of the maze by holding a magnet above the maze (Fig. 17.8.). Rensch and Döhl first taught her to put the ring in a slot machine that would reward her with fruit, and then she learned to move the ring along a path by means of a magnet. In the first series of definitive experiments, she chose between two simple paths of a small maze, one of which was blocked at various points. Her first movement led the ring to left or right off a small hump, so that she had to think backward from the goal to make the first choice of left or right correctly. In the next series, the paths became more and more complex through incorporation of angles and branched blind alleys. Next, the paths were interlaced, the size of the maze was doubled, the paths were made narrower, and false

FIGURE 17.8. *Julia starts a complex maze. She must first trace the path visually to the exit to make a correct initial choice of route. (Courtesy B. Rensch, and J. Döhl.)*

exits were added. "In the last 100 rather complicated mazes, the chimpanzee chose the correct path in 86 per cent of the trials . . . by watching the eye and head movements of the chimpanzee, we could in some cases see that the ape first looked to the exits and then to the path system near the starting point. Apparently, she combined the sensations of the latter with the mental images of the goal region, a form of behavior that is very similar to that of man" (Rensch and Döhl, 1968, p. 231). Six biology students tried the same complex mazes. On the average, they solved the maze in half the time that Julia did, but their scores overlapped so that sometimes Julia was the quickest.

Orangutans in captivity are even more manipulative than chimpanzees. Lethmate (1978, 1982) has taught young orangutans similar series of tool choices and openings as Julia's, as well as repeating Köhler's "insight" tests. Wright (1982) taught an orang to chip stones in imitation of early man. Orangutans seize, hoard, and use tools, clothes, and eating utensils. They lever off bars or pick loose fresh putty to escape from cages with far more skill and alacrity than any other ape. Birute Galdikas describes camp life in Borneo among ex-captive orangs. There are tooth marks on every object. One juvenile in the garden uses sticks to dig holes, to pull burning logs from the fire, or to poke the cat. Another is in the kitchen eating with spoons or stealing flour and sugar to throw in a glass, next finding the hidden eggs and breaking one into the glass, and then stirring vigorously like the cook making cakes (Galdikas, 1980). Ex-captive orangs steal dugout canoes and ferry them across a river with sticks as a kind of paddle to return to the comforts of camp (Galdikas, 1982).

All this is a paradox, behavior somehow uncorked by captivity. In 15,000 hours of observation, Galdikas has only once seen orang tool making in the wild, when an adult male broke off an ironwood branch and scratched his back with it (Galdikas, 1982).

Wild orangutans break off sticks and throw them, put leaves on their heads, and build nests. Adult males topple snags, the dead trees so common in the rain forest, and send them crashing noisily down, as a prelude to the male long calls. Two old males controlled the direction of fall, vibrating the snag or riding it partway down so that it fell toward the observer, even if it was initially leaning away. Orangs commonly use a similar technique in traveling, swinging one tree over until they can reach and transfer to the next. The most complex instrumental behavior by a wild orang was a female with an infant on her shoulder (who) crossed a narrow point in the Sekonyer Kanan river. Submerged up to her neck, she folded the large, stiff upright fronds of weeds lining the edge of the river out over the water. Then, standing on these, she caught hold of the fronds on the other side of the river, which she folded over onto the first fronds. Using this "nest" as a bridge, she crossed over to the other side. Like the backscratching, this is just one observation in eight and one half years. It does not seem to explain the ex-captives in their boats. It does, however, underline the laboratory finding that a new discovery, sudden insight, is rare and wonderful, whereas most complex behavior comes from the gradual accretion of experience (Galdikas, 1982b).

Köhler's original conclusion that insight is a qualitative leap in problem solving no longer seems so important. Hall (1963) pointed out in his famous review that there are very small steps from occasionally dislodging a stone to throwing one to the chimps' full-blown group attack on the stuffed leopard. All this could be explained by simple association. Birches' and Schiller's (1957) work showed how much part previous experience and inborn manipulative tendencies play in the sudden "insightful" solutions. Beck argued first for insight, but later observation of baboons solving problems first by luck, then more coherently, led him to think that insight is perhaps just skipping a step or combining two steps without acting them out.

One of the few accounts of a discovery of tool use by wild primates comes from Jane Goodall:

> A mature male, who was afraid to take a banana held out to him by hand, shook a clump of tall grass as a mild threat directed toward the human. When there was no response, he shook the grasses more violently so that one actually touched the banana. He stared at the fruit and suddenly released the grasses, pulled a long soft plant from the ground, dropped it, turned to break off a thicker stick, and then hit the banana to the ground and ate it. When a second banana was offered in this way, he hit it from the hand with no hesitation (Goodall, 1968b, p. 207).

What seems important is how and why wild or captive apes can associate so many steps of means to end, even if they are achieved by gradual accretion, not sudden insight.

ART

Chimpanzees, gorillas, orangutans, and cebus monkeys will paint or draw, given any materials at hand, even their own feces used on the wall of their cages. If one equips them with more orthodox pencil and paper or, even better, poster paints, they work with extraordinary concentration, to the extent of having tantrums if anyone tries to stop them before they are finished. They also seem to know when they are finished, ripping off a page to start a clean sheet (Kohts, 1923; Schiller, 1951; Rensch, 1957).

Desmond Morris (1962) has analyzed in detail the stylistic differences of primate painting. Some individuals make recognizable variations on a theme, not merely a whole arm circling motion but, for instance, a repeated fan pattern with the pencil drawn toward or away from the animal in converging lines. Schiller, Morris, and Rensch also find that the primates may attempt to complete or balance a preformed pattern, drawing around the outline of a square on a piece of cardboard or adding color on the opposite side from a large black blot. Julia does follow-the-dot pictures, copying square and triangle. Morris makes the tantalizing suggestion that his chimpanzee Congo stopped just before the stage when children can drawn "faces." Small children, after stages of random and then more coherent scribbling, will suddenly begin to make circular patterns with spots inside. From there, the child begins to claim that the spots are in fact a face, although there may at first be no assignment of eyes, nose, and mouth. Congo's final paintings (Fig. 17.9.) were of the form of a circle with marks inside it, but, although possessing far greater motor coordination than a child of comparable age, he never went on to recognizable representation.

Still more tantalizing, Moja, one of the Gardners' signing chimpanzees, names her pictures. In April 1956, she drew an atypical scribble with very few lines. Her companion, Tom Turney, put the chalk back in Moja's hand and signed to her to *try more*. She dropped the chalk and signed *All done*. The reply was unusual; so was the drawing. Turney signed *What that?* Moja placed the forefinger and thumb of her right hand to her lips: the sign for *bird*. (See Fig. 17.10.)

Since then, Moja has extended her oeuvre, labeling drawings consistently: radial shapes are *flower*, round shapes *berry*. She has also replied appropriately to questions, as in: Q. *What that?* A. *Grass.* Q. *Who draw that?* A. *Moja.* (Gardner and Gardner, 1978; Watson 1979).

FIGURE 17.9. *Congo choosing a color. (Courtesy D. Morris, from Morris, 1962.)*

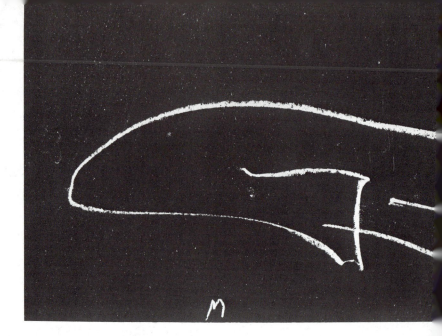

FIGURE 17.10. *Mojas "bird." (Courtesy R. A. and B. T. Gardner.)*

TOOL USE AND INTELLIGENCE

The mystery is not how much primates use tools in the wild, but why they do not do more when they clearly have such capacities. The answer seems to me to be that there is such a thing as general intelligence, which allows once to combine more than one perception or idea, and thus create new concepts, which are revealed in new patterns of behavior.

Such intelligence can be used to remember and map the locations of fruit trees. It can be used in deciphering and taking advantage of complex social interactions (Jolly, 1966; Humphrey, 1976; Beck, 1980). It can be used in feeding and social tactics, such as sea gulls dropping shellfish to break them, or wolves ambushing prey, which have little to do with what we define as tools. We can set up the discussion as an ''either/or,'' emphasizing either the importance of social life or the importance of manipulative foraging techniques. I suspect that both contributed to a general capacity for thought. Without some concept of general intelligence it seems to me impossible to explain the discrepancies between the object manipulation of wild and captive primates.

However, we also need to take account of different species' bias in their natural direction of attention. Here it is humans who are aberrant. Instead of ponderous statistics, let me quote Birute Galdikas (1980):

> After five years of living with orangutans, I had reached the point where the line between human and ape was getting somewhat blurred. Sometimes I felt that I was surrounded by wild, unruly children in orange suits who had not yet learned their manners. They used tools, liked to wear . . . clothing, loved . . . junk food and candies, were insatiably curious, wanted constant affection.
>
> But (our son) Bin's behavior in his first year highlighted the differences very clearly. At the same time I was hand-raising Princess, a one-year-old orangutan female. A one-year-old orangutan merely clings to its mother (me, in this case), showing little interest in things other than to chew them or put them on its head. For Princess the main interest in life seemed to be sustenance.

Bin, on the other hand, was not particularly food-oriented; in fact, unless he was very hungry he gave all his food to Princess. He was also fascinated by objects and implements and would watch with great concentration whenever Rod or I, or an orangutan for that matter, used one of them. He was constantly manipulating objects. Another major difference was that Bin babbled constantly, while Princess was silent except for squealing.

(Thus) many of the traits associated with the emergence of humankind were already expressed in Bin's development before the age of one: bipedal locomotion, food sharing, tool using, speech.

SUMMARY

Humans take far more interest in objects than do other primates. Wild primates' feeding determines much of their manipulation: insectivorous primates tend to have a longer attention span than vegetarians, and terrestrial, small object feeders have most fine control of the hand. There is little pretreatment of food in the wild or tool use except among chimpanzees, who fish for termites and ants, hit food down with sticks, and sop up fluids with leaves. Many primates dislodge or even throw stones and branches to discourage predators. Cebus, baboons, and all the apes wipe the body with objects, although cebus seem to wipe on odor and the others wipe off sticky substances. Great apes and some prosimians build nests.

Primates achieve far more complex object manipulation in captivity than in the wild. The earlier arguments about insight versus associative learning seem less important than the clear fact that the apes have some general capacity for intelligence, which turns toward tools in the odd environment provided by ape-oriented humans.

CHAPTER 18

COGNITION

COGNITIVE LEVELS

If manipulating tools and other objects is only one expression of cognitive abilities, where do we turn for a more comprehensive description of thought?

Köhler, as we have seen, was concerned with the number of objects his animals could combine as means to an end and whether they were out of sight; a relevant object had to be remembered or imagined as part of the ongoing situation. Piaget, whose work will be discussed later in this section, created a whole sequence of stages of logic that a child was supposed to pass through. The child progressed to solving Köhler-type chimpanzee problems and the first use of words and symbols in the *sensorimotor stage,* from infancy to about two and one half years. Then it developed logical concepts in the *preoperational stage,* from two to seven, and finally achieved fully *operational* logic after seven years.

Many other psychologists have attempted to describe such levels of cognition. Braine (1959) writes "It may be noted that the distinction between abstract and concrete thinking made by Goldstein and Scheerer (1941) appears to have much in common with Piaget's distinction between 'operational' and 'preoperational' thinking and with Lashley's (1938) between first- and second-order generalizations. All these distinctions seem to follow that made by logicians between 'class' and 'class of classes' " (Braine, 1968, p. 194). The importance of Braine's putting together all these pairs of words is that Goldstein and Scheerer were working with psychotics and mental defectives, Piaget with children, Lashley thinking principally of animal studies, and the logicians in the farther pastures of philosophy and symbolic logic. Braine is thus making a sweeping generalization across many of the fields that concern themselves with the nature of thought.

Lashley gives, as an example of first-order generalization, discriminating "triangularity *per se.*" Zimmerman has shown that rhesus infants can do this in the first weeks of life (Zimmerman and Torrey, 1965). A second-order generalization is the "oddity problem," in which the animal must choose the odd one of three objects, regardless of the particular characteristics of the two alike or the one odd. But one can go a bit further than saying that a first-order generalization is more directly linked to perception and a second-order generalization is more abstract. Second-order generalization means *consistently* applying some abstract criterion, that is, a mental concept, in the face of perceptual distraction by irrelevant factors.

Roger Thomas (1980, 1982) has offered a coherent scale in terms which can be applied through nonverbal tests to both children and animals. Table 18.1. gives his classification of cognitive levels. (See his original papers for descriptions of the similarities and differences between this classification and the many earlier ones.) The important point here is that the simple concepts of negation, conjunction, disjunction, and conditionality underly much of logical thinking. The ability to use these concepts is testable with animals; the same operations are also the basic repertoire of computers; and Whitehead and Russell's *Principia Mathematica* was an attempt to found all mathematics on these same logical functions.

Recently David Premack (1983) has proposed yet another level of abstraction. He suggests that there is a great difference in dealing with relations between elements and relations between relations. That is, it is fairly easy to train a chimpanzee, or a very young child, to reliably choose "the bigger thing" or "the odd thing," or in a matching-to-sample test, "the same kind of thing." It is not until children are around four years that they can deal with four-element

---

TABLE 18.1.   A Hierarchy of Mental Abilities

Non-concept learning

Level 1: Habituation. A learned decrement in responding to repeated presentation of the stimulus.
Level 2: Signal learning. Classical Pavlovian conditioning.
Level 3: Stimulus-response learning. Simple or discriminated operant conditioning.
Level 4: Chaining. A sequence of two or more stimulus-response connections.
Level 5: Concurrent discrimination learning. Two or more stimulus-response connections are learned simultaneously.

Concept learning

Level 6: Affirmative or negative class concepts.
   Such concepts produce consistent responses which do not depend on prior experience with the *specific* stimuli of the test trial. *Absolute concepts* do not necessarily involve comparing two or more objects, e.g., "a red thing, non-red things." *Relational concepts* demand comparison, e.g., "the bigger one" or "the odd one."
Level 7: Conjunctive, disjunctive, or conditional concepts. E.g., stimulus correct only if curved *and* blue (conjunction); correct if curved *or* blue (disjunction); or *if* the stimulus tray is white choose the blue stimulus, *if* the stimulus tray is black choose the curved stimulus (conditionality).
Level 8: Biconditional concepts. A is correct *if and only if* B is present (or true) where A and B are both class concepts.

(After R. K. Thomas, 1980, 1982).

TABLE 18.2. Piagetian Cognitive Levels

Stage	Age (months)	Description	Major Distinguishing Behavioral Parameters	Examples
Sensorimotor Period				
I Reflex	0–1	Unlearned, involuntary, stereotyped responses	Unlearned Involuntary Stereotyped	Rooting and sucking
II Primary Circular Reaction	1–4	Infant's action centered about his own body ("primary"), he learns to repeat actions to reinstate an event; First acquired adaptations to stimuli	Self-oriented Recognizes contexts Acquired adaptations to impinging stimuli	Reliably habituates so psychologists can test for novelty; Visually follows moving objects; Brings held objects to mouth and sucks; Smiles at faces and other visual stimuli; Contagious crying
III Secondary Circular Reaction	4–7	Repeated ("circular") attempts to reproduce environmental ("secondary") events first discovered by chance	Environment-oriented; Single behaviors toward a single object or person; Semi-intentional (initial act not intentional, but subsequent acts are); Establishes object- or person-action relationships through attempts to reproduce interesting environmental events	Reaches for and grasps objects reliably; Repeats gestures that have accompanied interesting spectacles, without attention to object connections; Does not uncover hidden objects; Gradual focusing of smiling on known people; Repeats own and others' sounds
IV Coordinations	8–12	Two or more behavioral acts are coordinated, one serving as instrument to another	Intentional; Goal established from outset; Establishes relationships (coordinations) between two objects; Coordinates or relates multiple aspects of single objects; Combines behaviors; Begins to attribute cause of environmental change to others	Uncovers hidden object; At first confused by two or more stage hiding ("A not B" or "Stage IV error"); Touches or combines two behaviors or two objects; Places parent's hand on object to restart spectacle; "social" causality; Fusses at strangers; Separation distress; Imitates with unseen body parts; Repeated dropping objects to watch them fall

V Tertiary Circular Reactions	12–18	The child becomes curious about the functions of objects, object-space-, gravity-, force relationships ("tertiary"); he repeats his behavior ("circular") with variation as he explores the potentials of objects through trial and error experimentation	Behavior becomes variable and nonstereotyped, as the infant invents new behavior patterns. Trial-and-error experimentation in play and to solve problems. Coordinates object-, person-, space-, gravity-, force-relationships	Can find toy hidden in series of places Will not check back through series if dump toy en route Trial and error causality—string-pulling, hoe problems, repeats what works using one object to obtain another Repeated throwing, catching, bouncing objects Putting objects in and out of containers Gradually lessening fear of strangers Imitates words and other new actions
VI Invention through mental combination (insight)	18–24	The solution is arrived at mentally and not through experimentation	Symbolically represents objects and events not present. Solves problems mentally	Checks through series of hiding places for toy dumped en route Fetches out-of-sight objects to solve problems Insight into object relations: e.g., box stacking without preceding trial and error Delayed imitation of people's acts, including language Symbolic and pretense play

Preoperational period: 2–7 years

Intuitive thought which groups like objects but does not deal with the relations of classes

Concrete operational period: 7–12 years

Can deal with class overlap and inclusion (some vs. all)
Seriation (if A > B > C, then A > C)
Analogy, metaphor
Conservation of mass or volume despite change of shape

Operational period: 12 years onward

Can explain logical operations verbally

(Modified after Chevalier-Skolnikoff, 1982.)

choices exemplifying relations. Is *XX* like *AA* or *CD?* Is *XY* like *AA* or *CD?*, which is much harder than the first problem. Is *Xx* like *Aa* or *aa?* The three chimpanzees in Premack's laboratory who could solve such relational problems had all been trained to use a "language" of plastic symbols on a magnetized board. Premack's concern has always been the study of logical thought, not a proliferation of vocabulary, so it may be that his apes have learned something more like symbolic logic than like language. There is, at any rate, a sharp distinction between the three chimpanzees who received such training and the three who did not, in their use of abstract relations. We shall return to Premack's "abstract codes" at the end of this chapter, after describing some of apes' abilities, and again in Chapter 20, "Language."

Probably the most influential attempt to construct such a hierarchy of thought has been the life work of Jean Piaget. Piaget developed a theory of the creation of logic in the human mind, based on empirical observation of children's growth—first his own three offspring, then larger numbers of children in Geneva, and now, through his disciples, widely across human cultures (see Table 18.2.). Piaget's work was promptly ignored by proponents of stimulus-response theory, particularly in America. Later it provoked many educators to fury, because Piaget insists that intelligence develops through discrete stages, so that training children to parrot rules they do not yet understand is no use. A number of educators have trained children to focus attention on the relevant parameters so as to accelerate parts of the Piagetian stages. Many of the experiments I will cite show that even untrained children understand more than Piaget gives them credit for, when tested in different manners. But I return to the Piagetian formulation, because whether or not one accepts his philosophy and its more didatic conclusions, Piaget has written a natural history of human intelligence (Piaget, 1951, 1954, 1960).

Ten years ago, in this book's first edition, the question was, Do primates go through the same intellectual developmental stages as human children? Today we can answer yes—at least for a handful of apes that have been studied and perhaps new-world cebus. Old-world monkeys seem to start in much the same way, but stop at different points on various series of tasks. At least it seems likely they do not skip to solving complex tasks while still unable to solve simpler tasks of the same form that are mastered earlier in our own species.

There follows a more interesting question. Piaget's descriptions encompass all aspects of thought: from finding hidden objects and simple tool use through the ability to imagine other people's points of view, both literally in physical space, and emotionally. When two traits that normally appear at the same age are dissociated in a particular child, he calls it a *decalage,* which means a displacement in space and time, or literally "an unwedging, or taking out the props." The word *decalage* implies that the two traits really *ought* to appear together but have fallen apart. The interesting question now for cross-species comparison is which traits do in fact appear together. We might now refine our comparisons from the crude observation that an adult chimpanzee has the strength of a human Olympic athlete but the tool using ability of a human child, to saying *which* of its abilities appear at different ages in humans and how these add up to an adaptive complex for the chimpanzee. Alternatively, if more and more of the chimp's cognition seems to center at a single mental age for the human, we might conclude that Piaget is right to think that abilities at each stage are logically connected. One answer would teach us more biology, the other, more philosophy.

Parker and Gibson (1979) have made a major attempt to correlate ontogeny and phylogeny in a Piagetian framework. In some respects prosimians resemble Stage 2 to Stage 3 human infants (under about eight months old). Macaques can find objects concealed in a succession of hiding places like Stage 5 humans up to about 18 months. Among primates only the apes progress beyond into the pre-operational stage, in which a mental image of concealed food and simultaneously or prospectively that of an extracting tool leads them on to using dipsticks or hammerstones. The paper is a landmark in its attempt to classify species' intelligence. It relates selection for formal intelligence to the evolution of feeding techniques in the wild as well as to neuroanatomy. However, it should have the same effect as Crook and Gartlan's seminal classification of primate social structures (Chapter 6). It should provoke others to break down categories into correlations: to find and then to understand the decalages in each species' mosaic of abilities.

INFANCY AND THE CONCEPT OF OBJECTS

Human Development: Visual and Manual Decalage

Piaget divides infancy, or the sensorimotor period, into six stages. In Stage 1 the child has only reflex actions, and in Stage 2, the first acquired adaptations. The baby can modify his sucking patterns, his crying, his cooing. He cannot, however, affect the outside world systematically unless some adult—mother or psychologist—arranges responses to his minute powers. At most he can repeat a few coordinations of parts of his own body, such as reliably bringing a thumb or bit of fist to his mouth.

Stage 3, the early manipulative stage, begins when the child voluntarily reaches out and grasps seen objects. Now his own hands and his own actions seem to cause external results:

> With a third level . . . which begins with the coordination of vision and prehension (between three and six months, usually around four and one half) new behavior appears that represents a transition between simple habit and intelligence. Let us imagine an infant in a cradle with a raised cover, from which hang a whole series of rattles and a loose string. The child grasps the string and so shakes the whole arrangement without expecting to do so, or understanding any of the detailed spatial or causal relations. Surprised by the result, he reaches for the string and carries out the whole sequence several times over. . . . If the child is confronted with a completely new situation, such as the sight of something moving several yards from his cot, he responds by seeking and pulling the same string, as though he were trying to restart the interrupted spectacle. . . . (Piaget, 1960, pp. 101–102).

> Laurent at seven months loses a cigarette box which he has just grasped and swung to and fro. Unintentionally he drops it outside the visual field. He then immediately brings his hand before his eyes and looks at it for a long time with an

expression of surprise, disappointment, something like an expression of its disappearance. But far from considering its loss irremediable, he begins again to swing his hand, although it is empty. After this he looks at it once more! For anyone who has seen this act and the child's expression, it is impossible not to interpret such behavior as an attempt to make the object come back. Such an observation . . . places in full light the true *nature of the object peculiar to this stage: a mere extension of action* (Italics mine) (Piaget, 1954, p. 22).

The fourth stage of infancy, for Piaget, begins at about eight to ten months when the child first uncovers hidden objects. Before this the baby may retrieve an object when a small part is left visible, but give up and stare or plead with the tester as soon as the object is gone completely, as though it somehow ceased to exist. The eight-to-ten-month child achieves "object permanence." However, he makes the "Stage 4 error"—the object is now continuous in time but not in space. The child defines a hiding place by its own action of uncovering the object.

Lucienne (ten months) is seated with a coverlet on her lap and a cloth spread on the floor, at her left. I hide her rubber doll under the coverlet, in A; without hesitation Lucienne raises the coverlet and searches. She finds the doll and sucks it. I immediately place the doll under the cloth in B, taking care to have Lucienne see me. She looks at me until the doll is entirely covered up again, then without hesitation looks at A and raises the coverlet. She searches for a while . . . same reaction with four sequential experiments . . . (then) once Lucienne has searched in A for the cloth hidden in B, I again raise the cloth at B in order to show her that the doll is still there, then I cover it up again; but Lucienne looks at the doll in B, and, as though moved by a new impetus, returns to A to pursue her search (Piaget, 1954, p. 52).

It may be that the errors and achievements of Stage 4 depend on the growth of a system of memory retrieval. One key experiment varied the length of delay before the child could pick up the cloth. As children begin to solve the Stage 4 or A, not B error, they do so with a three-second delay several weeks before they can cope with a seven-second delay (Kagan 1978). In stage 4, the child does not just drop objects, but follows the trajectory of the descending splodge of banana, and keeps throwing more from the highchair to watch them fall.

Stage 5, which lasts from 12 to 18 months, is one of vast advance in manipulative understanding. Piaget calls it the "trial-and-error strategy" stage. The child can not only track an object through successive displacements into different hiding places but it can use sticks, strings, and blankets to pull in food. Piaget notes that Stage 5 problems were solved by Köhler's chimpanzees.

Finally, Stage 6 begins with the first deferred imitation. From very early, a child will cry or laugh when others do. A few other actions, such as sticking out the tongue in response to a grownup doing so, start at a few weeks of age. Much more imitation seems to begin during Stage 4. Imitation involving unseen parts of the body, such as touching an ear, is generally labeled "Stage 4." In Stage 6 the model is not just out of sight, but copied after a lapse of time. Jacqueline Piaget at 16 months, deeply impressed by a little boy of 18 months, later copied both his loud laugh and, detail for detail, his tantrum. The first words learned and then used after a delay, not repeated directly after an adult, are the primary example of deferred imitation.

In this stage children also become upset when they cannot copy a model. In a series of experiments, psychologists played fairly complex games with toys; a doll might lay plates on a table and then "talk" on a toy telephone. Then the psychologist said to the child, "Now you play," *not* "Now you copy me." Beginning at about 14 months and peaking at 24 to 28 months, the children burst into tears (Kagan, 1978).

These stages of human infancy are defined (up to Stage 6) by achievements in the manipulative sphere. There is a rich background of speculation about the child's conceptions of object permanence and the child's construction of causal relations. At first he conjures causes by repeating his own actions in Stage 4, then he combines two objects to act on each other in Stage 5, and then he names and relates objects in words in Stage 6. A background of observation is necessary to show the child is solving each problem in human style, not mechanically. But is each stage a whole in human logic, or are there major "decalages" even in our own species?

When we turn to the visual sphere, object recognition and object permanence appear months before directed reaching. Neonates of six to ten days flap vaguely if an object is within their reach. If the object is not real, just a projected image, so their hand flaps through the space where a thing should be, they cry (Bower, 1970). Recent experiments show that they pay selective attention to human faces and even imitate expressions such as sticking out the tongues. Two-month-old infants can be conditioned to indicate two shapes look the "same" by turning their heads to touch a headrest, and thus they show they use binocular cues and parallax to judge shape and size much as adults do. (The reward for conditioning is that the experimenter pops up and says "peekaboo." A Martian observer, or even the baby, might misinterpret who is controlling whom.) Finally, five-month-old infants, who never uncover a toy hidden under an opaque cup, could reach for a dangling toy if the room was suddenly darkened, and they can (slowly) turn over a transparent cup to grasp the toy they could see beneath (Bower and Wishart, 1972). All this makes it seem that the *object concept* appears earlier as judged by vision than by reaching. But perhaps a more useful way to put the same idea is to return to Piaget's emphasis on the interrelation of idea and action. Visual discrepancies can focus a baby's attention on objects that behave strangely and help him learn, but only when his actions change the world, only when he can reach for or uncover objects, can he codify his concepts into reproducible results.

This leads us to a far more important general point. If we are trying to understand evolution, we are looking for robust structures that can be reliably applied to the environment. This is true for "I can uncover almost any hidden object" just as it is true for "I trust my hand's grasp to hold my weight on any given branch." An ability that appears in a highly restricted range of experimental conditions and disappears in all others is like an embryonic limb bud whose final function is not yet achieved. The Piagetian formulation outlines how bodily action and mental concept interact to produce flexible mental schema that apply to practical situations in the real world.

Do schema similar to humans appear in other primates?

Macaque Development: Locomotor and Manual Decalage

Wise, Wise, and Zimmerman (1974) directly addressed the comparative question of development, and turned up another major decalage. A human baby can reach months before he can crawl. A macaque reaches for objects at two weeks. It would normally not leave its mother for 10–14 days, but isolated laboratory infants crawl from birth with the reflexes that would pull them over their mothers' fur. Zimmerman and Torrey (1965) had already shown that a rhesus baby under two weeks old can learn brightness, color, and shape discriminations in two-choice maze (Fig 18.1.). Wise and colleagues gave two infant rhesus the Piagetian series of object problems to solve in three different manners: by crawling into a maze where the object disappeared out of sight round a corner or by pushing objects off a food well in the standard primate Wisconsin General Test Apparatus (Fig. 18.2.), or by the human test of picking up a soft cloth cover. In each case the babies were weeks ahead in solving the problems when they could crawl to the goal rather than picking up a cloth (Fig. 18.3A, B). Pushing the cover off a food well was solved at intermediate ages. Motivation may have played a role. The goal in the maze was the infants' cloth mother-surrogate, and mere food or toys in the other cases. Still, the infants could undoubtably find a hidden "mother" from two weeks of age (Fig. 18.4.). This is also the age when an infant stumptail macaque began the classic, "climb off mother and jump back on," the earliest game of baby primates from sifaka to chimpanzee (Parker, 1977). Perhaps we call this a game, unlike earlier rhythmic grasping or mouth opening, because it seems to bring a tiny portion of the world under control of the infant's own action, as it wobbles from safe to scary to safe again.

The earliest repetative actions of the human baby involve hand, eye, and

FIGURE 18.1. *A newborn rhesus can crawl to his goal and learn simple discriminations in the first week of life if tested in some form of maze. (Courtesy R. Zimmerman.)*

FIGURE 18.2. *An adult rhesus uncovers a food well in a Wisconsin General Test Apparatus, the traditional format for primate learning tests. (Courtesy Wisconsin Primate Laboratory.)*

FIGURE 18.3. *An infant stumptail macaque (A) tested for uncovering a hidden object and (B) playing with the cloth as his mother solves the problem.*

A

B

377   *Cognition*

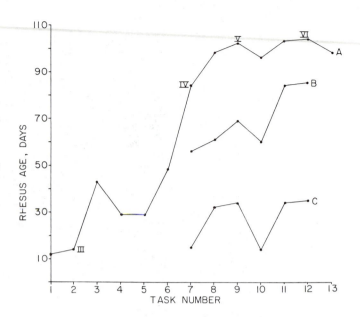

FIGURE 18.4. *The performance of two rhesus infants on Piagetian tasks presented in three testing modes. Curve C is finding the mother that disappeared round the corner of a maze; Curve B, displacing the cover of a food well; Curve A, the standard Piagetian test with a soft cover over hidden food or toys. The defining tasks for Stage IV, finding a hidden object, Stage V, finding an object in two successive hiding places, and Stage VI, invisible displacements, were solved with several weeks' decalage between different test conditions. (From data of Wise et al., 1974.)*

mouth coordination. In the stumptail they are locomotor play. This stumptail, like the rhesus, progressed through the normal stages of the "object series," although at a somewhat different rate. Also, from about four months on, he imitated the actions of an older juvenile, "sitting on his mother's head" (Parker, 1977). However, this macaque did not progress to Stage 5 of either imitation or sensorimotor series. It seems that macaques diverge before completing what for us are the six stages of infancy. Even in Stage 4, not all the human tasks are solved.

Chevalier-Skolnikoff (1982) has made a major contribution to primate Piagetian comparisons, in her analysis of the development of facial expressions. Table 18.3 shows her summary of expressions in stumptail macaques, a group of infant orangutans, and human babies.

There is a reservation about the form of the conclusions, in that she is taking the stage concept as the way of organizing her data, with the stages defined initially by object manipulation. Although she discusses fuzziness, overlap of stages, and incompletely present stages, and presents some quantitative data on the percentage of stage-specific facial expressions in infant orangutans, the anti-Piagetians will want rawer data, because they are not convinced that stages exist.

However, it is the *first* attempt to correlate facial expressions with cognitive development in a number of primate species. As such, it is an important mapping of new ground.

One point in the work is that there is decalage between modalities. Macaques may achieve complex behavior in locomotor activities before similar behavior shown manually or through facial expression, just as in the Wise experiment. Some aspects of a stage as it appears in humans may not appear in other primates, or may appear mainly in the locomotor and social modes. We have seen that rare individual macaques and baboons may use sticks to pull in food— Stage 5 object manipulation either by trial and error, or perhaps training through

TABLE 18.3. Facial Development in Primates
(With approximate ages of first appearance, in months)

Stage	Stumptailed macaque	Orangutan	Human
1. Reflex	0 Nursing: 　rooting 　sucking Crying face	0 Nursing: 　rooting 　sucking Crying-pout face	0 Nursing: 　rooting 　sucking Crying face REM smile
2. Self-oriented	0.5 Pucker-lips Self-mouthing	1 Pucker-lips Self-mouthing	1 Smile Self-mouthing
3. To single animal or environmental stimulus	0.75 Object-mouthing Emotional expressions: 　lipsmack (friendly) 　square-mouth (friendly) 　open-mouth eyelids-down (play) 　grimace (fear) 　open-mouthed stare (threat) 　teeth chatter (fear-affection)	3 Object-mouthing Emotional expressions: 　kiss 　smile 　open-mouth eyelids-down (play) 　open-mouthed bared-teeth (laughing face)	4 Object-mouthing Emotional expressions: 　smile 　laughing face
4. Establishes relations between animals	2 Emotional expressions: 　whispered chirl (subordinate threat) 　backing bared-teeth stare (subordinate threat) 　round-mouth stare (dominant threat)	7 Emotional expressions: 　grimace (fear) Novel behaviors: 　blows bubbles (once) 　purses out lower lip and looks down at it (once)	8 Emotional expressions: 　fear face 　angry face 　sad face 　surprise face Novel behaviors: 　spitting sound with lips
5. Coordinates relations between animals	4 Lipsmacks to enlist support	14 Novel behaviors: 　spits 　spitting sound with lips 　kissing sound with lips 　sticks out tongue 　sticks tongue in jar to get honey 　holds pen in mouth and draws 18+ Facial lying: 　pucker lips for subversive ends	12 Novel behaviors: 　makes bubbles 　sticks out tongue 18 Facial lying; teasing; offers toy and smiles, withdraws toy and laughs

(After Chevalier-Skolnikoff, in press.)

still simpler cognitive abilities. Perhaps the detour behavior that any macaque or baboon uses in traveling its home range is a far more complex combining of behavior, or even involves the "foresight" and "planning" of Stage 6.

In the social sphere, macaques make use of other animals to achieve their ends, for instance, by lipsmacking to solicit aid against a third party. We shall return to this in the final chapter on social learning. Attention to social cues is much commoner among mammals than attention to inedible objects. I suspect that humans are the only primate for whom "using" an object is in any way as easy as "using" a social companion.

Cats, Lemurs and Hummingbirds: Species' Cognitive Style

One devastating recent experiment on learning is Cole and associates' (1982) demonstration that hummingbirds learn more easily to visit a plastic flower where they have not just found sugar syrup than one where they have. This undercuts ordinary response-reward theory, which postulates repetition of a rewarded act. The evolutionary prediction was that hummingbirds in nature do not immediately revisit depleted flowers, so it should be relatively difficult for them to learn to do so in the laboratory.

Asking what animals are *adapted* to learn is still all too uncommon. It has long been known that monkeys are good at learning about junk objects they can touch, and rats are good at running mazes. This bedevils reports of success and failure on particular problems, including Piagetian problems. Laboratory cats failed to solve successive hiding problems in an experimental apparatus (Gruber et al., 1971). However, some pet cats and dogs could solve them and even solve successive invisible displacements. A piece of highly desired food was dropped from a cup under one of several hiding places, with the food removed by sleight of hand to avoid odor cues. The response was not manual lifting, but running to a hiding place and nosing off the cover. As well as being tested with appropriate response and reward, these pets presumably share much experience of retrieving balls from invisibility under the sofa (Troiana and Pasnak, 1981; Wilkerson and Rumlaugh, 1979).

Prosimians solve standard learning problems more slowly than most monkeys. They illustrate both the differences in species' approach to object problems and some Piagetian Stage 3 and 4 errors. Insectivorous lorisoids, like pottos and galagos, solved a series of simple object problems, including delayed response tests, more effectively than frugivorous lemurs. This was because an insectivore waits, watches, and then pounces, using a hand, but the fruit-eating species preferred to rush into a problem nose first. On the other hand, the lemurs succeeded more often on string-pulling tests, because the insectivores waited patiently without scaring the bait until it might crawl into reach. The lemurs succeeded by simply hooking in the nearest part of the setup with one hand, as they hook in fruit-bearing twigs in the wild (Jolly, 1964).

A far more detailed study of learning compares mouse lemurs, which are small, solitary, and insectivorous, with the forked lemur, which is also nocturnal but feeds primarily on gum and the secretions of insect colonies. Mouse lemurs respond relatively more to spatial cues, ignore irrelevant visual cues, retain learned discriminations for long periods, and have more difficulty with reversal and transfer of visual learning. This correlates with their activity at lower light levels (later at night, in dense understory), their home ranges scent-marked along known trails, and their opportunistic feeding on insects that happen to move in their field of vision. The forked lemur is a more visual animal entirely; it has a more definite *area centralis* in its retina (Pariente 1979); it moves high in the trees and finds new food sources by sight. Such an animal not only needs vision to find a new clump of colonial insects but needs the mental flexibility to learn where they are, as well as to forget old areas that have become unprofitable. Its social life also depends on vision as well as sound, which one might guess from the bold markings (Cooper, 1981).

Greater reliance on vision does not equal greater manipulative intelligence. Slender loris are said not to reach for an insect that crawls under a leaf, or to push open a cage door that stands ajar like a Stage 3 baby (Subramoniam, 1957). Tame lemurs reaching for a raisin may drop the raisin, yet still bring the hand to their mouths and sniff at the palm; so do baby rhesus. They may even reach for the raisin with one hand, stand on it, and then bring the *other* hand to their mouths. It seems incredible that an adult primate should not realize which hand has the raisin, but this is not a fair statement of the problem. What the lemur does know is that reaching-to-people-results-in-raisins. Lemurs, rewarded for reaching down into a bottle mouth, began by learning to reach downward, although not oriented to the bottle; those learning to lift a box lid repeated their upward gesture, but on box or lid. The result is very like the behavior of Lucienne and Laurent Piaget in their bassinets. The strategies are similar—to repeat an oriented action with little attention to physical connection of objects. This is not simple motor repetition. It has not only a perceptual component but also an emotional one—the lemur's own eagerness befuddles it (Jolly, 1964b).

New-world monkeys, like lemurs, have species differences in dealing with objects. Monogamous dusky titis manipulated novel objects more when tested with their mate than when alone; squirrel monkeys made more contact with novel objects than did titis; and female squirrel monkeys manipulated more when tested without males than with (Fragaszy and Mason, 1978). Squirrel monkeys are more active, more impulsive, and more distractable than titis, which affects their performance on both detour and object discrimination tasks. Titi monkeys were slower to move objects from foodwells, slower to make choices in an elevated maze, but more likely to see and use a new shortcut in the maze, and more persistent in solving difficult discrimination problems. All these differences relate to squirrel monkeys' high-energy, high-activity lives, in which they bounce into new problems as they bounce through a forest between giant fig trees, gleaning insects on the way. The more conservative titis keep under cover, hiding from aerial predators, and move without attracting notice (Wright, personal communication). They consider the situation visually, then map out an energy-saving route to a fruit tree, or slowly stalk an insect (Fragaszy, 1980, 1981). It would be interesting to compare titi and squirrel monkey's development of "object permanence."

The beginnings of object permanence have been compared in squirrel and cebus monkeys (Elias, 1977). Squirrel monkeys were consistently faster than cebus, both in locomotor development and response to objects. This may relate in part to slower growth because of cebus' large body size and in part to K-

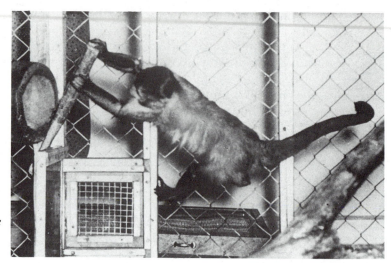

FIGURE 18.5. *The cebus monkey Pablo attempting to lever off his food box. Of the new-world monkeys, only Cebus have Stage 6 tool use. (Courtesy B. Rensch, from Rensch, 1973.)*

selection in this highly intelligent primate. Eight squirrel monkeys first uncovered a ring hidden under a cloth at an average of twelve and one half weeeks, eight cebus at 19 weeks. This contrasts with Vaughter and colleagues' (1972) tests in which a six-month squirrel monkey did not uncover objects and a nine-month one made Stage 4 errors. Like the adult lemurs, these animals were probably regressing to an earlier stage through excitement.

Among the new-world monkeys, *Cebus* is the chief tool user in captivity and the most manipulative in the wild (Fig. 18.5), easily solving the problems of Piagetian Stage 6 (Izawa, 1979; Chevalier-Skolnikoff, 1979; Mathieu, et al., 1976). Cebus' convergence with the great apes reinforces the idea that feeding strategy, rather than phylogeny, is the chief determinant of manipulative intelligence. On the whole, it is animals that have varied and opportunistic feeding in the wild who have most curiosity and most varied approaches to many problems offered in captivity (Parker, 1974, 1978).

The Great Apes: Symbolic Play and Others' Intentions

One of the most detailed comparisons of development in primate cognition (Fig. 18.6.) is Redshaw's (1978) analysis of four infant gorillas. She gave the gorillas the Usgiris and Hunt scales of object permanence, object relations, and causality. The gorillas followed the human sequence of success almost perfectly, or, rather, Redshaw's human controls followed the gorillas after several weeks' time lag. This is the clearest demonstration of the robustness of the Piagetian descriptions, as a hominoid (or perhaps primate) sequence, not just a human one. Redshaw's gorillas did not succeed in building a tower of two blocks, or in releasing a toy car down an inclined ramp. Object combinations are difficult for them (perhaps easier for chimps). Much more strikingly, they paid little attention to the person demonstrating a spectacle, but concentrated their attention on the toys themselves. Both chimp and child hand back toys to an experimenter; the gorillas at this age did not.

However, Redshaw's tests stopped when the gorillas reached two years of age. Chevalier-Skolnikoff (in press) continued, showing that although the apes

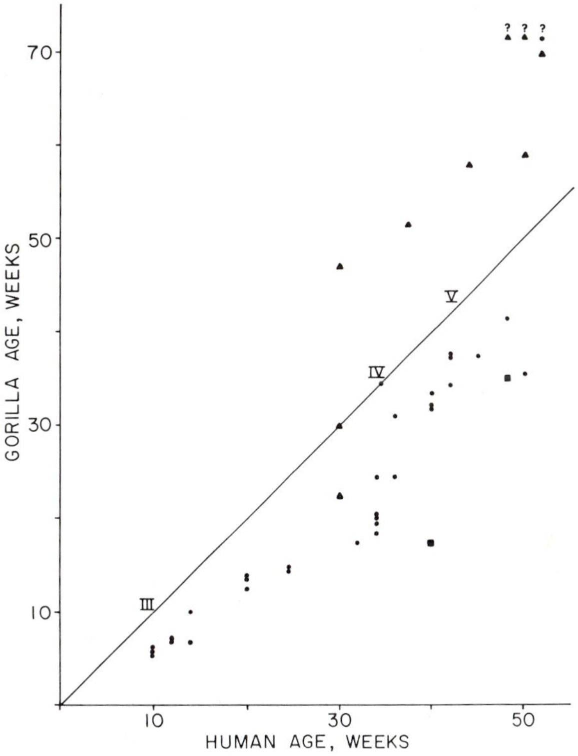

FIGURE 18.6. *Comparison of the average age that four gorillas solved object, spatial, and causal tasks with the age at which two human controls solved the tasks. The line shows the human norm, and the dots, the gorillas' performance. That is, three tasks that the humans solved at ten weeks, the gorillas solved at six weeks, and so forth. The gorillas matured consistently more quickly during their first year. Two tasks involving locomotor responses (square dots) were solved especially rapidly by gorillas, but tasks involving social responses (triangles) were generally solved more slowly by the gorillas. (After data in Redshaw, 1978.)*

speed through the early stages, they may not progress beyond Stage 6 until they are five to eight years old (Fig. 18.7). The gradual increase in orangutans' control of their environment is remarkably like humans if one substitutes "play pen" for "cage." (See Figs. 18.8. and 18.9A. and B.)

The ape's great social capacity appears in the neonates' highly mobile faces (Chevalier-Skolnikoff, 1982). Even in the first, reflex stage an ape or human infant, held up to face a human, will make a series of apparently random facial movements. A macaque newborn, held the same way, leaves its face perfectly blank. The facial mobility continues through the early months until it becomes the more intentional communication of Stage 4. For instance, a Stage 4 human-reared orangutan, after his "mother" tickled him, took her hand and placed it on his body again and laughed; an ape-reared orangutan looked at her father's face with a play face and grabbed at the big male's toes as he wiggled them.

In Stage 5, the apes, unlike the macaques, began to experiment with novel facial gestures—spitting, blowing bubbles, sticking out the tongue, and sometimes looking down at it. They also used "conventional" communication—the begging gesture of outstretched arm and upraised palm, which Chevalier-Skolnikoff believes analogous to learning sign language.

Mathieu and colleagues (1980) have tried to analyze a different question with two home-raised chimpanzees: the question of causality. To what extent did the chimps simply repeat their own actions (Stage 3), combine external event and their own action (Stage 4), combine objects to affect each other (Stage 5), or look for an invisible cause or tool (Stage 6)? The younger chimpanzee gave a majority of Stage 4 responses at age 22 months—pulling ahead on a harness

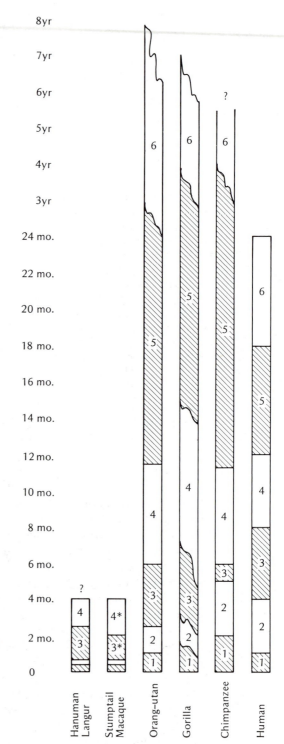

FIGURE 18.7. *Duration of Piagetian stages of object manipulation. The stumptail Stages 3 and 4 are only partially comparable to the ape and human stages. The apes reached Stage 4 more quickly than the humans, but then slowed, solving our 1½–2-year-old object problems only at ages 3–7 years. (After data in Chevalier-Skolnikoff, in press.)*

FIGURE 18.8. *One of the earliest adjustments to the world is attempting to fixate an object. (Courtesy P. Coffey and the Jersey Wildlife Preservation Trust.)*

or putting the experimenters' hand on a toy to restart any action. By the age of two and one half years, both chimps were capable of some Stage 6 responses: one looked for the hidden rubber bulb that made a squeeze-toy jump; they looked behind them to find an experimenter throwing paper airplanes; they looked for the missing ingredients to blow soap bubbles with jar and spatula; and, like Köhler's Sultan, they looked for a hidden rake with which to rake in food. All these feats involve memory or deduction of a missing cause, divorced from their own immediate action. This, according to Piaget, involves the kind of mental representation needed to label things with words.

Köhler first carried out similar experiments, just as he anticipated so much else. His Sultan went back to another room to fetch a ladder and a box to reach suspended fruit. Even Sultan, though, did not succeed at once. He only remembered the ladder or box when his attention was distracted from the bait—

FIGURE 18.9. *Repeating gestures that create a pleasant effect is classified as Stage 3. (Photos by Jeff Berner, and Stephen Longstreth, S. Chevalier-Skolnikoff 1982.)*

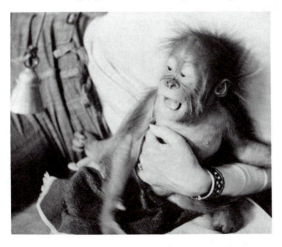 

once by an intervening door whose bolt he was trying to chew off, and once by a fracas among the other chimpanzees whom he was "trying to use as footstools."

It seems clear that chimpanzees achieve Stage 6 causality—looking for a hidden cause or tool as well as understanding mirrors as a representation of the world. They achieve Stage 6 social communication and social deceit. Do they have the final criterion of stage 6, symbolic representation, outside the special case of language training?

A few home-raised chimpanzees clearly do. Washoe, the Gardners' chimpanzee, had been regularly bathed. Sometime between the ages of one and a half and two years, she picked up her doll, filled the bathtub with water, dumped the doll in the tub, and then took it out and dried it with a towel. Sometimes in repetition she has even soaped her doll. This is the same age at which human girls begin such elaborate play. The most sustained pretense of this kind was the "imaginary pull toy" (Hayes, 1952). Viki, a home-raised chimp who could pronounce four "words," was just at the toddler stage when everything possible to drag along becomes a pulltoy. While playing in the bathroom she began to trail one arm behind her as though dragging a toy on a string that did not exist. She repeated the game over and over. One day, the "rope" apparently caught on the plumbing pipes; she fumbled, pulled, gave a jerk, and was off again. Still later, she sat on the potty "fishing" the object from the floor hand over hand. Then after several weeks, wrote Cathy Hayes, Viki's foster mother. "It was one of those days when Viki loves me to distraction. She had pattered along in my shadow [and] at every little crisis she called for 'Mama' . . . I was combing my hair before the bathroom mirror while Viki dragged the unseen pull toy around the toilet. I was scarcely noticing what had become commonplace, until she stopped once more at the knob and struggled with the invisible tangled rope. But this time she gave up after exerting very little effort. She sat down abruptly with her hands extended as if holding a taut cord. She looked up at my face in the mirror and then she called loudly, "Mama, Mama!"

Cathy Hayes, still half unbelieving, went through an elaborate pantomine of untangling the rope and handed it back to the little chimp. "Then I saw (her) expression—a look of sheer devotion—her whole face reflected the wonder in children's faces when they are astonished at a grownup's escape into make-believe. But perhaps Viki's look was just a good hard stare."

A few days later Cathy Hayes decided to improve on the game. She invented a pull toy of her own, which went clackety-clackety on the floor and squush-squush on the carpets. Viki stared at the point on the floor where the imaginary rope would have met the imaginary toy, uttered a terrified "oo-oo-oo," leaped into Cathy's arms, and never played that game again.

CHILDHOOD AND THE CONCEPT OF CLASSES

An adult's mental world consists of far more than individual objects, haphazardly juxtaposed. We are accustomed to thinking in classes, series, and invariances. It seems obvious to us that if apes are primates that lack tails, there must be more primates than there are tailed primates. If Morris is taller than Susan who is taller than Margaretta who is taller than Dickon, we know the relative heights of any pair of them. If I tip a whole coffeepot over my cup, there is just as much coffee in the cup and on the saucer, table, and floor as there was in the original pot. These seems to us necessary truths, barring quibbles about the coffee evaporating or Dickon jumping.

To the young child, these truths are far from obvious: he may deny them, even when demonstrating with the objects themselves. Piaget calls the mental processes *operations*. Not until early adolescence does a child think fully operationally in the sense that he can not only answer such questions but also explain his answers. From the ages of 2 to 11 or so, the child is gradually progressing from happy ignorance even of plurality to the ability to cope with this sort of problem concretely by internalized action sequences on objects present for demonstration, and to the formalized, mental operations explained in words.

As examples of such abstract thought we shall treat oddity and learning set problems, which have been extensively studied in primates, and then the Piagetian problems of classification, seriation, and conservation.

Harlow (1949) first codified primate learning set, working with rhesus monkeys in standard test conditions, in the Wisconsin General Test Apparatus. Harlow presented rhesus with two different objects for discrimination learning, one of which covered a reward, for six trials. He then started again with two new objects, one of which was always rewarded. Gradually, the rhesus monkey became more and more proficient. When first given two new objects, he would pick up one. If that one had the reward underneath, he would choose the same one again for the next five trials. If he had guessed wrong the first time, he then knew that the other object would be rewarded throughout the series.

However, the apparent supremacy of primates in learning set problems turns out to be largely the result of primate-oriented experimental design. Cats, rats, and raccoons achieve learning sets as fast as monkeys. The final blow comes from Menzel's saddlebacked tamarins. He gave them a series of household objects, some smeared with honey, some not, in a special mesh box in their home greenhouse. On a second trial, days later, the tamarins approached the objects that had had honey before, and ignored those that had no reward before. The "win-stay, lose-shift" strategy, painfully shaped over so many trials in an experimental design, is "natural" and "immediate" to primates in an approximation of the normal foraging situation with the reward on the stimulus itself, and with the group reinforcing or discouraging each other's approaches. Wild tamarins feed in small fruit trees, scattered fairly evenly through the range, whose fruits ripen a few at a time over several weeks. If they can keep track of what is currently in fruit, and if they return to a tree that had fruit a few days before, the more efficient their foraging. Squirrel monkeys, which gorge in a large fig tree until it is exhausted and then trek off, following the screech of circling parrotts to find another, should learn and forget in quite different rhythm (Menzel and Juno, 1982, Terborgh, in press).

What about unlearning? Given comparable designs, monkeys often excel in shifting to something new. Thus, they can switch from repeated reversal problems to classic learning set, or transfer from learning set to probability learning when the pattern of reward changes (Warren, 1974).

One suggestion of a qualitative difference in kinds of reversal learning is Rumbaugh's distinction between "association" and "mediational learning." He

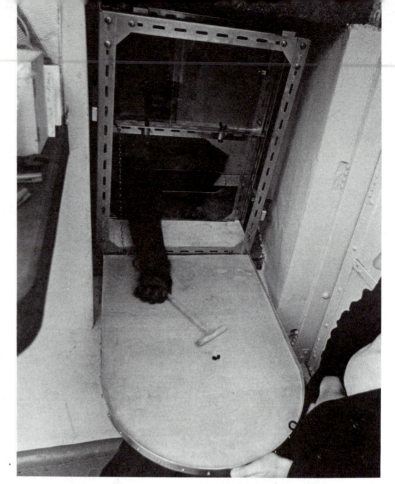

FIGURE 18.10. *An 18-month-old gorilla failing a rake test. Combining tool and reward by trial and error is a Stage 5 ability in human babies; around 18 months to two years, the Stage 6 human succeeds with immediate insight into the objects' relations. (Courtesy P. Coffey and Jersey Wildlife Preservation Trust.)*

trained a series of different primate species and then reversed the rewarded stimuli. The reversals were either total (instead of A^+, B^-, the rewards changed to A^-, B^+) or partial (A became minus, but was paired with a new stimulus, C^+, or else B became positive paired with a new negative C). Lemurs and young rhesus monkeys had very different success in the three new conditions, which Rumbaugh attributes to emotional valences associated with the stimuli. I would paraphrase this as "Nice A has gone horrid and empty but at least there is an interesting new C to approach"; versus "A has gone horrid and B always was horrid and I hate school." The adult rhesus and most apes succeeded about as well as on all new conditions. I would phrase this as "Which way round is it this time?" or even "Dratted psychologists keep playing with the stimuli." (Rumbaugh, in press).

We saw a similar difference earlier between mouse lemur and forked lemur, where it correlated with feeding differences. The forked lemurs that fed on insect colonies and nectar forgot old responses and learned new ones relatively quickly, because they need to revisit a series of feeding sites that change from week to week. The mouse lemurs that collect much of their food by encountering insects at random have less either to learn or to unlearn. This again recalls Piaget's insistence on the importance of flexibility and reversability in consolidating either mental or motor achievements to deal with the real world. (*See* Fig. 18.10.)

Oddity Problems

In oddity problems (Lashley's original example of second-order generalization), the primate is offered three or five stimuli and has to choose the one that does not resemble the others. Rhesus and chimpanzees can learn Weigl-type oddity problems, in which if the stimulus tray is one color they should choose the odd stimulus, or if it is another color they should choose one of the two matching stimuli that is on the end of the row of three stimuli.

Davis and colleagues (1967) have tested lemurs and seven species of monkeys on identical oddity problems, one of the few comparative studies done in a single laboratory. This measure of cognitive development reinforces the point that old-world and new-world monkeys have a parallel evolution toward complex and intelligent species and individuals in each phylogenetic line. However, Thomas and Boyd (1973) found no difference between squirrel monkey and cebus in speed of oddity learning.

Matching to Sample

In the simplest form of this test the primate is shown a sample, say, a triangle, and required to choose for reward between another triangle and a circle. This is a rudimentary step of classification: the triangle class as against the nontriangle, with the flexibility to choose the circle class if the sample is a circle. Weinstein (1945) took the process further. He taught two rhesus, Corry and Zo, to choose red objects when shown a red triangle and blue objects when shown a blue ellipse. As the number and variety of choice objects increase, Zo became disturbed and dropped out, but Corry eventually could choose all the red objects from a collection and none of the blue ones if he was shown an *unpainted* triangle, whereas, if shown an unpainted ellipse, he chose all the blue. Lehr, 1967, has similarly shown that primates can choose insect and flower classes.

Again these abilities are not unique to primates. Pigeons can pick out pictures that contain trees, or even a particular person in different poses and setting (Herrnstein, 1979). As with learning set problems, primate skills may be still more complicated: stumptails can match to a particular sample shown one or three days earlier. Primates may also be better than other mammals at second-order class concepts. Squirrel monkeys learned to choose green objects (or nongreen objects) in 600 learning trials. They also managed to learn to choose the odd stimulus versus the nonodd stimulus, but it took 1,600 trials. (Thomas and Crosby, 1977).

It is thus clear that primates can sort objects in consistent classes. They can do so to a symbolic cue, as detached perceptually as the verbal instruction, "Pick out the blue ones." This is the last step before true classification in which the child is told, "Pick out the ones that are alike," and the psychologist waits to see if the child chooses a criterion like "the blue ones."

Classification

But children do not so choose—not until the age of seven or so. Nor did Goldstein and Scheerer's (1941) schizophrenics. In one test, Golstein and Scheerer gave out skeins of wool to be sorted into "those that are alike." A conventional adult casually groups the reddish hues together; schizophrenics might start so, then add a bright yellow because it looks nice with red. Piaget gave small children geometric shapes; they made "edge-matched" collections by linking one shape to the next, not sorting to a single criterion. Olver and Hornsby (1966) gave children pictures of objects to sort. Children explained their groupings: "some are red, some are gold, and some are yellow." "Candle and clock are on a table and this lamp is round and this clock is." There is a logic, not the logic of classes, but one that relates anything in turn to anything else.

Viki, the Hayes' chimpanzee, spontaneously sorted objects into exclusive classes: forks against spoons, or buttons against screws. She could change her basis of sorting, dividing the same set of objects by form, size, color, or material. She sorted photographs of animals versus people, classing her own chimpanzee father in his cage as an animal, and herself (unclad, but in the Hayes' living room) unhesitatingly as a person (Hayes and Nissen, 1971).

Analogy

Problems which conserve functional, rather than perceptual, identity are also standard in children's testing. A 16-year-old language-trained chimpanzee, Sarah, solved a series of such problems: padlock and key are the same as closed can and (choose one) can opener or paintbrush? Some of her successful choices were remarkably abstruse: Cut paper is to scissors as cut, peeled apple is to knife or peeler? Then painted, sanded wood is to sandpaper as cut peeled apple is to knife or peeler? Sarah solved, as well, a series in which she chose words for same or different: wet wood to water sprayer is the same as wet, cut paper to bowl of water, but different from wet cut paper to scissors. Even harder, she had to complete abstract shape analogies that differed in one of four aspects: shape, size, color, and marking (Gillan, et al., 1981).

Seriation

Young children also have difficulty in placing objects in a series. They may take a row of sticks of various lengths and pair off one as longer and another as shorter, but to put a whole lot in a row with any consistent ordering mechanism is far more difficult. Even worse, when asked to insert another stick into their final arrangement, they generally begin by taking the whole thing apart and laboriously rematching pars of sticks together. The child must be able to cope with the transitive relation "If A is longer than B, and B is longer than C, then A is longer than C" to be able to perform such problems efficiently. The same sort of relation underlies the use of measuring sticks to determine which of two separated objects is taller. Object A must be related to measuring stick B and the stick to object C, the child retaining some notion of the transitive relation among the three things if he is to compare separated objects by measurement. Braine (1968) did nonverbal experiments on measuring and seriation. He concluded that these operations appear around five years of age, a good two years earlier than Piaget found children using verbal methods.

Some of the children's diffculties are attributable to memory, not reasoning per se: if extensively trained on each pair in a series, four-year-olds can make transitive inference (Bryant, 1981). Again, Piaget's sequence of solving problems seems valid, but subject to huge decalages when measured in different modes of response.

Primates can, of course, distinguish larger from smaller pieces of food. They can also be trained to choose signs like different-colored rings for different sizes of reward (Menzel, 1969b). Kapune (1966) taught a rhesus monkey to select from six different levels of reward, each represented by a different colored ring. This monkey, like Corry, could choose a category of objects or a series of objects, but perhaps neither could have arranged objects by series or category.

Three five- to six-year-old chimpanzees made transitive inferences in an experiment very like Bryant and Trabasso's. They had a series of five stimuli, such that B was positive when with A, negative when with C, and C positive when with B but negative with D, and so on. The crucial test is comparing D to B, because each had been rewarded in some combinations and wrong in others. Two of the three chimpanzees successfully chose D over B on test trials. Making the series circular, such that F was negative and A positive, disrupts learning just as it does with humans (Gillan, 1981b). Adult squirrel monkeys, as well, reproduce the human pattern of response (McGonigle and Chalmers, 1977). The squirrel monkeys, however, when they have all three stimuli present, give about a third of their responses to the middle one, which is consistent with the idea that they are still comparing pairs of stimuli rather than really integrating them into a transitive series. We now need this experiment repeated with children of different ages to see if the squirrel monkeys are really using a different approach. The capacities of squirrel monkeys when appropriately trained are impressive. Some individuals can even choose the largest stimulus on a white background, the smallest on black, and an intermediate size on gray (Thomas and Ingram, 1979).

Conservation

Piaget's most frequently quoted experiment is conservation of quantity. The experimenter or the child fills two identical glasses to the same level. The child is then asked if there is the same amount of water. He says (one hopes) yes. The experimenter then pours the water from one glass into another that is markedly taller and thinner or wider and squatter. The young child, under seven years old, says there is now a different quantity of water.

There are many forms of conservation experiment. Rows of buttons can be spread out longer or shorter and clay snakes rolled thinner or pummeled into pancakes. Culture produces decalages: Western children typically solve number conservation before volume; the sons of Mexican potters know all about clay snakes but are baffled by pouring liquid (Price-Williams et al., 1968). We may also be ignoring social influences. McGonigle and Donaldson found that four- to five-year-olds solemnly say there are more counters when an experimenter solemnly spaces a row out longer. The same children were not baffled at all if a playfully naughty teddy bear interrupted the session, and amidst other chaos kicked the counters into different spacings (Bryant, 1981). Again, nonverbal methods give different results: full verbal conservation starts at about seven years old, but toddlers of 18 months can adjust their grip to compensate for the expected weight of an object as if they are conserving mass (Bower, 1974).

Sarah, the language-trained chimpanzee, can certainly conserve liquid and solid quantity. She chose the word *same* reliably for transformations that changed dimension, and *different* for transformations that added or subtracted mass. She, like us, was confused if the transformation took place out of sight; she does not have super primate powers to calculate volume from perceived dimension. Rhesus monkeys, like Sarah, can distinguish processes that add or subtract materal from simple shape changes (Pasnak, 1979).

Sarah's performance is more impressive than seven-year-old children's, because she chose to respond to volume on her own initiative. We ask a child, "Is there more water now, or less, or the same?" To focus its attention on volume, not just height or width, Sarah apparently guessed what entity Woodruff (or Piaget) wanted her to describe (Woodruff et al., 1978).

Although Sarah's performance is dramatic, it is not unique. Even squirrel monkeys can indicate "sameness" of volume length when two sticks are equal or one displaced right or left, or "difference" when another is substituted of different length, even though the substitution involves sleight-of-hand (Thomas and Peay, 1976). They can also abstract volume cues from height and width though they achieve this through prolonged training (Czerny and Thomas, 1975).

Counting

Counting is a complicated process, which develops in children from reciting inaccurate strings of numbers through complex numerical operations. Some conception of number is widespread, for instance, among birds; Rensch showed that crows and ravens could match numerical samples.

Sarah, as well as dealing with conservation and analogy, can match number samples up to four and fractional samples of ¼, ½, and ¾ when the samples are as different perceptually as a partly-filled glass and a circle with a fraction cut out.

Sarah had difficulty with conservation problems involving four to seven buttons, so number concepts may be relatively harder for her than for children. Squirrel monkeys, though, succeeded on the relatively simpler number task of judging more, less, or the same on sequences up to seven versus eight (Thomas et al., 1980). As with volume, one individual could follow conditional instructions. If he was shown one light, he chose one of three cards with least spots (given a possible two to seven spots per card); if three lights, he chose the card with most spots; if two, the intermediate number. It is difficult to decide which of these numerical tasks are seriation, analogy, or conservation. Because (for us) number is a way of categorizing quantities, we may be dealing with an even more abstruse problem than size, or volume conservation.

Abstract Codes

Sarah's name keeps appearing as the star of many of these experiments. Two other language-trained chimpanzees in Premack's laboratory have also solved some of such abstract problems. Three other juvenile chimpanzees in the same laboratory seem to solve "imaginal" problems as well as the languaged-trained chimpanzees, but to fail on the abstract ones (Premack, 1983).

All the chimpanzees can match to sample. All could use natural reasoning to deduce that if a banana and an apple were put in two containers in a field, then the chimp was removed, and a trainer walked out of the field eating an apple, best to go look for the banana. (Children do not solve this one reliably at three and a half or four years.) Similarly, if a trainer walks toward one of two containers carrying fruit and comes back empty-handed, the chimp can infer where to go look for food. However, only the language-trained chimps solved matching like or unlike *pairs* of objects to a like or unlike paired sample, the problem mentioned at the start of the chapter. The language-trained chimps could "fill in the blanks" in such three-object "sentences" as apple-blank-cut apple, or apple-knife-blank. (This is Piagetian Stage 6 causality, at the further remove of a test sentence laid out before the animal, not a case where the animal is looking for the missing implement in a process it wants to do itself.) Sarah could even do problems distinguishing a fixed left-to-right order of cause and effect. Thus she appropriately chose between pencil and eraser, for paper-blank-marked paper, or marked paper-blank-clean paper. This led into the series of analogies already described.

Such an ability to deal with relations rather than objects, and simultaneous symbols for temporal sequences, seems to come only with particular symbolic training. Premack concludes that apes have this latent cognitive capacity, but only human training brings it into play. He defends himself against the doubt that evolution would produce "unused" capacity by suggesting that extraterrestrial beings who trained humans might well reveal capacities we do not yet suspect we possess. Closer to home, it seems that formal western-style schooling produces a formalization of thought not shared by illiterate adults in the same societies. (I do not mean to equate formal logical style with wisdom; the elders may well be wiser than their abstract-minded offspring.) However, it would be somewhat strange, after all the effort you the reader and I have put into our education, if it has not affected our forms of thought.

Premack's formulation is far from being accepted by everyone. Philosophers worry about the meaning of *abstract*, psychologists worry about underlying mechanisms and, in particular, whether the diverse tasks he cites are actually examples of a single ability. In the context of a textbook, the important point seems to me that yet again a psychologist is attempting to juxtapose primate behavior and a philosophical view of the nature of abstract thought—and finding that at least one ape will play his games.

Caveats: Ape Prodigies and Normal Humans

In all these experiments, both this text and the experimenters tend to dwell on the animals that succeed. The most tantalizing accounts are anecdotes, usually of home-raised apes. How much of this is not species-typical behavior but the feats of gifted infants uncommonly driven by their elders (the young Mozart, the young Brooke Shields)?

Mathieu (in press) reports that their three, much cosseted and much studied chimps have never made spontaneous classifications of objects or seriation of objects. They can, however, be taught. One chimp, Sophie, can pick small and larger sticks in order like three-year-old children. She can also pick a row with more pieces of food, up to six pieces, whether or not it is longer or denser than the row with less pieces. The other two chimpanzees just rely on visual length or density. It takes both extensive training and immediate reward to make the chimps pay attention to object poperties—the behavior so normal in humans.

Finally, we are still far from being *sure* that the thought processes are the same, even when we solve similar tasks. There are doubts about tightly designed experiments (Thomas, in press) and even more about spectacular anecdotes or sweeping assignments of an ape behavior to a particular cognitive stage. Although my bias is to give apes, when possible, the benefit of the doubt, it is still important to proceed with caution. This is especially crucial if we admit the possibility of primate awareness.

Self-concept

It is difficult or perhaps impossible to draw a boundary line for consciousness. Is an animal aware of sensations? Of images of external objects? Of itself as a separate entity? Or perhaps of multiple-step thoughts of the form, "I want you to believe that I trust you to love me . . . ?" (Griffin, 1981; Dennett, 1983). We may not be able to draw a boundary, but we can list some primate behaviors which we would accept in other humans as showing a self-concept, even though we would not be so charitable if they were shown by a computer.

Köhler, as usual, anticipated the current series of experiments on self-recognition. Chimpanzees, after a few days with a mirror, begin to grimace and pose at the mirror, treating it like their own reflection, not another animal. (See Figs. 18.11. and 18.12.) Gallup (1979, 1980) proved this experimentally by anaesthetizing chimpanzees and daubing their faces with paint spots. On recovery, the chimpanzees groomed the spots when they looked in the mirror. Orangutans react the same way, but macaques, wild vervets, and perhaps gorillas always treat the reflection as another animal, though not a very interesting one (Suarez and Gallup, 1981; Gallup et al., 1980). It should be added that pigeons can be conditioned to peck at blue dots on themselves in the presence of a mirror. The pigeons were first trained to peck at dots on themselves. Then they learned to peck at dots on one wall that they saw reflected on the other wall. Then the dots were flashed off as they turned away from the mirror, and they were rewarded for pecking the place where the dot had been. At last the birds were given bibs, so they could not see a spot on their own chests, and the mirror was uncovered. They at once pecked at the spot on the bib that corresponded to the dot's position (Epstein et al., 1981). This experiment raises doubts whether such mirror experiments are a sole criterion of self-awarness, and the probable thought that monkeys could be similarly trained. The pigeons do not quite compare with Köhler's chimps spontaneously making faces into the mirror, Belle Benchley's orang trying on hats while peering into its drinking pool, or the Köhler group using pieces of tin and glass to observe other objects while glancing back and forth intently from the real thing to its image. Koko, the language-trained gorilla, casts some doubts on the Gallup gorilla experiments by looking in a mirror and signing to herself, "eye, teeth, lip, pimple." (It is not quite clear whether she meant herself, though.) The wild moun-

FIGURE 18.11. *Koko, having made up her face with chalk, looks in a pocket mirror and signs* Lip. *(Courtesy Ron Cohn and the Gorilla Foundation.)*

FIGURE 18.12. *Gombe Stream chimpanzees grimace and reassure each other when confronted by a mirror. (Courtesy P. Marler.)*

tain gorillas of the Virungas imitate intrepid photographers who start uphill armed with telephoto lenses. The gorillas put their eyes to the lens, apparently peering at the reflections therein.

One aspect of awareness less abstruse than self-awareness is intention. Intentional communication or initiation of actions is a criterion of Stage 4 for Piaget. Most animal psychologists have been skirting the subject, as one whose existence may be assumed but not mentioned in polite society. Let me quote in full an episode that Chevalier-Skolnikoff believes shows Mom's intention to convey her facial expression, and Girl's intention to avoid receiving it (this description is based on motion picture film and simultaneous running narrative notes):

Stanford, April 16, 1968, 2 P.M.

Mom, the most subordinate adult female in the stumptail group, is sitting next to the dominant male, subadult female Girl, and subadult male Boy. Mom looks about (nervously) as the dominant male forcefully restrains and grooms her five-week-old infant. She looks directly at Girl, the most subordinate animal in the vicinity, but Girl is looking the other way and does not receive the look. Mom shoves her nose directly into Girl's face, but Girl ignores her and continues to look the other way with a neutral expression on her face. Mom retrieves her infant and again looks directly at Girl, grasps her chin, and turns her face toward her. As Mom starts to deliver an open-mouthed stare threat toward Girl (redirected aggression directed toward the most subordinate animal present), Girl, with head correctly oriented toward Mom, again averts her gaze, avoids receiving the threat, and maintains her neutral facial expression. Mom nips Girl on the shoulder and Girl draws away. Boy (who often supports Girl in dominance interactions) has been sitting behind Mom and now nips Mom on the head. Girl (now supported by Boy) turns toward Mom with an open-mouthed stare, and both animals rush at each other with attack faces. Just as the two animals are about to make contact, Mom leaps back and off the shelf with a shriek.

This sequence fulfills most of the criteria that Chevalier-Skolnikoff uses to recognize intentional behavior. First, the animal appears to study the situation before acting. Second, it initiates the behavior. Third, it is looking at the animal, person, or object toward whom the behavior is directed. Fourth, the behavior is part of a sequence or combination of behaviors, often toward an identifiable goal. Fifth, the behavior is repeated in several different bouts, often with modifications, but, sixth, it stops when the goal is achieved.

All these are criteria we use to decide when another human's or human baby's behavior is voluntary or if "he really didn't mean it." They would also, however, apply perfectly well to a robin building a nest, or a group of bees in the special circumstances of choosing the site for a new hive (Griffin, 1984). Within the limited scope of a book on primate behavior, perhaps we can beg the wider philosophical question of whether it is ever possible to define voluntary or intentional behavior. It is already a long step forward to say that there are such homolgous actions in macaques and humans—and in the species we know best we call such behavior "intentional."

Here again pygmy chimpanzees seem peculiarly like us. The one studied infant, Kanzi, points in the direction he wants to be carried, vocalizes while pointing, leads people by the hand, and "asks" by eye contact for help in moving. The impression of "voluntary" communication began over a two-week period when Kanzi was ten months old; the typical human age for the same transition is nine months. Human mothers train their children by responding as if their earlier gestures were communicative. The child strains to reach a toy; the human mother puts it in the baby's hand. Matata, the pygmy chimp who has reared Kanzi since birth, did the same. However, she has responded more in the locomotor than the manual mode, for instance, by helping him climb when he reached upward. As with human children, Kanzi's gestures after this transition became conventional—a reach for toy or high shelf while looking to the mother (Savage-Rumbaugh, in press).

Awareness of Others

All the language-trained apes use signs for their own names, and a sign for *me*. They learn such signs early and use them frequently, because most of their conversation seems to be self-centered requests. It is an even further step to ask how much they can imagine others' emotions. Of course, any primate must *react* to others' emotions.

Hamadryas baboons seem sometimes to calculate what each other knows. Bonded females were caged out of sight of their harem leader, behind a concrete wall, but still in earshot. Then another male was introduced. When the female began to groom the "wrong" male, he would suppress the loud call males usually give on being groomed. This was acoustic hiding. Visual hiding is far more frequent. Females of many species mate with subadults behind a rock or a bush. Kummer reports one very deliberate instance, when a female hamadryas spent 20 minutes edging herself behind a rock where her harem leader could see only the top of her head and back, while her hidden arms busily groomed a subadult male. The leader could check her presence, but not what she was doing (Kummer, 1982). The "use" of social companions and assessment of their reactions when reaching a further goal are much commoner in accounts of chimpanzee behavior than the multi-step use of objects. Two accounts from Mathieu and Bergeron (in preparation): Sophie, aged two and a half, who wanted to go out in a forbidden field but dare not without company, grabbed her companion Spock from behind. He screamed and hugged her, and she then led him into the field. Sophie at three and a half, while the caretaker was cleaning the floor and left the door open, played with the adult who was supposed to be watching her, and led him into a four-foot high play tunnel, where he had to move at a crouch. Sophie then raced out of the tunnel, out the door, and straight to the fridge.

De Waal (1982) gives an even more cogent example. "Luit, the dominant male, was being challenged by Nikkie, another adult male. After Luit and Nikkie had displayed in each other's presence for over ten minutes . . . Nikkie was driven into a tree, but a little later he began to hoot at the leader again. . . . Luit was sitting at the bottom of the tree with his back to his challenger. When he heard the renewed sounds of provocation he bared his teeth but immediately put his hand to his mouth and pressed his lips together. . . . I saw the nervous grin appear on his face again and once more he used his fingers to press his lips together. The third time Luit finally succeeded in wiping the grin off his face; only then did he turn round. A little later he displayed at Nikkie as if nothing had happened, and with Mama's help chased him back into the tree. Nikkie watched his opponents walk away. All of a sudden he turned his back and, when the others could not see him, a grin appeared on his face and he began to yelp very softly. I could hear Nikkie because I was not very far away, but the sound was so suppressed that Luit probably did not notice that his opponent was also having trouble concealing his emotions."

De Waal speculates that the shifting coalitions of the three males in this colony must have involved prediction of each other's actions in the short term. The oldest male, Yeroen, seemed to manipulate the other two and the females in ways we would ascribe to foresight in a human, changing sides whenever necessary to preserve the balance of power. Over the longer term, he points out that the jockeying for dominance status may be no more consciously calculated than the first adolescent squabbles with one's parents. One usually ascribes them to some momentarily irritating behavior on the other person's part, rather than to a forthcoming shift of status within the family.

Woodruff (1981) and Woodruff and Premack (1979) have done coherent experiments that test the chimps' estimate of others' capacities. They started from Menzel's observation that chimps can lead each other to food, when only one animal knows where the booty is hidden (Chapter 21). In these experiments the chimpanzee was shown food in a locked box, while an ignorant human caretaker wore the key on a chain round his neck. On half the trials the caretaker wore a cloth round his head as a headband; on the other half he wore it as a blindfold. When he was wearing the headband, the four chimps attracted his attention, then visually led him. When he was blindfolded, they tried touching him, then set off as usual, repeating the process 20 or 30 times when he did not follow, then sat down and gave up. After several trials, though, the chimpanzees solved their problem: three of four chimps pulled him along by the arm or the keychain. The fourth chimp, on her second and succeeding try, pulled down the blindfold enough for him to see out.

Premack has attempted to test his chimpanzee Sarah's prediction of others' actions. He showed Sarah videotapes of her favorite trainer attempting to solve familiar chimpanzee tasks. Sarah then chose the appropriate solution from a group of still photos—the plug to start the record player or her trainer removing concrete blocks from a box too heavy to shift. When Premack ran through the same series using a trainer Sarah disliked, she chose pictures of disasters, such as the trainer prone and strewn with concrete blocks.

In a succeeding experiment, Sarah and three younger chimpanzees were given two containers, one baited with food. They have a kind human companion or a liar. The kind one indicates which container holds food and the liar indicates the wrong one. Or, in the converse equipment, the chimpanzee knows which container is baited, but when the liar is shown where food is, he eats it himself. One chimpanzee misdirects the liar, one chooses the container he does not indicate, and one does both. The eldest, Sarah, does both, and hurled every available missile at the liar into the bargain. In her experience, humans always know the answer. As Premack points out, it may be much harder for Sarah to conceive of a human fool who merely guesses wrongly than a human who actually wants to deceive her (Premack and Woodruff, 1978).

At least one pygmy chimpanzee apparently makes allowances for intention. Common chimpanzee mothers react almost automatically if their babies scream while with another caretaker. They attack and bite the caretaker, or, if very subordinate, seize their infant and run away. Savage-Rumbaugh (in press) cites two extraordinary occasions. Matata is a pygmy chimp captured as an adult. She has nonetheless accepted human "friends," often draping an arm or leg over her teachers. She is not either imprinted as a child or fearful as an adult of the humans. Matata was attempting a sorting task, placing objects into bowls. Each time she was correct, the teacher cut her a slice of food. Meanwhile, Kanzi, her infant, was being rowdy and obnoxious, jumping about and even into the bowls. He was too young to understand, and both his mother and the teacher attempted to ignore him. Suddenly he jumped toward the cutting board just as the teacher was slicing downward. She could not check in time to stop the knife thumping him hard on the shoulder, although he was not cut. Kanzi screamed and tried to bite. The teacher looked at Matata in dismay, expecting attack. Matata who had watched the whole incident simply pulled Kanzi to herself, tried to quiet him, and then hugged the teacher.

On another occasion, "a teacher (who had her long hair tightly tied in a bun on her head) was holding Kanzi over her head and tickling him . . Kanzi began to playfully slap the teacher and to grab and tug laughingly at her hair. Suddenly his hand became entangled . . . He panicked and screamed and began to bite hard and repeatedly. The experimenter who did not realize that Kanzi's hand was caught (she thought Kanzi was grabbing her hair) began to scream and tried to shove him away . . . Matata rushed to Kanzi's aid, also began screaming at the experimenter, and was about to leap on her and bite her severely. At this point a second experimenter who . . . realized that Kanzi's hand was caught gesturally pointed this out to Matata and verbally told the experimenter who was being bitten. Immediately Matata restrained herself from attacking . . ." Eventually the second experimenter freed Kanzi's hand, and everyone hugged everyone except Kanzi, who still tried to scratch and threaten from the safety of his mother's ventrum. Reading the account, one objects that perhaps the second experimenter was so dominant she just frightened off Matata, but both humans believed the chimp was ready to attack when Kanzi seemed threatened. It seems as likely that Matata actually understood.

Young children grow from apparently considering the world an extension of themselves to awareness of their own identity and desires and to increasing ability to consider others' opinions. One such step comes toward the end of the second year, when children begin to talk of I-me-mine, when they smile to themselves in silent mastery play, and when they are distressed at being offered a model they know they cannot copy (Kagan, 1978). Over the following years of childhood, they gradually begin to imagine others' points of view (Piaget, 1951). Although, as usual, we cannot assign the primate's social performances to particular human ages, we can say that they are somewhere in the range of understanding we achieve in later childhood years. It is incredible that some scientists still discipline themselves to regard such creatures as unaware. It seems as though apes with some aspects of human mental ages of seven or ten are forced to deal with experimenters who have regressed to the self-centered age of two.

SUMMARY

Cognitive tasks may be ranked from simpler to more abstract problems. Children progress from simple learning, such as habituation, to dealing with concepts of class, conservation, and analogy. Piaget has provided a natural history of this development. However, he correlates social and cognitive development in many spheres into *stages* of infant growth. We now know that humans show *decalage,* or disjunction, between logically similar abilities as revealed in different spheres. Visual habituation or surprise at objects' appearance or disappearance appears months before manual reaching for a hidden object, and manual sorting, or seriation, years before verbal formulations of the same tasks. Similarly, macaques solve Piagetian locomotor problems younger than they solve the equivalent manual tasks. Social complexity and multi-step "use" of companions may be easier for most primates than "use" of objects.

Species bias affects styles of learning, including the solutions of Piagetian tasks. Insectivorous prosimians may not manipulate a task at all, or wait, watch, and pounce at the bait, not other parts of the apparatus. Frugivorous lemurs hook in test apparatus as they hook in fruit-bearing twigs in the wild. Squirrel monkeys that travel in gangs from fig to fig also bounce in and out of problems. Slow-moving titi monkeys, which hide from predators in the wild, observe a test slowly and calculate detours and shortcuts. Saddleback tamarins form learning sets of objects; rewarded from day to day; in the wild they return to small trees whose fruit ripens slowly over weeks. Cebus use tools in captivity and crack nuts by complicated means in the wild.

The great apes progress through the Piagetian stages of infancy in terms of handling objects. At first they mature more quickly, but tend not to combine objects like our 18- to 24-month children until they are in the two- to five-year-old age range. They also have increasing facial control, beginning with neonatal facial movements when facing a companion, and progressing to deliberate tongue protrusion, spitting, and imitation of others' expressions. Home-raised chimpanzees have rare, symbolic "pretend" play, as in the imaginary pull toy.

More abstract tasks include reversal, which may indicate differences in style of learning, oddity problems, classification, analogy, seriation, and conservation. Primates have some success on these problems, which humans solve in childhood, not infancy. However, it takes intensive training for primates either to solve or to pay attention to these tasks, which humans incorporate into play.

Self-concept, as measured by using mirrors to check one's own appearance, is unequivocal, at least in chimpanzees and orangutans. Intentionality, or voluntary control of action appears, by Piagetian criteria, at eight to nine months in the human baby; apes make a similar transition. Prediction of others' intentions is so much a part of primate social reactions that we find social awareness hard to define as the fish defining water. We have at least observations of chimpanzee, pygmy chimpanzee, and hamadryas allowance for others' states of knowledge, and perhaps even for others' intentions. Future studies should be framed in terms of primates' personal and social consciousness.

CHAPTER 19

PLAY

FUNCTIONS OF PLAY

Play is important. If it were not important in evolution, why would young animals spend so much time playing? (Bruner, 1974).

Some have argued that play is merely a way to get rid of surplus energy. It is true that if animals are short of food, they cut down, or even eliminate play (Southwick, 1967; Baldwin and Baldwin, 1972, 1976; Müller-Schwarze, 1978; Müller-Schwarze et al, 1982). Play also trades off with other energetic activity: if deprived of play, deer fawns do more nonplay running, or if forced to run they subsequently spend less energy playing (Müller-Schwarze, 1978). Young veverts play less in the dry season when they also feed less. (Lee, 1981). Play thus seems to be a kind of "behavioral fat" which disappears if there are more urgent needs. The question is if play is also like fat in that it sometimes permits survival and success when there are urgent needs.

Then what is play for? It seems to be for practice (Groos, 1898). It is practice in locomotor skills, as young spider monkeys chase each other round and round through the same two trees, swinging one after the other off a springy branch; practice in sex, motherhood, and object manipulation; and practice in fighting, when kittens, gorillas, or nursery school children roll over and over in rough-and-tumble wrestling.

Play may seem an inefficient way to practice the important tasks of life. Wouldn't rote drill, or gradually extending one's skills in an organized fashion, serve as well as all this ungainly activity? It would also be far less dangerous than crashing through the trees or teetering along the top of a high brick wall. Yellow baboons sometimes catch and eat young vervets that play away from the adults, and 66,000 American children are hurt each year using playground equipment (Hausfater, 1976; Fagen, 1981). The cost of play is not just time and energy, but sometimes injury or death.

The chief benefit seems to be that play is a way of learning to cope with extremes. Locomotor play is far more vigorous than any normal locomotion. It might seem safer to extend one's locomotor skills little by little, and never take chances with a flying leap. The day comes, though, when a branch breaks under you or a band of chimps is chasing you through the trees. A wild leap that connects then is the difference between life and death. Similarly, the flailing of rough-and-tumble prepares for real combat, when the opponent will lunge or parry with all his speed or strength. I believe that the analogy extends to mental play. Games of make-believe are far more complex than a child needs for survival, but they may develop adult ability to imagine alternate courses of action with different scenarios of future events.

These are likely extremes. The young male primate will probably have to fight, in its own species' manner. Any primate may well be driven to some desperate flight. Does play prepare principally for the predictable, or does it also allow new combinations and discoveries? Does play develop the flexibility to cope with the unexpected?

We intuitively think this is true in humans. It is clearly true for experimental rats. Rats that grow up in enriched environments and, above all, rats with normal play partners (not drugged, unresponsive ones) grow up versatile and flexible in learning behavior (Einon, Morgan, and Kibbler, 1978). However, it is so rare to observe any behavioral innovation in the wild that we can only speculate about its possible link with play, or quote a few anecdotes, as in the following sections. In these circumstances the most helpful (and extensive) speculation is Fagen's computer model of selection for innovation. Fagen lists and quantifies the factors that would favor the evolution of a genetically based tendency for innovative play. These include the transmission of the benefits of innovation to close genetic kin by cultural means. Nonplaying kin might guard the players from the risks of playfulness and later copy innovations. Useful discoveries would be most likely in small groups where the young had little opportunity for social play and so turned more attention to playing with objects. The first two conditions are common to most primate groups. The last, Fagen suggests, may have played a special role in the human lineage (Fagen, 1981).

A third function of play may be to prolong the period of learning. Swamp sparrows produce many variants of notes in subsong. Then they discard all but one version, which becomes their territorial singing. The subsong could be a way to keep behavioral flexibility long past the normal critical period for new learning, so the adult's options remain open (Marler and Peters, 1982).

Admittedly such functions are unproved. We deduce them from species comparison. More intelligent species play more. Species that deal as adults with a wide variety of food and live in a wide variety of habitats usually play as young animals, as Lorenz pointed out for parrots and his beloved jackdaws (1971). Groos concluded (1896), "At the moment when intelligence is sufficiently evolved to be more useful in the struggle for life than the most perfect instinct, then will selection . . . favor those animals which play. . . . The animals do not play because they are young; they are young because they must play."

Play is fun. It is a goal in itself, which implies that it has evolved to be so. By working without the pressure of immediate need, we sometimes achieve discoveries, like Sultan fiddling with his two sticks. So far the lesson of primate evolution is that hominid Tom, who messes about with barrels and mud and things, eventually outcompetes the grim professionalism of Captain Najork and his hired sportsmen (Hoban, 1974).

CHARACTERISTICS OF PLAY

Play is hard to define. Respectable psychologists have sternly concluded that play is undefinable and therefore is not a fit subject for study. There are some markers of play—the open-mouthed play face of monkeys, apes, and humans, and the bouncing gait which even lemurs use to invite a chase. They do not always appear though, and most types of locomotion and gesture occur in both play and serious contexts.

Gregory Bateson (1973) has given us the most influential description of play. He went to the zoo to look for signs of *metacommunication*—some indication that animals can communicate about communication, not just about their immediate mood. He saw two young monkeys playing. There it was—a play bite is not a real bite. Somehow they framed their game off from reality, and communicated the paradox, "This is not real." Even if a play bite happens to hurt, and has all the components of the real bite it represents, it does not denote the aggression that a real bite does.

Bateson did not suppose that the tumbling infants analyzed these signals in philosophical terms. Symons (1978) takes violent issue and says that play, like many other social signals, is ritualized communication. A play bite denotes just plain play to a rhesus monkey, not whatever layers of negation it may denote to Bateson. As usual, when there is an argument, I agree with both sides. Just plain play is evolved, as Symons says, to be actions that are ritualized in their own right. If one monkey nips too hard and the other bites back, we say the game has degenerated into a real fight with different motives, not that the monkeys have forgotten the logical frame.

However, as we wish to consider the evolution of consciousness, the possibility of mental framing or detachment becomes crucial. Rough-and-tumble has long-evolved signals that say, "This is not real!" The older and heavier partner must also restrain himself from hurting the little one. Isolated monkeys learn this slowly. The innate framing, symbols and learned restraints do distinguish play, whether or not the animals are aware of this. When we move on to consider pretense, we have explicit framing. Pretense, by definition, is conscious acting out what is not real. Somewhere in between is the infant chimp playing with termite sticks. He does not catch anything for the first two or three years; if it were "for real," that should discourage him. Bateson's analysis remains the chief logical link between all the diverse actions we call play.

Besides this logical form, play has three common attributes: exaggeration, repetition, and restraint. None is unique to play, nor do they always appear. However, they seem to be common factors for the spider monkey and its sapling-swing, and the child jumping over cracks on the sidewalk. Stephen Miller calls the combination of these three factors "galumphing." There is a restraint rule (don't just go the easy way round to arrive and don't bite too hard), an exaggeration of mental or locomotor effort, and the repetition that is usually what cues an observer that he is watching play in the first place. Repetition, as well, recalls Piaget's emphasis on "circular reactions" or repeated acts in mastering the environment. Repetition brings some aspect of the world under the child's control.

Finally, there is what Köhler called "serious play." This may be somewhat different—better called exploration or creation, not play. It is usually performed with deliberate movements, and the compressed mouth or protruded lower lip of concentration, not open-mouthed laughter and galumphing gait. We will discuss serious play at the end of this chapter, because it involves Bateson's "framing," because primates do it, and because whatever it is, it is not work.

MASTERY PLAY

Most mammals and many birds play. Fagen (1981), who has reviewed the literature to date, points out that only two reptiles are said to play, the Mississippi alligator and the Komodo dragon. One has extensive parental care, the other quasi-mammalian exercise physiology. Recent evidence that the dinosaurs had effective thermoregulation and brood care evokes the vision of them thunderously galumphing and at least the possibility that play evolved in our therapsid ancestors. Fish and invertebrate evidence is equivocal, although Fagen includes it along with Baron Munchausen's account of "the noble sphinx, gamboling like a huge leviathan."

All known primates indulge in locomotor play—swinging, sliding, leaping, bouncing, pirouetting (Fig. 19.1.). The practice incorporates physical subroutines into functioning wholes (Bruner, 1973; Chalmers, 1980). The first at-

FIGURE 19.1. *Most or all primates play at mastering locomotor skills; here the gorilla plays a species-specific gesture. (Courtesy P. Coffey, the Jersey Wildlife Preservation Trust.)*

tempts at solo locomotion usually look like play—the game of climb-off-mother-and-bounce-back-on. Later, more complicated, repeated actions become projects—climb the fence, jump to the branch, ricochet off the water trough with a bang, and so round again (Simpson, 1976).

Chimpanzees in the wild have projects too. Goodall (1968) reports:

> When the pattern involves a complicated series of swings and leaps from branch to branch, the infant is sometimes markedly more proficient during the final round than it was during the first, which suggests that locomotor play may function, at least in part, as an aid to muscular coordination. . . .
>
> Some infants fell when they were playing in the trees. One dropped some eight feet to the ground when a brittle branch on which she was swinging broke off. Subsequently, I saw her on three occasions holding firmly to a thick branch of the same tree while she tested a smaller branch, pulling downwards, and only gradually entrusting it with her whole weight. Thus, during locomotor play the infant becomes increasingly familiar with its environment.

SOCIAL PLAY

Mother-Infant Play

Social play varies widely among species. Again, apparently, all primates indulge in it.

The first game of bounce onto mother can turn into many further infant-mother games. Surprisingly, it often does not. Infant-mother peekaboo games, and all sorts of reciprocal teasing, are so frequent in our own species that psy-

FIGURE 19.2. *Most mothers tolerate their infants' playfulness; only a few reciprocate actively. A. Cleo at six months pulls Flossie's hair in a play session, Virunga Volcanoes. (Courtesy D. Fossey.) B. Flo tickles Flint. (Courtesy H. van Lawick.)*

A

B

405   *Play*

chologists see them as setting the basic timing and social taking turns that we need for language as well as all other human communication (Watson, 1974). In other primates, one tends to think of playful individual mothers, rather than playful species. (See Fig. 19.2A and B.) Sarah, one of Hinde and Simpson's macaques, was the only one who played approach-withdrawal games with her offspring. Flo, a Gombe chimp, played with and tickled her young more than any other. Both these females had particularly close bonds with their children, into adulthood.

Other chimpanzee mothers also play with their children, sometimes boisterously. Gorillas are less demonstrative. "Infants played on their mothers' bodies . . . sliding down their backs and abdomens, hair-pulling, or wrestling with the mother's arms and legs. The mothers sometimes reciprocated by gently nudging the infant or rocking one of their own arms or legs back and forth. The infant tackled the mother's extremities in much the same manner it would later tackle with and wrestle other infants. If the play appeared to become uncomfortable for the mother, she would often simply rest her arm or leg on top of the infant, thus hampering its movements" (Fossey, 1979).

Rough-and-Tumble

Chasing and wrestling with peers is ubiquitous. Every species has rough-and-tumble (Fig. 19.3A, B, C, D). Two unequivocal play-markers almost always go with rough-and-tumble: the open-mouthed play face and a bobbing invitation gait. Evolution of the play face was described in Chapter 10. The bouncing play invitation of young macaques is a ritualized intermediate gait, between the infant form of locomotion where the forefoot moves just after the hindfoot on the same side, and the adult primate pattern where the same-side forefoot moves just before the hindfoot. As an infant grows from the early stage, when its center of gravity lies forward, to the later gait with center of gravity near its hindquarters, it lurches from side to side without proper diagonal support. This ungainly, off-balance walk becomes the "bob" (Rollinson and Martin, 1980). This in turn is further ritualized to the rotating of head and shoulders common to so much wrestling play (Sade, 1973).

Rough-and-tumble play has been described and analyzed in a lovely book by Symons (1978). To telescope the results of this extensive filmed and written study, Symons concludes that rough-and-tumble is unequivocally practice fighting. It incorporates all the same components, and, as the infant matures, the subroutines of skill are linked into smooth final performance. Symons stresses that this is not practice for innovation or discovery. The point is not to invent a wrestling hold no rhesus ever used before. The point is to achieve speed and fluidity in adult aggression, which almost certainly will be needed. In every species of primate, the males spend far more time in rough-and-tumble than females do.

Other functions of rough-and-tumble have been proposed. Levy (quoted in Fagen, 1981) watched the same colony of rhesus on Cayo Santiago that Symons watched, and collected similar data. Levy concluded that play fighting develops social bonds and that young male sparring partners are likely to migrate to other troops together. The opposite suggestion is made by Geist (1978) that playing so hard you exhaust your future competitors may actually damage

FIGURE 19.3. *Rough-and-tumble wrestling and chasing are common primate play forms. A. Squirrel monkey infant and juvenile. (Courtesy R. Fontaine and Monkey Jungle.) B. Rhesus play king of the castle on a Nepalese temple. (Courtesy J. Teas.) C. Bonobos or pygmy chimpanzees. (Courtesy of Doris Sorby.) D. The baby at one year is just transforming his ventral clinging into play-wrestling. (A. Jolly.)*

their health and growth, leaving you a clearer field in the future. Still another suggestion is that young baboons sort out their prospects in a future dominance hierarchy, which avoids damaging fights later on (Hall and DeVore, 1965). This last has dropped out when we realized that many of these young males migrate to different troops at adolescence, but familiarity with each other's prowess may still help decide whom *not* to migrate with, or whom to rejoin when shifting again as an adult.

All these suggestions seem to me to apply to particular species or situations and to be testable only by following up the behavior of particular groups. A case which illuminates species differences is the play of twin Coquerel's mouse lemurs (Pages-Feuillade, in preparation). The species is monogamous. Adult male and female each forage alone, but meet during the second half of the night for a long bout of grooming and play. They hang by their feet and spar with their hands, twisting and play biting. Adult aggression is mostly vocal and olfactory, although quite possibly physical attack would also use the same motor components as play. Coquerel's mouse lemurs normally bear twin young, who have a 50 per cent chance of being unlike sexes. The young play wrestle with the adult repertoire of gestures, though using unstereotyped variants. The young twins play reciprocally, taking turns as rhesus do, without the need for self-limiting by heavier, older animals. It is this reciprocity and the mix of grooming

FIGURE 19.4. *Play-wrestling between twin Coquerels mouse lemurs has the reciprocal gestures of most rough and tumble. It does not mature into fighting, but into sexual pair-bonding play of the monogamous adults. (Courtesy E. Pages.)*

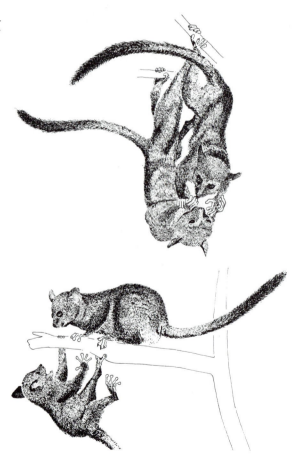

with play that persists to cement the adults' social pair-bond, (Fig. 19.4.), whereas in rhesus we see similar beginnings turn into thrust and parry.

Mummies and Daddies

Play parenting is a third kind of social play. We have seen that allomothering among adult females is equivocal. It often seems to turn into mistreatment of infants by their mother's competitors, not babysitting to give the mother relief. However, juveniles fascinated with young infants almost certainly practice adult caretaking, and improve the complicated skills of holding a baby so that it does not start squealing (Lancaster, 1971).

Again, the particular social structure of each species shapes play parenting. In most primates, including chimpanzees and humans, young females pay most attention to babies. In Barbary macaques, whose males use infants for pawns in the status game, it is young males who play more with babies. In common marmosets and lion tamarins, where the male carries growing infants much of the time, male and female juveniles seem to spend an equal amount of time carrying infants, as well as the same amount of time in rough-and-tumble play (Box, 1977; Hoage, 1977; Chalmers and Lock-Hayden, 1981). Experience with carrying young is one of the crucial factors in successful rearing for marmosets and tamarins. Until people realized this, captive juveniles were too often removed before the next infant's birth, and then they neglected or killed their own litters.

Adult Play

Adult play is rare. Fagen lists a few cases of like-sexed play and more of play in courtship, though in canids rather than primates. Humans seem to be the only primates that play in groups as adults.

OBJECT PLAY AND SERIOUS PLAY

Many of the examples of tool use and intelligent behavior that were quoted in the last two chapters were actually examples of play, such as the taboo pole or ape painting. Various general morals emerge.

Primates often solve difficult problems first in play, rather than when their attention is focused on an out-of-reach goal (Fig. 19.5.). Schiller (1967) showed this systematically by giving the chimpanzees problems without reward, but it also appears in Köhler's own protocols. The psychologist D. O. Hebb was once testing a chimpanzee on oddity problems when he ran out of banana slices. He noticed that the chimp had been hoarding, rather than eating the rewards, so he took a chance and continued testing. The chimp not only solved his problems but rewarded Hebb with a slice of banana. He ended 22 slices to the good. The lateral thinking in this experiment characterizes great psychologists; the chimp remains anonymous. The general moral is that play with objects often has its own motivation.

The imposition of another motivation on play turns it into work. It may console parents confronted by messy children's rooms that Köhler never suc-

FIGURE 19.5. *Rhesus is formally tested for curiosity. (Courtesy Wisconsin Primate Laboratory.)*

ceeded in teaching Sultan to pick up the cage; after the first few days the chimp turned as sullen as any teenager.

The next general point is that object play differs in different species, just as tool use does. Orangutans, chimps, cebus and, above all, people, seem drawn to playing with objects. Lowe's guenons repeatedly carry stones up a tree, drop them, and watch them fall (like a Piagetian Stage 4 baby in its highchair), and repeatedly scare a flock of noisy hornbills when they settle to roost. Perhaps this is just the playfulness of a particular troop, or perhaps, as a species of poke-and-pry insect catchers, they focus on their environment in play.

Third, there is a definite trade-off between social and object play. Marmosets whose twin dies, so they are raised as singletons, play more with objects than those who have a twin. In the wild, primates who have few play-mates through the accidents of demography play more alone—both in locomotor gymnastics and with objects in the environment (Fagen, 1981). This in part explains why captive animals play so much more than wild ones, when combined with the steady food supply that takes no time to find, and the lack of danger from predators. Altogether a human baby is very like a captive animal. Unless it has a twin, its siblings are at least two and often more years older, and its food and safety are provided by its keepers. The environment as well as the genes focus the baby on to objects.

Finally, though, object play usually does not look like social play. A child or primate approaches a new object gingerly, in fear combined with exploration. It then usually pokes and pulls, trying to find out "What does it do?"

A   B

FIGURE 19.6. *Object play begins with cautious exploration, but ends with boisterous destruction. A. Chimpanzee infant with stick, Gombe Stream; B. Chimpanzee in rough and tumble with "nest materials." (Courtesy I. Bernstein.)*

Later the child takes more active control: "What can I do with it?" Eventually the actions can turn boisterous—stand on it, throw it, break it! (Hutt, 1974). (See Fig. 19.6A,B.) Only in the last stage does the child put on the laughter of rough-and-tumble, and the gestures of one kind of social play. Up till then there is an intent look, and deliberate or rigidly precise gestures. This is what Köhler called "serious play." The description of ape "art" in Chapter 17 gave one set of examples. Let me add one more from Köhler, quoted in Bruner, 1972:

> On the playground a man has painted a wooden pole in white color. After the work is done he goes away leaving a pot of white paint and a beautiful brush. I observe the only chimpanzee who is present, hiding my face behind my hands as if I were not paying attention to him. The ape for a while gives much attention to me before approaching the brush and paint because he has learned that misuse of our things may have serious consequences. But very soon, encouraged by my attitude, he takes the brush, puts it into the pot of color and paints a big stone, which happens to be in the place, beautifully white. The whole time the ape behaved completely seriously.

Such play is seriously important in the wild. Young champanzees learn to fish for termites, in the Gombe stream, over four or five years. Little infants of one play with grass blades during the termite season. Two two and a half-year-olds attempt to fish with inappropriate tools, some of them tiny grass scraps about two inches long. They then jerk the tool from the hole, usually dislodging any termite that happens to cling, with occasional luck. Slowly they improve. Four-year-olds have achieved adult technique, although with only about a 15-minute attention span. One infant, Merlin, was orphaned in his third year and never progressed beyond the initial ineffective jerking. He fished with near-adult concentration, 40 minutes at a stretch, but only caught two termites while his age mates gobbled their catch beside him. Finally by age five, technique and attention have come together in one- to two-hour bouts. Not all adult activities take so long to learn; chimpanzees do not dip for biting ants until they are near adulthood and can stand an adult's arm's length from the nest. However, termiting clearly takes skill. When Teleki tried to duplicate the chimpanzees' performance, the chimps out-fished him every time (Goodall, 1968, 1971; McGrew, 1977; Teleki, 1974).

PLAY, PRETENSE, AND MENTAL FRAMING

Let us return to Bateson's formula that the essence of play is the frame which sets it off from reality. Leave aside the question of whether wrestling mouse lemurs or play mothering juvenile vervets have any need for mental detachment. Are there clear cases of pretense in primates?

We have already described Viki's imaginary pull toy. It disappeared when Cathy Hayes, the human "mother," pretended to have a toy of her own. Cathy speculated, "Perhaps Viki recognized that make-believe is fun for a baby, but dangerous in a parent." I find this easy to accept, knowing how easy it is for a parent to overdo stimulation at any level, from simply tickling too hard to telling too scary a ghost story. If you admit the imaginary pull toy in the first place, it is not too hard to see how Cathy inadvertently broke the frame.

Apes are certainly capable of social pretense. Figan, cleverest of the Gombe chimps, learned as an adolescent that he could lead adult males away from the banana pile by striding purposefully into the woods. He then returned, by another route, to gorge unmolested. On one occasion he "spotted a banana in a tree that the older chimps had overlooked, but Goliath was resting directly underneath it. After no more than a quick glance from the fruit to Goliath, Figan moved away and sat on the other side of the tent so he could no longer see the fruit. Fifteen minutes later, when Goliath got up and left, Figan, without a moment's hesitation, got up and collected the banana." He had apparently sized up the situation to the point where he realized both that he could not snatch the banana with Goliath there and that he could not help looking at the banana—which would guide Goliath's own gaze to the prize. Even Figan occasionally got his comeuppance. Once, when he had led one group of chimps from camp, he returned to find yet another high-ranking male installed at the banana pile. "Figan stared at him a few moments and then flew into a tantrum, screaming and hitting at the ground" (Goodall, 1971).

Laboratory experiments confirm such deliberate misdirection by chimpanzees. One pet pigtailed macaque clearly misdirected her owners, particularly by pretending a need to defecate at awkward moments (Bertrand, 1976). Even baboons can deceive. One of the female baboons at Gilgil grew particularly fond of meat, although the males do most hunting. A male, one who does not willingly share, caught an antelope. The female edged up to him and groomed him until he lolled back under her attentions. She then snatched the antelope carcass and ran. Later, the same male killed again. Again she groomed him. This time he kept a hand on the carcass. She left off grooming, and chased after his favorite female. He dithered, but at length went to his friend's defense. The first female promptly doubled back to snatch the antelope (Strum, personal communication). Sometimes, then, we do see primate pretense. Are there other cases of pretense in play besides Viki's pull toy? Among the language-trained chimpanzees, clearly yes. Koko, the signing gorilla, is terrified of alligators (she has never seen a real one). She plays, however, with little toy alligators, and pretends to sneak up and frighten her human friends with them. You are supposed to respond with an elaborate startle; as Patterson remarks, it is funny to think that a nearly full-grown gorilla supposes she needs a plastic alligator to frighten anyone. More to the point, Koko plays with "words," signing to herself and her dolls. Like many children and other home-raised apes, she does not much like to be watched while playing. Again, adults would intrude on the frame (Patterson and Linden, 1981).

It seems to me reasonable that this is the same sort of detachment children impose on pretense, and that this, in turn, is like rules that bind a game. The most often quoted example is Vygotsky's account of five- and seven-year-old Russian girls playing "sisters," which meant following rules they usually didn't; in fact, they were sisters. Or consider ten-year-old Susan who decided to get married to a toy frog. She assembled rings and a book, and then asked her father to perform the ceremony. The gleeful father couldn't resist adding red carpet, music, and an unctuous preacher's voice. At this point the child burst out in tears of fury, "Stop it, Daddy! You are just making a *game* of it!"

As adults, we operate within many different frames of rules, and usually manage not to break through between levels. Sometimes the theater drags them into consciousness, as in Jean Genet's play, "The Blacks." There, a group of simpleminded village Africans rehearse a play they will give for the colonial governor. They cavort in pink-cheeked, blond-thatched masks, comically misinterpreting Western ways. In the second act the governor and his entourage enter, with blank white masks and cruel eyes. The "innocent" Africans reveal their play is a only ruse to distract the white governor and his lady while revolutionaries assemble to massacre the hated rulers. The rustic comedy is performed with undertones of murder, though you are not sure which side will strike first. In the end, this frame too is broken. The actors, all blacks, strip off all their masks and announce the victims of the massacre—you, the audience, who came to the theater to titter at Africa's agony. Genet quite possibly hoped for a performance so convincing that the actors' closing move to block the exits would lead to real panic in the audience, and a stampede into real tragedy, breaking yet one more frame.

This kind of agreed detachment underlies the separation of a symbol and its meaning. Koko and Genet have more in common than some capacity for pretense. They both have some capacity for language. Language, in this sense, is a form of serious play.

SUMMARY

Play functions mainly to practice adult skills. Most play prepares for predictable aspects of adult life, particularly for extreme aspects, such as headlong flight or serious fighting. It occasionally allows for flexible or innovative behavior.

Almost all primates enjoy locomotor play. Rough-and-tumble is very common, particularly among young males. Play parenting is also common, particularly among young females. Mother-infant and adult play occur, but more rarely. Species differ in their play. For instance, where adult males do much infant-care, young males are likely to play with or carry babies. Similarly, the species that manipulate objects for reward do so also in play. Chimpanzee termite-fishing matures over three or four years' play before reaching adult efficiency.

Play involves logical framing, or detachment, which marks it off from "real" activities. There seems little need to suppose any conscious detachment in the evolved enjoyment of leaping and swinging, or rough-and-tumble or play mothering. However, more complicated pretense and symbolic play in apes or humans demand some such framing. Wild baboons and chimpanzees sometimes deceive each other over food; human-raised apes have a little true symbolic play.

CHAPTER 20

LANGUAGE

People have drawn one boundary line after another to distinguish themselves from brute beasts. Our supposedly unique attributes have ranged from the immortal soul to the humbler, but nearly as indestructible hand-axe. The most recent Rubicon was language. So far, two gorillas, more than a dozen chimpanzees, and perhaps three orangutans seem to have knuckle walked or brachiated toward the Rubicon, as a few pigeons in Skinner boxes cynically peck away in the background (review by Ristau and Robbins, 1982).

Of course, experiments that enlarge our understanding of the components of language do not diminish the importance of language in human life. Language shapes human emotional and cognitive growth, our societies, and our civilizations. Ape performance does not explain the total importance of language to humans any more than a termite stick explains the hydrogen bomb. But if we believe we evolved from some protohominid, our language somehow started from a primate signal system and a primate's mental abilities.

Summing up current research in this field is difficult, because the researchers are even more at odds among themselves than scientists in other fields. It may seem strange that people who communicate so well with alien species are frequently not on speaking terms with each other. The disputes are phrased in terms of distrust of each other's methods, and doubt that inappropriate methods can lead to valid or interesting results. Only a little below the surface lies an emotional split between those who treat the apes as much as possible like children and those who treat them as experimental animals. Proponents of the first school argue that the most effective way to encourage childlike behavior, including languagelike skills, is through the same rich mixture of love, training, cuing, parental response, and over-response that surrounds a human child. What, they ask, would you expect from a human one-year-old kept in a cage, taken to a bare teaching cell, and put through stereotyped training procedures by a succession of student volunteers or even by a machine? You might produce John Stuart Mill, but you would more likely produce a mental cripple. On the other hand, psychologists attempting to prove or elucidate the precise nature of ape skills argue that treating apes as children means that you have little idea what training procedures led to a given response, or what is the immediate context of a given response, and worst of all, how much the involved "parent" may be deluding himself about the skills of his own furry prodigy. In short, a chapter like this, that attempts to describe the claims of each research worker, will probably annoy them all.

The controversies between ape language workers also turn on the commitment to language as a reified principle, a unique attribute of our own species. It was not long ago that people doubted apes could use signs to mean categories of referents, as we use words. Umiker-Sebeok and Sebeok (1980) have reopened this question, in spite of the Gardners' extensive double-blind experiments on naming and the Rumbaughs' double-blind computer experiments on class names and inter-chimpanzee communication. I believe it is now clear that ape signs resemble human words. It is still doubtful whether, or in what sense, apes use either word-order grammar, like English, or a grammar of inflections, like American Sign Language (ASL). It is clear that apes combine several signs relevant to a single situation. It should be a solvable empirical question as to how much these combinations resemble small children's two- and three-word "sentences."

The Gardners compare "the ineffable qualities attributed to human language" to the nineteenth-century concept of a mysterious vital principle in organic chemistry. When Wöhler synthesized urea in the laboratory, there was furor and furious claims that urea was too simple a chemical to include the vital principle, even though by any other definition it is an organic compound. Synthesizing such a compound, identical to urea of organic origin, attacks such reified vitalism. It does not disprove the existence of life; it just illustrates the truism that the interesting properties of complex structures are a function of their complexity, not of their simplest parts. The fascination in studying apes' simple productions is that "the early utterances of children and chimpanzees have much in common, and that this agrees with the hypothesis that both obey common biological laws of intelligence" (Gardner and Gardner, in press). Only those who wish to reify human language need take this as an attack on human pomposity.

EXPERIMENTS AND EXPERIMENTERS

Nadia Kohts (1923), the first psychologist to raise an infant chimpanzee, was convinced that her charge understood some 50 to 60 phrases of Russian (Kellogg, 1968). Gua, the Kellogg's chimpanzee, understood 95 words and phrases of English (Kellogg and Kellogg, 1933). Viki, a home-raised chimp whom the Hayes tutored intensively, could actually pronounce four words: *Mama, Papa, cup,* and *up.* However, she pronounced them by holding her mouth shut and sometimes closing her nostrils with a hand, then hoarsely breathing out (Hayes, 1951). Apparently, primates have little voluntary control over their vocal sounds,

FIGURE 20.1.   *Washoe and Beatrice Gardner sign* drink. *(Courtesy R. A. and B. T. Gardner.)*

although they can slightly modify the frequency of giving coos and barks (Aitken and Wilson, 1979; Randolph and Brooks, 1967). Two orangutans have, like Viki, been trained to utter a few unvoiced words which they used as requests. Cody could ask for food *(fuh),* drink *(kuh),* to be picked up *(puh),* and to have his long red fur brushed *(thuh)* (Laidler, 1978).

R. Allen Gardner and Beatrice Gardner (1969, 1971, 1980) first began to teach American Sign Language for the deaf to an infant chimpanzee (Fig. 20.1.). They reasoned, as Yerkes had previously remarked, that a chimpanzee might accomplish by manual signing what it could not do in the vocal sphere. Washoe lived in a house-trailer in Washoe County, Nevada. She was raised with all the appurtenances of a middle-class American infant, and with as much as possible of continuous caretaking by devoted teachers. The Gardners call this *cross-fostering*—the raising of one species by foster parents of another. It is one way of attempting the supposed background of a controlled experiment, which keeps all other conditions the same, and changes one variable at a time—in this case, the species. Of course, the species' difference reverberates into many others, but the background was explicitly meant to be a human environment, not that of a caged creature, and certainly not that of a pet. The teaching methods were any that seemed to work—rewarding learning, molding Washoe's hands into shape, modeling for imitation, and much testing by asking questions.

An important aspect was that the humans used only ASL in Washoe's

FIGURE 20.2. *Moja at four months signs* drink. *This picture sums up many controversies: Do you see a signed word? A conditioned relfex? A baby putting its thumb in its mouth? Is the philosophy of teaching apes language in the messy, social way that human babies learn more or less revealing than teaching rigidly controlled symbolic systems? (Courtesy R. A. and B. T. Gardner.)*

presence. (See Fig. 20.2.) The Gardners did not want the chimpanzee to feel that babies are forced to perform in sign language, but grownups speak English. Washoe's first teachers were just learning ASL themselves, but in later stages the teachers were fluent signers—either deaf signers or hearing children of deaf parents. Now, another four young chimpanzees have been raised by the Gardners and their co-workers, and Washoe signs to her own adopted infant.

It is difficult to isolate hand-raised apes from spoken language. A further group of chimpanzees are being taught by Roger Fouts, an ex-student of the Gardners (Fouts and Couch, 1976). Lucy is the best-known, now being rehabilitated in Africa (Temerlin, 1975). These chimpanzees are "bilingual," hearing English and seeing signed ASL, although they can only respond with signing or the usual chimp behavioral repertoire. Terrace attempted to repeat the Gardners' methods with Nim, though there were more than 60 trainers, an emphasis on conditioning as a teaching method, and a background of spoken English (Terrace, 1979). Chantek, a young orangutan raised by Lynn Miles, makes the point that an orangutan may learn as many signs as a chimpanzee and use similar combinations of signs. These abilities seem to be a function of ape-level intelligence and are not confined to our closer relatives, the African apes (Miles, 1983). Koko and Michael are gorillas raised by Francine Patterson. Koko, who is now adult, is said to translate or even eavesdrop from one medium to the other (Patterson and Linden, 1981).

The strength of this approach is simultaneously its scientific weakness. These apes hear and see words used between adult humans, just as a baby does. Like a baby, they may abstract new phrases from the general flow of conversation, and add a nimbus of connotations to known phrases. Furthermore, they can use words when and how they like—mistakenly, irrelevantly, in play, or even in insult. In short, skeptical psychologists can never look at the whole corpus of training procedures and can never be sure of the animals' precise usage. One of Koko's favorite annoyance words is the ASL sign for *devil*. Just what does that mean to a gorilla? The more like language, the less like science.

There are quite different approaches that aim at convincing the psychologists. Premack trained, first, Sarah, and now several more chimps, to use plastic shapes for words, which Sarah placed in strings on a magnetic board (Premack, A. J. 1976; Premack, D., 1971; Premack, D., 1976; Premack and Premack, 1983). This meant that Sarah's training and responses could apparently be limited to maneuvers with specified objects. It is true that we do not know how much English she understood and whether she may use her trainer's spoken words for the shapes to help remember them. However, this is a very useful link with further tests of intelligence. The feats of analogical reasoning and the abstract ability to deal with relational concepts, which Sarah shows as an adult, may relate to her early education in symbolic logic (Chapter 18).

Finally, Lana, Sherman, and Austin at the Yerkes laboratories have learned to type out a computer language called Yerkish in which again the meanings and combinations of symbols can be explored with greater rigor (Rumbaugh, 1977). (See Fig. 20.3.)

Each school of training has spawned its challengers—linguists and grammarians, but also the trainers of animals. Terrace concluded that his chimp, Nim, signed but with an ungrammatical sequence of signs, often unconsciously cued by the trainers. Epstein and colleagues have taught pigeons to peck computer keys labeled with words that ape the apes. Sherman and Austin

FIGURE 20.3. *The chimpanzee Lana with the prototype computer. Note that the meanings of the signs must be deduced by context, as for ASL. However there are no ambiguous, improperly formed signs, and the computer records Lana's output. (Courtesy D. Rumbaugh.)*

use the computer to ask each other for named tools and food. The pigeons can mimic the same sequence by pecking keys marked with the following "conversation":

Jack: *What Color?*
Jill: Looks through curtain at red (or yellow or green) light.
  She pecks *R* (or *Y* or *G* as appropriate).
Jack: *Thank you.*
Jill is rewarded with food.
Jack pecks red (or yellow or green) disk, and is rewarded with food.

The controversy over ape language, then, centers around what our criteria are for language and how complex behavior must be before we call it language. Few dispute that the apes do what they are said to do, but a mixture of logic and hurt pride makes us question whether they mean what they seem to mean.

DESIGN FEATURES AND PRIMATE CALLS

Human beings have a vast range of nonlinguistic communication (Chapter 10). Most of our personal relations, the expression of our emotions and our feelings for each other, are communicated nonverbally. That is, we have language as a second system for talking about objects or refining our concepts, alongside the first system of communication that is so homologous to that of the chimpanzee.

Charles Hockett (Hockett and Ascher, 1964; Hockett and Altmann, 1968) gives a list (Table 20.1) of the design features of language. Note that the first item does not necessarily apply to language: language may be written or, among the deaf, gestural. However, human language as it evolved was vocal-auditory.

Design features 2 through 6 are shared with much animal communication. Hockett and Ascher (1964) discuss several of the attributes of human language that seem to differ from primate call systems or human nonverbal communication.

1. Openness

Language is an *open* system; call systems are *closed*. With language a speaker can say new things, utterances that have probably never been said before and certainly have not been said before by that speaker. This is true of grammar. It is even true of words. However, with most call systems one is limited to a predetermined set of expressions or vocalizations. The best one can do is mix these vocalizations. For instance, in a situation where food and danger are both represented, some intermediate call might indicate some intermediate situation. Hockett and Ascher suggested that "The constant rubbing together of whole utterances by (this) blending mechanism . . . generated an increasingly large stock of minimum meaningful signal elements —the premorphemes of pre-language."

It now seems unlikely that such blending produced new meanings. When

TABLE 20.1. Design Features of Language

DF 1 Vocal-auditory channel
DF 2 Broadcast transmission and directional reception
DF 3 Rapid fading (the sound of speech does not hover in the air)
DF 4 Interchangeability (adult members of any speech community are interchangeably transmitters and receivers of linguistic signals)
DF 5 Complete feedback (the speaker hears everything relevant of what he says)
DF 6 Specialization (the direct-energetic consequences of linguistic signals are biologically unimportant; only the triggering consequences are important)
DF 7 Semanticity (linguistic signals function to correlate and organize the life of a community because there are associative ties between signal elements and features in the world; in short, some linguistic forms have denotations)
DF 8 Arbitrariness (the relation between a meaningful element in a language and its denotation is independent of any physical or geometrical resemblance between the two)
DF 9 Discreteness (the possible messages in any language constitute a discrete repertoire rather than a continuous one)
DF 10 Displacement (we can talk about things that are remote in time, space, or both from the site of the communicative transaction)
DF 11 Openness (new linguistic messages are coined freely and easily, and, in context, are usually understood)
DF 12 Tradition (the conventions of any one human language are passed down by teaching and learning, not through the germ plasm)
DF 13 Duality of patterning (every language has a patterning in terms of arbitrary but stable meaningless signal-elements and also a patterning in terms of minimum meaningful arrangements of those elements)
DF 14 Prevarication (we can say things that are false or meaningless)
DF 15 Reflexiveness (in a language, we can communicate about the very system in which we are communicating)
DF 16 Learnability (a speaker of a language can learn another language)

(After Hockett and Altmann (1968) and Marler (1969b).)

two calls are strung together in tandem, the compound call often has a quite different meaning from either of its parts (Marler and Tenaza, 1977). The apparently graded series of Japanese macaque vocalizations has very specific meanings to the listeners (see Chapter 10). Vocalizations may have learned meanings such as alarm calls applied to particular predators, or, the large group of calls that convey the identities of individual birds and mammals. Rarely, if ever, are these *new* calls.

2. Discreteness

A graded series of calls, even if it were opened somewhat by the use of intermediate calls, would rapidly become overloaded. Such a graded series would be an analog system, in which the intensity of the call might correspond with the intensity of the situation, or the shading from aggressive to fearful call might indicate the degree of aggressiveness or timidity of the animal. This representation, in which the continued variation of some parameter of the call is equivalent to the continuous variation of what is to be expressed, could not attain the complexity of true language, unless there were a separate parameter for every single thing that varies in the world around. Language instead is digital. Discrete words represent discrete ranges of experience.

Not only words but also sounds are divided into discrete quantities. We produce a huge variety of sounds, and many more could be made that do not exist in any given language. However, in objectively recorded speech and in large part through the mind of the hearer, the continuum of sound is classified into a limited number of permitted single sounds called *phonemes*. We can thus process discrete bits of acoustic information without being swamped by the infinite continuum.

Chopping up the continuum of sound into discrete groups may be mammalian, rather than a human capacity. Chinchillas, as well as rhesus monkeys, seem to make distinctions between such phonemes as /b/ and /p/ about as humans do. A computer generates artificial speech syllables that say ba-ba-ba, with the slight unvoiced pause after the *b* growing longer by equal, imperceptible steps. Chinchillas, rhesus, small babies, and human adults abruptly begin to hear the sound as pa-pa-pa, though the step between ba and pa is no greater than the steps we classify as a single sound. Small babies apparently recognize the phonemic classifications used in any human language. Only later, by about a year or 18 months, do they begin to ignore the distinctions which their own language ignores (Kohl, 1979; Snowdon, 1979; Eimas and Tartter, 1979).

Sarles (1969) makes the point that "The history of phonetic listening in language description is long, well-documented, and exactly what we try to teach out and unlearn in introductory courses in linguistics . . . the idea of discreteness, as is well known to those who have any acquaintance with psycho-acoustics, is in the mind of the perceivers, not in the nature of language. Thus the different operations that are performed on animal and human language yield entirely different and noncomparable kinds of units; phonemic for human language, phonetic for animal language."

As discussed in Chapter 10, Japanese macaques categorize their coo vocalizations in similar fashion. Linguistic analysis of primate calls has begun since Sarles' remarks. The question now is how many of these perceptually categorized systems will appear now that people have begun to listen for them? The various discrete calls described so far are used in emotionally different situations; will we now find categorical perception of varying call components, which would be then used as neutral elements of a range of different calls (Snowdon, 1982; Petersen, 1982)?

3. Duality of Patterning

The two levels—of sounds and of words—give language a *duality of patterning*. Both levels are discrete: neither operates as a graded continuum. However, the phoneme level is closed. We do not suddenly invent a new sound in our language except by conscious or snob mimicking of foreign words. (The pleb uses foreign words with his familiar phonemes: "Gay Paree.") We can then combine the finite number of meaningless sounds into an open system with an infinite number of new but meaningful words: blitzkrieg, behaviorism, muppet.

Thus, as in so many natural hierarchies, the component parts (phonemes) of language are relatively few in number and simple in form, but their groupings (as words) are many and complex. We still do not know just how most primates group components, except in experiments like Zoloth and colleagues' (1979) in which rhesus did *not* distinguish calls that were clearly separate to Japanese macaques (see Chapter 10). There are, however, indications that separate call types may be grouped in sequences with internal hierarchical structure. This is a formal approximation to the grouping of words (Robinson, 1979).

4. Denotation

There is at least one clear case in which a primate call system denotes categories of objects in the environment. Vervet monkeys give different calls to major mammalian predators, flying predators, and to snakes (Struhsaker, 1967; Cheney and Seyfarth, 1980, 1981; Seyfarth et al., 1980a, b). They also respond differently to the calls. An eagle "rraup" makes the troop look up or dive for cover in low bushes; a threat-alarm bark or chirp can trigger a dash for the branches of high trees; and a snake chutter often stimulates vervets to descend from trees to the ground and join in chuttering at the serpent. It would be dangerous for a vervet to get its responses wrong—the response that protects it from one kind of predator makes it more vulnerable to a different kind.

The three calls, and their responses, are not simple shades of alarm. Cheney and Seyfarth played back the various calls to their vervet troop. Each call led to the appropriate response, or to looking at the speakers, or looking around. Volume and proximity changes did not matter; a loud snake call did not produce "eagle" reactions. What did matter was the original emitter of the call. If the animal whose recorded voice called the alarm was peacefully feeding in plain view, the others glanced up (or down as needed), and ignored the false alarm.

Infants and juveniles apparently learn the proper use of calls (Fig. 20.4.) At first they gave eagle "rraups" to any real startle, then to a variety of flying birds, and finally only to true raptors (Seyfarth and Cheney, 1980).

This is the clearest case of "semantic" primate calls, which denote classes of objects in the environment, not simply emotional states. Of course, we have the human (or modern Western human) bias that denoting objects is more important than denoting social situations; we usually do not analyze primate social calls in the same terms.

It seems likely that we have underplayed the cognitive or semantic component of animal communication systems in our effort to preserve language as a quite separate phenomenon. Much animal communication includes a component of glance or posture that directs the recipient's attention to some companion or object. Affective signals generally have some external referent; the vervet alarm calls are striking mainly because the referents are so precise. And even fairly unprecise signals, such as human smiling for happiness, are rarely given unless there is someone to smile *at*—in other words, the smile is a gesture that *conveys* happiness. "Symbolic and affective signaling can be viewed as extremes on a continuum . . . as such, they may prove to differ more in degree than in kind. . . . (This) implies that man did not invent a new symbolic mode of operation, but rather that he enriched and elaborated a system already operating in animals" (Marler, in press).

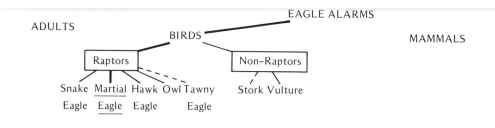

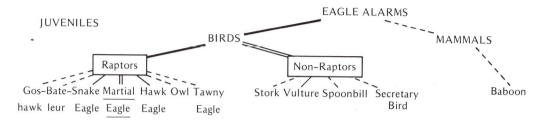

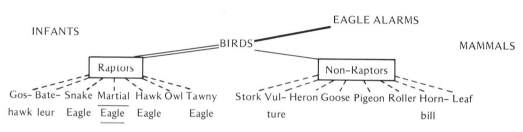

FIGURE 20.4. *The ontogeny of vervet eagle alarm calls. Infants called to overhead objects, including a falling leaf. Adults narrowed the calls' use to the truly dangerous species. (After Marler, in press.)*

5. Displacement

A fourth characteristic of language is displacement. An animal gives a call in the situation appropriate to the call, presumably when the appropriate emotion is evoked. Words, on the other hand, can be used at a distance in time and space from their referents. We will have much more to say on displacement in the section on symbols, for the major distinction between signal and symbol is that the signal is still tied to the situation and its accompanying emotion, whereas the symbol is divorced and independent.

However, it must be said here that, ontogenetically, infants' speech begins to develop from their babbling and comfort noises rather than from the higher-intensity distress calls that demand instant response from the mother. Marler (1969a) says that of all the chimpanzee calls, their "rough grunts" are most closely related to the phonetic characteristics of human speech. Rough grunts include food barks, greeting grunts, and the laughter of play. In all these situations the chimpanzees are fairly relaxed, and all these noises may have a rewarding quality to other chimpanzees.

It could be then that (making the leap to supposing that protohominid calls

424   *Intelligence*

were something like those of chimpanzees) such grunts were the raw material for the development of speech. This would at least be consistent with the fact that we have retained a full set of nonverbal, emotional communications while speech has evolved alongside.

Chimpanzee grunts resemble the pulsed structure of speech but lack the formants we then impress on the basic grunt. Andrew (1963) says there is a primate parallel for our vowels and our resonances of different parts of the upper respiratory tract. Baboon grunts, like human speech, have a low fundamental and many overtones, which resonate differently with movements of the mouth and thus with facial expression. Baboons have a stylized "greeting" grunt, in which lip smacking and/or tongue protrusion modifies the sound of the greeting. Andrew argues that baboons in a large, cohesive social group, under continual pressure to know each other's state of mind because of aggression within and predators round about the troop, may have evolved this refined means of signaling facial expression by sound. His arguments need not be confined to aggressive, large-group savannah animals, although the protohominids may have been such. What the signaling does imply is a need to keep in accurate touch at a distance or when the other animals' backs are turned. Thus, the subtleties that can be signaled in chimpanzees by actual touch, such as reassurance, may be shown in baboons as well as men by vowels!

One common factor in both Marler's and Andrew's derivations is the greeting grunt. This may be coincidence, or it might be a real clue to the earliest social functions of speech.

6. Arbitrariness and Tradition

Finally, language is *arbitrary*, learned, and traditional. This is a necessary corollary of the point that language is open in the sense that an infinite number of new statements are possible. Even if the structure of grammar turns out to be innate, and integral with the structure of the brain, the actual forms of language must be learned if it is to be a system of infinite capacity. A genetic program is indeed a program and not an inexhaustible store. Thus, a call system that contains only a few types of calls, or that has the calls grading analogically into each other through only a few parameters, may be biologically fixed in its content, and so might the deep structure of a grammar, but the Shorter *Oxford English Dictionary* is unlikely to be.

We again do not know how much of nonverbal communication is learned. Bruner (1968), for instance, points out that children's crying may be conventionalized after the first week or so of life. They can use only a part to indicate the whole, being fairly sure that a whimper will bring mother running. We know that looking at another person directly establishes an intensity of relationship and may often serve as a threat, if only by analogy with all the rest of the mammals. This probably is innately determined in us as well, but the frequency of glancing at other people in conversation and the distance at which we may stand to catch their eye is a highly cultural phenomenon. In England, a man who looks you straight in the eye is honest and true; in Madagascar, he is simply impolite. Not only the form of gestures may be learned but also contexts or responses. But even though troops of Japanese macaques differ in some of their communication (Green, 1975), and chimpanzees very likely do, this is a long way from the sheer quantity of arbitrary learning accomplished by a baby who learns to speak.

WORDS

Signals and Symbols

Symbols are distanced from the things they represent. They may be at a physical distance in space and time; they are also at an emotional distance. They may contain the pleasure of control, or an emotional satisfaction with naming an object; they need not form part of an immediate appeal to deal with the situation they describe. Signals, on the other hand, are tied to the situation in space, time, and emotional relevance.

The distinction between signal and symbol has been a major concern of philosophy, not just of animal behavior. In fact, study of any field in depth—art, mathematics, sense perception—tends to lead back to the nature of symbolization or abstraction itself. Philosophers also recognize that there are intermediate forms. When you bang your finger with a hammer, there is a gradient from the outright yell through the heartfelt curse through saying "ouch" in various tones of voice to the symbolic report, "I banged my finger horribly and it hurts." It is probably true that all symbols have some emotional overtones. For those who speak the language of mathematics, a feeling of beauty and purity can rise out of an equation on a blackboard. A few words are still somehow continuous with their referents: the magical names of God and the unmagical names of sexual intercourse and excrement. The distinction between symbol and signal divides word from invocation.

Signal and Symbol in Talking Apes

In what sense do ASL signs resemble spoken words, and how do apes use such signs? The Gardners write:

> As words can be analyzed into phonemes, so signs can be analyzed into what have been called cheremes. There are fifty-five cheremes, nineteen of which identify the configuration of the hand or hands making the sign; twelve, the place where the sign is made; twenty-four the action of the hand or hands. Thus the pointing hand configuration yields one sign near the forehead, another near the cheek, another near the chin, still another sign near the shoulder, and so on. At any given place, the pointing hand yields one sign if moved toward the signer, another if moved away, another if moved vertically . . . and so on, but, if a tapered hand is used instead of a pointing hand, then a whole new family of signs is generated (Gardner and Gardner, 1971).

Thus, ASL is an arbitrary language, not a set of pictorial or "natural" gestures. There is no international sign language for the deaf. The language is a formal set of symbols with duality of patterning, which must be learned like a spoken language.

Do apes use ASL symbolically, in the same way that humans do? They achieve "vocabularies" of a hundred or more signs, used (usually) in appropriate circumstances. However, critics have raised one criticism after another of ape symbol use as resembling humans'. Let us consider iconic symbols, conditioning, cuing, and finally return to affective signals as opposed to symbols.

As in a spoken language, ASL has some iconic intermediates between signal and symbol. The common primate open palm gesture of begging is formalized into the Ameslan sign *come-gimme*, much as the baby's anguished reaching for its mother becomes a conventionalized lifting of the arms, which means "Pick me up." As Bellugi and Klima (1976) demonstrate, the iconic or pictorial origins of signs are fairly irrelevant to their final use. Each deaf-language has its own symbol for *tree*, which originated from a mime of one aspect of a real tree (Fig. 20.5.). The resulting gestures are each so stylized and arbitrary they must be learned like any other symbol (Fig. 20.6.).

The most famous of Washoe's iconic gestures was *bib*, as reported by the Gardners:

> Washoe had begun to use (the napkin/wiper) sign appropriately for bibs, but it was still unreliable. One evening at dinner time, a human companion was holding up a bib and asking her to name it. Washoe tried *come-gimme* and *please*, but did not seem to be able to remember the bib sign we had taught her. Then, with the index finger of both hands, she drew an outline of a bib on her chest—starting from behind her neck where a bib should be tied, moving her index fingers down along the outer edge of her chest, and bringing them together again just above her navel.
>
> We could see that Washoe's sign for bib was at least as good as ours, and both were inventions. At the next meeting of the human participants in the project, we discussed the possibility of adopting Washoe's invention as an alternative to ours, but decided against it. The purpose of the project was, after all, to see if Washoe could learn a human system of two-way communication, and not to see if human beings could learn a system devised by an infant chimpanzee. We continued to insist on the napkin/wiper sign for bib, until this became a reliable item in Washoe's repertoire.
>
> Five months later, when we were presenting films on Washoe's signing to fluent signers at the California School for the Deaf in Berkeley, we learned that drawing an outline of a bib on the chest with both index fingers is the correct sign for bib (Gardner and Gardner, 1969).

Iconic signs no longer are a focus of doubt—most ASL signs are arbitrary to some extent. Furthermore, it seems to demonstrate a fairly high level of sophistication to make a sign like Washoe's *bib*.

Perhaps, then, the ape gestures are no more than an extensive set of conditioned reflexes. Apes chiefly use their words as requests for food, walks, and tickling. The training reinforces this tendency. A trainer holds an apple. The young ape may feel only desire. He learns a long list of contingencies—if the desired object is an apple, make one gesture, if banana another gesture, if milk a third. (Note even this supposes some mental categorization of apples versus bananas.) The trainer writes down the ape's gestures as three discrete symbols, but the ape might instead be making three conditioned reflexes for "gimme."

This is the basis for Skinnerian criticism of the ape work, and for Epstein's pigeons, conditioned to peck their "conversation" keys. It is not actually a distinction between ape and human language learning, in Skinner's view, because he supposes human language is also an extended chain of conditioned reflexes.

Sign   Drawing Showing
Iconic Relationship

American
Sign Language

Danish
Sign Language

Chinese
Sign Language

FIGURE 20.5. *There is no international sign language, although many signs have a pictorial or iconic origin. The sign for* tree *in three different sign languages. (Bellugi and Klima, 1976.)*

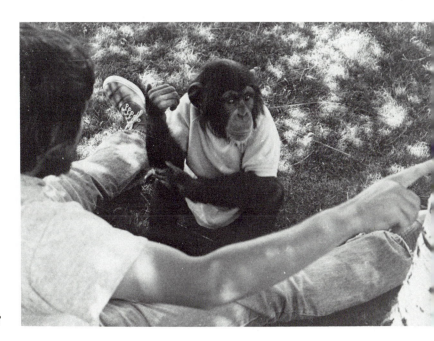

FIGURE 20.6. *Moja replies tree to the question,* what that?

In any case, Sarah, Premack's (1971) star pupil, seems to have disposed of that argument. In her "language" of plastic chips, the symbol for apple was a blue triangle. She identified the blue triangle as the-same-as red, round, and having a stem.

A far more serious set of doubts is raised by Sebeok and Umiker-Sebeok, and by Terrace. They suggest that the majority of ape signing is cued by human teachers. The cuing can be largely unconscious on the teacher's part, like that of Wilhelm von Osten and Clever Hans. Clever Hans was a horse in turn-of-the-century Austria. His trainer, von Osten, was transparently sincere in his belief that the horse could solve arithmetic problems, stamping with its hoof until it reached the correct sum. He was also transparent to the horse; as von Osten relaxed ever so slightly at the right answer, so Clever Hans stopped stamping. The cues were so small that neither von Osten nor the crowned heads of Europe were aware of what the horse saw clearly. At length, Oscar Pfungst, psychologist, tried giving different problems simultaneously to Hans and von Osten; Hans stamped out the answer his trainer expected. Although the shadow of Clever Hans lies over all animal psychology, von Osten was luckier than John Banks, whose bay horse Morocco could sum a throw of dice by pawing in just the same fashion. Banks and Morocco were apparently both burned at the stake for witches, about 1600. The Oscar Pfungst of the day, Samuel Rid, described the whole process of cuing a horse by slight signals in "The Art of Juggling," 1612, and percipiently concluded that the audience watched the horse when they should have watched the man (quoted in Sebeok, 1979).

Animals' sensitivity to nonverbal cues is surely involved in much "language" learning. The trainers' desire that their apes shall learn is openly admitted in all their accounts. It is also obvious that apes often will not concentrate at all on "schoolwork" unless there is a strong emotional component—a trainer they like, who is showing pleasure in the results. Terrace, in his analysis of videotapes of Nim, and even the *National Geographic* photographer who records Koko and Patterson, show the teachers frequently cuing the *particular* sign they expect the animal to make.

Experiments to exclude cuing have been carried out by the Gardners (Gardner and Gardner, 1971, 1974, in press). Washoe, and more recently Moja, Tatu, and Dar, are shown slides of various objects by back projection on a screen that only the chimp could see. An experimenter projects the slides; two different observers record the chimpanzees' signs. One observer is in the room with the chimpanzee, the other watching through a one-way window behind the screen. The two independent observers reported 54 per cent success in naming (for Moja) and 80 per cent or more for the other three, out of 16 to 35 vocabulary items on a test.

The Gardners then analyzed errors to see if there were consistent patterns. A disproportionate number of signed errors were correctly placed, that is with the hand inside lips, or mid face, or side of the head and so forth where the right sign should have been. However, hand shape or movement was wrong. That is, *cat* could be confused with *apple,* as one is pulling an imaginary whisker at the side of the face, the other is closed fist at the side of the face. Such errors also show familiarity with ASL. Only a language-trained ape would confuse *cat* with *apple.* (Patterson also notes such errors, though she considers some as manual "puns.")

A high proportion of other errors fell into the right conceptual categories. Washoe confused *pants* with *clothes,* and most of the apes had cats somehow mixed with dogs, cows, birds, bugs, or babies.

Premack and Premack (1983) also have done extensive "blind" tests. In one series, the trainer did not know the ape's language of plastic chips. Sarah's performance declined, but remained far above chance level. In another, Sarah marked the correct answer with masking tape while the trainer was out of the room. It is fascinating that the apes will solve such puzzles for their own sake, being, if anything, distracted by immediate food reward. It is also interesting that for the crucial blind tests the Gardners chose to study the generality of such words as *cat, shoe,* and *tree,* whereas the Premacks chose to teach a new logical relation: *similar* as distinct from either *same* or *different.*

This brings us to a broader question. Does the early vocabulary of child or chimpanzee indicate its concepts? What actually *concerns* a baby? Nelson (1973) recorded in detail the first 50 words of eight American children. She asked the mothers to write down the words as they appeared, with pronunciation, notes on context, and English gloss. She classified the ones with the same English referent as the same, that is, *cat, kitty,* and *meow* are different children's versions of the one referent, *cat.* By this criterion, the eight children had only 188 different words in their first 50, not 400; that is, there is great overlap, or even a common set of concerns of such middle-class American infants (Gardner and Gardner, in press).

Table 20.2. gives the list of the first 50 words used by the Gardners' five chimpanzees. They overlap greatly with each other and with the similarly raised children, including one deaf child of deaf parents who used the same sign language as the chimpanzees. The vast majority of nouns are things the child eats, wears, or acts upon, with the addition of a few animals (Gardner and Gardner, in press). Interestingly, 44 of Nim Chimpsky's first 50 words also fall on this list, allowing for *eat = food, listen = hear,* and *play = chase.* Nim's six different signs include one adjective, *red,* and five nouns for the same sorts of objects that appealed to the Gardners' chimps—Nim spoke of powder and handcream while the Gardners' proteges used lipstick. The apes, of course, were consciously taught many or most of their signs, but exposed to many more, as adults signed around them. To some unknown extent this is their "choice" of vocabulary, from what is available, like the children's choices. And to some extent, it is more revealing when they do what we do *not* expect. Sarah did not like using *no.* Given a red card on a blue one, and asked, *Query blue on red,* she sometimes rearranged the cards, then answered, correctly, *yes.* Even more telling, when instructed, *Sarah give apple Mary,* Sarah took the apple, and instead of obeying, or using her considerable physical strength to appropriate it, she took her own name-chip and stamped it all over the apple (Premack and Premack, 1983).

Whether this choice has much reference to what an adult ape wants to say is another question, as DeVore (personal communication) observes. We do not teach ASL signs for *lust, kill* or even *grovel, present, penile display.* The metaphors of four- and five-year-old apes, with their range of anal insults, or Koko's description of herself as "red rotten mad," may be attempts to find a way to communicate sentiments that are the real stuff of ape life, using the inadequate symbols we dole out to them.

Finally, we must admit that whether or not apes have the same meanings or concepts for their symbols, they rarely *use* the symbols as we do.

Children play with words. They name things to themselves. They name things to their parents, glancing from object to adult for confirmation. They also use words as requests, but "simple" naming appears even earlier. They point at objects, or point with nonsense syllables, glancing at an adult to confirm that they have directed the adult's attention. Besides, they share toys and food.

All this activity directed to others appears much later in apes, mainly with intensive human training. Apes babble, a little. ASL-trained apes play with gestures, a little. The drive to naming things for its own sake, that is, in play, seems to be far greater in humans. Washoe and Koko sometimes sit with picture books, signing to themselves. Do we want to call this naming? Or playing school with its associated gestures? Washoe commonly signed to herself in private, for instance up a tree, and stopped when she spotted people watching, again, like a child in play monologue. Sometimes her signs related to her actions, as in moving about a forbidden part of the yard signing *quiet* (Gardner and Gardner, 1974).

Two of the computer-trained apes unequivocally use signs to communicate:

> The ability to select a lexigram when shown an object . . . implies only that the chimpanzee can successfully produce the correct associations when the human teacher structures the word game. . . . Sherman and Austin, however, did something quite different. After three years of (complex) word games . . . Sherman and Austin suddenly took the game out of the teacher's hands . . . they began to name items and show them to the teacher. . . . For example, they say STRAW at the keyboard and point to the straw, then BLANKET and pick up the blanket. They would at times briefly touch an object before naming it, *without* looking at the teacher as they touched it, as though (they) were . . . singling it out from others to be named.

TABLE 20.2. Matches in the First 50 Words of Children and Chimpanzees

	Chimpanzees					Children	
	W	M	P	T	D	Words N=8	Signs N=1
Names and Pronouns							
Me	+	+	+	+	+	3	
You	+	+	+	+	+		
Own name	+	+	+	+		1	
Other names[a]	+	+	+	+		8	+
General Nominals							
Apple		+		+	+	1	+
Baby	+					6	+
Ball					+	5	+
Banana	+			+		1	
Bed	+					1	
Berry			+	+	+		
Bib		+	+	+	+		
Bird	+	+	+	+	+	1	
Blanket	+					2	
Book	+		+			2	
Brush	+		+	+	+		
Cat	+				+	5	+
Clothes	+						
Comb		+			+		
Actions							
Catch	+		+	+	+	1	
Chase		+	+	+	+		
Clean	+		+	+	+	1[c]	
Come	+	+	+	+	+	1	
Down	+	+	+	+	+	2	
Go	+	+	+	+	+	4	
Groom					+		
Hear	+	+	+	+	+		
Hug	+	+	+	+	+		
Hurry		+	+	+	+		
In	+		+			2	
Jump			+				
Kiss		+					
Open	+	+	+	+	+		
Out	+	+	+	+	+		
Peekaboo	+		+			3	
See	+	+	+	+		1	
Sleep			+	+		3	
Smell	+	+	+			1	+
Tickle	+	+	+	+	+	1	
Up	+	+	+	+	+	4	

This table lists vocabulary items with sign/symbol markings (+) across several columns and frequency counts.

Nouns							Count
Cookie		+				+	4
Cow	+	+	+			+	1
Diaper		+	+	+		+	
Dog	+	+	+	+		+	7
Drink	+	+	+	+		+	1
Flower	+	+	+	+		+	
Food	+		+	+		+	3
Glasses		+	+	+		+	1
Gum						+	
Hankie		+				+	
Hat	+	+	+	+		+	
Hurt	+	+	+	+		+	1[b]
Ice Cream			+				
Key	+			+		+	3
Light		+	+	+		+	2
Lipstick	+	+	+	+		+	
Milk	+	+	+	+		+	5
Nut	+	+	+	+		+	1
Oil	+		+				
Pants	+	+	+	+		+	
Pen	+	+	+	+		+	1
Potty		+	+	+		+	1
Shoes	+	+	+	+		+	3
Shirt		+					
Sweet	+	+					
Toothbrush	+	+	+	+		+	
Water	+	+	+	+		+	3
Wiper	+	+	+	+		+	

Modifiers

							Count
Dirty		+	+	+		+	2
Finished		+	+	+	+	+	1
Funny	+						
Hot	+	+	+	+	+	+	7
Mine	+	+	+	+	+	+	2
More	+	+	+	+	+	+	3
Quiet	+	+	+	+	+		
There		+	+	+	+	+	2

Personal

Social

							Count
Can't	+	+	+	+			1
Don't know		+	+	+			
Good	+	+	+	+	+		1
Good-bye	+	+	+	+			5
No	+	+	+	+	+	+	5
Please	+	+	+	+			
Refusal		+		+		+	3
Sorry	+						
Yes		+	+				5

Functions

That				+			

[a] Washoe had two and Moja had three other names.
[b] Classed as an action in Nelson sample.
[c] Classed as a modifier in Nelson sample.

After they named it, they would again point at the object, now deliberately and with a more expressive gesture, and they would look at the teacher or simply pick up the object and give it to the teacher. This behavior first appeared between the chimpanzees themselves in the food sharing context. . . . If Austin were asking Sherman for food and could not find the correct lexigram, Sherman began to make the selection for Austin and then give him the food" (Savage–Rumbaugh, in press).

After all, language as humans use it is *for* the communication of information and ideas, not just a kind of internal mental gymnastics. The fact that apes rarely use symbols to communicate confirms what we always suspected, that they are not human. The fact that some sometimes can do so confirms the power of culture to reveal latent mental capacities—in them or us (see Bruner, 1983).

Classification, Connotation, Combination

Washoe, the Gardners' chimpanzee, used keys to open the padlocked cupboard doors of her house trailer. She soon generalized the skill to using other keys, including ignition keys. Similarly, she learned the "word" for key, using the original padlock key for reference, and promptly generalized to using it for other keys, or to ask for a key when she wanted a door open.

Children's generalizations need not follow adult lines at all: "Charles Darwin's boy, when just beginning to speak, said 'Quack,' with reference to a duck and then applied the word to water, then to birds and insects on the one hand, and to liquids on the other. Later, having seen a representation of an eagle on a French coin, he named other coins quack" (Lewis, 1963, p. 49). A similar example is a child, Margaretta, one of whose words was *doggie*, applied to all small animals including dogs. One prized book displayed pictures of a doggie, a teddy bear, and daffodils. She called them all "doggie" and proceeded to generalize to include real flowers and circular geometric patterns. Either category of fuzzy animals or flower patterns makes sense to us, but it apparently was no problem for a child to use the one word for the lot.

Lewis analyzed in detail his child's learning animal names (Fig. 20.7.). The child progresses from "tee," the contraction of Timothy, their cat's name, through

FIGURE 20.7. *"Overgeneralization." A human infant's words for animals are only gradually narrowed to adult categories. Both ape signs and the vervet alarm calls of Fig. 20.4 follow the same pattern. (After M. M. Lewis, 1951.)*

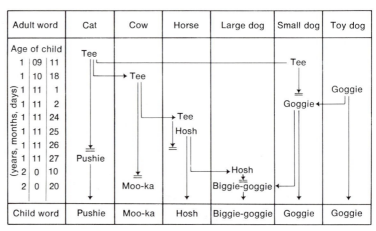

Key: Word is written at first use; = indicates final use

"goggie," "hosh," and so forth. Lewis points out that as well as the obvious generalization from one animal to another, some of the child's choices are perfectly logical. After all, the Saint Bernard is clearly more like a hosh than a toy goggie. However, as each distinct adult word was offered, the child dutifully incorporated it into his vocabulary, even accepting the compromise "biggie-goggie" for the Saint Bernard.

Brown (1958) argues that the referents of children's words are at first very large, almost undifferentiated categories, and then more concrete ones comparable with adults' objects. As the child matures, his vocabulary enlarges to include adult abstractions and groupings of classes. Thus, Lewis's child with his group of all quadrupeds, or Margaretta with her very unadult grouping of animals and flower patterns had formed infantile categories. The only requirement for membership seemed to be that one member shared *any* quality with an adjacent member, just like Bruner's series of objects or Piaget's graphic collections. Each member relates to an adjacent one, but no overall criterion that an adult would recognize applies to every member uniting the whole. As Brown points out, the fact that these collections take on simple names, such as "dog," with simple object meanings to an adult, tells us very little about the nature of the collection, merely that adults are likely to converse with their children about doggies.

Koko, the gorilla, similarly overgeneralized. There was a period when she used *bean* for cookies, shoes, artichokes, toy tigers, Jello, and a person, with Koko's toy glasses propped on his head. Did the word mean *anything?* Other overgeneralizations seemed reasonable—*straw* for long thin things, *tree* for asparagus, *corn* for small edible kernels, so a cut pomegranate became *red corn drink.*

This brings us back to the remark at the beginning that we have little idea what the signs mean to an ape. The Gardners' double-blind tests show that nouns such as *tree* and *shoe* are applied to the same variety of objects we would so class. *Shoes* are anything from sneakers to hiking boots. But what about *funny?* or Koko's *think* and *know?* It is highly doubtful that she distinguishes the two concepts; either sign, in the reported contexts, could simply be a reinforcer, like *sure is,* not anything so self-conscious as *know.*

Another equivocal sign is the back-of-hand on chin, which is coyly translated *dirty.* The connotations seem to be more like *shit* both in insult and as description of fecal-like substances. We have no idea if apes would associate feces and insult without a middle-class American upbringing. (I think this was first admitted in print by Crichton, 1979).

One formal approach is to study categories. We have seen that true classification appears fairly late in child development—age four to five if measured by nonverbal methods. Sherman and Austin could classify symbols (computer lexigrams) into two groups: edible versus nonedible objects. Another chimpanzee, Lana, had learned the same computer language. She had a different training technique, however, simply naming objects shown by her trainer rather than requesting and sharing tools and food like Sherman and Austin. Lana apparently can sort objects without difficulty, but cannot use a category symbol (Savage-Rumbaugh, et al., 1980; Savage-Rumbaugh, 1981).

The other clear indication of symbol use is invented combinations to describe new situations. The most famous was Washoe's *water bird*, to describe a

A

B

C

D

E

F

FIGURE 20.8. *How do you distinguish relevant meaning from semi-appropriate reflexes, or a string of linked words from separate gropings? A. The teacher molds Nim's hands into the sign* tea. *B. Nim approximates the sign. He then tries different signs relevant to the situation: C.* give. *D.* drink. *E.* more drink *(a combined sign). F.* more, *while the teacher insists* tea. *(Courtesy of H. Terrace.)*

swan. Skeptics, for example, Terrace, say we still do not know for sure that she was not referring to water, then the bird. Lucy's *cry hurt food* for a radish, and *drink fruit* for watermelon could be queried the same way (Temerlin, 1975). By the time one realises all the signing apes combine words, it may be simplest to assume the phrases describe the situation, if not naming the object. Recall this is just what animals and humans do *not* do with nonverbal sounds; sequential combinations are rare and, when they occur, seem to have new, evolved meanings unlike their components. (See Fig. 20.8A,B,C,D,E,F.)

GRAMMAR

Grammar, the classic symbol of dry-as-dust pedantry, has kept a few scholars through every century locked in mental combat, assailing each other with their inkhorns or floppy discs. For them, grammar is the study of how the mind encodes—or creates—reality.

Grammar is the internal structure of a statement, which assigns relations among the parts. One-word remarks can have no grammar, even though they may be relevant, or illuminating references to a situation. If a child says *Milk!* it is the listening adult who interprets the word to mean "I want more milk!" or "My milk, not yours!" or "The milk spilt!"

Two-word phrases are highly equivocal. The main grammatical device of English is word order. Other languages depend less on word order than in-flections—changes in the words themselves which tag a noun as subject or object, a verb as referring to the actions of a particular noun, and so forth. Latin, Chinese, and ASL are languages in which inflections matter more than in English. Children's two-word phrases have definite rules. Children combine words in preferred orders, although there are many exceptions. Children have certain "pivot words," which can be combined with many others. However, it is again the interpretation by adults in context that makes them seem grammatical. *Bite man* could have many meanings.

When one reaches three-word phrases, it is much clearer whether or not the phrase is grammatical. *Man bite dog* is different from *dog bite man*, even if the child leaves the inflected *s* off *bites*. One sentence or the other is true of a given action, not both. Now consider *bite dog man*. We instantly know this is ungrammatical. It may still evoke an important situation, but does not specify the relations within the situation. The doting parent can still think of expansions, such as "That's the biting dog, the one who bit the postman." This is no more farfetched an attempt to make the child phrase congruent with an adult phrase than the attempts we regularly make for two-word combinations. By this point, however, we expect a much closer approximation to adult word order. Note that ungrammatical as it is, it may still be useful communication. You might go and rescue the postman.

We have just been through a phase of mystic grammar. By mystic I do not mean vague and imprecise, but that the deep structure of grammar is the inner program of the human mind, the link between gene and soul, as Descartes used to consider the pineal gland. Noam Chomsky's teaching runs all the way from genetically given deep structure to the ideal of human freedom:

> The study of language . . . (is) a branch of theoretical human psychology. Its goal is to exhibit and clarify the mental capacities that make it possible for a human to learn and use a language. As far as we know, these capacities are unique to man, and have no significant analogue in any other organism. If the conclusions of this research are anywhere near correct, then humans must be endowed with a very rich and explicit set of mental attributes that determine a specific form of language on the basis of very slight and degenerate data. Furthermore, they make use of the mentally represented language in a highly creative way, constrained by its rules but free to express new thoughts. . . . If this is correct, there is no hope in the study of the control of human behavior. . . . (Chomsky, 1969, p. 84).

Chomsky's diametrical opposite, B. F. Skinner, believes that the organization of language is wholly given by conditioned response and reward. He can teach pigeons to play ping-pong or guide guided missiles, or to peck at a lighted spot in bafflingly complex sequence. He argues that surely the subtle, many-times-a-day rewards offered a growing baby could explain any complexity of speech. Furthermore, Skinner believes the child's responses are wholly determined by its history of conditioning. This is inverted Calvinism, with learning in the role of predestination. The recent "discovery" of "Motherese," that is, the fact that adults tend to speak clearly, repetitively and, above all, grammatically to very small children, supports Skinner's view that linguistic structure may be learned from models in the world around.

There is thus another controversy between one school that emphasizes structure as given, to be filled with a variety of different elements, and another that emphasizes the juxtaposition of learned elements, believing that the elements will in the end become a structure. The earliest two-word combinations of small children do not distinguish one theory from another. It is the rich interpretation offered by adults that confers grammatical structure on these sayings, and the obvious fact that children soon progress onward to adult word order and inflections.

One effect of the shifts of fashion or theory is that the linguists too often talk as though there were just one way of learning language. Even the four Jolly siblings joined rival theoretical camps in the cradle. The first was an imitator, with a good ear for sound and an average rate of development in words and grammar. At one and a half years she would babble to herself an "English" monolog, complete with the intonations of declarative and interrogative sentences, commas, and clauses. Rarely a real word intruded into these "sentences." Because she imitated sound rather than either meaning or grammar, and because no one is likely to argue the universality of mid-Atlantic English, she would have provided ample ammunition for the older theories of imitation learning. However, the second child was a Chomskyite. She began to speak on her first birthday, which is average, but promptly dropped all meaningless babbling in favor of a vocabulary of ten and then twenty words. At 23 months she made remarks like, "Mummy, can I take my apple all-the-way home?" The social psychologist might have seen her desperate to organize her world, but a linguist would have delightedly turned to studying the grammar rather than any other aspect of her speech. The third Jolly was and is a highly social creature. He had to be, born a teddy bear to two young sisters. Thus, his declarative babblings, like much of the rest of his behavior, were aimed at other people and frequently rewarded by an answer from one of the family. That baby would have confirmed the views of a hardened Skinnerian. Finally, the fourth went back to grammar with a side excursion into apelike signing. He reacted to the growing family cacaphony by refusing speech in spite of understanding much that was said. He simply mimed his desires with stereotyped gestures and a couple of personally invented words. Then he accelerated from the one-word stage to full sentences within about three months at two and a half years. Like his second sister, he needed to organize his world, but waited until he coped at a higher standard of his own.

Obviously, none of these children learned to speak by just one method. However, it seems worth pointing out that four siblings can differ enough in their apparent emphasis to "confirm" any theory in its own sense of importance. Further, the differences are not differences in some pure, single-factor, linguistic ability. The differences are rooted in each child's character and social surroundings.

This aspect of ape language learning is only beginning to receive attention, particularly in the work of Patterson and of Savage-Rumbaugh. We are only just beginning to ask what an ape *wants* to say, rather than how he says it.

What, then, are the ape data that bear on the use of grammar? Premack

(1976) has done most to test formal grammar. With the constrained "language" of plastic chips, his chimpanzee could deal with such commands as *Sarah put apple dish banana pail*, or *Sarah take apple in red dish, banana in blue dish*. As with the *same* and *different* chips, which parallel the Weigl-type oddity problems, Premack translated known chimpanzee capacities into what others called grammatical capacities. More recently (1983), the Premacks have established left to right as a criterion for three-object sequences: clean paper-pencil-marked paper *but* marked paper-eraser-clean paper (see Chapter 18). Only Sarah, an adult chimpanzee with years of training, seems to deal with such a problem. Premack concludes, however, there is no evidence that apes use anything like human grammar, and further that when we understand more about human use of words, we may find differences there, as well.

All the signing apes combine signs into strings of two or more. They have preferred pivot words, just as children do. Many of these strings combine words interpreted as different parts of speech, in appropriate semantic combinations such as noun and adjective and noun-verb, *berry red* or *you me tickle* (Table 20.3.). One of the few objections to the ape experiments that I have not seen in print is that the apes may not think of them as different parts of speech. If we write the phrases, *berry redness* or *hurrying cupboard opening,* they could be made to sound even less grammatical than usual. The combination of agent and action could just be that agents and actions are often juxtaposed in the mind, not that the chimpanzee has any inkling that we would distinguish *to bite* from *a bite*. The same is true for small children, of course.

Within each corpus, the apes have preferred word orders. However, these

---

TABLE 20.3. Examples of Nim's Two-sign Combinations

Action + Object (27%)	*Routine (6%)*
eat grape	out pants
drink tea	in pants
out shoe	
Object + Beneficiary (16%)	*Attribute + Entity (5%)*
food Nim	green color
ice Nim	red apple
baby Nim	orange balloon
yogurt Nim	hungry me
Two Propositions (16%)	*Action + Place (5%)*
eat tickle	clean there
dirty hug	out there
out open	tickle there
Entity + Place (6%)	*Agent + Action (3%)*
baby chair	Bill run
food there	Nim out
grape there	Nim wash
banana house	me open

(After Terrace, 1979.)

are not particularly consistent (subject does not always come before verb, for instance, but varies with each pivot word). The subject and object do not bear consistent relations—no firm distinction between *you me tickle* and *me you tickle*. ASL is an inflected language, as all concerned point out. The apes hold the end of questions longer, as humans do, and catch the recipient's eye, signing with a hand motion toward the other person for *you*, and so on. However, as Premack (personal comment) points out, wherever the trainers have not been fluent signers, they have had to fall back on pidgin English word order to convey grammatical relations, so one would expect the apes to have even more clearly modeled word order than if they had lived in fluent households.

In summary, although there is still much debate, the two-word phrases of young apes seem much like those of children, in pivot words, in preferred but not fixed word orders, and in impossibility of telling by clear-cut criteria whether there is any grammatical structure involved. When we consider longer strings, the situation changes.

The most publicized recent attack on the existence of ape grammar is that of Terrace and colleagues (1979). He raised a chimpanzee named Nim Chimpsky for four years. In that time Nim learned to make about 125 signs and to comprehend about 200. Nim clearly had preferred word orders. The pivot words *more* and *give* usually come before the thing demanded, and beneficiaries of an action would follow. Nim reached a mean phrase length of 1.6 signs, excluding directly repeated signs.

However, the phrase length did not increase over Nim's last two years of study. Sentences grew longer, but by repetition, which is also common for other apes, not elaboration of meaning. Nim's longest utterance of 16 signs went: *Give orange me give eat orange me eat orange give me eat orange give me you.* Such a string of words is apish, and most unlike the longer utterances of children. However, Nim's trainers tended to keep holding a reward just out of reach, to encourage further signs, which would provoke this kind of response. This string is also uncharacteristic of somewhat more coherent signing of cross-fostered apes, at least as reported in the literature.

Nim's pivot words were few in number: 76 per cent of agents were *you*, 99 per cent of beneficiaries of actions were *me* or *Nim* (used interchangeably), recurrence was nearly always *more*, place nearly always *point*. Further, when Terrace analyzed three and a half hours of videotapes, he found that Nim's teachers often unconsciously cued the chimpanzee. About 40 per cent of Nim's signs were imitations of a teacher's immediate signing. Only 12 per cent were spontaneous in that they were not preceded by a remark or question of the teachers. Terrace concluded that there is real doubt whether apes can create sentences.

Terrace does not minimize the shortcomings of Nim's rearing. In the little chimpanzee's four years, there were 60-odd different teachers and caretakers, and he was repeatedly separated from people he had loved. Jane Goodall's description of orphaned Merlin, who never developed the techniques necessary for termite fishing, recalls Nim's short phrase length. Further, at least some of his tapes show a very bored animal being lured or coerced into signing while his attention wandered (Ristau, personal communication). The sad remark that a mean phrase length of one and a half words seemed enough for what Nim had to say may be true—but only for that particular ape.

Miles (1983), for instance, makes a direct comparison between Nim and her three-year old orang-utan Chantek. Chantek is brought up in the rich environment of house trailer and yard festooned with ropes for brachiating, and by a few, long-term caretakers. The project stresses communication—that is, if Chantek seems thirsty but signs *food-eat,* he is given food not drink, and certainly not testing to a criterion of producing the right sign. Videotapes of relaxed communication, compared with Nim's show Chantek's mean phrase length as 1.9 words. Almost none of Chantek's signs imitated the teachers previous signs. In contrast, some 60 per cent of Chantek's remarks were spontaneous, in that the animal initiated the exchange, with no previous remark of its caretaker. The one measure where Chantek and Nim coincide is their low proportion of expansion. Young children frequently expand an adult's remark by taking one or two words from the adult's speech and then adding more words, which develops the adult's idea into something like a conversational sequence. Here the two apes resemble each other, and differ from children.

Patterson also compares Koko, the now near-adult gorilla, with four-year-old Nim. Koko's mean phrase length is 2.7 words, only 11 per cent of her utterances are direct imitations, and 36 per cent (compared to Chanteks or Nim's 7 per cent) of Koko's phrases are expansions on her teachers' statements. (It is almost impossible for an outsider to tell whether the various research workers mean quite the same thing by expansions and imitation.) Koko, like Nim and other signing apes, has preferred word orders. These are not always the same as her teachers'—adjectives generally follow the noun, for instance—which suggests she has formed some rules of her own.

Every one concerned adds we should pay more attention to the intrinsic grammar of ASL. Signs take longer to make than words, so they are often condensed by humans and by apes, with arms or the two hands signing simultaneously. Signs can be modulated, like tones of voice and like grammatical inflections. For instance, a cocked head and a glance, while holding the last hand sign slightly longer than usual, turns a statement into a question. Finally, in ASL (like Chinese and Latin) word order matters less than in English. All this means that it is easier to present a whole thought at once in ASL than in English, where one needs instead linear, sequential attention. However, we still do not know whether the signing apes' condensed thoughts are inflected as they would be in grammatically correct ASL, or if they resemble two-word baby ASL, and are just as equivocal as babies' two-word English

The upshot is that the existence or extent of ape grammatical competence remains in doubt. Apes certainly join words into meaningful combinations. They certainly have short, repetitive phrasing, unlike the three- and more word sentences of small children. We have yet to prove whether changes in the structure of ape phrases correspond or not with the structure of situations in the outside world.

TIME, LIES, AND COMMUNICATION

At intervals throughout the book we have come up against the question of negation, deception, and time sense. Chapter 17 discussed chimpanzees gathering and preparing tools for fishing long before they reach the termite nest, and the chimps who can leave bait to fetch a tool they saw in another room. In Chapter 5 there was the claim that chimpanzees, gorillas, and orangutans remember the phenology of their food, moving to new feeding grounds just before the fruit or bamboo begins to ripen or sprout.

In Chapter 17, Julia planned her sequence of opening locked boxes. There was also the deliberate deception of Figan, who led the other chimpanzees away from the bananas, or the female baboon who tricked a male away from his antelope carcass. There was Luit, wiping the fear grin from his face with his fingers. Then in Chapter 19 we discussed play as a formalized negation: "This is not real." Finally, in this chapter we have touched on displacement in two contexts: the derivation of speech sounds from generalized social grunts, and the divorce of arbitrary verbal symbols from the emotional urgency of whatever they denote.

Premack's account of chimpanzee's abstract codes may reveal yet one more step into the deliberate codification of reality. Sarah could choose *same as* rather than *different* in the test sentence *Apple is red* blank *red color of apple*. This, to our minds, is a metalinguistic comment, whatever it may be to the ape.

All these examples are loosely linked by an animal's capacity to detach itself from the immediate, and to accept some steps backward, or some framing, which in the end may lead by roundabout routes to the original goal. I am *not* claiming that all these diverse behaviors derive from one mechanism. Even starfish migrate to new feeding grounds, and the play of a baby is far from the conscious use of negation. What I do claim is that our own foresight, hindsight, and endless arguing must include such a capacity to step backward away from the present and to keep in mind chains of events, particularly what is not true, not true now, or only hypothetically true. What insight do the signing apes give into this sort of complex thought? Koko, Michael, and the Gardners' chimpanzees have all reported past events of which their trainers were unaware. This is a true communicative function of language.

For instance, when Washoe was 27 months old, she managed to insert into a hole in her trailer wall a toy that fell down between the double walls. When Allen Gardner arrived that evening, she signed *open, open* many times at the level of her bed, below the hole in the wall. He guessed the trouble and fished out the toy—then realized Washoe had used signing to convey new information and achieve her goal. This is, of course, a very simple communication—no grammar, word combination, or any criterion of language except that the symbol was learned and arbitrary. She could even have conveyed as much by staring and scratching at the wall. Still, this sort of account to me is more interesting than arguing over grammar, for it shows how a precursor of language can be *used*. (See Laguna, 1963, Bruner, 1983). (See Figs. 20.9 and 20.10.)

Several or perhaps all these apes use signs as insult—usually variants on *dirty* and *toilet*. Nim, Koko, and others use signs instead of aggressive actions, *bite*, for instance instead of actually biting. What is more significant, several—Lucy, Nim, Koko, Michael—have attempted to shift blame for misdemeanors onto human companions, for anything from feces on the carpet to a smashed sink that only a gorilla could break. Of course, these lies are somewhat invited by their teachers' rhetorical demand, "Now who did that?"

The most amazing developments are Koko's verbal humor and verbal metaphors. She sometimes describes herself as a *red mad gorilla* or *red rotten mad*.

FIGURE 20.9. *Inter-ape communication with signs: Tatu and Moja gaze at each other as Tatu signs* drink.

She once refused all day to make the *drink* sign, a sign she had used thousands of times before. When her exasperated tutor at last said, "Koko, please please sign *drink* for me," Koko leaned back against the counter, and grinning, signed *drink*—but into her ear, not her mouth. Washoe and other signing apes have a sign for *funny,* which they use in many situations that make them laugh.

Koko and Michael are also said to discuss abstract ideas such as death. There are many possible reasons that these are the only apes for whom such claims have been made—they are gorillas, not chimpanzees; they have lived in a human environment to adulthood, and so on. It is also possible their humans imagine what they wish to see. The ultimate interest, of course, would be to use language to do just what Patterson claims—to explore the mental world of an ape, its references as well as its grammatical architecture (if any). Even if it

FIGURE 20.10 *Communication with signs: Koko signs* visit. *Michael inside the cage may be signing* more.

is finally agreed that the structure of ape word combinations remains equivo-cal, and in need of interpretation and context to make them clear, what apes have to say is still fascinating.

One of Koko's conversations, one of the most amazing claimed for an ape, can serve as a test for credibility. Various people who have worked with ape intelligence flatly don't believe this. Patterson might reply that 25 years ago we defined man as the only toolmaker, and 15 years ago as the only user of word-like symbols, and in the last five years we learned an ape can solve formal problems of analogy. Why not now accept Koko's metaphors? Critics in turn point out that the gullible love to believe in mathematical horses or human te-lepathy. It is all too easy to delude oneself (or be deluded by one's assistants) into crediting metaphor in apes, especially after investing a lifetime's work and money in the teaching. The only immediate conclusion is that the ape workers must let each other watch unedited videotapes, and publish together, or the rest of us will have to decide by default: does this sound likely?

Koko nags for soft drinks, and, when too persistent, is given water and a straw to suck with. One day she was drinking through a fat rubber tube in-stead. Her teacher, Barbara Hiller, refused to give her yet another drink:

K. *Sad elephant*
B. *What mean?*
K. *Elephant*
B. *You sad elephant?*
K. *Sad . . . elephant me . . . elephant love thirsty*
B. *Thought you gorilla*
K. *Elephant gorilla thirsty*
B. *You gorilla or elephant?*
K. *Elephant me me . . . Time*
B. *Time for what?*
K. *Time know Coke elephant good me*
B. *Want drink, good elephant?*
K. *Drink fruit.*
B. (Showing Koko the rubber tube) *What this?*
K. *That elephant stink.*
B. *That why you elephant?*
K. *That nose.* (Points to the tube, and then wanders away laughing.)

SUMMARY

Apes can produce voluntary sounds only with great difficulty. One chimpan-zee and one orangutan have achieved four such different sounds, used in ap-propriate contexts. However, home-raised apes comprehend a good deal of spoken language.

Several chimpanzees have learned plastic chip and computer-based "lan-guages," and others, as well as two gorillas, use ASL, the American sign lan-guage of the deaf. They have achieved vocabularies of several hundred signs, which may be used in various combinations.

The design features of language that differentiate it from nonverbal communication include openness (the ability to say new things), discreteness (chopping the continuum of sound into limited segments), duality of patterning (sound versus word), denotation (the words have meaning), displacement (the meanings of words persist in the absence of what they denote), and arbitrariness (the meanings of words are learned and can be changed). There are few wild primate parallels to any of these. Vervet alarm calls apparently denote three different categories of predators. Juvenile vervets gradually learn to limit eagle alarms from general startles to situations when birds of prey appear, but the calls may not be arbitrary in the sense that eagle alarms might not be changed to apply to different situations.

The rough grunting of chimpanzees and baboons is the primate sound most like human speech. It is used in many low-key social situations and can be modified by changing facial expression and mouth shape.

True symbols can be displaced from their referents in space and time. ASL signs are such symbols. Much ape signing indicates "I want that," and could be skeptically viewed as an elaborate set of conditioned responses. However, Sarah described her blue plastic "word" for apple as red, round, with a stem, and the Gardners' chimpanzees pass extensive double-blind vocabulary tests. Signing apes look through picture books, naming objects to themselves in play. Sherman and Austin chose to name things with their computers before handing them to the teacher or to each other. This is unequivocal symbolism. It is still difficult to interpret the exact meaning of apes' or small children's symbols. They surely are not so polite or so abstract as their supposed translations. Child vocabularies of the first 50 words are much like those of apes in similar environments. Children and young apes may overgeneralize, using one symbol for an "edge matched" category that seems disparate to adults. Sherman and Austin can make one adultlike verbal classification, grouping symbols for food separately from symbols for tools. All the apes invent new combinations of words to describe unfamiliar objects.

Grammar is a surprisingly emotional subject. Apes can interpret "sentences" with several phrases, and spontaneously string words together in combinations. They have preferred word orders, which are not always those of their teachers. However, their phrases are short and much more prone to repetition of signs or close synonyms than children's phrases. Apes' conversation is mainly demands, so their corpus of speech is heavily weighted toward the pivot words *give, please, more, hurry,* which may be almost synonomous. This weighting is in part ape character, in part a consequence of training with rewards. The controversies over ape grammar are still on the boil. So far there is little difference between ape and human two-word phrases, but little definite grammar or "sentences" that appear in longer human phrases.

All signing apes use words as insults, naming aggression instead of just performing it. Several apes have lied, accusing others as the perpetrators of forbidden damage. Koko the gorilla is said to sign some jokes or metaphors. These abilities indicate combinations and detachment of thoughts and symbols, which are more clearly like humans' than is the apes' rudimentary grammar. Above all, the apes sometimes use their learned symbols in communicating new information to their teachers and to each other. It was presumably in communicating information useful to protohominids that true human language evolved.

CHAPTER
21

SOCIAL LEARNING

Evolutionary change comes through the selection of advantageous genes. Historical change is Lamarkian, not Darwinian. Advantageous historical changes can pass to the offspring directly, without needing to wait for chance mutation, or undergo the cruel process of natural selection.

History does not supersede evolution. There is ample feedback between the two levels of change. A chance mutation may lead to quicker or more practical learning ability, and so be selected for. Learning ability may buffer the rigors of the environment, favoring genes for bodily vigor at later ages, or allowing formally lethal genetic anomalies to live and make their own contribution to society.

There are many current attempts to quantify and explain these interactions (Lumsden and Wilson, 1981; Bonner, 1980; Pulliam and Dunford, 1980). One of the major themes is that of polymorphism, or variability. Does mammalian social life allow or depend on diversity of genotype and behavior and diverse roles within the group? Or is there an "ideal type" toward which all gnus or gorillas tend?

Other major questions are how knowledge accumulates within the social group and how quickly such knowledge can change with changing circumstances.

A final question is how the accumulation of knowledge may change an animal's own environment. That question often seems almost trivial in the time scale of species evolution. It has now become the central question in the time scale of human history. We have already transformed the biosphere at a speed unknown before—unless indeed a falling comet brought global winter which killed the dinosaurs. We have farmed, felled, bulldozed, and armed against our own species, through evolved desire to survive and protect our children. We do not yet know if our children, or the biosphere, will survive.

447

DIFFERENCES IN LEARNING IN THE GROUP

One of the dangerous tendencies of humans is that we so often picture ideal types. It may be the Aryan superman or the Hollywood goddess or the high school senior who scores perfectly on a standardized intelligence test. The great advances in biology instead seem to involve randomization. The genome does not congeal into a perfect type for each species. We go through all the confusion of sexual reproduction just to mix up our genes.

G. E. Hutchinson (1981) suggests that selection has favored some "noise," even in the creation of our proud brain. If the readout of genetic instruction or the recording of experience has a random element, this allows polymorphism of intellect within a kin group, which may in turn be adaptive. Differing powers of intelligence from innovative genius to complaisant follower, and different styles of intelligence from quick-witted hunter to tribal bard have allowed us to mix ideas, not merely genes.

Within the social group "the individuals who detect an object, and those who lead the approach, and those who initially test it out, and those who fool around with it the longest, and those who will best remember its nature and location need not necessarily be the same individuals; and in each of these aspects of group performance there might be transmission of information about the object from one animal to the others" (Menzel and Menzel, 1979). It is distinctly difficult to test out these ideas, because adults keep the young from dangerous or highly interesting objects, so with judicious choice of object and context, you might produce almost any pattern of group response.

On the whole, it is semi-independent infants and juveniles who approach and handle new objects most often and for the longest time. They thus perpetrate most of the useful innovations (Fig. 21.1.).

Menzel (1966) put out small plastic toys where a troop of wild Japanese macaques would find them. Infants and juveniles sometimes played, but older monkeys showed a studied indifference that was clear to the observer but hard to define. Sometimes a mother apparently ignored the object, but then pulled her infant sharply away from it. Of course, context mattered. Menzel placed nine innocuous toys on top of a favorite sitting rock, in the position usually taken by the most dominant animal. The troop began to avoid the rock area altogether. Menzel then left 20 toys, including the same objects, under an overturned crate in the woods all night. The next day the test materials had all been stolen except for two rubber snakes, and juveniles and infants appeared throughout the morning with a toy in hand.

The Japanese macaque troop of Koshima Island is famous for its cultural innovations and their spread. One infant female called Imo began to wash sweet potatoes to remove the grit when she was only one and a half years old. The innovation spread to her playmates and her mother, and from mother to baby sibling. Imo was likewise among the first to swim, and to "placer wash" wheat kernels by throwing a handful of wheat and sand in the water so the sand sinks and kernels float. Each of these discoveries also spread through the troop.

As usual, we will have to qualify any generalization as we learn more about transmission in different species. Cambefort (1981) tested a troop of wild chacma baboons and a troop of vervets with hidden foods marked by visible clues, such as yellow sticks stuck in the ground, or matchboxes that held a peanut. Among

FIGURE 21.1. *Young primates are more likely than adults to manipulate unfamiliar objects. If the object proves safe or profitable, adults then copy the juveniles, particularly within a kin group.*

the baboons, as among Japanese macaques, the juveniles were nearly always first to detect such food. Learning passed nearly instantly, though, to all the others, after an average of one demonstration by the discoverer. (This was a small troop of just ten animals.) The vervets behaved very differently. Adults and juveniles were equally likely to find the food. Use of the clues spread from an animal to its close associates, but it took several days and an average of 668 demonstrations for all 36 animals to acquire the new knowledge. This correlates with baboons' tight hierarchical structure, where the animals pay a great deal of attention to each other, as compared with the vervets' more loosely integrated group.

These baboons, vervets, and forest-living mandrills also differed in learning which foods to avoid (Cambefort, 1981; Jouventin et al., 1976). When they were offered two colors of banana slices, one dosed with bitter chemicals, mandrills who observed each other always avoided the bitter color. Observing baboons ate the right color first but sampled the wrong one later. Vervets seemed to learn nothing from each other's reactions, but had to make their own mistakes.

The infants may be most playful, but adolescents are perhaps particularly

449   *Social Learning*

quick to learn formal tasks. Tsumori (1966, 1967, Tsumori et al., 1965) did a variant of delayed response test with the entire Koshima troop of Japanese macaques. He buried peanuts in the beach sand, patiently arranging the situation so each troop member in turn had a chance to dig without too much interference from colleagues. Late adolescents were quickest and most successful at unearthing the peanuts. This agrees with the laboratory tests that show learning speed increases to near adulthood.

Older adults are perhaps slower to solve individual tasks, but they have a greater store of knowledge in which to integrate new facts. When Menzel (1969a) released a group of chimpanzees in a new enclosure, the adults glanced around and apparently accumulated a great deal of information, but the youngsters may have learned little more from bouncing several times up and down each tree.

The Menzels tested a family group of tamarins on problems with novel objects, comparing two-year-old twin infants, a four-year-old adolescent son, a six-year-old adult daughter, and the eight-year-old parents. The objects were neither edible nor especially fear-inspiring. An object was left in the greenhouse home for 30 minutes, then removed for five minutes, and then replaced. The infants investigated the object most, adults least, but in neither session could the infants play much until the older animals' interest began to decline. As Menzel says, if one could quantify the amount learned by the adults during their first 30 seconds of investigation, they might turn out to be most highly motivated of all.

Within two captive rhesus troops, the three most dominant males performed worse than other males on conditioning and reversal learning tasks, and a complex task requiring win-stay/lose-shift strategy (Bunnell et al., 1980a, 1980b). Their performance did not suggest explanations of this difference, such as that dominants do not inhabit responses or dominants are not so hungry. The age range of these animals was 4 to 16, so there may be an age effect as well. The performance of individuals shifted as their rank shifted, which suggests the effects relate to dominance, not just individual age or intelligence.

These laboratory tests confirm the observation in the wild that dominant males are less likely than others to take up new habits. On Koshima Island the dominants ate their sweet potatoes gritty, and would never dream of going swimming, but juveniles splashed into the water all around. This conservatism may in turn be useful to their offspring. One such dominant Japanese macaque kept his troop away from a novel object, which was, in fact, a trap. In another case, an adult male chacma baboon frustrated seven successive attempts to trap his troop with drugged oranges. Each time he approached, tasted, and discarded an orange first, then chased away the infants and juveniles who tried to eat (Fletemeyer, 1979).

We know that there are differences in male and female feeding patterns in several primates (see Chapter 3). Perhaps such differences underlay the evolution of food sharing and division of labor in our own line. Among orangutans, the huge males forage on the ground. They are more likely than females to find termites nesting in dead logs. Three times during eight and a half years of study, Birute Galdikas and her Indonesian assistants have seen males carry termite-infested logs into the trees and share with their consorts. These rare episodes seem similar to the more frequent sharing of vertebrate prey by hunting male chimpanzees, who sometimes offer pieces to females and young. If our own ancestors' foraging differed by sex, it could have been increasingly worthwhile to exchange food (Galdikas and Teleki, 1982). Furthermore, it could have been adaptive to evolve slightly different mentalities, initially for the particular blend of stamina, patience, and attention to detail appropriate to vertebrate hunting or to insect and fruit gathering. Then, as sharing food became normal, it could have been advantageous to both sexes and their jointly supported offspring, if the *other* sex was good at its tasks. All this is faintly foreshadowed by the female chimp squatting for hours by the termite mound while her baby plays beside (McGrew, 1979, 1981).

IMITATION AND TEACHING

How is a new idea communicated? We have seen that human babies grow through stages of contagious acts like crying, through what seems deliberate imitation of known acts like babbling and clapping, through imitation of novel actions including words. Great apes and some monkeys imitate novel actions (Fig. 21.2A,B,C). Galdikas' rehabilitant orangs, for instance, imitated using logs

FIGURE 21.2. *Young primates usually learn by observation rather than through teaching. Here a mother dips for biting driver ants in the Gombe Stream Reserve. Her infant watches, then reaches for the ant-laden tool during a pull-through, and eventually succeeds in mouthing the tool when his mother pauses. (Courtesy J. Moore and Anthro-Photo.)*

as bridges after a workman was in such a hurry to escape (they had sunk his boat and charged him) that he dragged a log to the river and scrambled over. He pulled the log up after him, but the two most aggressive orangs, among eight watching, dragged everything they could find toward the river and succeeded in crossing on a vine before the day's end. In the years before this incident no orangs made bridges. Thereafter, the scientists had to collect and destroy every usable log, and even then the orangs sometimes crossed to camp using vines they pulled down from the canopy (Galdikas, 1982).

There are transitions from vaguely contagious actions to such deliberate imitation. Contagion may be enough to transmit information, particularly if it is about biologically important, easily learned matters such as predation (Hall, 1963). Blackbirds can transmit fear of particular objects through a chain of at least six birds (Curio et al., 1978). Young vervets apparently learn just which predators are worthy of eagle alarms (Chapter 19), and baboons clearly learn whether or not to fear Land Rovers (DeVore, 1965).

The nicest experiment on such social tradition is again one of Menzel's (Menzel et al., 1972). Three young chimpanzees (designated A, B, C) were tested with two initially unfamiliar toys: a swing and a motorized satellite that moved erratically round the floor beeping. Each week one chimp in the test group (A) was removed, and a new one (D) brought in. The question was whether the two linking animals would transmit a group attitude to their new companions.

These were mainly chimps raised in isolation in their early years, so even a swing frightened them somewhat. Playing with the swing did not take off until the third trio, but carried on thereafter. The far more frightening satellite was not even approached before the fifth trio, but then each group grew bolder until they demolished a satellite every session. Individual scores were far more erratic than the group totals, because in each trio the boldest animals played most, but who was boldest changed within the different social groups.

In wild troops, animals tend to learn from close kin or associates, and subordinates are more likely to copy dominants (Fig. 21.3.). This may just be a result of the subordinates' keeping a wary eye on the dominants' moves in general (Miyadi, 1967).

Although it is clear that the observing monkeys are actively learning, it has long been supposed there is no active teaching. However, even this is a difficult distinction. Many mammals place their young in situations conducive to learning (Ewer, 1969). Chimpanzees communicate both contents and techniques.

Menzel (1971) kept a group of eight young chimpanzees in an observation house with access to a ten acre field. He would take out one chimpanzee at a time, and show it a hidden pile of fruit, then return it to the group and let them all into the field. The apes were too young to venture out alone. Instead, the guiding ape led its companions to the trove. If dominant, it might stride off confidently, sure that the others would follow. If subordinate, it begged, and tugged the others' hands and fur. Sometimes when the others paid no attention the guide flung itself on the ground in a tantrum.

Once the group knew what to expect, Menzel elaborated his experiments. The chimpanzees could distinguish which of two guides had seen the bigger reward. They could march past an apple on a stick, following a guide with more to offer. Then, Menzel hid frightening objects such as rubber snakes. Again the

FIGURE 21.3. *Brothers. Social learning may occur through direct imitation, often in family lines. The subordinate commonly observes and copies the dominant. (Courtesy J. Thompson.)*

group followed the guide, but with bristling fur, and tentative approach, unearthing the reptile with sticks and slapping. When Menzel removed his rubber snake before the group found it, they searched the area around its hiding place, slapping piles of leaves and poking along the boundary fence.

None of this communication is mysterious. It used bodily cues which are obvious to everyone. I might have told these experiments in Chapter 10, as an example of non-verbal communication applied to the environment. I might also have raised it in Chapter 18, as the keeping in mind of goals which are out of sight. It would fit in Chapter 20: this is the kind of communication for which we use language, a message about a fact removed in space and time. It seems most important to raise here. This is the kind of social control of the environment for which we invented language, and through which our intellect evolved. We should not be too surprised that Washoe has actually shown her adopted infant a chair and modeled the sign *chair* five times over, while looking at him, or, on another occasion, signed *food*, then molded his hands in hers into the sign for *food* (Chevalier-Skolnikoff, 1981).

LOCAL CULTURES

The Japanese macaques of Koshima and Takasakiyama first convinced the world of primate protoculture. Their innovations, though, came through contact with scientists and their new foods. Do fully wild primates have local traditions? (See Figs. 21.4. and 21.5.)

They have traditions about which foods to eat in each area and which predators to avoid. This is usual in mammals. Primates as well have local food-processing habits. Marais (1969) described troops of baboons who hammered baobab fruit with stones and a troop that cooled the water from a hot spring by scooping drainage channels in the soft mud beside.

FIGURE 21.4. *Belle cautiously uncovers a hidden model of a snake that had previously been shown to Bandit (at right). Bandit led the group to the site and communicated that the snake is scary. Hidden food would have been snatched with the hand, not levered with a stick. (Courtesy E. Henzel.)*

FIGURE 21.5. *Only one local population of chimpanzees grooms with this gesture of hands clasped above the head. (Courtesy C. Tutin.)*

FIGURE 21.6. *Chimpanzees in the Mahale mountains, only 150 km south of the Gombe Stream, fish for different species of termites with slightly different tools, inserting a long segment in the termite hole. (Courtesy S. Vehara.)*

Chimpanzees in some West African populations use large sticks for insect fishing (Fig. 21.6.), and peel off the bark, unlike their East African counterparts (McGrew et al., 1979). Even adjacent populations in East Africa use different tools and techniques to deal with different ant species (Nishida and Uehara, 1980; Uehara, 1982). Only West African chimpanzees apparently use hammer-stones and anvils (Sugiyama). Are there no hard nuts in East Africa? (See Chapter 3 for descriptions).

As we look for more local differences, we shall find more, and in many other primates besides chimpanzees. One group of vervet guenons have invaded clearings in true rain forest, far from their normal habitat in savannah or forest fringe. In these man-made clearings the vervets forage in unpredictable patterns and communicate by the softest of their calls. If dogs approach, the monkeys suppress all noise and hide, whereas in the savannah they would give loud alarms. All these local habits keep them safe from humans, including treating dogs as presages of humans (Kavanagh, 1980).

So far in this chapter we have seen that the roles of members of a primate group differ in respect to learning about their environment, as in other aspects of their lives. We have seen that primates learn from each other. This leads to local traditions in foods, and predator defense, occasionally in tool uses. Let us conclude with two examples that show how tool using, the growth of knowledge, and the ability to use knowledge combine within the social group.

The chimpanzee group of Bossou, Guinea, had a huge fig tree in its range. Its trunk was too thick to climb. There were no adjacent trees allowing direct access, only a thorny kapok tree that extended below. From mid-February to early March the chimpanzees climbed the kapok and grappled with the problem. A protocol of March 3 tells of 51 minutes' effort before the first chimpanzee climbed into the fig. The dominant and the third-ranking male alternated positions on the kapok branch. They broke kapok branches, peeled off the thorny bark, and flailed at the fig branch with a total of nine different stick tools. Some of the sticks had hooked projections that might have caught the fig, but they were either too short or too heavy for the chimps to control. Whenever a stick caught momentarily, all the watching chimpanzees called and barked. At last, the third-ranking male began quickly to break branches from the limb he was standing on and drop them to the ground without any attempt to use them. This so lightened the limb that it rose slightly, until by bouncing and stretching upright he seized a twig of the fig tree. He climbed up, and then the second-ranked male succeeded in hanging a stick on the dangling fig. Taking advantage of each other's weight, the whole group swarmed upward (Sugiyama and Koman, 1979). (See Figs. 21.7. and 21.8A,B,C.)

An even more complex account of group achievement is the invention and use of poles as ladders (Menzel, 1972, 1973). His eight little chimpanzees lived together for five years. In the first months of captivity, Shadow, a young male, began to stand a food tray on one end and jump off the top. He and his two companions were given brooms to play with. All three became skilled pole climbers and vaulters in two or three days. Then, as opportunities came, the behavior spread to the other chimpanzees.

It was not until roughly four and a half years later that the chimps began to use pole vaulting for anything other than play. The chimpanzees were so fascinated by humans in the two-story observation house that they persistently attempted to climb the house corners and peep in. The humans installed an electric grid on the house corners. Rock, then seven years old, leaned a tree branch about three meters long against the house, climbed up it, and slapped the window. The humans chased him off, but he repeated his performance, so they confiscated his stick. There was a week's quiet in spite of at least 15 poles lying about the compound. Then, one morning the chimps were found sleeping in the observation room, having spread several hundred dollars' worth of equipment and test stimuli round the field. When chased from one window, they swarmed up poles and in another. The humans boarded up the windows and chimp-watched thereafter from the roof.

The chimps' field cage had several trees, but not enough to satisfy the climbing and bark-eating tastes of the group. Eventually the trees had to be electric-wired as well, to preserve any shade, and a runway system of 15-cm-wide planks, two and a half meters above-ground, provided some outlet for the chimps' climbing. In 15 months the young chimpanzees had passed the shock wires only two or three times during power failures. Two weeks after the laboratory break-in, all eight chimps were up in the live trees. Like Köhler's chimps, they made many mechanical errors in setting up their ladders. And like Savage-Rumbaughs' chimps, they also cooperated, soliciting close com-

FIGURE 21.7. *Aid may be one-sided. Only a few of the squirrel monkeys can scale the fence; the majority cannot. Most are able to leave the compound by scaling tails. The helper's position rotates through the group, so only three or four monkeys climb any one monkey's tail. (Courtesy R. Fontaine and Monkey Jungle.)*

panions to help hold the ladders. Rock was apparently the chief innovator, but with four months' practice seven of the eight chimps could efficiently set up ladders alone. At last, Menzel tested them with extrinsic lures in a tree too far from the runway to use a ladder directly. The chimpanzees looked at the lure, then went straight to a different tree whose branches touched the first, set up their ladders on the runway, and climbed up and round to their goal. The chimpanzees had made an invention, diffused it through the group, shared in its use, and applied it to new goals.

The end of the story could be a parable for man. The chimpanzees eventually destroyed their environment, killing all the trees. Rock fell from a dead branch, and broke his ankle. One tree fell and smashed a section of runway. At last, all the trees had to be cut down.

A

B

FIGURE 21.8. *Wild chimpanzees in Guinea. A day after the joint success described in the text, the group returned. A., An adolescent male broke a branch from the Kapok and stripped its bark with his teeth; B., flailed until he caught the fig tree above; and C., again the group shared the spoils. (Courtesy Y. Sugiyama.)*

C

FIGURE 21.9. *Having erected a pole as a ladder, Bandit grimaces and gives a little scream as he nears his goal, which is escape into waiting humans' arms. (Courtesy E. W. Menzel.)*

While Rock was still hospitalized, Gigi used a ladder to escape the compound. Again, the group followed suit, even little Bandit who mainly tried to reach the top of the observation tower to hug the humans (Fig. 21.9.) The chimpanzees did not in the end escape to a new planet outside the rules of the old. The group was first caged, then translocated to become a breeding group for a medical laboratory.

CONQUEST OF THE ENVIRONMENT?

What are the conclusions of this summary of primate behavior?

There are grim conclusions about the way we treat other primates. Too often we cage them without regard for their sociability and their manipulative curiosity. We breed them without regard for their own preferences for mates, or their long-evolved incest taboos. We rear their infants in isolation, which was fascinating when we scarcely imagined the results, but now can only be seen as wanton cruelty. We experiment on them, attempting to be humane, too often failing, when we think of animals as merely inferior "medical models" of a human, bodies without minds.

These are sins of commission. The sins of omission are worse. We destroy the habitats of wild primates and all other species that cannot survive as human weeds. We have not yet set firm limits to the human reserve, even in the United States, which pioneered the preservation of wilderness land. When we omit to guard tropical rain forest or mountain watershed, we are condemning not just individuals but species and ecosystems.

There are heroes and heroines of conservation, inspired scientific visionaries, and people of great kindness. So far, though, for every endangered species that breeds in captivity, every park protected, far more species and wild lands are lost.

How has it come about that our own kind has erupted in such a plague over the earth? The technology that gives us this control is a qualitatively new thing, and yet it is not unlike other major advances in evolution.

In each great forward step of evolution, living organisms have gained internal information and control. Claude Bernard called this the *milieu interieur*—the internal environment. They have gained this control by sucking in energy, and by creating structures to surround the kernel of reproducing substance. The original genes, coalescing to live like modern viruses within the prebiotic soup did just that. So did later genes, which actually assembled a cell around themselves. Then, after another billion years, these cells managed to build a multicellular body that supported their gametes and fed them, protected them, and carried them about. Perhaps a part of those advances came from symbiosis—a division of labor between organelles or cells that had been independent entities before.

Each great advance has required a greater input of energy. Each has required construction or coordination of a vast external sphere, a new environment, to clothe the naked reproducing creature within.

In this light, current human technology lies in the tradition of other biological innovations. We have made a quantum leap in knowledge of how to control our environment. In doing so, we suck energy out of the environment.

In developing countries, people turn the new food supplies and disease control into more humanity. In the richer countries, the rate of population growth has slowed but consumption still accelerates. A matrix of technology surrounds each individual, supporting us as our bodies support our gametes. We believe our few rich offspring will have better lives—and be more sure to survive and successfully support their children—much as the original amoebae evolved to accept the energy investment and the time delay of building organic bodies.

It would be hubris to claim that no other organism has so transformed the earth's environment. After all, the plants and their offspring created our oxygenated atmosphere. We are, however, approaching the same power. We have tilled the land, felled the forests, and polluted at least the smaller oceans. We have suspended a nuclear sword of Damocles over the head of the biosphere, hung up on human aggression and faulty computer chips. We do not know yet if our technological venture will succeed or fail. Even if it succeeds, what room will be left for other species besides our own?

The leap to being alive, to being cellular, to being multicelled were all successful. Still there are intermediate creatures left in the world today: viruses, oozing slime molds, sponge cells, which, when sieved and separated, creep together to reconstruct their communal form.

The living primates today are such transitional forms, instructing us with their alien or all-too-familiar minds, their rich societies, their rudimentary tools. The primates stand at the hinge of evolution.

Will they continue to do so?

That depends on us.

BIBLIOGRAPHY

Ainsworth, M. D. S., 1967, *Infancy in Uganda,* Johns Hopkins Press, Baltimore, 471 pp.

Aitken, P. G., and Wilson, W. A., Jr., 1979, Discriminative vocal conditioning in rhesus monkeys: Evidence for volitional control?, *Brain and Lang.* 8, 227–240.

Albignac, R., in press, The carnivores, in *Madagascar,* A. Jolly, P. Oberlé, and R. Albignac, eds., Pergamon, Oxford.

Aldrich-Blake, F. P. G., 1980, Long-tailed macaques, in *Malayan Forest Primates,* D. J. Chivers, ed., Plenum, New York, pp. 147–165.

Aldrich-Blake, F. P. G., Bunn, T. K., Dunbar, R. I. M., and Headley, P. M., 1971, Observations on baboons, *Papio anubis,* in an arid region in Ethopia, *Folia primat.* 15, 1–35.

Alexander, R. D., 1974, The evolution of social behavior, *Ann. Rev. Ecol. Systematics* 5, 325–383.

Alexander, R. D., Hoagland, J. L., Howard, R. D., Noonan, K. M., and Sherman, P. W., 1979, Sexual dimorphisms and breeding systems in pinnipeds, ungulates, primates, and humans, in *Evolutionary Biology and Human Social Behavior,* N. A. Chagnon and W. Irons, eds., Duxbury Press, North Scituate, Mass., pp. 402–435.

Alexander, R. D., and Noonan, K. M., 1979, Concealment of ovulation, parental care, and human social evolution, in *Evolutionary Biology and Human Social Behavior,* N. A. Chagnon, and W. Irons, eds., Duxbury Press, North Scituate, Mass., pp. 436–453.

Ali, R., 1981, *Ecology and behavior of the bonnet macaque (Macaca radiata) in Mundanthurai Sanctuary, Tamil Nadu,* Ph.D. thesis, University of Bristol.

Allen, M. L., and Lemmon, W. B., 1981, Orgasm in female primates, *Amer. J. Primatol.* 1, 15–34.

461

Alley, T. R., 1980, Infantile colouration as an elicitor of caretaking behaviou in Old World primates, *Primates* 21, 416–429.

Altmann, J., 1979, Age cohorts as paternal sibships, *Behav. Ecol. Sociobiol.* 6, 161–164.

Altmann, J., 1980, *Baboon Mothers and Infants,* Harvard, Cambridge, Mass., 242 pp.

Altmann, J., Altmann, S. A., Hausfater, G., and McCuskey, S. A., 1977, Life history of yellow baboons: Physical development, reproductive parameters, and infant mortality, *Primates* 18, 315–330.

Altmann, S. A., 1962, Social behavior of anthropoid primates: analysis of recent concepts, in *Roots of Behavior,* E. L. Bliss, ed., Harper, New York.

Altmann, S. A., 1974, Baboons, space, time and energy, *Amer. Zool.* 14, 221–248.

Altmann, S. A., 1979, Baboon progressions: Order or chaos. Study of one-dimensional group geometry, *Anim. Behav.* 27, 46–80.

Altmann, S. A., and Altmann, J., 1970, *Baboon Ecology,* Bibl. Primatol. 12, S. Karger, Basel, 220 pp.

Altmann, S. A., and Altmann, J., 1979, Demographic constraints on behavior and social organization, in *Primate Ecology and Human Origins,* I. S. Bernstein and E. O. Smith, eds., Garland STPM Press, New York, pp. 47–63.

Ambrose, J. A., 1963, The concept of a critical period for the development of social responsiveness, in *Determinants of Infant Behavior,* B. M. Foss, ed., Methuen, London, 2, pp. 201–226.

Andersson, M., 1982, Female choice selects for extreme tail length in a widowbird, *Nature* 299, 818–820.

Andrew, R. J., 1962, Evolution of intelligence and vocal mimicking, *Science* 137, 585–589.

Andrew, R. J., 1963, The origins and evolution of the calls and facial expressions of the primates, *Behav.* 20, 1–109.

Andrew, R. J., 1963, Trends apparent in the evolution of vocalization in the Old World monkeys and apes, *Symp. Zool. Soc. Lond.* 10, 89–101.

Anthoney, T. R., 1968, The ontogeny of greeting, grooming and sexual motor patterns in captive baboons (supersp. *Papio cynocephalus*). *Behaviour* 31, 358–372.

Argyle, M., 1967, *The Psychology of Interpersonal Behaviour,* Penguin, Harmondsworth, 223 pp.

Attenborough, D., 1961, *Zoo Quest to Madagascar,* Lutterworth, London, 144 pp.

Bachmann, C., and Kummer, H., 1980, Male assessment of female choice in hamadryas baboons, *Behav. Ecol. Sociobiol.* 6, 315–321.

Badrian, A., and Badrian, N. L., 1977, Pygmy chimpanzees, *Oryx* 15, 463–468.

Badrian, N. L., Badrian, A., and Susman, R. L., 1981, Preliminary observations on the feeding behavior of *Pan paniscus* in the Lomako Forest of Central Zaire, *Primates* 22, 173–181.

Bajpai, R. K., 1980, The effect of early social deprivation on rhesus monkeys: social responsiveness, *Primates* 21, 510–514.

Baldwin, J. D., and Baldwin, J. I., 1972, Population density and use of space in howling monkeys (*Alouatta villosa*) in southwestern Panama, *Primates* 13, 371–379.

Baldwin, J. D., and Baldwin, J. I., 1976, Vocalizations of howler monkeys (*Alouatta palliata*) in southwestern Panama, *Folia primat.* 26, 81–108.

Baldwin, P. J., Sabater Pi, J., McGrew, W. C., Tutin, C. E. G., 1981, Comparisons of nests made by different populations of chimpanzees *(Pan troglodytes)*, *Primates* 22, 474–486.

Balzama, E., Guillon, R., Forni, C., Fady, J. C., and Bert, J., 1973, Modifications du comportement du babouin *Papio papio* dans son milieu naturel par l'apport d'aliments, *Folia primat.* 19, 404–408.

Bateson, G., 1973, *Steps to an Ecology of Mind*, Paladin, St. Albans, 510 pp.

Bauchop, T., 1978, Digestion of leaves in vertebrate arboreal folivores, in *The Ecology of Arboreal Folivores*, G. G. Montgomery, ed., Smithsonian, Washington, pp. 193–205.

Baxter, M. J., and Fedigan, L. M., 1979, Grooming and consort partner selection in a troop of Japanese monkeys *(Macaca fuscata)*, *Arch. sex. Behav.* 8, 445–458.

Beach, F. A., 1976, Sexual attractivity, proceptivity and receptivity in female mammals. *Hormones and Behav.* 7, 105–138.

Bearder, S. K., and Martin, R. D., 1979, The social organization of a nocturnal primate revealed by radio tracking, in *A Handbook of Biotelemetry and Radiotracking*, C. J. Amlaner, Jr., and D. W. MacDonald, eds., Pergamon, Oxford, pp. 633–648.

Bearder, S. K., and Martin, R. D., 1980, Acacia gum and its use by bushbabies, *Galago senegalensis* (Primates: Lorisidae), *Intl. J. Primatol.* 1, 103–128.

Beck, B. B., 1980, *Animal Tool Behavior*, Garland Press, New York, 307 pp.

Beck, B. B., and Tuttle, R., 1972, The behavior of gray langurs at a Ceylonese waterhole, in *The Functional and Evolutionary Biology of Primates*, R. Tuttle, ed., Aldine-Atherton, Chicago, pp. 351–377.

Beecher, M. D., Petersen, M. R., Zoloth, S. R., Moody, D. B., and Stebbins, W. C., 1979, Perception of conspecific vocalizations by Japanese macaques: Evidence for selective attention and neural lateralization, *Brain Behav. Evol.* 16, 443–460.

Bellugi, U. and Klima, E. S., 1976, Two faces of sign: iconic and abstract, *Ann. N.Y. Acad. Sci.* 280, 514–538.

Berenstain, L., Rodman, P. S., and Smith, D. G., 1981, Social relations between fathers and offspring in a captive group of rhesus monkeys *(Macaca mulatta)*, *Anim. Behav.* 29, 1057–1063.

Berkson, G., and Shusterman, R. J., 1964, Reciprocal food sharing of gibbons, *Primates*, 5, 1–10.

Bernstein, I. S., 1962, Response to nesting materials of wild-born and captive chimpanzees, *Anim. Behav.*, 10, 1–6.

Bernstein, I. S., 1966, Analysis of a key role in a capuchin (*Cebus albifrons*) group, *Tulane Studies Zool.* 13, 49–54.

Bernstein, I. S., 1974, Taboo or toy?, in *Play*, J. S. Bruner, A. Jolly, and K. Sylva, eds., Penguin, Harmondsworth, pp. 194–198.

Bernstein, I. S., 1981, Dominance: the baby and the bathwater, *J. Brain Behav. Sci.* 4, 419–457.

Bertram, B. C., 1978, Living in groups: predators and prey, in *Behavioural Ecology*, J. R. Krebs, and N. B. Davies, eds., Blackwell, Oxford, pp. 64–96.

Bertrand, M., 1976, Acquisition by a pigtail macaque of behavior patterns beyond natural repertoire of species, *Z. Tierpsychol.* 42, 139–169.

Bertrand, M., 1969, *The Behavioral Repertoire of the Stumptail Macaque*, Karger, Basel, 273 pp.

Birch, H. G., 1945, The relation of previous experience to insightful problem solving, *J. Comp. Physiol. Psychol.* 38, 367–383.

Birdsell, J. B., 1979, Ecological influences on Australian Aboriginal social organization, in *Primate Ecology and Human Origins,* I. S. Bernstein, and E. O. Smith, eds., Garland, New York, pp. 117–151.

Bishop, A., 1962, Control of the hand in lower primates, *Ann. N.Y. Acad. Sci.* 102, 316–337.

Bishop, A., 1964, Use of the hand in lower primates, in *Evolutionary and Genetic Biology of the Primates,* J. Buettner-Janusch, ed., Academic, New York, 2, pp. 133–226.

Blaustein, A. R., and O'Hara, R. K., 1981, Genetic control for sibling recognition? *Nature* 290, 246–248.

Blurton-Jones, N. G., 1972, Non-verbal communication in children, in *Non-Verbal Communication,* R. A. Hinde, ed., Cambridge, U.P., Cambridge, pp. 271–295.

Boaz, N. T., 1979, Early hominid population densities: new estimates, *Science* 206, 592–594.

Boesch, C., and Boesch, H., 1981, Sex differences in the use of natural hammers by wild chimpanzees: A preliminary report, *J. Hum. Evol.* 10, 585–593.

Boggess, J., 1979, Troop male membership changes and infant killing in langurs *(Presbytis entellus), Folia primat.* 32, 65–107.

Boggess, J., 1980, Intermale relations and troop male membership changes in langurs *(Presbytis entellus)* in Nepal, *Int. J. Primatol.* 1, 233–232.

Boggess, J., 1982, Immature male and adult male interactions in bisexual langur *(Presbytis entellus)* troops, *Folia primat.* 38, 19–38.

Boggess, J., in press, Infant killing and male reproductive strategies in langurs *(Presbytis entellus),* in *Infanticide,* G. Hausfater and S. Hrdy, eds., Aldine, New York.

Bolwig, N., 1959, A study of the behavior of the chacma baboon, *Papio ursinus, Behaviour* 14, 136–163.

Bongaarts, J., 1980, Does malnutrition affect fecundity? A summary of evidence, *Science* 208, 564–569.

Bonner, J. T., 1980, *The Evolution of Culture in Animals,* Princeton University Press, Princeton, 280 pp.

Boskoff, K. J., 1978, Estrous cycle of brown lemur, *Lemur fulvus, J. Reprod. Fertil.* 54, 313–318.

Bourlière, F., 1964, *The Natural History of Mammals,* 3rd ed., Knopf, New York, 387 pp.

Bourlière, F., 1975, Mammals, small and large: the ecological implications of size, in *Small Mammals: Their Productivity and Population Dynamics,* International Biological Programme 5, 1–8, Cambridge U.P., Cambridge.

Bourlière, F., 1979, Significant parameters of environmental quality for non-human primates, in *Primate Ecology and Human Origins,* I. S. Bernstein, and E. O. Smith, eds., Garland Press, New York, pp. 23–46.

Bourlière, F., in press, Primate communities: their structure and role in tropical ecosystems, *Int. J. Primatol.*

Bowen, R. A., 1981, The behavior of three hand-reared lowland gorillas with emphasis on the response to a change in accommodation, *Dodo* 17, 63–78.

Bower, T. G. R., 1974, *Development in Infancy*, Freeman, San Francisco.

Bowlby, J., 1969, *Attachment and Loss 1: Attachment*, Hogarth, London, 428 pp.

Box, H. O., 1975, A social developmental study of young monkeys *(Callithrix jacchus)* within a captive family group, *Primates* 16, 419–435.

Box, H. O., 1977, Observations on spontaneous hand use in the common marmoset *(Callithrix jacchus)*, *Primates* 18, 395–400.

Box, H. O., 1977, Social interactions in family groups of captive marmosets *(Callithrix jacchus)* in *The Biology and Conservation of the Callitrichidae*, D. G. Kleiman, ed., Smithsonian, Washington, pp. 281–292.

Brace, C. L., 1979, Biological parameters and pleistocene hominid life-ways, in *Primate Ecology and Human Origins,* I. S. Bernstein, and E. O. Smith, eds., Garland STPM, New York, pp. 263–290.

Braine, M. D. S., 1959, The ontogeny of certain logical operations: Piaget's formulation examined by nonverbal methods, *Psychol. Monogr.* 475.

Brainerd, C. J., 1978, The stage question in cognitive-developmental theory, *Behav. Brain Sci.* 1, 173–214.

Brandon-Jones, D., 1979, The zoogeography of living primates, *Primate Eye* 13, 2–3.

Brandt, E. M., and Mitchell, G., 1971, Parturition in primates: Behavior related to birth, in *Primate Behavior*, L. A. Rosenblum, ed., Academic, New York 2, pp. 177–223.

Brown, C. H., Beecher, M. D., Moody, D. B., and Stebbins, W. C., 1978, Localization of primate calls by Old World monkeys, *Science* 201, 753–754.

Brown, C. H., Beecher, M. D., Moody, D. B., and Stebbins, W. C., 1979, Locatability of vocal signals in Old World monkeys: Design features for the communication of position, *J. comp. physiol. Psychol.* 93, 806–819.

Brown, J. L., 1964, The evolution of diversity in avian territorial systems, *Wilson Bull.* 76, 160–169.

Brown, J. L., and Orians, G. H., 1970, Spacing patterns in mobile animals, *Ann. Rev. Ecol. and System* 1, 239–262.

Brown, K., and Mack, D. S., 1978, Food sharing among captive *Leontopithecus rosalia, Folia primatol.* 29, 268–290.

Brown, R., 1958, How shall a thing be called?, *Psychol. Rev.* 65, 14–21.

Bruner, J. S., 1968, *Process of Cognitive Growth: Infancy*, Clark U. P., Williamstown, Mass., 75 pp.

Bruner, J. S., 1972, in 1974, Nature and uses of immaturity, in *Play,* J. S. Bruner, A. Jolly, and K. Sylva, eds., Penguin, Harmondsworth, pp. 28–64.

Bruner, J. S., 1973, *Beyond the Information Given*, Norton, New York, 502 pp.

Bruner, J.S., 1983. *Child's Talk*, W. W. Norton, New York, 144.

Bryant, P. E., 1981, Training and logic, comment on Magali Bovet's paper, in *Intelligence and Learning*, M. P. Freidman, J. P. Das, and N. O'Connor, Plenum Press, New York, pp. 203–211.

Budnitz, N., 1978, Feeding behavior of *Lemur catta* in different habitats, *Perspect. Ethol.* 3, 85–108.

Bunnell, B. N., and Perkins, M. N., 1980, Performance correlates of social behavior and organization: Social rank and complex problem solving in crab-eating macaques *(M. fascicularis)*, *Primates* 21, 515–523.

Bunnell, B. N., Gore, W. T., Perkins, M. N., 1980, Performance correlates of social behavior and organization: Social rank and reversal learning in crab-eating macaques *(M. fascicularis)*, *Primates* 21, 376–388.

Burley, N., 1979, The evolution of concealed ovulation, *Amer. Nat.* 114, 835–858.

Burt, W. H., 1943, Territoriality and home range concepts as applied to mammals, *J. Mammal.* 24, 346–352.

Burton, F. D., 1971, Sexual climax in the female *Macaca mulatta, Proc. 3rd Intl. Congr. Primatol.,* Karger, Basel, 3, 190–191.

Burton, F. D., 1972, The integration of biology and behavior in the socialization of *Macaca sylvana* of Gibraltar, in *Primate Socialization,* F. E. Poirier, ed., Random House, New York, pp. 29–62.

Burton, J. J., and Fukada, F., 1981, On female mobility: The case of the Yuga-wara—T group of *Macaca fuscata, J. Hum. Evol.* 10, 318–386.

Busse, C. D., 1977, Chimpanzee predation as a possible factor in the evolution of red colobus monkey social organization, *Evolution* 31, 907-911.

Busse, C., in press, Infant killing by male baboons, in *Infanticide,* G. Hausfater, and S. B. Hrdy, eds., Aldine, New York.

Busse, C., and Hamilton, W. J., III, 1981, Infant carrying by male chacma baboons, *Science* 212, 1281–1283.

Butynski, T. M., 1982, Vertebrate predation by primates: a review of hunting patterns and prey, *J. Hum. Evol.* 11, 421–430.

Butynski, T. M., 1982, Harem male replacement and infanticide in the blue monkey *(Cercopithecus mitis stuhlmanni)* in the Kibale Forest, Uganda, *Amer. J. Primatol.* 3, 1–22.

Bygott, J. D., 1972, Cannibalism among wild chimpanzees, *Nature* 238, 410–411.

Bygott, J. D., 1979, Agonistic behavior, dominance and social structure in wild chimpanzees of the Gombe National Park, in *The Great Apes,* D. A. Hamburg, and E. R. McCown, eds., Benjamin/Cummings, Menlo Park, 405–428.

Cambefort, J. P., 1981, A comparative study of culturally transmitted patterns of feeding habits in the chacma baboon *Papio ursinus* and the vervet monkey *Cercepithecus aethiops, Folia primat.* 36, 243–263.

Candland, D. K., Blumer, E. S., and Mumford, M. D., 1980, Urine as a communicator in New World primates, *Saimiri sciureus, Anim. Learn. Behav.* 8, 468–480.

Cant, J. G. H., 1978, Population survey of the spider monkey, *Ateles geoffroyi* at Tikal, Guatemala, *Primates* 19, 525–535.

Cant, J. G. H., 1980, What limits primates?, *Primates* 21, 538–544.

Carpenter, C. R., 1934, A field study of the behavior and social relations of howling monkeys, in *Naturalistic Behavior of Nonhuman Primates,* C. R. Carpenter, Penn. State U.P., University Park, Penn., 1964, 3–92.

Carpenter, C. R., 1942, Sexual behavior of free-ranging rhesus monkeys *(Macaca mulatta)* in *Naturalistic Behavior of Nonhuman Primates,* C. R. Carpenter, Penn State U.P., University Park, 1964, 289–318.

Cartmill, M., 1974a, Rethinking primate origins, *Science* 184, 436–443.

Cartmill, M., 1974b, Daubentonia, Dactylopsia, woodpeckers and klinorhynchy, in *Prosimian Biology*, R. D. Martin, G. A. Doyle, and A. C. Walker, eds., Duckworth, London, pp. 655–672.

Casimir, M. J., 1975, Feeding ecology and nutrition of an eastern gorilla group in the Mt. Kahuzi Region (Republique du Zaire), *Folia primat.* 24, 81–136.

Caughley, G., 1977, *Analysis of Vertebrate Populations,* Wiley, New York, 234 pp.

Chalmers, N., 1972, Comparative aspects of early infant development in captive Cercopithecines, in *Primate Socialization,* F. Poirier, ed., Random House, New York, pp. 63–82.

Chalmers, N. R., 1980, Developmental relationships among social, manipulatory, postural and locomotor behaviors in olive baboons, *Papio anubis, Behaviour* 74, 22–37.

Chalmers, N. R., 1980, The ontogeny of play in feral olive baboons *(Papio annubis), Anim. Behav.* 28, 570–585.

Chalmers, N. R., and Locke-Haydon, J., 1981, Temporal patterns of play bouts in captive common marmosets, *(Callithrix jacchus), Anim. Behav.* 29, 1229–1238.

Chapais, B., and Schulman, S. R., 1980, An evolutionary model of female dominance relations in primates, *J. Theoret. Biol.* 82, 47–89.

Chapman, M., and Hausfater, G., 1980, The reproductive consequences of infanticide in langurs: a mathematical model, *Behav. Ecol. Sociobiol.* 5, 227–240.

Chance, M. R. A., and Jolly, C. J., 1970, *Social Groups of Monkeys, Apes and Men,* Jonathan Cape, London, 224 pp.

Charles-Dominique, P., 1977, *Ecology and Behavior of Nocturnal Primates,* Duckworth, London, 277 pp.

Charles-Dominique, P., 1978, Ecological position of the family Lorisidae compared to other mammalian families, *Bull. Carnegie Mus. Nat. Hist.* 30, 26–30.

Charles-Dominique, P., and Hladik, C. M., 1971, Le Lépilémur du sud de Madagascar: écologie, alimentation et vie sociale. *Terre et Vie* 1, 3–66.

Charles-Dominique, P., and Martin, R. D., 1972, Behaviour and ecology of nocturnal prosimians, *Z. f. Tierpsychol.* Suppl. 9, 89 pp.

Cheney, D. L., 1978, Interactions of immature male and female baboons with adult females, *Anim. Behav.* 26, 389–408.

Cheney, D. L., Lee, P. C., Seyfarth, R. M., 1981, Behavioral correlates of non-random mortality among free-ranging female vervet monkeys, *Behav. Ecol. Sociobiol.* 9, 153–161.

Cheney, D. L., and Seyfarth, R. M., 1980, Vocal recognition in free-ranging vervet monkeys, *Anim. Behav.* 28, 362–367.

Cheney, D. L., and Seyfarth, R. M., 1981, Selective forces affecting the predator alarm calls of vervet monkeys, *Behaviour* 76, 25–61.

Chepko-Sade, B. D. and Sade, D. S., 1979, Patterns of group splitting within matrilineal kinship groups, *Behav. Ecol. Sociobiol.* 5, 67–86.

Chevalier-Skolnikoff, S., 1981, The Clever Hans phenomenon, cueing and ape signing: a Piagetian analysis of methods for instructing animals, in *The Clever Hans Phenomenon,* T. A. Sebeok, and R. Rosenthal, eds., *Ann. N.Y. Acad. Sci.* 364, 60–94.

Chevalier-Skolnikoff, S., 1982, A cognitive analysis of facial behavior in Old World monkeys, apes, and humans, in *Primate Communication*, C. Snowdon, C. Brown, and M. Peterson, Cambridge University Press, Cambridge, pp. 308–368.

Chevalier-Skolnikoff, S., in press, Sensorimotor development in orangutans and other primates.

Chevalier-Skolnikoff, S., Galdikas, B. M. F., and Skolnikoff, A. Z., 1982, The adapative significance of higher intelligence in wild orangutans: a preliminary report, *J. Hum. Evol.* 11, 639–652.

Cheverud, J. M., Buettner-Janusch, J., and Sade, D., 1978, Social group fission and the origin of inter-group genetic differentuation among the rhesus monkeys of Cayo Santiago, *Amer. J. Phys. Anthrop.* 48, 449–456.

Chism, J., Olson, D. K., and Rowell, T. E., 1983, Diurnal births and perinatal behavior among wild patas monkeys: evidence of an adaptive pattern, *Int. J. Primatol.* 4, 167–184.

Chism, J., Rowell, T. E., and Richards, S. M., 1978, Daytime births in captive patas monkeys, *Primates* 19, 765–767.

Chivers, D. J., 1969, On the daily behavior and spacing of free-ranging howler monkey groups, *Folia primat.* 10, 48–103.

Chivers, D. J., 1977, The lesser apes, in *Primate Conservation*, Prince Rainier III and G. H. Bourne, eds., Academic Press, New York, pp. 539–599.

Chivers, D. J., and MacKinnon, J., 1977, On the behavior of siamang after playback of their calls, *Primates* 18, 943–948.

Chivers, D. J., and Raemaekers, J. J., 1980, Long-term changes in behaviour, in *Malayan Forest Primates*, D. J. Chivers, ed., Plenum, New York, pp. 209–260.

Clark, A. B., 1978, Sex ratio and local resource competition in a prosimian primate, *Science* 201, 163–165.

Clark, T. W., 1979, Food adaptations of a transplanted Japanese macaque troop (Arashiyama West), *Primates* 20, 399–410.

Clark, W. E. Le Gros, 1960, *The Antecedents of Man*, Quadrangle, Chicago, 374 pp.

Clutton-Brock, T. H., 1973, Feeding levels and feeding sites of red colobus (*Colobus badius tephrosceles*) in the Gombe National Park, *Folia primat.* 19, 368–379.

Clutton-Brock, T. H., 1975a, Feeding behaviour of red colobus and black-and-white colobus in East Africa, *Folia primatol.* 23, 165–207.

Clutton-Brock, T. H., 1975b, Ranging behaviour of red colobus (*Colobus badius terphrosceles*) in the Gombe National Park, *Anim. Behav.* 23, 706–722.

Clutton-Brock, T. H., 1977, Some aspects of interspecific variation in feeding and ranging behaviour in primates, in *Primate Ecology*, T. H. Clutton-Brock, ed., Academic, London, 539–556.

Clutton-Brock, T. H., and Harvey, P. H., 1976a, Primate ecology and social organization, *J. Zool.*, London, 183, 1–39.

Clutton-Brock, T. H., Guinness, F. E., and Albon, S. D., 1982, *Red Deer*, Univ. of Chicago Press, Chicago, 378 pp.

Clutton-Brock, T. H., and Harvey, P. H., 1979, Home range size, population density and phylogeny in primates, in T. S. Bernstein and E. O. Smith, eds., *Primate Ecology and Human Origins,* Garland STPM Press, New York, pp. 201–214.

Clutton-Brock, T. H., and Harvey, P. H., 1980, Primates, brains and ecology, *J. Zool. Lond.* 190, 309–321.

Clutton-Brock, T. H., Harvey, P. H., and Rudder, B., 1977, Sexual dimorphism, socionomic sex ratio and body weight in primates, *Nature* (Lond.) 269, 797–800.

Cochran, C. G., 1979, Proceptive patterns of behavior throughout the menstrual cycle in female rhesus monkeys, *Behav. Neural Biol.* 27, 342–353.

Cody, M. L., 1975, Towards a theory of continental species diversities: Bird distributions over Mediterranean habitat gradients, in *Ecology and Evolution of Communities,* M. L. Cody, and J. M. Diamond, eds., Harvard University Press, Cambridge, Mass., pp. 214–257.

Coe, C. L., and Rosenblum, L. A., 1978, Annual reproductive strategy of the squirrel monkey *(Saimiri sciureus), Folia primatol.* 29, 19–42.

Coelho, A. M., Jr., 1974, Socio-bioenergetics and sexual dimorphism in primates, *Primates* 15, 263–269.

Coelho, A. M., Jr., Bramblett, C. A., Quick, L. B., and Bramblett, S. S., 1976, Resource availability and population density in primates: a socio-bioenergetic analysis of the energy budgets of Guatemalan howler and spider monkeys, *Primates* 17, 63–80.

Coelho, A. M., Bramblett, C. A., and Quick, L. B., 1977a, Social organization and food resource availability in primates: A socio-bioenergetic analysis of diet and disease hypotheses, *Amer. J. phys. Anthrop.* 46, 253–264.

Coelho, A. M., Coelho, L. S., Bramblett, C. A., Bramblett, S. S., and Quick, L. B., 1977b, Ecology, population characteristics, and sympatric association in primates: a sociobioenergetic analysis of howler and spider monkeys in Tikal, Guatemala, *Yrb. Phys. Anthrop.* 20, 96–135.

Coelho, A. M., Bramblett, C. A., and Quick, L. B., 1979, Activity patterns in howler and spider monkeys: an application of socio-bioenergetic methods, in *Primate Ecology and Human Origins,* I. S. Bernstein, and E. O. Smith, eds., Garland STPM Press, pp. 175–199.

Cohen, J. E., 1971, *Casual Groups of Monkeys and Men,* Harvard University Press, Cambridge, Mass., 175 pp.

Cohen, M. N., 1977, *The Food Crisis in Prehistory.* Yale University Press, New Haven, 341 pp.

Coimbra-Filho, A. F., and Mittermeier, R. A., 1977, Conservation of the Brazilian lion tamarins *(Leontopithecus iosalia),* in *Primate Conservation,* Prince Rainier III and G. H. Bourne, eds., Academic Press, New York, pp. 60–94.

Cole, S., Hainsworth, F. R., Kamil, A. C., Mercier, T., and Wolf, L. L., 1982, Spatial learning as an adaptation in hummingbirds, *Science* 217, 655–657.

Connell, J. H., 1978, Diversity in tropical rain forest and coral reefs, *Science,* 199, 1302–1310.

Cooper, H. M., 1980, Ecological correlates of visual learning in nocturnal prosimians, in *Nocturnal Malagasy Primates,* P. Charles-Dominique et al., eds., Academic, New York, pp. 191–204.

Coss, R. G., 1977, Perceptual determinants of gaze aversion by the lesser mouse lemur *(Microcebus murinus)* the role of two facing eyes, *Behaviour* LXIV, 3–4.

Cowgill, U. M., 1964, Recent variations in the season of birth in Puerto Rico, *Proc. Nat. Acad. Sci.* 152, 1149–1151.

Crichton, M., 1979, *Congo*, Knopf, New York, 348 pp.

Crockett, C. M., in press, Emigration by female red howler monkeys and the case for female competition, in *Females on Females,* M. F. Small, ed., Alan R. Liss, New York.

Crockett, C., and Sekulic, R., 1982, in press, Infanticide with and without rapid male replacement in red howler monkeys *(Alouatta seniculus),* in *Infanticide,* G. Hausfater and S. B. Hrdy, eds., Aldine, New York.

Crook, J. H., 1965, The adaptive significance of avian social organizations, *Symp. Zool. Soc. Lond.* 18, 237–258.

Crook, J. H., 1980, *The Evolution of Human Consciousness,* Clarendon Press, Oxford, 445 pp.

Crook, J. H., and Gartlan, J. S., 1966, On the evolution of primate societies, *Nature* 210, 1200–1203.

Cross, J. F., and Martin, R. D., 1981, Calculation of gestation period and reproductive parameters for primates, *Dodo* 18, 30–43.

Curio, E., Ernst, U., and Vieth, W., 1978, Cultural transmission of enemy recognition: One function of mobbing, *Science* 202, 899–900.

Curtin, S. H., 1980, Dusky and banded leaf-monkeys, in *Malayan Forest Primates,* D. J. Chivers, ed., Plenum, New York, pp. 107–146.

Curtin, R. A., 1981, Strategy and tactics in male gray langur competition, *J. Hum. Evol.* 10, 245–253.

Curtin, R. A., 1982, Females, male competition and gray langur troop structure, *Folia primat.* 37, 216–227.

Daly, M., 1979, Why don't male mammals lactate?, *J. Theor. Biol.* 78, 325–345.

Dare, R., 1974, Food-sharing in free-ranging *Ateles geoffroyi* (Red spider monkeys), *Lab. Primate Newsletter* 13, 19–20.

Darwin, C. R., 1859, *On the Origin of Species by Means of Natural Selection,* J. Murray, London, 502 pp.

Davies, N. B., 1978, Ecological questions about territorial behaviour, in *Behavioural Ecology,* J. R. Krebs, and N. B. Davies, eds., Blackwell, Oxford, pp. 317–350.

Davis, R. T., Leary, R. W., Stephens, D. A., and Thompson, R. F., 1967, Learning and perception of oddity problems by lemur and seven species of monkey, *Primates* 8, 311–323.

Dawkins, R., and Krebs, J. R., 1977, Animal signals: Information or manipulation?, in *Behavioural Ecology,* J. R. Krebs and N. B. Davis, eds., Blackwell, Oxford, pp. 282–313.

Dawson, G. A., 1978, Composition and stability of social groups of the tamarin, *Saguinus oedipus geoffroyi,* in Panama: Ecological and behavioral implications, in *The Biology and Conservation of the Callitrichidae,* D. G. Kleimann, ed., Smithsonian, Washington, pp. 23–38.

Deag, J. M., 1973, Intergroup encounters in the wild Barbary macaque *Macaca sylvanus,* in *Comparative Ecology and Behaviour of Primates,* R. P. Michael, and J. H. Crook, eds., Academic Press, London and New York, pp. 315–374.

Deag, J., and Crook, J. H., 1971, Social behavior and agonistic buffering in the wild Barbary macaque, *Macaca sylvana L., Folia primat.* 15, 183–200.

Deevey, E., 1960, The human population, *Sci. Amer.* 204, 194–204.

DeCasper, A. J., and Fifer, W. P., 1980, Of human bonding: Newborns prefer their mothers' voices, *Science* 208, 1174–1176.

Defler, T. R., 1980, Notes on interactions between the tayra *(Eira barbara)* and the white-fronted capuchin *(Cebus albifrons), J. Mamm.* 61, 156.

Demarest, W. J., 1977, Incest avoidance among human and non-human primates, in *Primate Biosocial Development,* S. Chevalier-Skolnikoff and F. Poirier, eds., Garland, New York, pp. 323–342.

Dennett, D. C., in press, Intentional systems in cognitive ethology: the Panglossian Paradigm defended, *Behav. Brain Sci.*

de Pelham, A., and Burton, F. D., 1976, More on predatory behavior in non-human primates, *Curr. Anthrop.* 17, 512–513.

de Pelham, A., and Burton, F. D., 1977, Still more on predatory behavior in nonhuman primates: Reply, *Curr. Anthrop.* 18, 108–109.

Deputte, B. L., 1982, Duetting in male and female songs of the white-cheeked gibbon *(Hylobates concolor leucogenys),* in *Primate Communication,* C. T. Snowdon, C. H. Brown and M. R. Petersen, eds., Cambridge U.P., Cambridge, 67–93.

Deputte, B., and Goustard, M., 1978, Etude du repertoire vocal du Gibbon a favoris blancs *(Hylobates concolor—leucogenys):* Analyse du structure des vocalisations, *Z. Tierpsychol.* 48, 225–250.

DeVore, I., and Hall, K. R. L., 1965, Baboon ecology, in *Primate Behavior,* I. DeVore, ed., Holt, New York, pp. 20–52.

DeVore, I., and Washburn, S. L., 1960, *Baboon Behavior* (16-mm film), University of California, Berkeley.

Diamond, J. M., 1975, Assembly of species communities, in *Ecology and Evolution of Communities,* M. L. Cody, and J. M. Diamond, eds., Harvard University Press, Cambridge, Mass., pp. 342–444.

Dittus, W. P. J., 1975, Population dynamics of the toque monkey, *Macaca sinica,* in *Socioecology and Psychology of the Primates,* Prince Rainier and G. H. Bourne, eds., Mouton, the Hague, pp. 125–152.

Dittus, W. P. J., 1977, The socioecological basis for the conservation of the toque monkey *(Macaca sinica)* of Sri Lanka (Ceylon), in *Primate Conservation,* Prince Rainier III, and G. H. Bourne, eds., Academic Press, New York, pp. 238–267.

Dittus, W. P. J., 1977, The social regulation of population density and age-sex distribution in the toque monkey, *Behaviour* 63, 281–322.

Dittus, W. P. J., 1979, The evolution of behaviors regulating density and age-specific sex ratios in a primate population, *Behaviour* 69, 265–302.

Dittus, W. P. J., 1980, The social regulation of primate populations: a synthesis, in *The Macaques,* D. G. Lindburg, ed., Van Nostrand, New York, pp. 263–286.

Dixson, A. F., and Fleming, D., 1981, Parental behaviour and infant development in owl monkeys *(Aotus trivirgatus griseimembra, J. Zool.* 194, 25–39.

Dixson, A. F., Gardner, J. S., and Bonney, R. C., 1980, Puberty in the male owl monkey *(Aotus trivirgatus griseimembra):* A study of physical and hormonal development, *Int. J. Primatol.* 1, 129–140.

Dixson, A. F., 1977, Observations on the displays, menstrual cycles and social behavior of the "Black ape" of Celebes, *Macaca nigra*, *J. Zool. Lond.* 182: 63–84.

Dixson, A. F., 1983, Observations on the evolution and behavioral significance of sexual skin in female primates, *Advances in the Study of Behavior*, Academic, New York 13: 63–106.

Döhl, J., 1966, Manipulier Fähigkeit und "einsichtiges" Verhalten eines Schimpansen bei Komplizierten Handlungsketten, *Z. Tierpsychol.* 23, 77–113.

Dolhinow, P. J., 1977, Normal monkeys?, *Amer. Sci.* 65, 266.

Dolhinow, P. J., 1980, An experimental study of mother loss in the Indian lemur monkey *(Presbytis entellus), Folia primat.* 33, 77–128.

Doty, R. L., Ford, M., Prett, G., and Huggins, G. R., 1975, Changes in the intensity and pleasantness of human vaginal odors during the menstrual cycle, *Science* 190, 1316–1318.

Drickamer, L. C., 1974, A ten-year summary of reproductive data for free-ranging *Macaca mulatta, Folia primatol.* 21, 61–80.

Dunbar, R. I. M., 1977, The gelada baboon: status and conservation, in *Primate Conservation*, Prince Rainier III, and G. H. Bourne, eds., Academic Press, New York, pp. 63–419.

Dunbar, R. I. M., 1979, Structure of gelada baboon reproductive units. I. Stability of social relationships, *Behaviour* 69, 72–87.

Dunbar, R. I. M., 1979, Population demography, social organization, and mating strategies, in *Primate Ecology and Human Origins*, I. S. Bernstein and E. O. Smith, eds., Garland STPM Press, New York, pp. 65–88.

Dunbar, R. I. M., 1980, Determinants and evolutionary consequences of dominance among female gelada baboons, *Behav. Ecol. Sociobiol.* 7, 253–265.

Dunbar, R. I. M., 1980, Demographic and life history variables of a population of gelada baboons *(Theropithecus gelada), J. Anim. Ecol.* 49, 485–506.

Dunbar, R. I. M., and Dunbar, P., 1974, Ecology and population dynamics of *Colobus guereza* in Ethiopia, *Folia primatol.* 21, 188–208.

Dunbar, R. I. M., and Dunbar, P., 1975, *Social Dynamics of Gelada Baboons*, Karger, Basel.

Dunbar, R. I. M., and Sharman, M., 1977, Dominance and reproductive success among female gelada baboons, *Nature* 266, 351–352.

Dunn, J., 1976, How far do early differences in mother-child relations affect later development?, in *Growing Points in Ethology*, P. P. G. Bateson, and R. A. Hinde, eds., Cambridge University Press, Cambridge, pp. 481–496.

Dyson, T., and Crook, N., 1980, Seasons of births and deaths in developing countries, in *Seasonal Dimensions to Rural Poverty*, R. Chambers, ed., Institute of Development Studies, Brighton, Sussex.

Eibl-Eibesfeldt, I., 1972, Similarities and differences between cultures in expressive movements, in *Non-Verbal Communication*, R. A. Hinde, ed., Cambridge University Press, Cambridge.

Eibl-Eibesfeldt, I., 1973, The expressive movement of the deaf-and-blind-born, in *Social Communication and Movement*, M. von Cranach, and I. Vine, eds., Academic Press, London.

Eibl-Eibesfeldt, I., 1982. Warfare, man's indoctrinability, and group selection. *Z. Tierpsychol.* 60: 177–198.

Eimas, P. D., and Tartter, V. C., 1979, On the development of speech perception: Mechanisms and analogies, in *Advances in Child Development and Behavior,* Vol. 13, H. W. Reese, and L. P. Lipsitt, eds., Academic Press, New York.

Eisenberg, J. F., 1966, The social organization of mammals, *Handbuch Zool.* 8, 39, 1–92.

Eisenberg, J. F., 1979, Habitat, economy, and society: some correlations and hypotheses for the neotropical primates, in *Primate Ecology and Human Origins,* I. S. Bernstein and E. O. Smith, eds., Garland, New York, pp. 215–262.

Eisenberg, J. F., 1980, The density and biomass of tropical mammals, in *Conservation Biology,* M. E. Soulé and B. A. Wilcox, eds., Sinauer, Sunderland, Mass., pp. 35–56.

Eisenberg, J. F., 1981, *The Mammalian Radiations,* University of Chicago Press, Chicago, 610 pp.

Eisenberg, J. F., and Lockhart, M., 1972, An ecological reconnaissance of Wilpattu National Park, Ceylon, *Smithsonian Contrib. Zool.* 101, 1–118.

Eisenberg, J. F., and Thorington, R. W., Jr., 1973, A preliminary analysis of a neotropical mammal fauna, *Biotropica* 5, 150–161.

Elias, M. F., 1977, Relative maturity of cebus and squirrel monkeys at birth and during infancy, *Develop. Psychobiol.* 10, 519–528.

Ellefson, J. O., 1968, Territorial behavior in the common white-handed gibbon, *Hylobates lar* Linn., in *Primates: Studies in Adaptation and Variability,* P. C. Jay, ed., Holt, New York, pp. 180–199.

Elliot Smith, G., 1927, *Essays on the Evolution of Man,* 2nd ed., Oxford University Press, Oxford.

Emmons, L. H., Gautier-Hion, A., and DuBost, G., 1983, Community structure of the frugivorous-folivorous mammals of Gabon, *J. Zool.* (Lond.) 199, 209–222.

Epple, G., 1970, Maintenance, breeding and development of marmoset monkeys (Callithricidae) in captivity, *Folia primatol.* 12, 56–76.

Epple, G., 1975, Parental behavior in *Saguinus fuscicollis* ssp. (Callithricidae), *Folia primatol.* 24, 221–238.

Epple, G., 1975, The behavior of marmoset monkeys (Callithricidae), in *Primate Behavior,* Vol. 4, L. A. Rosenblum, ed., Academic Press, New York.

Epple, G., 1980, Relationships between aggression, scent marking and gonadal state in a primate, the tamarin *Saguinus fuscicollis,* in *Chemical Signals: Vertebrates and Aquatic Invertebrates,* D. Mueller-Schwarze, and R. M. Silverstein, eds., Plenum, New York, pp. 87–105.

Epple, G., 1981, Effect of pair-bonding with adults on the ontogenetic manifestation of aggressive behavior in a primate, *Saguinus fuscicollis, Behav. Ecol. Sociobiol.* 8, 117–123.

Epple, G., and Conroy, V. A., 1979, Effects of castration and social change on scent-marking behavior of *Saguinus fuscicollis* (Callithricidae), *Folia primatol.* 32, 252–263.

Epple, G., and Katz, Y., 1980, Social influences on first reproductive success and related behaviors in the saddle-back tamarin (*Saguinus fuscicollis,* Callitrichidae), *Int. J. Primatol.* 1, 171–184.

Epstein, R., Lanza, R. P., and Skinner, B. F., 1980, Symbolic Communication between Two Pigeons (*Columba livia domestica*), *Science* 207, 343–545.

Epstein, R., Lanza, R. P., and Skinner, B. F., 1981, "Self-awareness" in the pigeon, *Science* 212, 695–696.

Estrada, A., and Estrada, R., 1976, Establishment of a free-ranging colony of stumptail macaques *(Macaca arctoides):* Relations to the ecology 1. *Primates* 17, 337–355.

Etter, H. F., 1973, Terrestrial adaptations in the hands of Cercopithecinae, *Folia primatol.* 20, 331–350.

Ewer, R. F., 1968, *The Ethology of Mammals,* Logos, London, 418 pp.

Ewing, L. L., 1982, Seasonal variation in primate fertility with emphasis on the male, *Amer. J. Primat.* Suppl. 1, 145–160.

Fagen, R., 1981, *Animal Play Behavior,* Oxford University Press, New York, 684 pp.

Fairbanks, L. A., and Bird, J., 1979, Ecological correlates of interindividual distance in the St. Kitts vervet *(Cercopithecus aethiops sabaeus), Primates* 19, 605–614.

Fleagle, J. G., 1978, Size distributions of living and fossil primate faunas, *Paleobiology* 4, 67–76.

Fleagle, J. G., Kay, R. F., and Simons, E. L., 1980, Sexual dimorphism in early anthropoids, *Nature* 287, 328–330.

Fleagle, J. G., and Mittermeier, R. A., 1980, Locomotor behavior, body size, and comparative ecology of seven Surinam monkeys, *Amer. J. Phys. Anthrop.* 52, 301–314.

Fletemeyer, J. R., 1978, Communication about potentially harmful foods in free-ranging chacma baboons, *Papio ursinus, Primates* 19, 223–226.

Foerg, R., 1982, Reproduction in *Cheirogaleus medius, Folia primatol.* 39, 49–62.

Fooden J., 1980, Classification and distribution of living macaques *(Macaca, Lacepede, 1799),* in *The Macaques,* D. G. Lindburg, ed., pp. 1–9.

Ford, S. M., 1980, Callitrichids as phyletic dwarfs, and the place of Callitrichidae in Platyrrhini, *Primates* 21, 31–43.

Fossey, D., 1979, Development of the mountain gorilla *(Gorilla gorilla beringei):* the first 36 months, in *The Great Apes,* D. A. Hamburg, and E. R. McCown, Benjamin/Cummings, Menlo Park, pp. 139–186.

Fossey, D., 1981, The imperiled mountain gorilla, *National Geographic* 159, 501–523.

Fossey, D., in press, Infanticidal episodes of mountain gorilla *(Gorilla gorilla beringei)* and chimpanzees *(Pan troglodytes schweinfurthii),* in *Infanticide,* G. Hausfater, G., and S. B. Hrdy, eds., Aldine, New York.

Foster, R. B., 1982, The seasonal rhythm of fruitfall on Baro Colorado Island, in *The Ecology of a Tropical Forest,* E. G. Leigh, A. S. Rand and D. M. Windsor, eds., Smithsonian, Washington, pp. 151–172.

Fouts, R. S., and Couch, J. B., 1976, Cultural evolution of learned language in signing chimpanzees, in *Communicative Behavior and Evolution,* M. E. Hahn and E. C. Simmel, eds., Academic Press, New York, pp. 141–162.

Fox, G. J., and Taub, D. M., 1980, Odor cues in vaginal pheromones, *Amer. J. Phys. Anthrop.* 52, 227.

Fragaszy, D. M., 1980, Comparative studies of squirrel monkeys *(Saimiri)* and titi monkeys *(Callicebus)* in travel tasks, *Z. Tierpsychol.* 54, 1–36.

Fragaszy, D. M., 1981, Comparative performance in discrimination learning tasks in two New World primates, *(Saimiri sciureus* and *Callicebus moloch), Anim. Learning and Behav.* 9, 127–134.

Fragaszy, D. M., and Mason, W. A., 1978, Response to novelty in Saimiri and Callicebus: Influence of social context, *Primates* 19, 311–331.

Frankel, O. H., and Soule, M. M., 1981, *Conservation and Evolution,* Cambridge U.P., Cambridge, 327 pp.

Freeland, W. J., 1976, Pathogens and evolution of primate sociality, *Biotropica* 8, 11–24.

Freeland, W. J., 1979, Primate social groups as biological islands, *Ecology* 60, 719–728.

Freeland, W. J., and Janzen, D. H., 1974, Strategies in herbivory by mammals: the role of plant secondary compounds, *Amer. Nat.* 108, 269–289.

Freese, C. H., 1978, The behavior of white-faced capuchins *(Cebus capucinus)* at a dry-season water hole, *Primates* 19, 275–286.

French, J. A., 1981, Individual differences in play in *Macaca fuscata:* The role of maternal status and proximity, *Intl. J. Primatol.* 2, 237–246.

Froelich, J. W., Thorington, R. W., Jr., and Otis, J. S., 1981, The demography of howler monkeys *(Alouatta palliata)* on Barro Colorado Island, Panama, *Int. J. Primatol.* 2, 207–236.

Galdikas, B. M. F., 1978, Orangutan death and scavenging by pigs, *Science* 200, 68–70.

Galdikas, B. M. F., 1979, Orang adaptation at Tanjung Puting Reserve: mating and ecology, in *The Great Apes,* D. A. Hamburg and E. R. McCown, eds., Benjamin/Cummings, Menlo Park, CA, pp. 195–234.

Galdikas, B. M. F., 1980, Living with the great orange apes, *National Geographic Magazine* 157, 830–853.

Galdikas, B. M. F., 1981, Orangutan reproduction in the wild, in *Reproductive Biology of the Great Apes,* C. E. Graham, ed., Academic Press, New York, pp. 281–300.

Galdikas, B. M. F., 1982a, An unusual instance of tool use among wild orangutans in Tanjung Puting Reserve, Indonesian Borneo, *Primates* 23, 138–139.

Galdikas, B. M. F., 1982b, Orangutan tool use at Tanjung Puting Reserve, Central Indonesian Borneo (Kalimantan Tengah), *J. Hum. Evol.* 11, 19–33.

Galdikas, B. M. F., and Teleki, G., 1981, Variations in subsistence activities of female and male pongids: New perspectives on the origins of hominid labor division, *Curr. Anthrop.* 22, 241–256.

Galef, B. G., Jr., Mittermeier, R. A., and Bailey, R. C., 1976, Predation by the tayra *(Eira barbara), J. Mammal.* 57, 760–761.

Gallup, G. G., 1979, Self-awareness in primates, *Amer. Scientist* 67, 417–421.

Gallup, G. G., Jr., 1980, Chimpanzees and self-awareness, in *Species Identity and Attachment: A Phylogenetic Evolution,* M. A. Roy, ed., Garland STPM Press, New York, pp. 223–243.

Gallup, G. G., Jr., Wallnau, L. B., and Suarez, S. D., 1980, Failure to find self-recognition in mother-infant and infant-infant rhesus monkey pairs, *Folia primatol.* 33, 210–219.

Gandini, G., and Baldwin, P. J., 1978, An encounter between chimpanzees and a leopard in Senegal, *Carnivore* 1, 107–109.

Garber, P. A., Moya, L., and Malaga, C. 1984. A preliminary field study of the moustached tamarin monkey (*Saguinus mystax*) in northeastern Peru: questions concerned with the evolution of a communal breeding system. *Folia prmat.* 42, 17–33.

Gardner, B. T., and Gardner, R. A., 1971, Two way communication with an infant chimpanzee, in *Behavior of Nonhuman Primates,* A. M. Schrier and F. Stollnitz, eds., Academic Press, New York, pp. 117–185.

Gardner, B. T., and Gardner, R. A., 1980, Two comparative psychologists look at language acquisition, in *Children's Language,* K. E. Nelson, ed., Halsted Press, New York, Vol. 2, pp. 331–369.

Gardner, R. A. and Gardner, B. T., 1978, Comparative psychology and language acquisition, *Ann. New York Acad. Sci.* 309, 37–76.

Gardner, R. A., and Gardner, B. T., in press, Early signs of reference in children and chimpanzees.

Gardner, B. T., and Wallach, L., 1965, Shapes of figures identified as a baby's head, *Percept. Mot. Skills* 20, 135–142.

Gartlan, J. S., 1968, Structure and function in primate society, *Folia primatol.* 8, 89–120.

Gartlan, J. S., 1975, Adaptive aspects of social structure in *Erythrocebus patas,* in *Proc. Symp. 5th Congr. Int'l Primat. Soc.,* S. Kondo, M. Kaway, A. Ehara, and S. Kawamura, Japan Science Press, Tokyo, pp. 161–171.

Gaulin, S. J. C., and Kurland, J. A., 1976, Primate predation and bioenergetics, *Science* 191, 314–315.

Gaulin, S. J. C., and Schlegel, A., 1980, Paternal confidence and parental investment: A cross-cultural test of a sociobiological hypothesis, *Ethol. Sociobiol.* 1, 301–310.

Gaull, G. E., 1979, What is biochemically special about human milk?, in *Breastfeeding and Food Policy in a Hungry World,* Academic Press, New York, pp. 217–227.

Gause, G. F., 1947, Problems of evolution, *Trans. Conn. Acad. Arts. Sci.,* 37, 17–68.

Gautier-Hion, A., 1970, L'organization sociale d'une bande de talapoins (*Miopithecus talapoin*) dans le Nord-est du Gabon, *Folia primatol.* 12, 116–141.

Gautier-Hion, A., 1971, L'ecologie du talapoin du Gabon, *Terre Vie* 25, 427–490.

Gautier-Hion, A., 1978, Food niches and coexistence in sympatric primates in Gabon, in *Recent Advances in Primatology, I. Behaviour,* D. J. Chivers, ed., Academic Press, London, pp. 269–286.

Gautier-Hion, A., and Gautier, J. P., 1974, Associations polyspecifiques de cercopitheques sur le plateau de M'Passa, Gabon, *Folia primatol.* 22, 134–177.

Gautier-Hion, A., and Gautier, J. P., 1978, Le singe de Brazza: Une strategie originale, *Z. Tierpsychol.* 46, 84–104.

Gautier, J. P., and Gautier-Hion, A., 1969, Les associations polyspecifiques chez les cercopithecidae du Gabon, *Terre Vie* 2, 164–201.

Geist, V., 1978, On weapons, combat and ecology, in *Aggression, Dominance and Individual spacing,* L. Krames, P. Pliner, and T. Alloway, eds., Plenum, New York, pp. 1–30.

Gillan, D. J., 1981, Reasoning in the chimpanzee: II. Transitive inference *J. Exp. Psychol.: Anim. Behav. Processes* 7, 150–164.

Gillan, D. J., Premack, D., and Woodruff, G., 1981, Reasoning in the chimpanzee: I. Analogical reasoning *J. Exp. Psychol.: Anim. Behav. Processes* 7, 1–17.

Gittins, S. P., and Raemaekers, J. J., 1980, Siamang, lar and agile gibbons, in *Malayan Forest Primates*, D. J. Chivers, ed., Plenum, New York, pp. 63–106.

Gittleman, J. L., and Harvey, P. H., 1980, Why are distasteful prey not cryptic?, *Nature* 286, 149–150.

Glander, K. E., 1978a, Howling monkey feeding behavior and plant secondary compounds: a study of strategies, in *Ecology of the Arboreal Folivores*, G. G. Montgomery, ed., Smithsonian, Washington, pp. 561–574.

Glander, K. E., 1978b, Drinking from arboreal water sources by mantled howling monkeys *(Alouatta palliata Gray)*, *Folia primatol.* 29, 206–217.

Glander, K. E., 1980, Reproduction and population growth in free-ranging mantled howling monkeys, *Amer. J. Phys. Anthrop.* 53, 25–36.

Glander, K. E., 1982, The impact of plant secondary compounds on primate feeding behavior, *Yrb. Phys. Anthrop.* 25, 1–18.

Goldfoot, D. A. 1981, Olfaction, sexual behavior, and the pheromone hypothesis in rhesus monkeys: A critique, *Amer. Zool.* 21, 153–164.

Goldfoot, D. A., Westerborg-Van Loon, Groeneveld, W., and Koos Slob, A., 1980, Behavioral and physiological evidence of sexual climax in the female stump-tailed macaque *(Macaca arctoides)*, *Science* 208, 1477–1479.

Goldstein, K., and Scheerer, M., 1941, Abstract and concrete behavior: An experimental study with special tests, *Psychol. Monogr.* 58, 1–51.

Goodall, A. G., 1977, Feeding and ranging behaviour of a mountain gorilla group *(Gorilla gorilla beringei)* in the Tshibinda-Kahuzi Region (Zaire), in *Primate Ecology*, T. H. Clutton-Brock, ed., Academic Press, London, pp. 450–479.

Goodall, A. G., 1979, *The Wandering Gorillas*, Collins, London, 253 pp.

Goodall, J., 1962, Nest building behavior in the free-ranging chimpanzee, *Ann. New York, Acad. Sci.*, 102, 455–467.

Goodall, J., 1963, Feeding behavior of wild chimpanzees, a preliminary report. *Symp. Zool. Soc. Lond.* 10, 39–47.

Goodall, J. van Lawick, 1967, *My Friends the Wild Chimpanzees*, National Geographic, Washington, 204 pp.

Goodall, J. van Lawick, 1968, The behaviour of free-living chimpanzees in the Gombe Stream Reserve, *Anim. Behav. Monogr.*, 1, 161–311.

Goodall, J. van Lawick, 1971, *In the Shadow of Man*, Collins, London, 256 pp.

Goodall, J., 1977, Infant killing and cannibalism in free-living chimpanzees, *Folia primatol.* 28, 259–282.

Goodall, J., 1983, Population dynamics during a 15-year period in one community of free-living chimpanzees in the Gombe National Park, Tanzania, *Z. Tierpsychol.* 61, 1–60.

Goodall, J., and Athumani, J., 1980, An observed birth in a free-living chimpanzee *(Pan troglodytes schweinfurthii)* in Gombe National Park, Tanzania, *Primates* 21, 545–549.

Goodall, J., Bandora, A., Bergmann, E., Busse, C., Matama, H., Mpongo, E., Pierce, A., and Riss, D., 1979, Intercommunity interactions in the chimpanzee population of the Gombe National Park, in *The Great Apes*, D. A. Hamburg, and E. R. McCown, eds., Benjamin/Cummings, Menlo Park, CA, pp. 13–54.

Goodall, J. van Lawick, van Lawick, H., and Packer, C., 1973, Tool-use in free-living baboons in the Gombe National Park, *Nature* 241, 212–213.

Gordon, T. P., 1981, Reproductive behavior in the rhesus monkey: social and endocrine variables, *Amer. Zool.* 21, 185–195.

Goss-Custard, J. D., Dunbar, R. M., and Aldrich-Blake, F. P. G., 1972, Survival mating and rearing structures in the Evolution of Primate Society structure, *Folia primat.* 17, 1–19.

Gould, S. J., 1978, *Ever Since Darwin,* Burnett, London, 285 pp.

Gouzoules, H., 1980, The alpha female: Observations on captive pigtail monkeys, *Folia primatol.* 33, 46–56.

Gouzoules, H., 1980, A description of geneological rank changes in a troop of Japanese monkeys *(Macaca fuscata), Primates* 21, 262–267.

Gouzoules, H., Gouzoules, S., and Fedigan, L., 1981, Japanese monkey group translocation: Effects on seasonal breeding, *Int. J. Primat.* 2, 323–334.

Graham, C. A., and McGrew, W. C., 1980, Menstrual synchrony in female undergraduates living on a coeducational campus, *Psychoneuroendocrinology* 5, 245–252.

Graham, C. E., 1979, Reproductive function in aged female chimpanzees, *Amer. J. Phys. Anthrop.* 50, 291.

Graham, C. E., 1981, Menstrual cycle in the great apes, in *Reproductive Biology of the Great Apes,* C. E. Graham, ed., Academic Press, London, 1–44.

Grant, E. C., 1969, Human facial expression, *Man* 4, 525–536.

Green, S., 1975, Variation of vocal pattern with social situation in the Japanese monkey *(Macaca fuscata):* A field study, in *Primate Behavior,* Vol. 4, L. A. Rosenblum, ed., Academic Press, New York, pp. 1–102.

Green, S., 1981, Sex differences and age gradations in vocalizations of Japanese and lion-tailed monkeys *(Macaca fuscata* and *Macaca silenus), Amer. Zool.* 21, 165–183.

Green, S., and Marler, P., 1979, The analysis of animal communication, in *Handbook of Behavioral Neurology,* Vol. 3, *Social Behavior and Communication,* P. Marler and J. G. Vandenbergh, eds., Plenum, New York, pp. 73–158.

Green, S., and Minkowski, K., 1977, The lion-tailed monkey and its South Indian rain forest habitat, in *Primate Conservation,* Prince Rainier III, and G. H. Bourne, Academic Press, New York pp. 290–338.

Greenwood, P. J., 1980, Mating systems, philopatry, and dispersal in birds and mammals, *Anim. Behav.* 28, 1140–1162.

Grewal, B. S., 1980, Social relations between adult central males and kinship groups of Japanese monkeys at Arashiyama with some aspects of organization, *Primates* 21, 161–180.

Griffin, D. R. 1981, *The Question of Animal Awareness,* 2nd ed., The Rockefeller University Press, New York, 209 pp.

Griffin, D. R., 1984, *Animal Thinking,* Harvard University Press, Cambridge, Mass.

Groos, K., 1898, *The Play of Animals,* D. Appleton, New York.

Gruber, H. E., Girgus, J. S., and Banuazizi, A., 1971, The development of object permanence in the cat, *Devel. Psychol.* 4, 9–15.

Haimoff, E. H., Chivers, D. J., Giffins, S. P., and Whitten, A. J., 1982, A phylogeny of gibbons *(Hylobates* spp.) based on morphological and behavioral characters, *Folia primatol.* 39, 213–237.

Hall, K. R. L., 1962, Numerical data, maintenance activities and locomotion of the wild chacma baboon *(Papio ursinus), Proc. Zool. Soc. Lond.,* 139, 181–220.

Hall, K. R. L., 1963, Tool using performances as indications of behavioral adaptability, *Curr Anthrop.* 4, 479–494.

Hall, K. R. L., 1965, Behavior and ecology of the wild patas monkey, *Erythrocebus patas,* in Uganda, *J. Zool.,* 148, 15–87.

Hall, K. R. L., and DeVore, I., 1965, Baboon social behavior, in *Primate Behavior,* I. DeVore, ed., Holt, New York, pp. 53–110.

Halle, F., Oldeman, R. A. A., and Tomlinson, P. B., 1978, *Tropical Trees and Forests: An Architectural Analysis,* Springer-Verlag, New York, 442 pp.

Hamden, C. E., and Rodman, P. S., 1980, Social development of bonnet macaques from six months to three years of age: a longitudinal study, *Primates* 21, 350–356.

Hamilton, W. D., 1970, Selfish and spiteful behavior in an evolutionary model, *Nature* 228, 1218–1220.

Hamilton, W., Buskirk, R., and Buskirk, W., 1975, Defensive stoning by baboons, *Nature* 256, 488–489.

Hamilton, W. J., Buskirk, R. E., and Buskirk, W. H., 1976, Defense of space and resources by chacma *(Papio ursinus)* baboon troops in an African desert and swamp, *Ecology* 57, 1264–1272.

Harcourt, A. H., 1979, Social relationships among adult female mountain gorillas, *Anim. Behav.* 27, 251–264.

Harcourt, A. H., 1981, Intermale competition and the reproductive behavior of the great apes, in *Reproductive Biology of the Great Apes,* C. E. Graham, ed., Academic Press, New York, pp. 301–318.

Harcourt, A. H., Fossey, D., and Sabater Pi, J., 1981, Demography of *Gorilla gorilla, J. Zool.* 195, 215–233.

Harcourt, A. H., and Stewart, K. G., 1978, Coprophagy by wild mountain gorilla, *E. Afr. Wildl. J.* 16, 223–225.

Harcourt, A. H., and Stewart, K. J., 1981, Gorilla male relationships: can differences during immaturity lead to contrasting reproductive tactics during adulthood?, *Anim. Behav.* 29, 206–210.

Harcourt, A. H., Harvey, P. H., Larson, S. G., and Short, R. V., 1981, Testis weight, body weight, and breeding system in primates, *Nature* 293, 55–57.

Harcourt, C., 1981, An examination of the function of urine washing in *Galago senegalensis, Z. Tierpsychol.* 55, 119–128.

Harding, R. S. O., 1973, Predation by a troop of olive baboons *(Papio anubis), Amer. J. Phys. Anthrop.* 38, 587–592.

Harding, R. S. O., 1975, Meat-eating and hunting in baboons, in *Socioecology and Psychology of Primates,* Tuttle, R. H., ed., Mouton, the Hague, pp. 245–258.

Harding, R. S. O., 1976, Ranging patterns of a troop of baboons *(Papio anubis)* in Kenya, *Folia primatol.* 25, 143–185.

Harding, R. S. O., 1977, Patterns of movement in open country baboons, *Amer. J. phys. Anthrop.* 47, 349–353.

Harlow, H. F., 1949, The formation of learning sets, *Psychol. Rev.* 56, 51–96.

Harlow, H. F., 1951, Primate learning, in *Cooperative Psychology,* C. P. Stone, ed., Prentice-Hall, New York, 3rd ed., pp. 183–238.

Harlow, H. F., and Harlow, M. K., 1965, The affectional systems, in *Behavior of Nonhuman Primates,* A. M. Schrier, H. F. Harlow, and F. Stollnitz, eds., Academic Press, New York pp. 287–334.

Harrington, J. E., 1977, Discrimination between males and females by scent of *Lemur fulvus, Anim. Behav.* 25, 147–151.

Harrington, J. E., 1979, Responses of *Lemur fulvus* to scents of different subspecies of *Lemur fulvus* and to scents of different species of Lemuriformes, *Z. Tierpsychol.* 49, 1–9.

Hartung, J., 1981, Genome parliaments and sex with the Red Queen, in *Natural Selection and Social Behavior,* R. D. Alexander, and D. W. Tinkle, eds., Chiron, New York, pp. 382–404.

Hartung, J., 1981, Paternity and the inheritance of wealth, *Nature* 291, pp. 652–654.

Harvey, P., Kavanagh, M., and Clutton-Brock, T., 1978, Canine tooth size in female primates, *Nature* 276, 817–818.

Hasegawa, T., and Hiraiwa, M., 1980, Social interactions of orphans observed in a free-ranging troop of Japanese monkeys, *Folia primatol.* 33, 129–158.

Hausfater, G., 1975, *Dominance and Reproduction in Baboons (Papio cynocephalus),* Karger, Basel, 150 pp.

Hausfater, G., 1976, Predatory behavior of yellow baboons, *Behaviour* 56, 44–68.

Hausfater, G., 1977, Tail carriage in baboons *(Papio cynocephalus):* relationship to dominance rank and age, *Folia primatol.* 27, 41–59.

Hausfater, G., Altmann, J., and Altmann, S., 1982, Long-term consistency of dominance relations among female baboons *(Papio cynocephalus), Science* 217, 252–255.

Hausfater, G., Aref, S., and Cairns, S. J., 1982, Infanticide as an alternative male reproductive strategy in langurs: A mathematical model. *J. Theoret. Biol.* 94, 391–412.

Hausfater, G., and Bearce, W. H., 1976, Acacia tree exudates: their composition and use as a food source by baboons, *E. Afr. Wildl. J.* 14, 241–243.

Hausfater, G., and Meade, B. J., 1982, Alternation of sleeping groves by yellow baboons *(Papio cynocephalus)* as a strategy for parasite avoidance, *Primates* 23, 287–297.

Hausfater, G., Saunders, C. D., and Chapman, M., 1981, Computer models of primate life-histories, in *Natural Selection and Social Behavior,* R. D. Alexander and D. Tinkle, eds., Chiron, New York, pp. 345–362.

Hausfater, G., and Watson, D. F., 1976, Social and reproductive correlates of parasite ova emissions by baboons, *Nature* 262, 688–689.

Hayes, C., 1951, *The Ape in our House,* Harper, New York.

Hayes, C. and Nissen, C. H., 1971, Higher mental functions of a home-raised chimpanzee, in *Behavior of Nonhuman Primates,* A. M. Schreier and F. Stollnitz, eds., Academic Press, New York, 59–115.

Hediger, H., 1950, *Wild Animals in Captivity,* Butterworth, London.

Hennessy, M. B., Mendoz, S. P., Coe, C. L., Lowe, E. L., and Levine, S., 1980, Androgen-related behavior in the squirrel monkey: An issue that is nothing to sneeze at. *Behav. Neural Biol.* 30, 103–108.

Henwood, K., and Fabrick, A., 1979, A quantitative analysis of the dawn chorus: temporal selection for communicatory optimization, *Amer. Nat.* 114, 260–274.

Herrnstein, R. J., 1979, Acquisition, generalization and discrimination reversal of a natural concept, *J. Exptl. Psychol: Anim. Behav. Processes* 5, 118–129.

Hess, J. P., 1973, Some observations on the sexual behavior of captive lowland gorillas, in *Comparative Ecology and Behavior of Primates*, R. P. Michael and J. H. Crook, eds., Academic Press, London, pp. 507–581.

Hildegarda, St., Abatissa, col. 1329, *Physica*, Book 7, *De animalium*, 1385 pp.

Hill, J. P., 1932, The developmental history of the primates, *Phil. Trans. Roy. Soc.* Series B, 221, 45.

Hinde, R. A., 1966, *Animal Behavior*, McGraw-Hill, New York, 544 pp.

Hinde, R. A., 1977, Mother-infant separation and the nature of inter-individual relationships: experiments with rhesus monkeys, *Proc. Roy. Soc. Lond. B. Biol. Sci.* 196, 29–50.

Hinde, R. A., Rowell, T. E., and Spencer-Booth, Y., 1964, Behavior of socially living rhesus monkeys in their first six months, *Proc. Zool. Soc. Lond.* 143, 609–649.

Hiraiwa, M., 1981, Maternal and alloparental care in a troop of free-ranging Japanese monkeys, *Primates* 22, 309–329.

Hladik, A., 1978, Phenology of leaf production in the rain forest of Gabon: distribution and composition of food for folivores, in *The Ecology of the Arboreal Folivores*, G. G. Montgomery, ed., Smithsonian, Washington, pp. 51–72.

Hladik, A., 1980, The dry forest of the west coast of Madagascar: climate, phenology, and food available for prosimians, in *Nocturnal Malagasy Primates*, P. Charles-Dominique et al., eds., Academic Press, New York, pp. 3–40.

Hladik, A., and Hladik, C. M., 1969, Rapports trophiques entre végétation et primates dans la forêt de Barro Colorado (Panama), *Terre et Vie* 1, 25–117.

Hladik, C. M., 1973, Alimentation et activité d'un groupe de chimpanzés réintroduits en forêt Gabonaise, *Terre et Vie* 27, 343–413.

Hladik, C. M., 1974, La vie d'un groupe de chimpanzes dans la forêt du Gabon, *Sci. Nat. Environ.* 121, 5–14.

Hladik, C. M., 1975, Ecology, diet, and social patterning in Old and New World primates, in *Socioecology and Psychology of Primates*, R. H. Tuttle, ed., Mouton, The Hague, pp. 3–35.

Hladik, C. M., Charles-Dominique, P., and Petter, J. J., 1980, Feeding strategies of five nocturnal prosimians in the dry forest of the west coast of Madagascar, in *Nocturnal Malagasy Primates*, Charles-Dominique et al, eds., Academic, New York, 41–74.

Hoage, R. J., 1977, Parental care in *Leontopithecus rosalia rosalia*, in *The Biology and Conservation of the Callitrichidae*, D. Keliman, ed., Smithsonian, Washington, pp. 293–306.

Hoban, R., 1974, *How Tom beat Captain Najork and his Hired Sportsmen*, Jonathan Cape, London.

Hockett, C. F., and Altmann, S. A., 1968, A note on design features, in *Animal Communication*, T. A. Sebeok, ed., Indiana University Press, Bloomington, pp. 61–72.

Hockett, C. F., and Ascher, R., 1964, The human revolution, *Curr. Anthrop.* 5, 135–168.

Hoff, M. P., Nadler, R. D., and Maple, T. L., 1982, Control role of an adult male in a captive group of lowland gorillas. *Folia Primatol.* 38, 72–85.

Hofstadter, D. R., 1979, *Gödel, Escher, Bach,* Vintage, New York, 777 pp.

Holmes, W. G., and Sherman, P. W., 1983, Kin recognition in animals, *Amer. Sci.* 71, 46–55.

Homewood, K. M., 1978, Feeding strategy of tana mangabey *(Cercocebus galeritus galeritus)* (Mammalia-Primates), *J. Zool.* 186, 375–391.

Hoof, J. A. R. A. M. van, 1967, The facial displays of the catarrhine monkeys and apes, in *Primate Ethology,* D. Morris, ed., Weidenfeld and Nicholson, London, pp. 7–68.

Hoof, J. A. R. A. M. van, 1972, The phylogeny of laughter and smiling, in *Non-Verbal Communication,* R. A. Hinde, ed., Cambridge University Press, Cambridge, pp. 209–241.

Hooley, J. M., and Simpson, M. J. A., 1981, A comparison of primiparous and multiparous mother-infant dyads in *Macaca mulatta, Primates* 22, 379–392.

Hopf, S., 1981, Conditions of failure and recovery of maternal behavior in captive squirrel monkeys (Saimiri). *Int. J. Primatol.* 2, 335–349.

Horn, R. N. van, and Eaton, G. G., 1979, Reproductive physiology and behavior in prosimians, in *The Study of Prosimian Behavior,* Doyle, G., and Martin, R. D., eds., Academic Press, New York, pp. 79–122.

Howell, N., 1979, *Demography of the Dobe !Kung.* Academic Press, New York.

Hrdy, S. B., 1976, The care and exploitation of non-human primate infants by conspecifics other than the mother. *Adv. Study Behav.* 6, 101–151.

Hrdy, S. B., 1977, *The Langurs of Abu,* Harvard University Press, Cambridge, Mass., 361 pp.

Hrdy, S. B., 1979, Infanticide among animals: A review, classification and examination of the implications for the reproductive strategies of females, *Ethology and Sociobiology* 1, No. 1.

Hrdy, S. B., 1981, *The Woman That Never Evolved,* Harvard University Press, Cambridge, Mass., 256 pp.

Hrdy, S. B., and Hrdy, D. B., 1976, Hierarchical relations among female hanuman langurs, *Science* 193, 913–915.

Hrdy, S. B., in press, Commentary on "Infant killing and male reproductive strategies in langurs, by J. Boogess," in *Infanticide,* G. Hausfater and S. B. Hrdy, eds., Aldine, New York.

Hrdy, S. B. and Hausfater, G., in press, Comparative and evolutionary perspectives on infanticide: an introduction, in *Infanticide,* G. Hausfater and S. B. Hrdy, eds., Aldine, New York.

Huebner, D. K., Lentz, J. L., Wooley, M. J., and King, J. E., 1979, Responses to snakes by surrogate-reared and mother-reared squirrel monkeys, *Bull. Psychon. Soc.* 14, 33–36.

Huffman, S. L., Chowdhury, A. K. M. A., and Moseley, W. H., 1978, Postpartum amenorrhea: how is it affected by maternal nutritional status?, *Science* 200, 1155–1156.

Humphrey, N. K., 1976, The social function of intellect, in *Growing Points in Ethology,* Bateson, P. P. G., and Hinde, R. A., eds., Cambridge University Press, Cambridge, pp. 303–317.

Hunter, A. J., and Dixson, A. F., 1980, Olfactory communication in the owl monkey *(Aotus trivirgatus), Primate Eye* 14, 5–6.

Hunter, J., Martin, R. D., Dixson, A. F., and Rudder, B. C. C., 1979, Gestation and inter-birth intervals in the owl monkey *(Aotus trivirgatus griseimembra)*, *Folia primatol.* 31, 165–175.

Hutchinson, G. E., 1959, Homage to Santa Rosalia, or why are there so many kinds of animals?, *Am. Nat.* 93, 145–159.

Hutchinson, G. E., 1961, The paradox of the plankton, *Am. Nat.* 95, 137–144.

Hutchinson, G. E., 1978, *An Introduction to Population Ecology*, Yale University Press, New Haven, Conn., 280 pp.

Hutchinson, G. E., 1981, Random adaptation and imitation in human development, *Am. Sci.* 69, 161–165.

Hutchinson, G. E., and MacArthur, R. H., 1959, A theoretical ecological model of size distributions among species of mammals, *Am. Nat.* 93, 133–134.

Hutt, C., 1974, Exploration and play in children, in *Play*, J. S. Bruner, A. Jolly, and K. Sylva, eds., Penguin, Harmondsworth, pp. 202–215.

Huxley, A., 1939, *After Many a Summer Dies the Swan*, Chatto, London, 314 pp.

Huxley, T. H., 1863, *Man's Place in Nature*, Macmillan, London, 328 pp.

Ingram, J. C., 1978, Parent-infant interaction in the common marmoset (Callithrix jacchus), in *The Biology and Conservation of the Callitrichidae*, D. Kleiman, ed., Smithsonian, Washington, pp. 281–292.

Itani, J., 1972, A preliminary essay on the relationship between social organization and incest avoidance in nonhuman primates, in *Primate Socialization*, F. E. Poirier, ed., Random House, New York, pp. 165–171.

Itoigawa, N., Negayama, K., and Kondo, K., 1981, Experimental study on sexual behavior between mother and son in Japanese monkeys *(Macaca fuscata)*, *Primates* 22, 494–502.

Izawa, K., and Mizuno, A., 1977, Palm-fruit cracking behavior of wild black-capped capuchin *(Cebus apella)*, *Primates* 18, 773–792.

Izawa, K., 1979, Foods and feeding behavior of wild black-capped capuchin *(Cebus apella)*, *Primates* 20, 57–76.

Janetos, A. C., and Cole, B. J., 1981, Imperfectly optimal animals. *Behav. Ecol. Scoiobiol.* 9, 203–210.

Janos, D. P., 1980, Vesicular-arbuscular mycorrhizae affect tropical rain forest plant growth, *Ecology* 61, 151–162.

Janson, C. H., in press, Female choice and mating system of the brown capuchin monkey *Cebus apella*.

Janson, H. W., 1952, *Apes and Ape Lore in the Middle Ages and the Renaissance*, Warburg Inst. Studies, London 20, 384 pp.

Janzen, D. H., 1978, Complications in interpreting the chemical defenses of trees against tropical arboreal plant-eating vertebrates, in *Ecology of the Arboreal Folivores*, G. G. Montgomery, ed., Smithsonian, Washington, pp. 73–84.

Janzen, D. H., 1979, How to be a fig, *Ann. Rev. Ecol. and Syst.* 10, 13–52.

Jay, P. C., 1963, Aspects of maternal behavior among langurs, *Ann. New York Acad. Sci.* 102, 468–476.

Jay, P. C., 1965, The hanuman langur of north India, in *Primate Behavior*, I. Devore, ed., Holt, Rinehart, Winston, New York, pp. 197–249.

Jolly, A., 1964, Prosimians' manipulation of simple object problems, *Anim. Behav.* 12, 560–570.

Jolly A., 1964, Choice of cue in prosimian learning, *Anim. Behav.* 12, 517–577.

Jolly, A., 1966, *Lemur Behavior*, Chicago University Press, Chicago, 187 pp.

Jolly, A., 1966, Lemur social behavior and primate intelligence, *Science* 153, 501–506.

Jolly, A., 1967, Breeding synchrony in wild *Lemur catta*, in *Social Communication among Primates*, S. A. Altmann, ed., Chicago University Press, Chicago, pp. 3–14.

Jolly, A., 1972, Hour of birth in primates and man, *Folia primatol.* 18, 108–121.

Jolly, A., 1973, Primate birth hour, *Internat. Zoo Yrb.* 13, 391–397.

Jolly, A., 1972, Troop continuity and troop spacing in *Propithecus verreauxi* and Lemur catta at Berenty (Madagascar), *Folia primatol.* 17, 335–362.

Jolly, A., 1980, *A World Like Our Own: Man and Nature in Madagascar*, Yale, New Haven, 272 pp.

Jolly, A., Oberle, P., and Albignac, R., 1984, Madagascar, Pergamon, Oxford.

Jolly, C. J., 1970, The seed-eaters: a new model of hominid differentiation based on a baboon analogy, *Man* 5, 5–26.

Jones, C. B., 1980, The functions of status in the mantled howler monkey, *Alouatta palliata Gray:* Intraspecific competition for group membership in a folivorous neotropical primate, *Primates* 21, 389–405.

Jones, C. B., 1980, The evolution and socioecology of dominance in primate groups: a theoretical formulation, classification and assessment, *Primates* 22, 70–83.

Jones, C., and Sabater Pi, J., 1969, Sticks used by chimpanzees in Rio Muni, West Africa, *Nature* 223, 100–101.

Johnson, J. M., 1974, Status, ecology and behaviour of the liontailed macaque, Dissertation. Forest Research Institute, Dehra Dun.

Jouventin, P., Pasteur, G., and Cambefort, J. P., 1976, Observational learning of baboons and avoidance of mimics, *Evolution* 31, 214–218.

Kagan, J., 1978, *The Second Year*, Harvard University Press, Cambridge, Mass.

Kagan, J., 1978, Continuity and stage in infant development, In *Perspectives in Ethology*, Vol. 3, P. P. G. Bateson, and P. H. Klopfer, Plenum, New York, pp. 67–84.

Kano, T., 1980, Social behaviors of wild pygmy chimpanzees *(Pan paniscus)* of Wamba, a preliminary report. *J. Hum. Evol.* 9, 243–260.

Kano, T., 1982, The social group of pygmy chimpanzees, *Pan paniscus* of Wamba, *Primates* 23, 171–188.

Kaplan, J. N., 1981, Effect of surrogate-administered punishment on surrogate contact in infant squirrel monkeys, *Develop. Psychobiol.* 14, 523–532.

Kaplan, J. N., Cubicciotti, D. D., III, and Redican, W. K., 1979, Olfactory and visual differentiation of synthetically scented surrogates by infant squirrel monkeys, *Develop. Psychobiol.* 12, 1–11.

Kapune, T., 1966, Untersuchungen zur Bildung eines ''Wertbegriffs'' bei neideren Primaten, *Z. Tierpsychol.* 23, 324–363.

Kaufman, I. C., and Rosenblum, L. A., 1967, Depression in infant monkeys separated from their mothers, *Science* 155, 1030–1031.

Kavanagh, M., 1972, Food-sharing behavior within a group of douc monkeys *(Pygathrix nemaeus nemaeus)*, *Nature* 239, 406–407.

Kavanagh, M., 1978, The diet and feeding behavior of *Cerceopithecus aethiops tantalus*, *Folia primatol.* 30, 30–63.

Kavanagh, M., 1980, Invasion of the forest by an African savannah monkey: behavioral adaptations, *Behaviour* 73, 238–260.

Kavanagh, M., 1981, Variable territoriality among tantalus monkeys in Cameroon, *Folia primatol.* 36, 76–98.

Kawai, M., 1958, On the system of social ranks in a natural troop of Japanese monkeys, 1: Basic rank and dependent rank, *Primates* 1, 111–130.

Kawai, M., 1960, A field experiment on the process of group formation in the Japanese monkey (Macaca fuscata) and the releasing of the group at Ohirayama, *Primates* 2, 181–255.

Kawai, M., ed., 1979, *Ecological and Sociological Studies of Gelada Baboons*, Karger, Basel.

Kawai, M., Dunbar, R., Ohsawa, H., and Mori, U., 1983, Social organization of gelada baboons: social units and definitions, *Primates* 24, 13–24.

Kellogg, W. N., 1968, Communication and language in the home-raised chimpanzee, *Science* 162, 423–427.

Kellogg, W. N., and Kellogg, L. A., 1933, *The Ape and the Child*, McGraw-Hill, New York, 341 pp.

Keverne, E. B., 1980, Olfaction in the behaviour of non-human primates, *Symp. Zool. Soc. Lond.* 45, 313–327.

King, G. E., 1980, Alternative uses of primates and carnivores in the reconstruction of early hominid behavior, *Ethol. Sociobiol.* 1, 99–109.

Kingdon, J. S., 1980, The role of visual signals and face patterns in African forest monkeys (guenons) of the genus *Cercopithecus*, *Trans. Zool. Soc. Lond.* 35, 425–475.

Kingsley, S., 1979, Social and endocrine condition of secondary sexual development in male orangutans, paper, *Primate Soc.*, Great Britain.

Kinzey, W. G., and Gentry, A. H., 1979, Habitat utilization in two species of Callicebus, in *Primate Ecology*, R. W. Sussman, ed., Wiley, New York, pp. 89–100.

Kinzey, W. G., and Robinson, J. G., 1981, Intergroup loud calls serve two different roles in two species of *Callicebus*, *Am. J. Phys. Anthrop.* 54, 240.

Kinzey, W. G., and Robinson, J. G, 1983, Intergroup loud calls, range size, and spacing in *Callicebus torquatus*, *Am. J. Phys. Anthrop.* 60, 539–544.

Kitamura, K., 1983, Pygmy chimpanzee association patterns in ranging, *Primates* 24, 1–12.

Kleiman, D. G., 1977, Monogamy in mammals, *Q. Rev. Biol.* 52, 29–69.

Kleiman, D. G., and Mack, D. S., 1980, Effects of age, sex, and reproductive status on scent marking frequencies in the golden lion tamarin, *Leontopithecus rosalia*, *Folia primatol.* 33, 1–14.

Kleiman, D. G., and Malcolm, J. R., 1981, The evolution of male parental investment in mammals, in *Parental Care in Mammals*, D. J. Gubernick and Klopfer, P. J., ed., Plenum, New York.

Klein, L. L., and Klein, D. J., 1975, Social and ecological contrasts between four taxa of neotropical primates, in *Socioecology and Psychology of Primates*, R. H. Tuttle, ed., Mouton, The Hague, pp. 59–86.

Klein, L. L., and Klein, D. B., 1977, Feeding behaviour of the Colombian spider monkey, in *Primate Ecology*, T. H. Clutton-Brock, ed., Academic Press, London, pp. 153–182.

Klopfer, P. H., 1970, Discrimination of the young in galagos, *Folia primatol.* 13, 137–143.

Klopfer, P., and Boskoff, K. J., 1982, Maternal behavior in prosimians, in *The Study of Prosimian Behavior*, G. Doyle and R. D. Martin, eds., Academic Press, New York.

Koford, C. B., 1965, Population dynamics of rhesus monkeys on Cayo Santiago, in *Primate Behavior*, I. DeVore, ed., Holt, New York, pp. 160–174.

Köhler, W., 1921, *Intelligenzprüfungen an Menschenaffen*, Springer, Berlin, 194 pp.

Köhler, W., 1927, *The Mentality of Apes*, 2nd ed., Routledge and Kegan Paul, London.

Kohts, N., 1923, *Infant Ape and Human Child*, Scientific Memoirs of the Museum Darwinianum in Moscow, 1935.

Konner, M, and Worthman, C., 1980, Nursing frequency, gonadal function and birth spacing among !Kung hunter-gatherers, *Science* 207, pp. 788–791.

Kortlandt, A., 1967, Experimentation with chimpanzees in the wild, in *Neue Ergebnisse der Primatologie*, D. Starck, R. Schneider, and H.-J. Kuhn, eds., Gustau Fischer Verlag, Stuttgart.

Kortlandt, A., 1978, The ecosystem in which the incipient hominines could have evolved, in *Recent Advances in Primatology*, Vol. 3, D. J. Chivers, and K. A. Joysey, eds., Academic Press, London, pp. 503–506.

Kortlandt, A., 1980, How might early hominids have defended themselves against large predators and food competitors?, *J. Hum. Evol.* 9, 79–112.

Kortlandt, A., and Kooiji, M., 1963, Protohominid behavior in primates, *Symp. Zool. Soc. Lond.* 10, 61–88.

Kleiber, M., 1932, Body size and metabolism, *Hilgardia* 6, 315–353.

Krebs, J. R., 1978, Optimal foraging: decision rules for predators, in *Behavioral Ecology*, J. R. Krebs, and N. B. Davies, eds., Blackwell, Oxford, pp. 23–63.

Kress, J. H., Conly, J. M., Eaglen, R. H., and Ibanez, A. E., 1978, The behavior of *Lemur variegatus*, Kerr, 1792, *Z. Tierpsychol.* 48, 87–99.

Kuhl, P. K., and Miller, J. D., 1975, Speech perception by the chinchilla: voiced-voiceless distinction in alveolar plosive consonants, *Science* 190, 69–72.

Kummer, H., 1968, *Social Organization of Hamadryas Baboons*, Karger, Basel, 189 pp.

Kummer, H., 1982, Social knowledge in free-ranging primates, in *Animal Mind-Human Mind*, Dahlem Konferenzen, 1982, D. R. Griffin, ed., Springer-Verlag, Berlin, pp. 113–130.

Kurland, J. A., 1980, Kin selection theory: a review and selective bibliography, *Ethol. Sciobiol.* 1, 255–274.

Kuroda, S., 1980, Social behavior of the pygmy chimpanzees, *Primates* 21, 181–197.

Kurup, G. U., 1975, Status of the Nilgiri langur, *Presbytis johni* in the Anamalai, Cardamon and Nilgiri hills of the Western Ghats, India, *J. Bombay Nat. Hist. Soc.* 72, 21–29.

Laguna, G. A. de, 1963, *Speech, Its Function and Development*, Indiana University Press, Bloomington (orig. publ. 1927).

Laidler, K., 1980, *The Talking Ape*, Stein and Day, New York, 181 pp.

Lancaster, J. B., 1972, Play-mothering: the relations between juvenile females and young infants among free-ranging vervet monkeys, in *Primate Socialization*, F. E. Poirer, ed., Random House, New York, pp. 83–104.

Lancaster, J. B., and Lee, R. B., 1965, The annual reproductive cycle in monkeys and apes, in *Primate Behavior*, I. DeVore, ed., Holt, New York, pp. 486–513.

Lashley, K. S., 1938, The mechanism of vision: XV Preliminary studies of the rats' capacity for detail vision, *J. Genet. Psychol.* 18, 123–193.

Lauer, C., 1980, Seasonal variability in spatial defense by free-ranging rhesus monkeys *(Macaca mulatta)*, *Anim. Behav.* 28, 476–482.

Lawick-Goodall, J. van *see* Goodall, J.

Le Gros Clark, W. E., 1960, *The Antecedents of Man*, Quadrangle Books, Chicago, 374 pp.

Lee, P. C., 1981, Ecological influence on social development of vervet monkeys *(Cercopithecus aethiops)*, *Primate Eye* 17, 4–5.

Lee, P. C., and Olover, J. I., 1979, Competition, dominance and the acquisition of rank in juvenile yellow baboons *(Papio cynocephalus)*, *Anim. Behav.* 27, 576–585.

Lee, R. B., 1980, Lactation, ovulation,infanticide, and women's work: a study of hunter-gatherer population regulation, in *Biosocial Mechanisms of Population Regulation*, M. N. Cohen, R. S. Malpass, and H. G. Klein, eds., Yale University Press, New Haven, Conn., pp. 321–48.

Lehr, E., 1967, Experimentelle Untersuchungen an Affen und Halbaffen über Generalisation von Insekten- und Blütenabbildungen, *Z. Tierpsychol.* 24, 208–244.

Leigh, E. G., 1971, *Adaptation and Diversity*, Freeman, Cooper and Co., San Francisco, 288 pp.

Leigh, E. G., Jr., 1975, Structure and climate in tropical rain forest, *Ann. Rev. Ecol. Syst.* 6, 67–86.

Leigh, E. G., 1977, How does selection reconcile individual advantage with the good of the group?, *Proc. Nat. Acad. Sci.* 74, 4542–4546.

Leigh, E. G., Jr., and Smyth, N., 1978, Leaf production, leaf consumption and the regulation of folivory on Barro Colorado Island, in *The Ecology of the Arboreal Folivores*, G. G. Montgomery, ed., Smithsonian, Washington, pp. 33–50.

Leigh, E. G. Jr., Rand, A. S., and Windsor, D. M. eds., 1982, *The Ecology of a Tropical Forest*, Smithsonian, Washington, 468.

Lethmate, J., 1978, Versuche zum "vorbedingten" Handeln mit einem jungen orangutan, *Primates* 19, 727–736.

Lethmate, J., 1982, Tool-using skills of orangutans, *J. Hum. Evol.* 11, 46–64.

Leutenegger, W., 1973, Maternal-fetal weight relationships in primates, *Folia primatol.* 20, 280–293.

Leutenegger, W., 1976, Allometry of neonatal size in eutherian mammals, *Nature* 263, 229–230.

Leutenegger, W., 1978, Scaling of sexual dimorphism in body size and breeding system in primates, *Nature* (London) 272, 610–611.

Leutenegger, W., 1980, Monogamy in callitrichids: a consequence of phyletic dwarfism?, *Int. J. Primatol.* 1, 95–98.

Leutenegger, W., 1980, Encephalization and obstetrics in primates, *Am. J. Phys. Anthrop.* 52, 248.

Leutenegger, W., 1982, Scaling of sexual dimorphism in body weight and canine size in primates, *Folia primatol.* 37, 163–176.

Leutenegger, W., and Kelly, J. T., 1977, Relationship of sexual dimorphism in canine size and body size to social, behavioral, and ecological correlates in anthropoid primates, *Primates* 18, 117–136.

Lewes, M. M., 1963, *Language, Thought, and Personality*, Harrap, London.

Lindburg, D. G., 1971, The rhesus monkey in North India: an ecological and behavioral study, in *Primate Behavior*, L. A. Rosenblum, ed., Academic Press, New York, pp. 2–106.

Lindburg, D. G., 1977, Feeding behaviour and diet of rhesus monkeys *(Macaca mulatta)* in a Siwalik Forest in North India, in *Primate Ecology*, T. H. Clutton-Brock, ed., Academic Press, London, pp. 223–249.

Lindburg, D. G., 1982, Primate obstetrics: the biology of birth, *Am. J. Primatol.* Suppl. 1, 193–200.

Linnaeus, C., 1758, Systema naturae per regna tri naturae, secundum classes, ordines, genera, species cum caracteribus, synonymis locis, 10th ed., rev. Stockholm, Laurentii Salvii 1 (2), 1–824.

Lorenz, K., 1935, Companions as factors in the birds' environment, in *Studies in Animal and Human Behaviour*, K. Lorenz, ed., Methuen, London, 1970, Vol. 1, pp. 101–258.

Lorenz, K., 1971, *Studies in Animal and Human Behaviour*, Methuen, London, Vol. II, 366 pp.

Lorenz, K., 1975, *Evolution and Modification of Behavior*, Chicago University Press, Chicago, 125 pp.

Lovejoy, C. O. 1981, The origin of man, *Science*, 211, 341–350.

Loy, J., 1981, The reproductive and heterosexual behaviors of adult patas monkeys in captivity, *Anim. Behav.* 29, 714–726.

Lumsden, C. J., and Wilson, E. O., 1981, *Genes, Mind and Culture*, Harvard University Press, Cambridge, Mass., 428 pp.

Lunn, S. F., and McNeilly, A. S, 1982, Failure of lactation to have a consistent effect on the interbirth interval of the common marmoset, *Callithix jacchus jacchus, Folia primatol.* 37, 99–105.

MacArthur, R. H., and Levins, R., 1964, Competition, habitat selection, and character displacement in a patchy environment, *Proc. Nat. Acad. Sci. U.S.A.* 51, 1207–1210.

MacArthur, R. H., and Wilson, E. O., 1967, *The Theory of Island Biogeography*, Princeton University Press, Princeton, N.J., 203 pp.

Mace, G. H., Harvey, P. H., and Clutton-Brock, T. H., 1981, Brain size and ecology in small mammals, *J. Zool. Lond.*, 193, 333–354.

Mace, G. M., Harvey, P. H., and Clutton-Brock, T. H., in press, Vertebrate home range size and energetic requirements, in *Animal Movement*, P. J. Greenland and I. Swingland, eds.

MacKinnon, J., 1974, The behavior and ecology of wild orangutans, *Anim. Behav.* 22, 3–74.

MacKinnon, J. R., 1977, A comparative ecology of the Asian apes, *Primates* 18, 747–772.

MacKinnon, J., 1978, *The Ape Within Us*, Collins, London, 287 pp.

MacKinnon, J., 1979, Reproductive behavior in wild orangutan populations, in *The Great Apes*, D. A. Hamburg, and E. R. McCown, eds., Benjamin/Cummings, Menlo Park, CA, pp. 257–274.

MacKinnon, J. R., and MacKinnon, K. S., 1980a, Niche differentiation in a primate community, in *Maylayan Forest Primates*, D. J. Chivers, ed., Plenum, New York, pp. 167–190.

MacKinnon, J. R., and MacKinnon, K. S., 1980b, The behavior of wild spectral tarsiers, *Int. J. Primatol.* 1, 361–380.

Makwana, S. C., and Pirta, R. S., 1978, Field ecology and behavior of rhesus macaque, *Macaca mulatta*, IV. Development and behavior of an independent and stable all juvenile group, *Biol. Behav.* 3, 163–167.

Mallow, G. K., 1981, The relationship between aggressive behavior and menstrual cycle stage in female rhesus monkeys *(Macaca mulatta)*, *Hormones Behav.* 15, 259–269.

Maples, W. R., Maples, M. K., Greenhood, W. F., and Walek, M. L., 1976, Adaptations of crop-raiding baboons in Kenya, *Am. J. Phys. Anthrop.* 45, 309–316.

Mathieu, M., in preparation, Intelligence without language: Piagetian assessment of cognitive development in the chimpanzee, IPS Congress, Atlanta, 1982.

Mathieu, M., and Bergeron, G., in preparation, Piagetian assessment of cognitive development in chimpanzee.

Mathieu, M., Daudelin, N., and Dagenais, Y., 1980, Piagetian causality in two house-reared chimpanzees *(Pan troglodytes)*, *Can. J. Psychol.* 34, 179–186.

Marais, E., 1969, *The Soul of the Ape*, Anthony Blond, London, 226 pp.

Marler, P., 1965, Communication in monkeys and apes, in *Primate Behavior*, I. DeVore, ed., Holt, New York, pp. 544–584.

Marler, P., 1967, Animal communication signals, *Science* 157, 769–774.

Marler, P., 1969, Vocalizations of wild chimpanzees, in *Proc. 2nd Int. Congr. Primatol.* 1. *Behavior*, C. R. Carpenter, ed., Karger, Basel, pp. 94–100.

Marler, P., 1970, Vocalizations of East African monkeys, I. Red colobus, *Folia primatol.* 13, 81–89.

Marler, P., 1972, Vocalizations of East African monkeys, II. Black and white colobus, *Behaviour* 42, 175–197.

Marler, P., 1973, A comparison of vocalizations of redtailed monkeys and blue monkeys, *Cercopithecus ascanius* and *C. mitis*, in Uganda, *Z. Tierpsychol.* 33, 223–247.

Marler, P., 1976, Social organization, communication, and graded signals: the chimpanzee and the gorilla, in *Growing Points in Ethology*, P. P. G. Bateson, and R. A. Hinde, eds., Cambridge University Press, Cambridge, pp. 239–280.

Marler, P., 1977, The structure of the animal communication sounds, in *Recognition of Complex Acoustic Signals*, T. H. Bullock, ed., West Berlin, Dahlem Konferenzen, pp. 17–35.

Marler, P., 1977, Primate vocalization: affective or symbolic?, in *Progress in Ape Research*, G. H. Bourne, ed., Academic Press, New York, pp. 85–96.

Marler, P., in press, Animal communication: affect or cognition?

Marler, P., and Hamilton, W. J., III, 1966, *Mechanisms of Animal Behavior*, Wiley, New York, 771 pp.

Marler, P., and Hobbet, L., 1975, Individuality in long-range vocalization of wild chimpanzees, *Z. Tierpsychol.* 38, 97–109.

Marler, P. and Peters, S., 1982, Subsong and plastic song: their role in the vocal learning process, in *Acoustic Communication in Birds*, D. E. Kroodsma and E. H. Miller, eds., Academic, New York, 2,25–50.

Marler P., and Tenaza, R., 1977, Signaling behavior of apes with special reference to vocalization, in *How Animals Communicate*, T. A. Sebeok, ed., Indiana University Press, Bloomington, Indiana, pp. 965–1033.

Mariott, B. M., and Salzen, E. A., 1978, Facial expressions in captive squirrel monkeys *(Saimiri sciureus), Folia primatol.* 29, 1–18.

Marsh, C. W., 1979, Comparative aspects of social organization in the Tana River red colobus, *Colobus badius rufomitratus, Z. Tierpsychol.* 51, 337–362.

Marsh, C. W., 1981, Ranging behaviour and its relation to diet selection in Tana River red colobus *(Colobus badius rufomitratus, J. Zool.* 195, 473–492.

Marten, K., Quine, D., and Marler, P., 1977, Sound transmission and its significance for animal vocalizations: II. Tropical forest habitats, *Behav. Ecol. Sociobiol.* 2, 291–302.

Martin, R. D., 1968, Toward a new definition of primates, *Man* 3, 377–401.

Martin, R. D., 1979, Phylogenetic aspects of Prosimian behavior, in *The Study of Prosimian Behavior*, G. A. Doyle, and R. D. Martin, eds., Academic Press, New York, pp. 45–78.

Martin, R. D., 1980, Sexual dimorphism and the evolution of higher primates, *Nature* 287, 273–275.

Martin, R. D., 1981, Field studies of primate behaviour, *Symp. Zool. Soc. Lond.* 46, 287–336.

Martin, R. D., 1981b, Well-groomed predecessors, *Nature* 289, 536.

Martin, R. D., in press, Human brain evolution in an ecological context, *Bull. Am. Mus. Nat. Hist.*

Maslow, A. H., 1936, The role of dominance in the social and sexual behavior of infra-human primates III: A theory of the sexual behavior of infra-human primates, *J. Genet. Psychol.* 48, 310–338.

Mathieu, M., Bouchard, M. A., Granger, L. and Herscovitch, J., 1976, Piagetian object-permanence in *Cebus capucinus, Laogthrix flavicauda* and *Pan troglodytes, Anim. Behav.* 24, 585–588.

Mathieu, M., Daudelin, N., Dagenais, Y., Decarie, T., 1980, Piagetian causality in two house-reared chimpanzees *(Pan troglodytes).* Can J. Psychol. 34:179–186.

Mayer, P. J., 1982, Evolutionary advantages of the menopause. *Human Ecology* 10: 477–494.

Maynard Smith, J., 1974, The theory of games and the evolution of animal conflicts, *J. Theor. Biol.* 47, 209–221.

Maynard Smith, J., 1976, Group selection, *Q. Rev. Biol.* 5, 277–283.

Maynard Smith, J., 1978, *The Evolution of Sex*, Cambridge University Press, London.

Maynard Smith, J., and Parker, G. A., 1976, The logic of asymmetric contests, *Anim. Behav.* 24, 159–175.

Maynard Smith, J., and Price, G. R., 1973, The logic of animal conflict, *Nature* 246, 15–18.

McClintock, M. K., 1971, Menstrual synchrony and suppression, *Nature* 229, 244.

McClintock, M. K., 1981, Social control of the ovarian cycle and the function of estrous synchrony, *Am. Zool.* 21, 243–256.

McClintock, M. K., and Adler, N. T., 1978, Induction of persistent estrus by airborne chemical communication among female rats, *Hormones and Behav.* 11, 414–418.

McGeorge, L. W., 1978, Influences on the structure of vocalizations of three Malagasy lemurs, in *Recent Advances in Primatology,* D. J. Chivers, and K. A. Joysey, eds., Academic Press, London, Vol. 3, pp. 103–109.

McGonigle, B. O., and Chalmers, M., 1977, Are monkeys logical?, *Nature* 267, 694–696.

McGrew, W. C., 1974, Tool use by wild chimpanzees in feeding upon driver ants, *J. Hum. Evol.* 3, 501–508.

McGrew, W. C., 1975, Patterns of plant food sharing by wild chimpanzees, in *Contemporary Primatol.,* S. Kondo, M. Kawai, and A. Ehara, eds., Karger, Basel, pp. 304–309.

McGrew, W. C., 1977, Socialization and object manipulation of wild chimpanzees, in *Primate Biosocial Development,* S. Chevalier-Skolnikoff and F. E. Poirier, eds., Garland STPM, New York, pp. 261–288.

McGrew, W. C., 1979, Evolutionary implications of sex differences in chimpanzee predation and tool use, in *The Great Apes,* D. A. Hamburg and E. R. McCown, eds., Benjamin/Cummings, Menlo Park, CA, pp. 441–464.

McGrew, W. C., 1981, The female chimpanzee as a human evolutionary prototype, in *Woman the Gatherer,* F. Dahlberg, ed., Yale University Press, New Haven, Conn., pp. 35–72.

McGrew, W. C., Baldwin, P. J., and Tutin, C. E. G., 1981, Chimpanzees in a savannah habitat: Mt. Assirik, Senegal, West Africa, *J. Hum. Evol.* 10, 227–244.

McGrew, W. C., Tutin, C. E. G., and Baldwin, P. J., 1979, New data on meat eating by wild chimpanzees, *Curr. Anthrop.* 20, 238–239.

McGrew, W. C., Tutin, C. E. G., and Baldwin, P. J., 1979, Chimpanzees, tools and termites: cross-cultural comparisons of Senegal, Tanzania and Rio Muni, *Man* 14, 185–214.

McGrew, W. C., Tutin, C. E. G., Baldwin, P. J., Sherman, M. J., and Whiten, A., 1978, Primates preying upon vertebrates: new records from West Africa, *Carnivore* 1, 41–45.

McGrew, W. C., and Tutin, C. E. G., 1973, Chimpanzee tool use in dental grooming, *Nature* 241, 477–478.

McKenna, J. J., 1979, Aspects of infant socialization, attachment, and maternal caregiving patterns among primates: A cross-disciplinary review, *Yrb. Phys. Anthrop.* 22, 250–286.

McKenna, J. J., 1979, The evolution of allomothering behavior among colobine monkeys: Function and opportunism in evolution, *Am. Anthrop.* 81, No. 4.

McKenna, J. J., 1981, Primate infant caregiving behavior: origins, consequences, and variability with emphasis on the common Indian langur monkey, in *Parental Care in Mammals,* D. J. Gubernick, and P. H. Klopfer, eds., Plenum, New York, pp. 389–416.

McKey, D., 1978, Soils, vegetation and seed-eating by black colobus monkeys, in *The Ecology of the Arboreal Folivores,* G. G. Montgomery, ed., Smithsonian, Washington, pp. 423–438.

McMillan, C. A., 1981, Synchrony of estrus in macaque matrilines at Cayo Santiago, *Am. J. Phys. Anthrop.* 54, 251.

McMillan, C., and Duggleby, C., 1981, Interlineage genetic differentiaton among rhesus macaques on Cayo Santiago, *Am. J. Phys. Anthrop.* 56, 305–312.

McNab, B. K., 1963, Bioenergetics and the determination of home range size, *Am. Nat.* 97, 113–140.

Menzel, E. W., 1966, Responsiveness to objects in free-ranging Japanese monkeys, *Behaviour* 26, 130–150.

Menzel, E. W., 1969, Responsiveness to food and signs of food in chimpanzee discrimination learning, *J. Comp. Physiol. Psychol.* 56, 78–85.

Menzel, E. W., 1969, Chimpanzee utilization of space and responsiveness to objects: age differences and comparison with macaques, in *Proc. 2nd Int. Congr. Primatol. 1. Behav.,* Carpenter, C. R., ed., Karger, Basel, pp. 72–80.

Menzel, E. W., 1971, Communication about the environment in a group of young chimpanzees, *Folia primat.* 15:220–232.

Menzel, E. W., 1972, Spontaneous investion of ladders in a group of young chimpanzees, *Folia primatol.* 17, 87–106.

Menzel, E. W., 1973, Further observations on the use of ladders in a group of young chimpanzees, *Folia primatol.* 19, 450–457.

Menzel, E. W., Davenport, R. K., and Rogers, C. M., 1963, The effects of environmental restriction on the chimpanzee's responsiveness to objects, *J. Comp. Physiol. Psychol.* 56, 78–85.

Menzel, E. W., Davenport, R. K., and Rogers, C. M., 1963a, The effects of environmental restriction on the chimpanzee's responsiveness in novel situations, *J. Comp. Physiol. Psychol.* 56, 329–334.

Menzel, E. W., Davenport, R. K., and Rogers, C. M., 1970, The development of tool use in wild-born and restriction-reared chimpanzees, *Folia primatol.* 12, 273–283.

Menzel, E. W., Davenport, R. K., and Rogers, C. M., 1972, Proto cultural aspects of chimpanzees' responsiveness to novel objects, *Folia primatol.* 17, 161–170.

Menzel, E. W. and Juno, C., 1982, Marmosets *(Saguinus fusciocollis)* are learning sets learned? *Science* 217, 750–752.

Menzel, E. W., and Menzel, C. R., 1970, Cognitive developmental and social aspects of responsiveness to novel objects in a family group of marmosets, *(Saguinus fuscicollis), Am. Behav.* 70, 251–279.

Mertl, A. S., 1977, Habituation to territorial scent marks in the field by *Lemur catta, Behav. Biol.* 21, 500–507.

Mertl-Milhollen, A. S., 1979, Olfactory demarcation of territorial boundaries by a primate *Propithecus verreauxi, Folia primatol.* 32, 35–42.

Mertl-Hilhollen, A. S., Gustafson, H. L., Budnitz, N., Dainis, K., and Jolly, A., 1979, Population and territory stability of the *Lemur catta* at Berenty, Madagascar, *Folia primatol.* 31, 108–122.

Meyer, D. R., Treichler, F. R., and Meyer, P. M., 1965, Discrete training techniques and stimulus variables, in *Behavior of Nonhuman Primates,* A. M. Schreier, H. F. Harlow, and F. Stollnitz, eds., Academic Press, New York, pp. 1–50.

Michael, R. P., and Bonsall, R. W., 1977, Chemical signals and primate behavior, in *Chemical Signals in Vertebrates,* D. Muller-Schwarze, and M. M. Mozell, eds., Plenum, New York, pp. 251–271.

Michael, R. P., and Keverne, E. B., 1968, Pheromones in the communication of sexual status in primates, *Nature* 218, 746–749.

Michael, R. P., and Zumpe, D., 1970, Aggression and gonadal hormones in captive rhesus monkeys, *Anim. Behav.* 18, 1–19.

Michael, R. P., and Zumpe D., 1976, Environmental and endocrine factors influencing annual changes in sexual potency in primates, *Psychoneuro-endocrinology* 1, 303–313.

Michael, R. P., Zumpe, D., Richter, M., and Bonsall, R. W., 1977, Behavioral effects of a synthetic mixture of aliphatic acids in rhesus monkeys, *(Macaca mulatta), Horm. Behav.* 9, 296–308.

Miles, H. L., 1983, Apes and Language: the search for communicative competence, in *Language in Primates,* J. de Luce and H. T. Wilder, eds., Springer-Verlag, New York, 43–62.

Miles, W. R., 1963, Chimpanzee behavior: removal of foreign body from companion's eye, *Proc. Nat. Acad. Sci.* 49, 840–843.

Miller, R. E., Caul, W. F., and Mirsky, I. A., 1967, Communication of affect between feral and socially isolated monkeys, *J. Pers. Soc. Psychol.* 7, 231–239.

Milton, K., 1978, Behavioral adaptation to leaf-eating by the mantled howler monkey *(Alouatta palliata),* in *The Ecology of the Arboreal Folivores,* G. G. Montgomery, ed., Smithsonian, Washington, pp. 535–550.

Milton, K., 1980, *The Foraging Strategy of Howler Monkeys,* Columbia University Press, New York, 165 pp.

Milton, K., 1982, Dietary quality and demographic regulation in a howler monkey population, in *The Ecology of a Tropical Forest,* E. G. Leigh, A. S. Rand and D. M. Windsor, eds., Smithsonian, Washington, pp. 273–289.

Milton, K., and May, M. L., 1976, Body weight, diet and home range area in primates, *Nature* (Lond.), 259, 459–462.

Mineka, S., Keir, R., and Price, V., 1980, Fear of snakes in wild- and laboratory-reared rhesus monkeys *(Macaca mulatta), Anim. Learn. Behav.* 8, 653–663.

Mitani, J. C., and Rodman, P. S., 1979, Territoriality: the relation of ranging pattern and home range size to defendability, with an analysis of territoriality among primate species, *Behav. Ecol. Sociobiol.* 5, 241–251.

Mitchell, G., 1979, *Behavioral Sex Differences in Nonhuman Primates,* van Nostrand, New York, 515 pp.

Mittermeier, R. A., and Coimbra-Filho, A. F., 1977, Primate conservation in Brazilian Amazonia, in *Primate Conservation,* Prince Rainier III, and G. F. Bourne, eds., Academic Press, New York, pp. 117–167.

Mittermeier, R. A., and Coimbra-Filho, A. F., 1981, Systematics: Species and subspecies, in *Ecology and Behavior of Neotropical Primates,* Academia Brasileira de Ciencias Rio de Janeiro, pp. 9–28.

Mittermeier, R. A., Macedo-Ruiz, H. de, Luscombe, B. A., and Cassidy, J., 1977, Rediscovery and conservation of the Peruvian yellow-tailed woolly monkey *(Lagothrix flavicauda),* in *Primate Conservation,* Prince Rainier III and G. H. Bourne, eds., Academic Press, New York, pp. 95–116.

Mittermeier, R. A., and van Roosmalen, M. G. M., 1981, Preliminary observations on habitat utilization and diet in eight Surinam monkeys, *Folia primatol.* 36, 1–39.

Miyadi, D., 1967, Differences in social behavior among Japanese macaque troops, in *Neue Ergebnisse der Primatologie*, D. Stark, R. Schneider, and H. J. Kuhn, eds., Fisher, Stuttgart.

Mohnot, S. M., 1971, Some aspects of social changes and infant-killing in the hanuman langur, *Presbytis entellus* (Primates, Cercopithecidae) in Western India, *Mammalia* 35, pp. 175–198.

Mohnot, S. M., 1980, Intergroup kidnapping in hanuman langur, *Folia primatol.* 34, 259–277.

Montgomery, G. G., and Sunquist, M. E., 1978, Habitat selection and use by two-toed and three-toed sloths, in *The Ecology of the Arboreal Folivores*, G. G. Montgomery, ed., Smithsonian, Washington, pp. 229–359.

Moore, J., in press, Female transfer in primates, *Int. J. Primatol.*

Mori, A., 1979, Analysis of population changes by measurement of body weight in the Koshima troop of Japanese monkeys, *Primates* 20, 371–398.

Morris, D., 1957, "Typical intensity" and its relation to the problem of ritualization, *Behavior* 11, 1–12.

Morris, D., 1962, *The Biology of Art*, Methuen, London, 176 pp.

Morris, D., 1967, *The Naked Ape*, Jonathan Cape, London, 267 pp.

Moynihan, M., 1969, Some adaptations which help to promote gregariousness, in *Proc. 12th Internat. Ornith. Congr.*, pp. 523–541.

Moynihan, M., 1967, Comparative aspects of communication in the New World primates, in *Primate Ethology*, D. Morris, ed., Weidenfeld and Nicolson, London, pp. 236–266.

Moynihan, M., 1970, Control, suppression, decay, disappearance and replacement of displays, *J. Theor. Biol.* 29, 85–112.

Moynihan, M., 1976, *The New World Primates*, Princeton University Press, Princeton, N.J., 262 pp.

Mukherjee, R. P., in press, Further observations on the golden langur *(Presbytis geei* Khajuria, 1956) with a note on the capped langur *(Presbytis pileatus* Blyth, 1843) of Assam.

Mukherjee, R. P., and Saha, S. S., 1978, The tail and its display behaviour in the golden langur, *Presbytis geei* Khajuria, *Bull. Zool. Surv. India* 1, 305–307.

Müller-Schwarze, D., ed., 1978, *Evolution of Play Behavior*, Dowden, Hutchinson and Ross, Stroudsburg, Pa.

Müller-Schwarze, D., Stagge, B., and Müller-Schwarze, C., 1982, Play behavior: persistence, decrease and energetic compensation during food shortage in deer fauns, *Science* 215, 85–87.

Myers, N., 1979, *The Sinking Ark*, Pergamon, Oxford, 307 pp.

Nadler, R. D., 1980, Child abuse: evidence from nonhuman primates, *Devel. Psychobiol.* 13, 507–512.

Nagy, K. A., and Milton, K., 1979, Energy metabolism and food consumption by wild howler monkeys, *Ecology* 60, 475–480.

Napier, J. R., 1961, Prehensility and opposability in the hands of primates, *Symp. Zool. Soc. Lond.* 5, 115–132.

Napier, P. H., 1976, *Catalogue of primates in the British Museum (Natural History):* I. *Families Callitrichidae and Cebidae*, London, British Museum (Natural History), 121 pp.

Napier, J. R., and Napier, P. H., 1967, *A Handbook of Living Primates*, Academic Press, London, 465 pp.

National Research Council, Committee on Non-Human Primates, 1981, *Techniques for the Study of Primate Population Ecology*, National Academy Press, Washington, 233 pp.

Nelson, K., 1973, Structure and strategy in learning to talk. *Monogr. Soc. for Res. in Child Devel.* 38.

Nigi, H., Tiba, T., Yamamoto, S., Floescheim, Y., and Ohsawa, N., 1980, Sexual maturation and seasonal changes in reproductive phenomena of male Japanese monkeys *(Macaca fuscata)* at Takasakiyama, *Primates* 21, 230–240.

Nishida, T., 1973, The ant-gathering behaviour by the use of tools among wild chimpanzees of the Mahali Mountains, *J. Hum. Evol.* 2, 357–370.

Nishida, T., and Uehara, S., 1980, Chimpanzees, tools, and termites: Another example from Tanzania, *Curr. Anthrop.* 21, 671–672.

Nishida, T., Uehara, S., and Nyundo, R., 1979, Predatory behavior among wild chimpanzees of the Mahale Mountain, *Primates* 20, 1–20.

Noë, R., de Waal, F., and van Hooff, J., 1980, Types of dominance in a chimpanzee colony, *Folia primatol.* 34, 90–110.

Noyes, S., 1981, Behavioral strategies associated with pregnancy among female Japanese monkeys, *Am. J. Phys. Anthrop.* 54, 259.

Oates, J. F., 1977a, The guereza and its food, in *Primate Ecology,* T. H. Clutton-Brock, ed., Academic Press, London, pp. 384–414.

Oates, J. F., 1977b, Social life of a black-and-white colobus monkey, *Colobus guereza, Z. Tierpsychol.* 45, 1–60.

Oates, J. F., 1978, The status of the South Indian black leaf-monkey *(Presbytis johnii)* in the Palni Hills, *J. Bombay Nat. Hist. Soc.* 75, 1–12.

Oliver, T. J., Ober, C., Buettner-Janusch, J., and Sade, D. S., 1981, Genetic differentiation among matrilines in social groups of rhesus monkeys, *Behav. Ecol. and Sociobiol.* 8, 279–285.

Olver, R. R., and Hornsby, J. R., 1966, On equivalence, in *Studies in Cognitive Growth,* J. S. Bruner et al., eds., Wiley, New York, pp. 68–85.

Oppenheimer, J. R., 1968, Behavior and ecology of the whitefaced cebus monkey, Cebus capucinus on Barro Colorado Island, Ph.D. thesis, University of Illinois.

Owen, O. J., 1979, Tool use in free-ranging baboons of Nairobi National Park, *Primates* 20, 595–597.

Owen-Smith, N., 1977, On territoriality in ungulates and an evolutionary model, *Q. Rev. Biol.* 52, 1–38.

Packer, C., 1977, Reciprocal altruism in *Papio anubis, Nature* 265, 441–443.

Packer, C., 1979a, Inter-troop transfer and inbreeding avoidance in *Papio anubis, Anim. Behav.* 27, 1–36.

Packer, C., 1979b, Male dominance and reproductive activity in *Papio anubis, Anim. Behav.* 37, 37–45.

Packer, C., 1980, Male care and exploitation of infants in *Papio anubis, Anim. Behav.* 28, 512–520.

Pages-Feuillade, E., in–press, Jeu et socialization: aspect theorique et descriptive de l'ontogenese du jeu chez un prosimien, *Microcebus coquereli.*

Pariente, G., 1980, The role of vision in prosimian behavior, in *The Study of Prosimian Behavior,* G. A. Doyle, and R. D. Martin, eds., Academic Press, New York, pp. 411–460.

Parker, C. E., 1978, Opportunism and the rise of intelligence, *J. Hum. Evol.* 7, 596–608.

Parker, S. T., 1977, Piaget's sensorimotor series in an infant macaque: a model for comparing unstereotyped behavior and intelligence in human and non-human primates, in *Primate Biosocial Development,* S. Chevalier-Skolnikoff, and F. E. Poirier, eds., Garland, New York, pp. 43–112.

Parker, S. T., Gibson, K. R., 1979, A developmental model of the evolution of language and intelligence in early hominids, *Brain Behav. Sci.* 2, 367–408.

Parra, R., 1978, Comparison of foregut and hindgut fermentation in herbivores, in *The Ecology of Arboreal Folivores,* G. G. Montgomery, ed., Smithsonian, Washington, pp. 205–230.

Pasnak, R., 1979, Acquisition of prerequisites to conservation by macaques, *J. Exp. Psychol.: Anim. Behav. Proc.* 5, 194–210.

Passingham, R. E., 1982, *The Human Primate,* Freeman, Oxford, 390 pp.

Patten, B. M., 1958, *Foundations of Embryology,* McGraw-Hill, New York, 578 pp.

Patterson, F., and Linden, E., 1981, *The Education of Koko,* Holt, Rinehart, and Winston, New York, 224 pp.

Pereira, M. E., in press, Abortion following the immigration of an adult male in yellow baboons, in *Infanticide,* G. Hausfater and S. B. Hrdy, eds., Aldine, New York.

Perret, M., 1982, Influence du groupement social sur la réproduction de la femelle de *Microcebus murinus, Z. Tierpsychol.* 60, 47–65.

Peters, R., and Mech, L. D., 1975, Behavioral and intellectual adaptations of selected mammalian predators to the problem of hunting large animals, in *Socioecology and Psychology of the Primates,* R. H. Tuttle, ed., Mouton, The Hague, pp. 279–300.

Petersen, M. R., 1982, The perception of species—specific vocalizations by primates: a conceptual framework, in *Primate Communication,* C. T. Snowdon, C. H. Brown, and M. R. Petersen, eds., Cambridge U.P., Cambridge, 171–211.

Petter, J. J., and Petter-Rousseaux, A., 1979, Classification of the prosimians, in *The Study of Prosimian Behavior,* G. A. Doyle, and R. D. Martin, eds., Academic Press, New York, pp. 1–44.

Petter, J. J., and Peyrieras, A., 1979, Nouvelle contribution à l'étude d'un lemurien malgache, le aye-aye (*Daubentonia madagascariensis* E. Geoffroy), *Mammalia* 34, 167–193.

Petter, J. J., Schilling, A., and Pariente, G., 1975, Observations on the behavior and ecology of *Phaner furcifer,* in *Lemur Biology,* I. Tattersall, and R. W. Sussman, eds., Plenum, New York, pp. 209–218.

Petter-Rousseaux, A., 1962, Recherches sur la biologie de la réproduction des primates inferieurs, Faculté des Sciences de l'Universite de Paris, A, 3794, 87 pp.

Petter-Rousseaux, A., 1968, Cycles genitaux saisonniers des lémuriens malgaches, in *Cycles Genitaux Saisonniers de Mammiferes Sauvages,* R. Canivenc, ed., Masson, Paris, pp. 11–22.

Piaget, J., 1951, *Play, Dreams and Imitation in Childhood,* Routledge and Kegan Paul, London, 296 pp.

Piaget, J., 1954, *The Construction of Reality in the Child*, Basic Books, New York, 386 pp.

Piaget, J., 1960, *The Psychology of Intelligence*, Littlefield, Adams, Peterson, N.J., 182 pp.

Pitcairn, T. K., and Eibl-Eibesfeldt, I., 1976, Concerning the evolution of non-verbal communication in man, in *Communicative Behavior and Evolution*, M. E. Hahn, and E. C. Simmel, eds., Academic Press, New York, pp. 81–113.

Pitelka, F. A., 1949, Numbers, breeding schedule, and territoriality in pectoral sandpipers in Northern Alaska, *Condor* 61, 233–264.

Plooij, F. X., 1978, Tool use during chimpanzees' bushpig hunt, *Carnivore* 1, 103–106.

Plooij, F., 1979, How wild chimpanzee babies trigger the onset of mother-infant play—and what the mother makes of it, in *Before Speech: The Beginning of Interpersonal Communication*, M. Bullowa, ed., Cambridge University Press, New York, pp. 223–243.

Poirier, F. E., 1977, Introduction, in *Primate Bio-social Development*, S. Chevalier-Skolnikoff, and F. E. Poirier, eds., Garland, New York.

Pollock, J. I., 1975, Field observations on *Indri indri:* a preliminary report, in *Lemur Biology*, I. Tattersall, and R. W. Sussman, eds., Plenum, New York, pp. 287–312.

Pollock, J. I., 1977, The ecology and sociology of feeding in *Indri indri*, in *Primate Ecology*, T. H. Clutton-Brock, Academic Press, London, pp. 38–71.

Pollock, J. I., 1979, Spatial distribution and ranging behavior in lemurs, in *The Study of Prosimian Behavior*, G. A. Doyle and R. D. Martin, eds., Academic Press, New York, pp. 359–410.

Pollock, J. I., 1979, Female dominance in *Indri indri*, *Folia primatol.* 31, 143–164.

Portmann, A., 1965, Uber die Evolution der Tragzeit bei Saügetieren, *Rev. Suisse Zool.* 72, 658–666.

Post, D. G., Hausfater, G., and McCuskey, S. A., 1980, Feeding behavior of yellow baboons *(Papio cynocephalus):* relationship to age, gender and dominance rank, *Folia primatol.* 34, 170–195.

Premack, A. J., 1976, *Why Chimps Can Read*, Harper and Row, New York, pp. 118.

Premack, D., 1976, *Intelligence in Ape and Man*, Erlbaum, Hillsdale, N.J.

Premack, D., 1983, The codes of man and beasts, *Behav. Brain Sci.* 6, 125–167.

Premack, D., and Premack, A. J., 1983. *The Mind of an Ape*, Norton, New York, 165 pp.

Price-Williams, D. R., Gordon, W., and Ramirez, M., 1968, Manipulation and conservation: a study of children from pottery-making families in Mexico, in *Eleventh Inter-Amer. Congr. Psychol. Proceedings*, Mexico City.

Pulliam, H. R., and Dunford, C., 1980, *Programmed to Learn: An Essay on the Evolution of Culture*, Columbia University Press, New York, 138 pp.

Quris, R., 1975, Écologie et organisation sociale de *Cercocebus galeritus agilis* dans le nord-est du Gabon, *Terre Vie* 29, 333–398.

Raemaekers, J. J., Aldrich-Blake, F. P. G., and Payne, J. B., 1980, The forest, in *Malayan Forest Primates*, D. J. Chivers, ed., Plenum, New York, pp. 29–62.

Rajecki, D. W., Lamb, M. E., and Obmascher, P., 1979, Toward a general theory of infantile attachment: A comparative review of aspects of the social bond, *Behav. Brain Sci.* 1, 417–464.

Randolph, M. C., and Brooks, B. A., 1967, Conditioning of a local response in a chimpanzee through social reinforcement, *Folia primatol.* 5, 70–79.

Ralls, K., 1976, Mammals in which females are larger than males, *Q. Rev. Biol.* 51, 245–276.

Ralls, K., and Ballou, J., 1982, Effects of inbreeding on infant mortality in captive primates, *Int. J. Primatol.* 3, 491–506.

Ralls, K., Brugger, K., and Ballou, J., 1979, Inbreeding and juvenile mortality in small populations of ungulates, *Science* 206, 1101–1103.

Ransom, T. W., 1981, *Beach Troop of the Gombe*, Bucknell University Press, Lewisburg, 319 pp.

Ransom, T. W., and Rowell, T. E., 1972, Early social development of feral baboons, in *Primate Socialization*, F. E. Poirier, ed., Random House, New York, pp. 105–143.

Rasmussen, D. R., 1979, Correlates of patterns of range use of a troop of yellow baboons *(Papio cynocephalus)*, 1. Sleeping sites, impregnable females, births and male emigrations and immigrations, *Anim. Behav.* 4, 1098–1112.

Rasmussen, D. R., 1981, Communities of baboon troops *(Papio cynocephalus)* in Mikumi National Park, Tanzania, A preliminary report, *Folia primatol.* 36, 232–242.

Rathbun, C. D., 1979, Description and analysis of the arch display in the golden lion tamarin, *Leontopithecus rosalia rosalia, Folia primatol.* 32, 125–148.

Redican, W. K., 1975, Facial expressions in nonhuman primates, in *Primate Behavior*, L. A. Rosenblum, ed., Academic Press, London.

Redshaw, M., 1978, Cognitive development in human and gorilla infants, *J. Hum. Evol.* 7, 133–141.

Rensch, B., and Döhl, J., 1967, Spontanes Offnen vershiedener Kistenverschlüsse durch eine Schimpansen, *Z. Tierpsychol.* 25, 216–231.

Rensch, B., and Döhl, J., 1968, Wahlen zwischen zwei überschaubaren Labyrinthwegen durch einen Schimpansen, *Z. Tierpsychol.* 25, 216–231.

Rhine, R. J., 1975, The order of movement of yellow baboons *(Papio cynocephalus)*, *Folia primatol.* 23, 72–104.

Rhine, R. J., Forthman, D. L., Stillwell-Barnes, R., Westlund, B. J., and Westlund, H. D., 1979, Movement patterns of yellow baboons *(Papio cynocephalus)*, the location of subadult males, *Folia primatol.* 32, 241–251.

Rhine, R. J., Hendy, H. M., Stillwell-Barnes, R., Westlund, B. J., and Westlund, H. D., 1980, Movement patterns of yellow baboons *(Papio cynocephalus)*, Central positioning of walking infants, *Am. J. Phys. Anthrop.* 53, 159–167.

Rhine, R. J., and Westlund, B. J., 1981, Adult male positioning in baboon progressions: Order and chaos revisited, *Folia primatol.* 35. 35, 77–116.

Richard, A. F., 1978, *Behavioral Variation*, Bucknell University Press, Lewisburg, 213 pp.

Rijksen, H. D., 1978, A field study on Sumatran orangutans *(Pongo pygmaeus abelii* Lesson 1827*)*, H. Veenman and Zonen B. V., Wageningen, 420 pp.

Ripley, S., 1967a, Intertroop encounters among Ceylon gray langurs *(Presbytis entellus)*, in *Social Communication Among the Primates*, S. A. Altmann, ed., Chicago University Press, Chicago, pp. 237–254.

Ripley, S., 1967b, The leaping of langurs; a problem in the study of locomotor adaptation, *Am. J. Phys. Anthrop.* 26, 149–170.

Ripley, S., 1980, Infanticide in langurs and man: Adaptive advantage or social pathology?, in *Biosocial Mechanisms of Population Regulation,* M. N. Cohen, R. S. Malpass, and H. G. Klein, eds., Yale University Press, New Haven, Conn., pp. 349–390.

Riss, D., and Goodall, J., 1977, The recent rise to alpha-rank in a population of free-living chimpanzees, *Folia primatol.* 27, 134–151.

Ristau, C., and Robbins, D., 1982, Language in the great apes: A critical review, in *Advances in the Study of Behavior,* J. Rosenblatt, R. A. Hinde, C. Beer, and M.-C. Busnel, eds., Academic Press, New York, Vol. 12.

Robinson, J. G., 1979, Vocal regulation of use of space by groups of titi monkeys *Callicebus moloch, Behav. Ecol. Sociobiol.* 5, 1–15.

Robinson, J. G., 1979, An analysis of the organization of vocal communication in the titi monkey *Callicebus moloch, Z. Tierpsychol.* 49, 381–405.

Robinson, J. G., 1979, Correlates of urine washing in the wedge-capped capuchin *Cebus nigrivittatus,* in *Vertebrate Ecology in the Northern Neotropics,* J. F. Eisenberg, ed., Smithsonian, Washington, pp. 137–143.

Robinson, J. G., 1981, Spatial structure in foraging groups of wedge-capped capuchin monkeys *Cebus nigrivittatus, Anim. Behav.* 29, 1036–1056.

Rodman, P. S., 1969, Individual activity patterns and the solitary nature of orangutans, in *The Great Apes,* D. A. Hamburg, and E. R. McCown, eds., Benjamin/Cummings, Menlo Park, CA, pp. 235–256.

Rogel, M. J., 1978, A critical evaluation of the possibility of higher primate reproductive and sexual pheromones, *Psychol. Bull.* 85, 810–830.

Rollison, J., and Martin, R. D., 1981, Comparative aspects of primate locomotion with special reference to arboreal cercopithecines, *Symp. Zool. Soc. Lond.* 48, 377–427.

Rose, M. D., 1978, The roots of primate predatory behavior, *J. Hum. Ecol.* 7, 179–189.

Rose, K. D., Walker, A., and Jacobs, L. L., 1981, Function of the mandibular tooth comb in living and extinct mammals, *Nature* 289, 583–585.

Rosenblum, L. A., and Alpert, S., 1977, Response to mother and strangers: a first step in socialization, in *Primate Bio-Social Development,* S. Chevalier-Skolnikoff, and F. E. Poirier, eds., Garland, New York, pp. 463–477.

Rosenblum, L. A., and Kaufman, I. C., 1967, Laboratory observations of early mother-infant relations in pigtail and bonnet macaques, in *Social Communication among Primates,* S. A. Altmann, ed., Chicago University Press, Chicago, pp. 33–41.

Rosenblum, L. A., Kaufman, I. C., and Stynes, A. J., 1966, Some characteristics of adult social and autogrooming patterns in two species of macaque, *Folia primatol.* 4, 438–451.

Rosenfeld, S. A., and Van Hoesen, G. W., 1979, Face recognition in the rhesus monkey, *Neuropsychologia* 17, 503–509.

Rowell, T. E., 1962, Agonistic noises of the rhesus monkey, *(Macaca mulatta), Symp. Zool. Soc. Lond.* 8, 91–96.

Rowell, T. E., 1963, The social development of some rhesus monkeys, in *Determinants of Infant Behavior,* B. M. Foss, ed., Methuen, London, 1, 35–49.

Rowell, T. E., 1966, Forest-living baboons in Uganda, *J. Zool., Lond.* 149:344–364.

Rowell, T. E., 1966, Hierarchy in the organization of a captive baboon group, *Anim. Behav.* 14, 430–433.

Rowell, T. E., 1967, Female reproductive cycles and the behavior of baboons and rhesus macaques, in *Social Communication among Primates,* S. A. Altmann, ed., Chicago University Press, Chicago, pp. 15–32.

Rowell, T. E., 1967, A quantitative comparison of the behavior of a wild and a caged baboon group, *Anim. Behav.* 15, 499–589.

Rowell, T. E., 1979, How would we know if social organization were *not* adaptive?, in *Primate Ecology and Human Origins,* I. S. Bernstein, and E. O. Smith, eds., Garland, New York, pp. 1–22.

Rowell, T. E., and Hartwell, J. M., 1978, The interaction of behavior and reproductive cycles in patas monkeys, *Behav. Biol.* 24, 141–167.

Rowell, T. E., and Richards, S. M., 1979, Reproductive strategies of some African monkeys, *J. Mammal.* 60, 58–69.

Roy, M. A., 1981, Abnormal behaviors in nursery-reared squirrel monkeys (*Saimiri sciureus*), *Am. J. Primatol.* 1, 35–42.

Rudder, B. C. C., 1979, The allometry of primate reproductive parameters, Ph.D. thesis, University of London.

Rudran, R., 1973, Adult male replacement in one-male troops of purple-faced langurs (*Presbytis senex senex*) and its effect on population structure, *Folia primatol.* 19, 166–192.

Rudran, R., 1973, The reproductive cycles of two subspp. of purple-faced langurs (P.S.) with relation to environmental factors, *Folia primatol.* 19, 41–60.

Rudran, R., 1978, Socioecology of the blue monkeys (*Cercopithecus mitis stuhlmanni*) of the Kibale Forest, Uganda, *Smithsonian Contr. Zool.* 249, 1–88.

Rudran, R., 1979, The demography and social mobility of a red howler (*Alouatta seniculus*) population in Venezuela, in *Vertebrate Ecology in the Northern Neotropics,* J. F. Eisenberg, ed., Smithsonian, Washington, pp. 107–126.

Rumbaugh, D., 1977, *Language Learning by a Chimpanzee,* Academic Press, New York, 312 pp.

Rumbaugh, D. M., and Pate, J. L., in press, Primate learning by levels.

Russell, M. J., 1976, Human olfactory communication, *Nature,* London, 260, 520–522.

Russell, R. J., in press, The population densities of nocturnal lemurs, *Bull. de l'Academie Malgache.*

Saayman, G. S., 1971, Baboons' responses to predators, *African Wild Life,* 25, 46–49.

Sackett, G. P., 1966, Monkeys reared in isolation with pictures as visual input: evidence for an innate releasing mechanism, *Science* 154, 1470–1473.

Sackett, G. P., Holm, R. A., Davis, A. E., and Fahrenbruch, E. E., 1975, Prematurity and low birth weight in pigtail macaques: incidence, prediction, and effects on infant development, in *Proc. Symp. 5th Intl. Congr. Primatol.* S. Kondo, M. Kawai, A. Ehara, and S. Kawanuraeds, Japan Science Press, Tokyo, pp. 189–206.

Sackett, G. P., Rappenthal, G. C., Fahrenbruch, C. E., Holm, R. A., Greenough, W. T., 1981, Social isolation rearing effects in monkeys vary with genotype, *Develop. Psychol.* 17, 313–318.

Sade, D. S. 1972, A longitudinal study of social behavior of rhesus monkeys, in *The Functional and Evolutionary Biology of Primates*, R. Tuttle, ed., Aldine, Chicago.

Sade, D. S., 1974, An ethogram for rhesus monkeys: antithetical contrasts in posture and movement, in *Play*, J. S. Bruner, A. Jolly, and K. Sylva, eds., pp. 146–152.

Sade, D. S., 1980, Population biology of free-ranging rhesus monkeys on Cayo Santiago, Puerto Rico, in *Biosocial Mechanisms of Population Regulation*, M. N. Cohen, R. S. Malpass, H. G. Klein, eds., Yale University Press, New Haven, Conn., pp. 171–187.

Sade, D. S., Cushing, K., Cishing, P., Dunaif, J., Figueroa, A., Kaplan, J. R., Lauer, D., and Schneider, J., 1977, Population dynamics and its relation to social structure on Cayo Santiago Island, *Am. J. Phys. Anthrop.* 20, 253–262.

Sarles, H. B., 1969, The study of language and communication across species, *Curr. Anthrop.* 211–220.

Savage-Rumbaugh, E. S., 1981, Can apes use symbols to represent their world?, *Ann. New York Acad. Sci.* 364, 35–59.

Savage-Rumbaugh, E. S., in press, *Pan paniscus* and *Pan troglodytes:* contrasts in preverbal communicative competence.

Savage-Rumbaugh, E. S., Rumbaugh, D. M., Smith, S. T., and Lawson, J., 1980, Reference: the linguistic essential, *Science* 210, 922–925.

Savage-Rumbaugh, S., and Wilkerson, B. J., 1978, Socio-sexual behavior in *Pan paniscus* and *Pan troglodytes:* a comparative study, *J. Hum. Evol.* 7, 327–344.

Schaffer, H. R., 1966, The onset of fear of strangers and the incongruity hypothesis, *J. Child Child Psychol. Psychiat.* 7, 95–106.

Schaffer, H. R., and Emerson, P. E., 1964, The development of social attachments in infancy, *Monogr. Soc. Res. Child Devel.*, 29, 3, no. 94.

Schaller, G. B., *The Mountain Gorilla*, Chicago University Press, Chicago, 431 pp.

Schessler, T., and Nash, L. T., 1977, Food sharing among captive gibbons *(Hylobates lar)*, *Primates* 18, 677–689.

Schiller, P. H., 1957, Manipulative patterns in the chimpanzee, in *Instinctive Behavior*, C. H. Schiller, ed., International University Press, New York, pp. 264–287.

Schilling, A., 1979, Olfactory communication in prosimians, in *The Study of Prosimian Behavior*, G. A. Doyle and R. D. Martin, eds., Academic Press, New York, pp. 461–542.

Schilling, A., 1980, The possible role of urine in territoriality of some nocturnal prosimians. *Symp. Zool. Soc. Lond.* 45, 165–193.

Schleidt, M., 1980, Personal odor and non-verbal communication, *Ethol. Sociobiol.* 1, 225–232.

Schlichte, H.-J., 1978, A preliminary report on the habitat utilization of a group of howler monkeys *(Alouatta villosa pigra)* in the National Park of Tikal, Guatemala, in *The Ecology of Arboreal Folivores*, G. G. Montgomery, ed., Smithsonian, Washington, pp. 551–560.

Schoener, T. W., 1968, Sizes of feeding territories among birds, *Ecology* 49, 123–141.

Schulman, S. R., and Chapais, B., 1980, Reproductive value and rank relations among macaque sisters, *Am. Nat.* 115, 580–593.

Schultz, A. H., 1969, *The Life of Primates*, Weidenfeld and Nicolson, London, 281 pp.

Schwartz, G. G., and Rosenblum, L. A., 1980, Novelty, arousal, and nasal marking in the squirrel monkey, *Behav. Neural Biol.* 28, 116–122.

Scollay, P. A., and Judge, P., 1981, The dynamics of social organization in a population of squirrel monkeys *(Saimiri sciureus)* in a seminatural environment, *Primates* 22, 60–69.

Sebeok, T. A., and Rosenthal, R., eds., 1979, The Clever Hans Phenomenon, *Ann. New York Acad. Sci.* 364.

Seitz, E., 1969, Die Bedeutung geruchlicher orientierung beim Plumpori *Nycticebus coucang* Boddaert 785 *(Prosimii*, Lorisidae), *Z. Tierpsychol.* 26, 73–103.

Serpell, J. A., 1981, Duetting in birds and primates: A question of function, *Anim. Behav.* 29, 963–965.

Seyfarth, R. M., 1976, Social relationships among adult female baboons, *Anim. Behav.* 24, 917–938.

Seyfarth, R. M., 1978a, Social relationships among adult male and female baboons, II. Behaviour throughout the female reproductive cycle, *Behaviour* 64, 227–247.

Seyfarth, R. M., 1978b, Social relations among adult male and female baboons, I. Behaviour during consortship, *Behaviour* 64, 204–247.

Seyfarth, R. M., and Cheney, D. L., 1980, The ontogeny of vervet monkey alarm calling behavior: A preliminary report, *Z. Tierpsychol.* 54, 37–56.

Seyfarth, R. M., Cheney, D. L., and Marler, P., 1980a, Vervet monkey alarm calls: Semantic communication in a free-ranging primate, *Anim. Behav.* 28, 1070–1094.

Seyfarth, R. M., Cheney, D. L., and Marler, P., 1980b, Monkey responses to three different alarm calls: evidence of predator classification and semantic communication, *Science* 210, 801–804.

Short, R. V., 1976, The evolution of human reproduction, *Proc. R. Soc. Lond.* B, 195, 3–24.

Short, R. V., 1981, Sexual selection in man and the great apes, in *Reproductive Biology of the Great Apes*, C. E. Graham, ed., Academic Press, New York, pp. 319–342.

Silk, J. B., 1978, Patterns of food sharing among mother and infant chimpanzees at Gombe National Park, Tanzania, *Folia primatol.* 29, 129–141.

Silk, J. B., 1979, Feeding, foraging, and food sharing behavior of immature chimpanzees, *Folia primatol.* 31, 123–142.

Silk, J. B., 1980, Kidnapping and female competition among captive bonnet macaques, *Primates* 21, 100–110.

Silk, J. B., Clark-Wheatley, C. B., Rodman, P. S., and Samuels, A., 1981, Differential reproductive success and facultative adjustment of sex ratios among captive female bonnet macaques *(Macaca radiata)*, Anim. Behav. 29, 1106–1120.

Simonds, P. B., 1965, The bonnet macaque in South India, in *Primate Behavior*, I. DeVore, ed., Holt, New York, pp. 175–196.

Simpson, M. J. A., 1974, The social grooming of male chimpanzees, in *Comparative Ecology and Behavior of Primates*, Academic Press, London, pp. 411–506.

Simpson, M. J. A., 1976, The study of animal play, in *Growing Points in Ethol-*

ogy, P. P. G. Bateson and R. A. Hinde, eds., Cambridge University Press, Cambridge, pp. 385–400.

Simpson, M. J. A., and Simpson, A. E., 1982, Birth sex ratios and social rank in rhesus monkey mothers, *Nature* 300, 400–441.

Simpson, M. J. A., Simpson, A. E., Hooley, J., and Lunz, M., 1981, Infant-related influences on birth intervals in rhesus monkeys, *Nature* 290, 49–51.

Singh, S. D., 1965, The effects of human environment upon the reactions to novel situations in the rhesus, *Behavior* 26, 243–250.

Singh, S. D., 1966, Effect of human environment on cognitive behavior in the rhesus monkey, *J. Comp. Phys. Psychol.* 61, 280–283.

Skinner, B. F., 1981, Selection by consequences, *Science* 213, 501–504.

Small, M. F., and Smith, D. G., 1981, Brief report: Interactions with infants by full siblings, paternal half-siblings, and nonrelatives in a captive group of rhesus macaques *(Macaca mulatta), Am. J. Primatol.* 1, 91–94.

Smith, C. C., 1977, Feeding behavior and social organization in howling monkeys, in *Primate Ecology,* T. H. Clutton-Brock, ed., Academic Press, London, pp. 97–126.

Smith, W. J., 1968, Message-meaning analysis, in *Animal Communication,* T. Sebeok, ed., Indiana University Press, Bloomington.

Smith, W. J., 1977, *The Behavior of Communicating,* Harvard University Press, Cambridge, Mass., 545 pp.

Smith, W. J., 1981, Referents of animal communication, *Anim. Behav.* 29, 1273.

Smith, W. J., Chase, J., and Leiblich, A. K., 1974, Tongue showing: a facial display of humans and other primate species, *Semiotica* 11, 201–246.

Snowdon, C. T., 1979, Response of nonhuman animals to speech and to species-specific sounds, *Brain, Behav. Evol.* 16, 409–429.

Snowdon, C. T., 1982, Linguistic and psycholinguistic approaches to primate communication, in *Primate Communication,* C. T. Snowdon, C. H. Brown, and M. R. Petersen, eds., Cambridge U.P., Cambridge, 212–238.

Soini, P., 1982, Ecology and population dynamics of the pygmy marmoset, *Cebuella pygmaea, Folia primatol.* 39, 1–21.

Southwick, C. H., 1980, Rhesus monkey populations in India and Nepal: Patterns of growth, decline and natural regulation, in *Biosocial Mechanisms of Population Control,* M. N. Cohen, R. S. Halpass, and H. G. Klein, eds., Yale University Press, New Haven, Conn., pp. 151–170.

Spitz, R. A., and Woolf, K. M., 1946, The smiling response: a contribution to the ontogeny of social relations, *Gen. Psychol. Monogr.* 34.

Steklis, H. D., and King, G. E., 1978, The craniocervical killing bite: toward an ethology of primate predatory behavior, *J. Hum. Evol.* 7, 567–581.

Stephens, W. N., 1963, *The Family in Cross-Cultural Perspective,* Holt, New York, 460 pp.

Stoltz, L. P., and Saayman, G. S., 1970, Ecology and behavior of baboons in the northern Transvaal, *Ann. Transvaal Mus.* 26, 99–143.

Struhsaker, T. T., 1967a, Ecology of vervet monkeys *(Cercopithecus aethiops)* in the Masai-Amboseli Game reserve, Kenya, *Ecology* 48, 891–904.

Struhsaker, T. T., 1967b, Social structure among vervet monkeys *(Cercopithecus aethiops), Behav.* 29, 83–121.

Struhsaker, T. T., 1974, Correlates of ranging behavior in a group of red colobus monkeys *(Colobus badius tephrosceles), Am. Zool.* 14, 177–184.

Struhsaker, T. T., 1975, *The Red Colobus Monkey,* Chicago University Press, Chicago.

Struhsaker, T. T., 1976, Further decline in numbers of Amboseli vervet monkeys, *Biotropica* 8, 211–214.

Struhsaker, T. T., 1977, Infanticide and social organization in redtail monkey *(Cercopithecus ascanius schmidti)* in Kibale Forest, Uganda, *Z. Tierpsychol.* 45, 75–84.

Struhsaker, T. T., 1978, Interrelations of red colobus monkeys and rain forest trees in the Kibale Forest, Uganda, in *Ecology of the Arboreal Folivores,* G. G. Montgomery, ed., pp. 397–492.

Struhsaker, T. T., 1981, Vocalizations, phylogeny and palaeogeography of red colobus monkeys *(Colobus badius),* *African J. Ecol.* 19, 265–283.

Struhsaker, T. T., 1981, Polyspecific association among tropical rain forest primates, *Z. Tierpsychol.* 57, 268–304.

Struhsaker, T. T., and Gartland, J. S., 1970, Observations on the behavior and ecology of the patas monkey *(Erythrocebus patas)* in the Waza Reserve, Cameroon, *J. Zool. Lond.* 161, 49–63.

Struhsaker, T. T., and Hunkeler, P., 1971, Evidence of tool using by chimpanzees in the Ivory Coast, *Folia primatol.* 15, 212–219.

Struhsaker, T. T., and Leland, L., 1979, Socioecology of five sympatric monkey species in the Kibale Forest, Uganda, *Advance Study Behav.* 9, 159–228.

Strum, S. C., 1975, Primate predation: Interim report on the development of a tradition in a troop of olive baboons, *Science* 187, 755–757.

Strum, S. C., 1976, Primate predation and bioenergetics, *Science* 191, 315–317.

Strum, S. C., 1982, Agonistic dominance in male baboons: An alternative view, *Intl. J. Primatol.* 3, 175–202.

Suarez, S. D., and Gallup, G. G., Jr., 1981, Self-recognition in chimpanzees and orangutans, but not gorillas, *J. Hum. Evol.* 10, 175–188.

Subramoniam, S., 1957, Some observations on the habits of the slender loris, *Loris tardigradus* (Linnaeus), *J. Bombay Nat. Hist. Soc.* 54, 387–398.

Sugardjito, J., and Nurhuda, N., 1981, Meat-eating behaviour in wild orangutans, *Pongo pygmaeus, Primates* 22, 414–416.

Sugiyama, Y., 1965, On the social change of hanuman langurs *(Presbytis entellus)* in their natural condition, *Primates* 6, 381–418.

Sugiyama, Y., 1966, An artificial change in a hanuman langur troop, *(Presbytis entellus), Primates* 7, 41–73.

Sugiyama, Y., and Koman, J., 1979, Tool-using and -making behavior in wild chimpanzees at Bossou, Guinea, *Primates* 20, 513–524.

Sugiyama, Y., and Ohsawa, H., 1982, Population dynamics of Japanese monkeys with special reference to the effect of artificial feeding, *Folia primatol.* 39, 238–263.

Suomi, S. J., Harlow, H. F., and Novak, M. A., 1974, Reversal of social deficits produced by isolation rearing in monkeys, *J. Hum. Evol.* 3, 527–534.

Suomi, S. J. and Ripp, C., 1983, A history of motherless mother monkey mothering at the University of Wisconsin Primate Laboratory, in *Child Abuse: The Nonhuman Primate Data,* eds. M. Reite and N. G. Caine, Alan R. Liss, New York, 49–78.

Susman, R. L., Badrian, N., Badrian, A., and Handler, N., 1981, Pygmy chimpanzees in peril, *Oryx* 16, 179–183.

Sussman, R. W., 1974, Ecological distinctions in sympatric species of *Lemur*, in *Prosimian Biology*, R. D. Martin, G. A. Doyle, and A. C. Walker, Duckworth, London, pp. 75–108.

Sussman, R. W., 1977, Feeding behavior of *Lemur catta* and *Lemur fulvus*, in *Primate Ecology*, T. H. Clutton-Brock, ed., Academic Press, London, pp. 1–37.

Sussman, R. W., 1977, Socialization, social structure and ecology of two sympatric species of Lemur, in *Primate Biosocial Development*, S. Chevalier-Skolnikoff, and F. E. Poirier, eds., Garland, New York, pp. 515–528.

Sussman, R. W. and Kinzey, W. G., in press, The ecological role of the Callitrichidae: a review, *Amer. J. Phys. Anthrop.*

Sussman, R. W. and Raven, P. H., 1978, Pollination by lemurs and marsupials: an archaic coevolutionary system, *Science*, 200, 731–736.

Suzuki, A., 1966, On the insect-eating habits among wild chimpanzees living in the savannah woodland of western Tanzania, *Primates* 7, 481–487.

Suzuki, A., 1975, The origin of hominid hunting: a primatological perspective, in *Socioecology and Psychology of Primates*, R. H. Tuttle, ed., Mouton, The Hague, pp. 259–278.

Suzuki, A., 1979, The variation and adaptation of social groups of chimpanzees and black-and-white colobus monkeys, in *Primate Ecology and Human Origins*, I. S. Bernstein and E. O. Smith, eds., Garland, New York, pp. 153–174.

Swayamprabha, M. S., and Kadam, K. M., 1980, Mother-infant relationship in the slender loris, *Primates* 21, 561–566.

Symmes, D., and Newman, J. D., et al., 1979, Individuality and stability of isolation peeps in squirrel monkeys, *Anim. Behav.* 27, 1142–1152.

Symons, D., 1978, *Play and Aggression: A Study of Rhesus Monkeys*, Columbia University Press, New York, 246 pp.

Szalay, F., 1975, Hunting-scavenging protohominids: a model for human origins, *Man* 10, 420–429.

Szalay, F. S., and Seligsohn, D., 1977, Why did the strepsirhine tooth comb evolve?, *Folia primatol.* 27, 75–82.

Tanner, J. M., 1962, *Growth at Adolescence*, Blackwell, Oxford, 325 pp.

Tattersall, I., 1977, Ecology and behavior of *Lemur fulvus mayottensis* (Primates, Lemuriformes) *Anthrop. Papers, Am. Mus. Nat. Hist.* 54, 421–482.

Tattersall, I., 1982, *The Primates of Madagascar*, Columbia U.P., New York, 382 pp.

Tattersall, I. and Sussman, R. W., 1975, Observations of the ecology and behavior of the mongoose lemur *Lemur mongoz mongoz* Linnaeus (Primates, Lemuriformes), at Ampijoroa, Madagascar, *Anthrop. Papers Amer. Mus. Nat. Hist.* 52, 195–216.

Taub, D. M., 1980, Testing the "agonistic buffering" hypothesis, I. The dynamics of participation in the triadic interaction, *Behav. Ecol. Sociobiol.* 6, 187–197.

Teas, J., Richie, T. L., Taylor, H. G., Siddiqi, M. F., and Southwick, C. H., 1981, Natural regulation of rhesus monkey populations in Kathmandu, Nepal, *Folia primatol.* 35, 117–124.

Teleki, G., 1973a, *The Predatory Behavior of Wild Chimpanzees*, Bucknell University Press, Lewisburg, Pa., 232 pp.

Teleki, G., 1973b, Group response to the accidental death of a chimpanzee in Gombe National Park, Tanzania, *Folia primatol.* 20, 81–94.

Teleki, G., 1974, Chimpanzee subsistence technology: Materials and skills, *J. Hum. Evol.* 3, 575–594.

Teleki, G., 1977, Still more on predatory behavior in nonhuman primates. *Curr. Anthrop.* 18, 107–108.

Teleki, G., and Baldwin, L., 1980, Disaster for chimpanzees, *Oryx* 15, 317–318.

Teleki, G., Hunt, E. E., Jr., Pfifferling, J. H., 1976, Demographic observations (1963–1973) on the chimpanzees of Gombe National Park, Tanzania, *J. Hum. Evol.* 5, 559–598.

Temerlin, M. K., 1975, *Lucy: Growing Up Human.* Souvenir Press, London, 216 pp.

Tenaza, R. R., 1976, Songs, choruses and counter-singing of Koss' gibbons (*Hylobates klossii*) in Siberut Island, Indonesia, *Z. Tierpsychol.* 41, 37–52.

Tenaza, R. R., and Hamilton, W. J., III, 1977, Preliminary observations of the Mentawai Islands gibbon, *Ho klossii, Folia primatol.* 51, 201–211.

Terborgh, J., in press, *Five New World Primates: A Study in Comparative Ecology,* Princeton University Press, Princeton.

Terrace, H. S., 1979, *Nim,* Knopf, New York, 303 pp.

Terrace, H. S., Petitto, L. A., Sanders, R. J., and Bever, T. G., 1979, Can an ape create a sentence?, *Science* 206, 891–902.

Thomas, R. K., 1980, Evolution of intelligence: an approach to its assessment, *Brain, Behav. Evol.* 17, 454–472.

Thomas, R. K., 1982, The assessment of primate intelligence, *J. Hum. Evol.* 11, 247–255.

Thomas, R. K., in preparation, Using the Piagetian approach with primates: a critical analysis, IPS Congress, Atlanta, 1982.

Thomas, R. K., and Boyd, M. G., 1973, A comparison of *Cebus albifrons* and *Saimiri sciureus* on oddity performance, *Anim. Learn. Behav.* 1, 151–153.

Thomas, R. K., and Crosby, T. N., 1977, Absolute and relative class conceptual behavior in squirrel monkeys (*Saimiri sciureus*), *Anim. Learn. Behav.* 5, 265–271.

Thomas, R. K., Fowlkes, D., and Vickery, J. D., 1980, Conceptual numerousness judgments by squirrel monkeys, *Am. J. Psychol.* 93, 247–257.

Thomas, R. K., and Ingram, D. K., 1979, Conceptual volume judgment by squirrel monkeys, *Am. J. Psychol.* 92, 33–43.

Thomas, R. K., and Peay, L., 1976, Length judgments by squirrel monkeys: evidence for conservation?, *Devlop. Psychol.* 12, 349–352.

Thorington, R. W., 1967, Feeding and activity of *Cebus* and *Saimiri* in a Colombian forest, in *Neue Ergebnisse der Primatologie,* D. Stark, R. Schneider, and H. J. Kuhn, eds., Fischer, Stuttgart.

Tilford, B., 1981, Nondesertion of a post-reproductive rhesus female by adult male kin, *J. Mamm.* 62, 638–639.

Tilson, R. L., 1981, Family formation strategies of Kloss's gibbons, *Folia primatol.* 35, 259–287.

Tinbergen, N., 1951, *The Study of Instinct,* Oxford University Press, Oxford.

Triana, E., and Pasnak, R., 1981, Object permanence in cats and dogs, *Anim. Learn. Behav.* 9, 135–139.

Trivers, R. L., 1972, Parental investment and sexual selection, in *Sexual Selection and the Descent of Man, 1871–1971,* B. Campbell, ed., Aldine, Chicago, pp. 136–179.

Trivers, R. L., 1974, Parent-offspring conflict, *Am. Zool.* 14, 249–264.

Tsumori, A., 1966, Delayed response of wild Japanese monkeys by the sand-digging method, II. Cases of the Takasakiyama troops and the Ohiragama troop, *Primates* 7, 363–380.

Tsumori, A., 1967, Newly acquired behavior and social interactions of Japanese monkeys, in *Social Communication among Primates,* S. A. Altmann, ed., Chicago University Press, Chicago, pp. 207–220.

Tsumori, A., Kawai, M., and Motoyoshi, R., 1965, Delayed response of wild Japanese monkeys by the sand-digging method, *Primates* 6, 195–212.

Turner, F. B., Jeinrich, R. I., and Weintraub, J. D., 1969, Home ranges and body size of lizards, *Ecology* 50, 1076–1081.

Tutin, C. E. G., 1979, Responses of chimpanzees to copulation, with special reference to interference by immature individuals, *Anim. Behav.* 27, 845–854.

Tutin, C. E. G., and McGinnis, P. R., 1981, Chimpanzee reproduction in the wild, in *Reproductive Biology of the Great Apes,* C. E. Graham, ed., Academic Press, New York, 239–264.

Uehara, S., 1982, Seasonal changes in the techniques employed by wild chimpanzees in the Mahale Mountains, Tanzania, to feed on termites *(Pseudacanthus spiniger), Folia primatol.* 37, 44–76.

Umiker-Sebeok, J., and Sebeok, T. A., 1980, Questioning apes, in *Speaking of Apes*, T. A. Sebeok, and J. Umiker-Sebeok, eds., Plenum, New York, pp. 1–60.

Vandenberg, J. G., and Drickamer, L. C., 1974, Reproductive coordination among free-ranging rhesus monkeys, *Physiol. Behav.* 13, 373–376.

Van Horn, R. N., 1980, Seasonal reproductive patterns in primates, *Prog. Reprod. Biol.* 5, 181–221.

Van Horn, R. N., and Eaton, G. G., 1979, Reproductive physiology and behavior in prosimians, in *The Study of Prosimian Behavior*, G. A. Doyle and R. D. Martin, eds., Academic Press, New York, pp. 79–122.

Vaughter, R. M., Smotherman, W. and Ordy, J. M., 1972, Development of object permanence in the infant squirrel monkey, *Devel. Psychol.* 7, 34–38.

Veit, P. G., 1982, Gorilla society, *Nat. Hist.* 91, 48–59.

Vogt, J. L., Carlson, H., and Menzel, E., 1978, Social behavior of a marmoset *(Saguinus fusciollis)* group, I. Parental care and infant development, *Primates* 19, 715–726.

de Waal, F., 1982, *Chimpanzee Politics*, Harper and Row, New York, 223 pp.

de Waal, F. B. M., and Van Hoof, J. A. R. A. M., 1981, Side-directed communication and agonistic interactions in chimpanzees, *Behaviour* 77, 164–198.

de Waal, F. B. M., and van Roosmalen, A., 1979, Reconciliation and consolation among chimpanzees, *Behav. Ecol. Sociobiol.* 5, 55–66.

Wada, K., and Ichiki, Y., 1980, Seasonal home range use by Japanese monkeys in the snowy Shiga heights, *Primates* 21, 468–483.

Wade, T. D., 1979, Inbreeding, kin selection and primate social evolution, *Primates* 20, 355–370.

Wagenen, G. van, 1972, Vital statistics from a breeding colony, Reproductive and pregnancy outcome in Macaca mulatta, *J. Med. Primatol.* 1, 3–28.

Walker, A., 1969, The locomotion of lorises, with special reference to the potto, *E. Afr. Wildl. J.,* 7, 1–60.

Walters, J., 1981, Inferring kinship from behaviour: maternity determinations in yellow baboons, *Anim. Behav.* 29, 126–136.

Warren, J. M., 1974, Possibly unique characteristics of learning by primates, *J. Hum. Evol.* 3, 445–454.

Waser, P. M., 1975, Experimental playbacks show vocal mediation of inter-group avoidance in a forest monkey, *Nature (Lond.)* 255, 56–58.

Waser, P. M., 1977a, Feeding, ranging and group size in the mangabey *Cercocebus albigena*, in *Primate Ecology*, T. H. Clutton-Brock, ed., Academic Press, London, pp. 183–222.

Waser, P. M., 1977b, Individual recognition, intra-group cohesion and inter-group spacing: Evidence from sound playback to forest monkeys, *Behaviour* 60, 28–74.

Waser, P. M., 1978, Postreproductive survival and behavior in a free-ranging female managabey, *Folia primatol.* 29, 142–160.

Waser, P. M., 1980, Polyspecific associations of *Cercocebus albigena*: geographic variation and ecological correlates, *Folia primatol.* 33, 56.

Waser, P. M., 1982, Primate polyspecific associations: do they occur by chance?, *Anim. Behav.* 30, 1–8.

Waser, P. M., and Floody, O., 1974, Ranging patterns of the mangabey *Cercocebus albigena* in the Kibale Forest, Uganda, *Z. Tierpsychol.* 35, 85–101.

Waser, P. M., and Homewood, K., 1979, Cost-benefit approaches to territoriality: A test with forest primates, *Behav. Ecol. Sociobiol.* 6, 115–119.

Waser, P. M., and Waser, M. S., 1977, Experimental studies of primate vocalization: Specializations for long distance propagation, *Z. Tierpsychol.* 43, 239–263.

Waser, P. M., and Wiley, R. H., 1979, Mechanisms and evolution of spacing in animals, in *Handbook of Behavioral Neurobiology*, 3: Social Behavior and Communication, P. Marler, and J. G. Vandenbergh, eds., Plenum, New York, pp. 159–223.

Washburn, S. L., and Lancaster, C. S., 1968, The evolution of hunting, in *Man the Hunter*, I. DeVore and R. Lee, eds., Aldine, Chicago.

Watson, J., 1974, Smiling, cooing, and "the game," in *Play*, J. S. Bruner, A. Jolly, and K. Sylva, eds., Penguin, Harmondsworth, pp. 268–276.

Watson, P., 1979, How Moja the talking chimp learned to draw, *Sunday Times Magazine*, Nov. 18, pp. 62–75.

Weinstein, B., 1945, The evolution of intelligent behavior in rhesus monkeys, *Genet. Psychol. Monogr.* 31, 3–48.

Welles, J. F., 1975, The anthropoid hand: a comparative study of prehension, *Proc. 5th Int. Congr. Primatol.*, Karger, Basel, pp. 30–33.

Whitten, A. J., 1982, *The Gibbins of Siberut*, Dent, London.

Whitten, A. J., in press, The trilling handicap in Kloss gibbons, in *The Lesser Apes*, H. Preuschoft, D. J. Chivers, N. Creel, and W. Brockelman, eds., Edinburgh University Press, Edinburgh.

Wickler, W., 1967, Sociosexual signals and their intraspecific imitation among primates, in *Primate Ethology*, D. Morris, ed., Weidenfeld and Nicolson, London, pp. 69–147.

Wiley, R. H., 1981, Social structure and individual ontogenies: problems of description, mechanism and evolution, in *Perspectives in Ethology*, P. P. G. Bateson, and P. H. Klopfer, eds., Plenum, London 4, pp. 105–133.

Wiley, R. H., and Richards, D. G., 1978, Physical constraints on acoustic communication in the atmosphere: implications for the evolution of animal vocalizations, *Behav. Ecol. Sociobiol.* 3, 69–94.

Wilkerson, B. J., and Rumbaugh, D. M., 1979, Learning and intelligence in prosimians, in *The Study of Prosimian Behavior*, G. A. Doyle, and R. D. Martin, eds., Academic Press, New York, pp. 207–246.

Wilson, E. O., 1975, *Sociobiology, The New Synthesis*, Harvard University Press, Cambridge, Mass. 697 pp.

Wise, K. L., Wise, L. A., and Zimmerman, R. R., 1974, Piagetian object permanence in the infant rhesus monkey, *Devel. Psychol.* 10, 429–437.

Wolf, K. E., and Fleagle, J. G., 1977, Adult male replacement in a group of silvered leaf monkeys *(Presbytis cristata)* at Kuala Selangor, Malaysia, *Primates* 18, 949–956.

Wolff, P., personal commun. A study of mother-infant pair interactions in *Lemur variegatus* at Jersey Wildlife Preservation Trust.

Woodruff, G., 1981, Methods for studying cognition in the chimpanzee, in *Self-awareness in domesticated animals*, D. G. M. Wood-Gush, M. Dawkins, R. Ewbank, eds., Universities Federation for Animal Welfare, Potters Bar, pp. 29–36.

Woodruff, G., and Premack, D., 1979, Intentional communication in the chimpanzee: The development of deception, *Cognition* 7, 333–362.

Woodruff, G., and Premack, D., 1981, Primitive mathematical concepts in the chimpanzee: proportionality and numerosity, *Nature* 293, 568–570.

Woodruff, G., Premack, D., and Kennel, K., 1978, Conservation of liquid and solid quantity by the chimpanzee, *Science* 202, 991–994.

World Wildlife Fund Indonesia, 1980, *Saving Siberut: A Conservation Master Plan*, WWF, Bogor, 134 pp.

Wrangham, R. W., 1974, Artificial feeding of chimpanzees and baboons in their natural habitat, *Anim. Behav.* 22, 83–93.

Wrangham, R. W., 1977, Feeding behaviour of chimpanzees in the Gombe National Park, Tanzania, in *Primate Ecology*, T. H. Clutton-Brock, ed., Academic Press, London, pp. 504–538.

Wrangham, R. W., 1980, An ecological model of female-bonded primate groups, *Behaviour* 75, 262–300.

Wrangham, R., and Dunbar, R. I. M., 1979, Species designated as threatened with extinction, *Primate Eye* 13, Nov. 15–20.

Wrangham, R. W., 1981, Drinking competition in vervet monkeys, *Anim. Behav.* 29, 904–910.

Wright, R. V. S., 1972, Imitative learning of a flaked-tool technology—the case of an orangutan, *Mankind* 8, 296–306.

Wu, H. M. H., Holmes, W. G., Medina, S. R., and Sackett, G. P., 1980, Kin preference in the infant *Macaca nemestrina*, *Nature* 285, 225–227.

Young, G. H., and Hankins, R. J., 1979, Infant behaviors in mother-reared and harem-reared baboons *(Papio cynocephalus)*, *Primates* 20, 87–93.

Zimmerman, R., and Torrey, C. C., 1965, Ontogeny of learning, in *Behavior of Nonhuman Primates*, A. M. Schreier, H. F. Harlow, and F. Stollnitz, eds., Academic Press, New York, pp. 405–445.

Zoloth, S., and Green, S., 1979, Monkey vocalizations and human speech: Parallels in perception?, *Brain Behav. Evol.* 16, 430–442.

Zoloth, S. R., Petersen, M. R., Beecher, M. D., Green, S., Marler, P., Moody, D. B., and Stebbins, W., 1979, Species-specific perceptual processing of vocal sounds by monkeys, *Science* 204, 870–873.

Zucker, E. L., and Kaplan, J. R., 1981, Allomaternal behavior in a group of free-ranging patas monkeys, *Am J. Primatol.* 1, 57–64.

Zuckerman, S., 1932, *The Social Life of Monkeys and Apes* K. Paul, Trench, Trubner, London, 357 pp.

Zumpe, D., and Michael, R. P., 1968, The clutching reaction and orgasm in the female rhesus monkey, *Macaca mulatta, J. Endocrinol.* 40, 117–123.

INDEX

*Asterisk indicates illustration.

511

Gunung Leuser Reserve, Sumatra, Indonesia, 175
Guyana, 31

Habitat destruction, 3–5, 28–30, 33
Hacienda Bargueta, Panama, 31
Haimoff, E. H., 216
Hairy-eared dwarf lemur, 6, 18, 22, 24
Haldane, J. B. S., 182
Hall, K. R. L., 60, 67, 76, 125, 147, 254, 256, 337, 341, 355, 364, 408, 452
Halle, F., 155
Hamden, C. E., 335
Hamilton, W. D., vii, 182, 186
Hamilton, W. J., 76, 148, 330, 355
Hands, 5, 48, 51, 59, 352–354, 367
Hankins, R. J., 337
Hapalemur. See Hapalemur
Hapalemur, 6, 18, 100, 117
    gray, 6, 24, 353*
    simus, 6, 22, 24
Haplorhini, 7, 12, 38, 39, 40
Harcourt, A. H., 45, 124, 125, 223, 251, 252, 264, 287, 330
Harcourt, C., 204
Hardin, G., 34
Harding, R. S. O., 62, 69, 76
Harem, 15, 115, 122, 124, 130, 133, 134, 255. *See also* One-male troops
Harlow, H. F., 188, 308, 313, 315, 354, 387, 410
Harlow, M. K., 313
Harrington, J. E., 202
Hartung, J., 184, 331
Hartwell, J. M., 254
Harvey, P. H., 47, 54, 70, 89–92, 109, 115, 120, 122, 134, 251
Hasegawa, T., 317, 318
Hausfater, G., 56, 62, 63, 79, 81–83, 125, 209, 256, 260, 271, 333, 402
Hayes, C., 386, 390, 412, 416
Hebb, D. O., 409
Hediger, H., 205
Height in trees, 99, 113, 173
Hennessy, M. B., 204
Henwood, K., 215
Herrnstein, R. J., 389
Hess, J. P., 321
Hildegard of Bingen, St., ii
Hill, J. P., 303
Hinde, R. A., 216, 270, 315, 321, 331, 406
Hiraiwa, M., 317, 318
Hladik, A., 49, 91, 99, 156–157, 160, 165
Hladik, C. M., 12, 45, 49, 55, 58, 60, 61, 74, 90, 91, 146, 165, 186, 356
Hoage, R., 328, 409
Hoban, R., 403
Hobbet, L., 217
Hockett, C. F., 420, 421
Hoff, M. P., 245
Hofstadter, D. R., 348
Holmes, W. G., 182
Home range, 37, 69, 71, 87, 90–92, 93, 97–99, 109–111, 113, 138, 149–151, 162–164
Homewood, K. M., 96, 144, 152
Hominidae, 11
Hominoidea, 11, 119, 129
Homo. See Human

Hooff, J. A. R. A. M. van, 192, 199, 209, 211
Hooley, J. M., 321
Hopf, S., 322
Horn, R. N. van, 205, 234
Hornsby, J. R., 390
Howell, N., 235
Howler, 9, 13, 18, 100, 207, 215, 278, 285, 355, 358
    black, 9, 25, 118, 129
    black-and-red, 9
    brown, 9, 25
    Guatemalan, 9, 93–95, 99, 108–109, 113, 118, 129
    mantled, 4, 9, 25, 47, 54, 57, 67, 68*, 70, 75*, 79, 89, 90, 91, 92–95, 104*, 108–109, 113, 118, 125, 126, 129, 134, 144–145, 146*, 149, 150, 161, 162, 168, 173, 225–227, 228*, 238, 245, 271, 280
    red, 9, 47, 81, 118, 125, 126, 129, 131, 139, 161, 164, 168, 175, 261, 263, 338
Hrdy, D. B., 20, 271
Hrdy, S. B., 27, 81–83, 125, 132, 133, 143, 250, 252, 253, 254, 258, 259, 260, 263, 269, 271, 272, 273, 275, 277, 280, 289, 331, 332, 340
Huebner, D. K., 73
Huffman, S. L., 235
Human, 3–5, 11, 15, 23, 27, 29, 33, 40, 61–62, 66, 83, 100, 110, 111, 114, 120, 135, 137–138, 139, 177, 202–203, 205, 210*, 211, 214*, 227, 229, 235–236, 237, 251, 252, 260, 273, 279, 288–290, 293, 303, 308, 311, 312, 315, 325, 331, 351, 352, 353*, 359, 360, 361, 367, 370–375, 382–384, 385*, 389–390, 399, 401, 402, 407*, 409, 410, 425, 428, 431–435, 434*, 439, 449*, 453*
Humphrey, N. K., 366
Hunkeler, P., 53
Hunter, A. J., 205, 231
Hunting. *See* Meat-eating
Hutchinson, G. E., v, vi, 34, 36, 37, 41, 42, 222, 223, 224, 304, 448
Hutt, C., 411
Huxley, A., 196, 294
Huxley, T. H., 3, 5
Hylobates. See Gibbon
Hylobatidae, 11, 17, 18, 26

Ichiki, Y., 93
Idenau, Cameroon, 174
Imitation, 374, 400, 439, 448–449, 451–453
Imprinting, 309–313, 325
Inbreeding, 28, 33, 139, 239–242
Incest, 239–242
Incisors, 51
India, 19, 23, 259
Indonesia. *See* Borneo, Gunung Leuser Reserve, Kutai, Mentawai Islands, Siberut
Indri. See Indri
Indri, 7, 18, 25, 40, 45, 47, 70, 89, 100, 109–110, 117, 128, 135*, 136, 143, 144, 149, 150, 151, 215, 216, 252, 279, 327
Indriidae, 6, 17, 18, 25, 54, 152
Infanticide, 81–83, 86, 126, 259–264, 273, 327
Infants, 224–227, 292–293, 304–326, 369, 370–371, 373–386, 405–406, 448
Ingram, D. K., 391
Ingram, J. C., 328